D0141631

BIOTECHNOLOGY:
A Textbook of
Industrial Microbiology

Wulf Crueger
Anneliese Crueger

BIOTECHNOLOGY:
A Textbook of Industrial Microbiology

SECOND EDITION

Editor of the English edition:
Thomas D. Brock

Sinauer Associates, Inc. • Sunderland, MA 01375

Dr. W. Crueger
Technical Microbiology Center

Dr. A. Crueger
Biochemical Process Development

Bayer AG
Friederich-Ebert Strasse 217
5600 Wuppertal 1
Federal Republic of Germany

Library of Congress Cataloging-in-Publication Data

Crueger, Wulf.
[Lehrbuch der angewandten Mikrobiologie. English]
Biotechnology : a textbook of industrial microbiology / Wulf Crueger, Anneliese Crueger : editor of the English edition. Thomas D. Brock. -- 2nd ed.
p. cm.
Translation of: Lehrbuch der angewandten Mikrobiologie.
Includes bibliographical references.
ISBN 0-87893-131-7
1. Industrial microbiology. I. Crueger, Anneliese. II. Brock, Thomas. III. Title.
QR53.C7813 1990
660 ' .62--dc20

89-26191
CIP

Printed in the United States of America.

Original translation by Caroline Haessly. Editorial supervision and new translation by Thomas D. Brock. Production supervision by Katherine M. Brock (Science Tech Publishers, Madison, Wisconsin).

Address orders and correspondence to
Sinauer Associates, Inc., Sunderland, MA 01375, USA

5 4 3 2 1

Foreword

Biotechnology deals with the use of living organisms or their products in large-scale industrial processes. It is an old field that has been rejuvenated in recent years because of the development of genetic engineering techniques. At present, biotechnology is in an amazing growth phase whose end is nowhere in sight. Industrial microbiology, a central part of biotechnology, matured as a science in the antibiotic era, and the large-scale manufacture of microbial products became a multibillion dollar industry. Genetic engineering has now made possible the directed construction of microorganisms that will do almost anything, and new products are being announced almost daily. Not only are new classes of substances being sought for human therapy, but cost-effective processes are being developed for major organic chemicals. Yet the mere engineering of a microbe is not enough. Large-scale, economically viable production must be attained. The industrial microbiologist knows that there are vast difficulties in the transfer of a laboratory process to the production plant. There are, however, some well-established principles of industrial microbiology, and it is the purpose of this textbook to enunciate them.

Although there are a number of advanced textbooks and research monographs dealing with industrial microbiology, there has been no up-to-date book suitable for use in universities and colleges. The first edition of this book has been used widely in university classes, not only in North America but throughout the world. The book obviously met a real need and I am pleased that a new edition can now be published. This second edition is based on a new German-language edition published in 1989. In the years since the first edition appeared, a number of important advances in industrial microbiology have taken place; all of these advances are described in some detail in this book. Among the new material is, of course, an extensive update of the material on genetic engineering, including the use of recombinant DNA (rDNA) techniques for strain selection and for the production of pharmaceutically useful proteins, enzymes, and amino acids. A whole series of new rDNA products, whose development makes use of the principles laid out in this book, are appearing on the market. Other major new developments include the use of immobilized enzymes and cells for large-scale processes. Biochemical reactors employing new membrane technology are discussed, as are membrane-based biosensors used in medicine and laboratory analysis. Some of the older industrial processes for the production of organic acids, alcohols, aldehydes, and flavor and aroma ingredients have been greatly improved in recent years, and these new developments are covered. The purification of industrial metabolites is almost a special branch of biotechnology, and new or improved chromatographic and extraction methods are presented.

As in the first edition, this book has been carefully edited to make it accessible to the English-speaking student. I have translated the new material myself, taking pains to ensure that the vocabulary is appropriate for an undergraduate student. In order to bring out this new edition as rapidly as possible, the translation was made directly from the German manuscript.

Students will certainly find this text useful. Research scientists and technicians already employed in industrial microbiology will also benefit from this practical overview of the field. In addition to industrial microbiologists, others who may find this book valuable include food scientists, environmental scientists, chemical engineers, and organic chemists.

The authors and I are pleased that we are able to offer this modern, attractive, accessible book, which should be of great value to both students and practicing researchers.

November 1989

Thomas D. Brock
Department of Bacteriology
University of Wisconsin
Madison, Wisconsin, USA

Contents

1 Introduction

Biotechnology is the use of microbiology, biochemistry, and engineering in an integrated fashion with the goal of using microorganisms and cell and tissue cultures (or their parts) to manufacture useful products. Biotechnology can be divided into two categories which are sometimes called "traditional biotechnology" and "new biotechnology". The major products of the traditional biotechnology industry are food and flavor ingredients, industrial alcohol, antibiotics, and citric acid. These products amount, on a worldwide basis, to about 300 billion dollars annually. The new biotechnology, which involves the use of the newer techniques of genetic engineering and cell fusion to produce organisms capable of making useful products, provides at present products with a total value of less than a billion dollars. In the future, however, it is predicted that the new biotechnology will account for a much larger fraction of the total biotechnology industry.

Industrial microbiology, the major foundation of biotechnology, arose out of empirical developments in the production of wine, vinegar, beer, and sake, and with the traditional fungal fermentations used in Asia and Africa for the production of food. An experimental approach to the production of microbial metabolites only began at the beginning of the 20th century. Up until the time of World War II, the main microbial products that had developed from this experimental approach were enzymes such as proteases, amylases, and invertase.

A major breakthrough in biochemical and microbial engineering occurred after World War II as a result of the large-scale production of the first antibiotic, *penicillin*. In order to produce this antibiotic economically, important engineering developments had to be made, including the development of techniques for large-scale sterilization, aeration, and growth of microorganisms. In addition, genetic methods for microbial strain improvement were perfected.

From World War II up until about 1960, the major new biotechnology products were antibiotics. Through intense efforts of the pharmaceutical industry, numerous new antibiotics were discovered and of these around 20 were put into commercial production. In addition, in this early post-World-War-II period, processes were developed for the chemical transformation of steroids, and the culture of animal cells for the production of virus vaccines was perfected.

In the period from 1960 through 1975, new microbial processes for the production of amino acids and 5′-nucleosides as flavor enhancers were developed, primarily in Japan. In addition, numerous processes for enzyme production for industrial, analytic, and medical purposes were perfected. During this same period, successful techniques for the **immobilization** of enzymes and cells were developed. During this time a further development was the use of **continuous fermentation** for the production of *single-cell protein* from yeast and bacteria for use as human and animal food. Single-cell protein processes were developed using microorganisms capable of using petroleum-based starting materials such as gas oil, alkanes, and methanol. In this same period, microbial biopolymers such as xanthan and dextran, used as food additives, were also developed into commericial processes. Somewhat distinct processes that were advanced during this period were the use of microorganisms for tertiary oil recovery (an aspect of geomicrobiology) and the perfection of techniques for anaerobic cultivation of microorganisms, derived out of studies on the sewage treatment process.

Since 1975 biotechnology has entered some important new phases. First was the development of the **hybridoma technique** for the production of monoclonal antibodies, of interest primarily in the medical diagnosis field. Soon after was the production of human proteins using genetically engineered *Escherichia coli*. The first product, **human insulin** was introduced in 1982, followed soon by Factor VIII, human growth hormone, interferons, and urokinase. At present, a vast array of human proteins are in the development stage.

Although the production of human proteins by engineered bacteria is generally recognized as the major "highlight" of the period since 1975, in actuality other products are economically more important. For instance, the production of ethanol by immobilized cells has become a major process. The enzyme *glucose isomerase* has become a 27 million dollar industry and is used to produce high-fructose syrup which itself has a value of 2.5 billion dollars. *Aspartame*, a major artificial sweetener, is produced microbially. Many new antibiotics have been introduced. Cheap fats are being increased in value by enzymatic esterification, the enzymes being microbial products. The biodegradation of persistent chemicals using specially developed microbial strains as starter cultures is being field-tested.

Table 1.1 Patent applications in 1984 for three countries with major biotechnology industries.

Country	Molecular biology patents	Fermentation patents (including enzymes)
USA	100	160
Japan	90	700
Federal Republic of Germany	12	55

To place in perspective research activities in traditional biotechnology and the "new" biotechnologies (using genetic engineering, etc.), Table 1.1 provides a comparison of the number of patent applications in the whole field of biotechnology for three major industrial countries, U.S.A., Japan, and the Federal Republic of Germany. As can be seen, traditional biotechnologies still dominate, especially in Japan.

REFERENCES

Demain, A.L. and N.A. Solomon. 1986. Manual of Industrial Microbiology and Biotechnology. American Society for Microbiology, Washington, D.C.

Moo-Young, M. (editor). 1985. Comprehensive Biotechnology. The Principles, Applications and Regulations of Biotechnology in Industry, Agriculture, and Medicine. Volume 1, The Principles of Biotechnology: Scientific Fundamentals; Volume 2, The Principles of Biotechnology: Engineering Considerations; Volume 3, The Practice of Biotechnology: Current Commodity Products; Volume 4: The Practice of Biotechnology:

Specialty Products and Service Activities. Pergamon Press, Oxford.

Rehm, H.J. and Reed, G. (editors). 1981-1988. *Biotechnology. A Comprehensive Treatise.* Volume 1, Microbial Fundamentals; Volume 2, Fundamentals of Biochemical Engineering; Volume 3, Biomass, Microorganisms for Special Applications, Microbial Products I; Volume 4, Microbial Products II; Volume 5, Food and Feed Production with Microorganisms; Volume 6a, Biotransformations; Volume 6b, Special Microbial Processes; Volume 7a, Enzyme Technology; Volume 7b, Gene Technology; Volume 8, Microbial Degradations. VCH Publishers, Weinheim, Germany and Deerfield Beach, Florida.

2

Screening for new metabolites

2.1 GENERAL

The biochemical capabilities of microorganisms are vast, and a wide variety of new or unusual compounds may be produced by various microbial isolates. One of the main tasks of the industrial microbiologist is to develop procedures for obtaining new microbial metabolites. There are five distinct approaches:

1. **Screening** for the production of new metabolites with new isolates and/or new test methods. This is the only way to obtain completely new classes of substances.
2. **Chemical modification** of known microbial substances.
3. **Biotransformation** (Chapter 15), which results in change in a chemical molecule by means of a microbial or enzymatic reaction.
4. **Interspecific protoplast fusion** (Section 3.4), which is a means of recombining genetic information from rather closely related producer strains. New or hybrid substances are expected and the method is widely used in the antibiotic industry.
5. **Gene cloning** (Section 3.6), in which genes may be transferred between unrelated strains which are producers of known substances. Alternatively, transfer may be to nonproducers which contain "silent" genes, leading to the generation of modified or even new substances. This will undoubtedly be the method of choice in the future.

In order to be successful, screening must be an interdisciplinary activity, combining the activities of microbiology, chemistry, biochemistry, engineering, and bibliographics. Microbiology is involved in the isolation and identification of microorganisms, strain preservation, testing for biological activity, and fermentation practice. Biochemistry provides the analytical procedures needed as well as the approaches to purification of the biologically interesting molecules whereas synthesis of substrates and inhibitors falls under the purview of the chemist. The bibliographic specialist searches the literature as well as the various computerized data bases. The engineer's activity focuses on the development of technical equipment needed for the successful process.

Screening can never be considered a routine activity, since the methods must always be adapted to the newest techniques and knowledge. The goal is always to detect and identify new substances of commercial interest and to separate them in the quickest possible way from the numerous easily detected substances that are of no commercial interest. Some of the most intelligent screening methods that are currently used are the products of Japanese scientific activity, particularly the group of Omura. For instance, 42 completely new compounds were found by Omura (1986) using systems that detected microbially produced substances with antibacterial, antimycoplasmal, antianaerobe, antifungal, antiparasite, and antitumor activity. Substances were also found that acted as herbicides and as inhibitors of penicillin, elastase, and adenosine deaminase.

Further overviews of the newer screening methods are given by Vandamme (1984), Box (1985), Verrall (1985), Cheetham (1987), and Elander (1987).

2.2 PRIMARY AND SECONDARY METABOLITES

Bu'Lock borrowed the term "secondary metabolite" from plant physiology in 1961 and applied it to microbiology. While **primary metabolites** are essential for life and reproduction of cells, and primary metabolism functions similarly in all microorganisms, the following is true for secondary metabolites:

- Every secondary metabolite is formed by only a few organisms.
- Secondary metabolites are seemingly not essential for growth and reproduction.
- Their formation is extremely dependent on environmental conditions.
- Some secondary metabolites are produced as a group of closely related structures; one strain of *Streptomyces*, for example, produces 32 different anthracyclines.
- Some organisms form a variety of different classes of substances as secondary metabolites.
- The regulation of the biosynthesis of secondary metabolites differs significantly from that of the primary metabolites.

There are several hypotheses about the role of secondary metabolites, of which Hans Zähner's, illustrated in Figure 2.1, is the most elegant. Besides the five phases of the cell's own metabolism (intermediary metabolism, regulation, transport, differentiation and morphogenesis), secondary metabolism is considered a "playing field" for the evolution of further biochemical development, which can proceed without damaging primary metabolism. Genetic changes leading to the modification of secondary metabolites would be expected not to have any major effect on normal cell function. If a genetic change leads to the formation of a compound that in some way is beneficial, then this genetic change would be fixed in the cell's genome, perhaps becoming essential. In this way the former secondary metabolite would be converted into a primary metabolite.

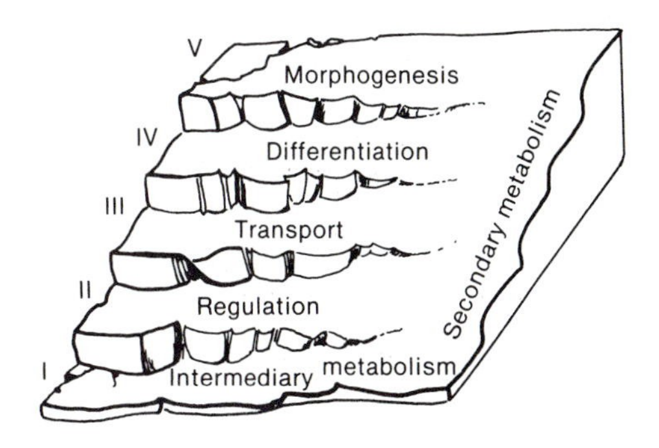

Figure 2.1 The five levels of primary metabolism with the "playing field" of secondary metabolism (Zähner, 1979).

Screening for new metabolites

There are no universal screening methods. The success of a screening program depends upon the selection of appropriate tests as well as appropriate microorganisms to be tested. The capacity of an industrial screening group for isolation of microorganisms and thorough testing is around 1000–2000 strains per year.

Today, most screening programs focus on chemotherapeutically useful products for the following areas: activity against antibiotic-resistant strains, tumors, and viruses, as well as a search for enzyme inhibitors and pharmacologically active substances (hormones, etc.). Better starter cultures for the food industry as well as microorganisms that are capable of degrading hazardous and persistent chemicals are also sought.

Of the 10,000 antibiotically active compounds known in the late 1980's, 67% are produced by microorganisms (67% by actinomycetes, 9% by other bacteria, and 15% by fungi). Additionally, about 2000 other biologically active secondary metabolites are known, as well as a large number of enzymes.

2.3 STRAINS USED IN SCREENING

The success of a screening program depends on both the kinds of organisms used and the methods for detection of activity. Currently, the choice of strain has a 30–40% influence on the outcome, the test procedure a 60–70% influence.

A gram of soil contains between 10^6–10^8 bacteria, 10^4–10^6 actinomycete spores, and 10^2–10^4 fungal spores. Less than 1% of the world's microorganisms have been intensively studied. Above all, the approximately 100,000 known fungi have been poorly studied, so that a vast number of new natural products can be expected from this group in the future.

In the isolation of new metabolic products, researchers try to isolate strains from extreme or unusual environments in the hope that such strains may be capable of producing new metabolites. For instance, microorganisms from high altitudes, cold habitats, sea water, deep sea, deserts, geysers, and petroleum fields are being examined. Depending on the inoculum source and enrichment procedure, specific groups of organisms may be isolated. Table 2.1 presents some examples of the kinds of organisms that can be isolated with various enrichment methods. The isolation of strains can be carried out with the following scheme: 1) The soil or water sample is suspended in a definite amount of sterile water to which Tween has been added as an emulsifying agent. The sample is vigorously agitated. 2) The supernatant is diluted 10^{-1}–10^{-10}. 3) Samples from this dilution series are plated on various culture media and then incubated. 4) Single colonies from the plates are picked and purified by restreaking. 5) The pure strains are maintained as agar cultures in test tubes.

The screening procedure can often be speeded up by testing the initial isolates directly for biological activity. Some examples of procedures that can be carried out directly on agar plates are given in Table 2.2. The soil or water samples are diluted directly onto the test plates and only those colonies showing activity are isolated.

The variability of metabolites produced by individual genera is somewhat limited except in the streptomycetes. For instance, except for industrial enzymes, *Bacillus* strains almost exclusively produce peptide antibiotics.

In one extensive study, 20,000 *Actinoplanes* strains were isolated, of which 13,000 were screened for the formation of antibiotics. Within

Table 2.1 Enrichment of microorganisms by selection of appropriate culture conditions

Enrichment methods	Type of isolate
Extreme pH values (pH 2–4)	Acidophiles
Low temperatures (4–15°C)	Psychrotrophs
High temperatures (42–100°C)	Thermophiles
High NaCl concentrations	*Nocardia*, halophiles
N_2 atmosphere	Anaerobes
Chitin as growth substrate	*Lysobacter*
Wood bark, roots	Myxobacteria
Pollen grains	Actinoplanes

Table 2.2 Test systems for screening of metabolites

Product sought	Test system
Antibiotics	Agar plates with strains of test microorganisms[a]. Inhibition zones as indicator of activity
β-Lactamase-resistant antibiotics	Agar plates with test microorganisms to which β-lactamase has been added
Proteases	Agar plates with casein, selection of colonies which produce clear zones on the turbid plates
Amylases	Agar plates with starch, selection of colonies after staining with iodine
Lipases	Agar plates with oil emulsion, selection of colonies after precipitation of free fatty acids with Ca^{2+}
Phosphatases	Agar plates with phenolphthalein-diphosphate and pH indicator, selection based on color change
NAD	Agar with microorganism auxotrophic for NAD

[a]For example, *Staphylococcus aureus, Proteus vulgaris, Candida albicans, Penicillium avellaneum,* bacteria resistant to or hypersensitive to aminoglycoside, macrolides, or β-lactam antibiotics.

a ten year span, 41 new antibiotics were isolated and characterized from these strains. These antibiotics turned out to be almost all either acetylmalonyl or amino acid derivatives.

2.4 TEST SYSTEMS

The success of a screening procedure is quite dependent on the development of "intelligent" tests with which known or undesirable antibiotics can be eliminated and compounds with the required properties can be recognized. For instance, in one screening program (Omura et al., 1979), a procedure was set up to discover inhibitors of cell wall biosynthesis, because all known antibiotics with this mode of action have low toxicity. In the first phase, screening was done for metabolites which inhibited *Bacillus subtilis,* but did not inhibit *Acholeplasma laidlawii* (which lacks a cell wall). In the second phase, substances were sought which inhibited the synthesis of *meso*-diaminopimelic acid, a component of the cell wall of bacteria, but did not inhibit the incorporation of leucine, an indicator of protein synthesis. In the third phase, substances whose molecular weights were greater than 1000 were eliminated by use of membranes, because larger molecules often elicit undesirable side effects when used therapeutically. With these three procedures, culture filtrates of 10,000 strains (fungi, bacteria and actinomycetes) were screened. A new antibiotic, azureomycin, was discovered, and six known antibiotics were also reisolated.

The antibiotic penicillin is a β-lactam, and one mode of resistance is through the production of β-lactamase, an enzyme which splits the β-lactam ring. Inhibitors of β-lactamase might thus prove useful in permitting penicillin therapy against resistant organisms. For the screening of microbial β-lactamase inhibitors, supernatants of the cultures were placed on agar plates containing penicillin or cephalosporin and one of the β-lactamase-producing microorganisms. Thus it could be determined during the first screening whether different β-lactamases could be inhibited.

Nozaki et al. (1987) used a screening procedure with β-lactam-hypersensitive mutants to isolate the antibiotic lactivicin, an antibiotic which combines with the penicillin-binding site of gram-negative and gram-positive bacteria even though it lacks a β-lactam ring.

Fleck and Strauss (1975) used molecular biology tests to discover an antitumor metabolite.

Continuous fermentation can provide a method to isolate from mixed cultures strains with improved properties. By raising the temperature or the alcohol concentration, microorganisms can be selected that are either thermophilic or alcohol tolerant. Similar tests can be used to isolate strains that produce temperature-stable extracellular enzymes.

REFERENCES

Box, S.J. 1985. Approaches to the isolation of an unidentified microbial product. pp. 32–51. In Verrall, M.S.

(editor), Discovery and Isolation of Microbial Products. Ellis Horwood Publishers, Chichester, U.K.

Cheetham, P.S.J. 1987. Screening for novel biocatalysts. Enzyme and Microbial Technology. 9: 194–213.

Elander, R.P. 1987. Microbial screening, selection and strain improvement. pp. 217–251. In: Bu'Lock, J. and B. Kristiansen (editors). Basic Biotechnology. Academic Press, London.

Fleck, W. and D. Strauss. 1975. Leukaemomycin, an antibiotic with antitumor activity, vol. 1, Screening, fermentation, and biological activity. Z. Allg. Mikrobiol. 15:495–503.

Lancini, C. 1980. Screening for new antibiotics. Lecture. Int. School of General Genetics: Microbial Breeding II, June 3-13, 1980. Erice/Italy.

Nozaki, Y., N. Katayama, H. Ono, S. Tsubotani, S. Harada, O. Okazaki, and Y. Nakao. 1987. Binding of a non-β-lactam antibiotic to penicillin-binding proteins. Nature 325: 179–180.

Omura, S. 1986. Philosophy of new drug discovery. Microbiological Reviews 50: 259–279.

Omura, S., H. Tanaka, R. Oiwa, T. Nagai, Y. Koyana, and Y. Takahashi. 1979. Studies on bacterial cell wall inhibitors, vol. VI, Screening method for the specific inhibitors of peptidoglycan synthesis. J. Antibiotics 32:978–984.

Vandamme, E.J. 1984. Antibiotic search and production: an overview. pp. 3–31. In Vandamme, E.J. (editor). Biotechnology in industrial antibiotics. Marcel Dekker, New York.

Verall, M.S. (editor). 1985. Discovery and isolation of microbial products. Ellis Horwood, Chichester.

Zähner, H. 1979. What are secondary metabolites? Folia Microbiol. 24:435–443.

3

Strain development

3.1 GENERAL

With the exception of the food industry, only a few commercial fermentation processes use wild strains isolated directly from nature. Mutants or recombinants which are specifically adapted to the fermentation process are used in the production of antibiotics, enzymes, amino acids, and other substances. The objective of a genetic strain improvement program depends on the process.

In general, the major motivation for industrial strain development is **economic**, since the metabolite concentrations produced by wild strains are usually too low for economical processes. Through an extensive strain development program (which may require several years), yield increases up to 100 times or more can usually be attained. The success of these programs depends greatly on the substance to be examined. For instance, the yield from products involving the activity of one or a few genes, such as enzymes, can be increased simply by raising the gene dose. However, with secondary metabolites, which are frequently the end result of complex, highly regulated biosynthetic processes, a variety of changes in the genome may be necessary to permit the selection of high-yielding strains.

For a cost-effective process, strains with improved fermentation properties may also be needed. Depending on the system, it may be desirable to isolate strains which require shorter fermentation times, which do not produce undesirable pigments, which have reduced oxygen needs, with lower viscosity of the culture so that oxygenation is less of a problem, which exhibit decreased foaming during fermentation, which are able to metabolize inexpensive substrates, with tolerance to high concentrations of carbon or nitrogen sources, or with resistance to bacteriophage.

Wild strains frequently produce a mixture of chemically closely related substances. Mutants which synthesize **one component** as the main product are preferable, since they make possible a simplified process for product recovery. Changes in the genotype of microorganisms can lead to the biosynthesis of new metabolites. Thus, mutants which synthesize modified antibiotics may be selected.

One of the most significant approaches to strain improvement can be anticipated from the use of recombinant DNA techniques. Bringing together in one organism genes from several organisms has the potential for not only increasing yields but also for producing entirely new substances. Perhaps of even greater significance is the use of recombinant DNA techniques for the production by microbes of nonmicrobial products, such as insulin, somatostatin, human growth hormone, virus vaccines, and interferon.

However, one difficulty in applying the newer genetic techniques to improvement of existing processes is that the organisms of widest use in industry are unfortunately not the organisms for which the greatest amount of basic genetic information is available. There is an apparent gap between basic research and industrial application. Biosynthesis and regulation, along with the genetic fundamentals of industrially important microorganisms, must be understood before the combination of empirical procedures currently used can be replaced by appropriate new approaches. In the last ten years in several industrial firms, steps to bridge the gap between basic knowledge and industrial application have been made, and we can anticipate a marked increase in the effectiveness of strain improvement programs. Protoplast fusion, site-directed mutagenesis, or recombinant DNA methods are examples of the use of newer technologies which have been especially useful in the production of primary metabolites such as amino acids, but are also finding increasing use in strain development programs for antibiotics.

In the present chapter we present the fundamental genetic approaches to strain improvement and show how these approaches are used in practice.

3.2 MUTATION

Introduction

Changes in the genotype are caused by **mutation** and **genetic recombination**. In a balanced strain development program each method should complement the other. In the past, procedures for achieving genetic recombination in industrially important production strains hardly existed, so that the spectacular successes of strain development in industry are basically due to the extensive application of mutation and selection. Table 3.1 shows that the success of this method has been impressive. However, current data on the production levels of industrial high-performance mutants are rarely published, so that the data in Table 3.1 may not show the full extent of development.

Although yields of penicillin in the early days of industrial production were less than 100 units/ml, today the penicillin yield is around 85,000 units/ml (approx. 50 g/l). Because of the low yield per weight of substrate used (weight of penicillin produced per weight of glucose used is around 0.12), continued increases can be expected in the future. The success of these purely empirical strain development programs depends on an optimal use of mutagenesis procedures in combination with an effective system for selecting high-yielding strains.

Spontaneous and induced mutations

Mutations occur in vivo spontaneously or after induction with mutagenic agents. Mutations can also be induced in vitro by the use of genetic engineering techniques. The **rate of spontaneous mutation** depends on the growth conditions of the organism and is between 10^{-10} and 10^{-5} per generation and per gene; usually the mutation

Table 3.1 Increase in antibiotic production through mutation and selection between 1943–1961

Antibiotic	Productivity at time of discovery units/ml	Productivity of high yield mutants units/ml
Penicillin	20 (1943)	8000 (1955)
Streptomycin	50 (1945)	5000 (1955)
Erythromycin	100 (1955)	2000 (1961)
Chlortetra-cycline	200 (1948)	4000 (1959)
Oxytetracycline	400 (1950)	6000 (1959)

(Alikhanian, 1962)

rate is between 10^{-7} and 10^{-6}. All mutant types are found among spontaneous mutations, although deletions are relatively frequent. The causes of spontaneous mutations which are thus far understood include integration and excision of transposons, along with errors in the functioning of enzymes such as DNA polymerases, recombination enzymes, and DNA repair enzymes. Because of the low frequency of spontaneous mutations, it is not cost-effective to isolate such mutants for industrial strain development. The mutation frequency (proportion of mutants in the population) can be significantly increased by using **mutagenic agents** (mutagens): it may increase to 10^{-5}–10^{-3} for the isolation of improved secondary metabolite producers or even up to 10^{-2}–10^{-1} for the isolation of auxotrophic mutants.

Spontaneous and induced mutants arise as a result of structural changes in the genome:

- **Genome mutation** may cause changes in the number of chromosomes.
- **Chromosome mutation** may change the order of the genes within the chromosome, e.g. by deficiency, deletion, inversion, duplication, or translocation.
- **Gene or point mutations** may result from changes in the base sequence in a gene.

Most information is available about point mutations. Through base substitution, a base pair in the wild type allele may be replaced by another base in the mutant allele. Several kinds of changes are recognized. A **transition** is an exchange of a purine with another purine or a pyrimidine with another pyrimidine. A **transversion** refers to the substitution of a pyrimidine with a purine or vice versa. One characteristic of point mutants is that they can revert. Another category includes those called **frameshift mutations**, which result when one nucleotide or more is inserted or deleted, thus altering the reading frame in the following transcription and translation processes, and leading to a changed amino acid sequence in the resulting protein.

Although genome mutations are important in plant genetics, mutations used in microbial strain improvement usually are point mutations; chromosome mutations also occur (e.g. deletions, duplications) but are of minor significance.

Repair mechanisms

Mutant formation is a complex process. Premutational lesions of DNA occur spontaneously or through the action of mutagenic agents; only part of these result in stable mutants during subsequent replication. The following structural changes occur in DNA:

- Pyrimidine dimers, in which two adjacent pyrimidines on a DNA strand are coupled by additional covalent bonds and thus lose their ability to pair.
- Chemical changes of single bases, such as alkylation or deamination, thus causing changes in the pairing properties of the DNA.
- Crosslinks between the complementary DNA strands, which prevent their separation in replication.
- Intercalation of mutagenic agents into the DNA, causing frameshift mutations.
- Single-strand breaks.
- Double-strand breaks.

These **premutational structural changes** can lead to mutations directly by causing pairing errors in replication or indirectly by error-prone repair in the next following round of DNA replication.

Repair systems play a significant role in the mutation process. As a result of repair, potentially lethal changes in the DNA may be eliminated. If the repair systems function in an error-free manner, potentially mutagenic lesions are eliminated before they can be converted into final mutations. A number of repair systems have thus far been discerned in microorganisms, particularly bacteria. In the following, these repair mechanisms are discussed briefly.

Photoreactivation Short-wavelength ultraviolet irradiation (254 nm) affects DNA in a number of ways, but a well-established action is the formation of thymine dimers, a state in which two adjacent thymine molecules are chemically joined, so that replication of the DNA cannot occur. When such an ultraviolet-irradiated population is subsequently exposed to visible light of a wavelength of 300 to 450 nm, the survival rate increases and the frequency of mutation decreases. This is due to the activation of a photoreactivating enzyme (photolyase) which splits thymine dimers. In the dark, this enzyme bonds with thymine dimers; in the presence of light the enzyme splits the dimers into monomer pyrimidines. Up to 80% of the thymine dimers existing in the genome can be photoreactivated. UV-induced DNA crosslinks can also be photoreactivated. This repair system functions in an error-free manner and thus does not allow mutations to occur.

Excision repair In contrast to photoreactivation, which is possible with single-stranded DNA, the complementary strand is required for excision repair. In a dark reaction, damages to the DNA, such as ultraviolet-induced pyrimidine dimers, alkylation, or deamination, are recognized by specific DNA endonucleases. Different repair paths are involved according to each DNA lesion. Excision repair can be partially prevented by inhibitors such as caffeine, acriflavin and 8-methoxypsoralen.

The mechanism of excision repair for **ultraviolet induced lesions** is different than that for lesions induced by alkylation or deamination. In one mechanism of repair of ultraviolet lesions, a **nucleotide-excision** repair mechanism is operable. In effect, defective nucleotides are cut out and replaced, according to the mechanism illustrated in Figure 3.1. For this mechanism to operate, the normal DNA replication process is not required.

In some bacteria, an ATP-dependent endonuclease has been demonstrated in ultraviolet-exposed cells. This endonuclease is controlled by three genes, *uvrA*, *uvrB* and *uvrC*. The endonuclease splits a phosphodiester bond on the 5′ side

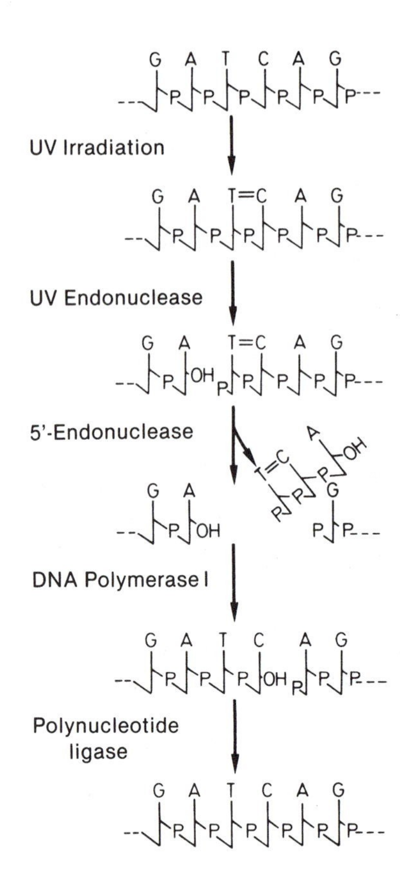

Figure 3.1 Repair of DNA containing pyrimidine dimers by nucleotide excision

of the pyrimidine dimer and 3′ hydroxyl and 5′ phosphoryl termini are produced. With the help of a 5′ exonuclease and DNA Polymerase I, a 6- to 7-nucleotide-long oligonucleotide is cut out along with the dimer and the resulting gap is expanded to approximately 30 nucleotides. The missing nucleotides are filled by DNA polymerase I *(polA)* starting from the 3′ end and are connected by a polynucleotide ligase *(lig)*. This repair mechanism is almost error-free and mutation is thus avoided.

Another mechanism for repair of ultraviolet damage is termed *recA*-dependent repair. This second repair path requires DNA replication and seems to be mutagenic since large gaps of up to 1500 nucleotides are cut out. For this repair mechanism to function *recA, recB, recC, lexA, uvrD* and *polC* genes are needed. This type of repair is a form of the so-called SOS response (see later).

Excision repair of **alkylated and deaminated DNA** operates in a different manner. In this repair path, modified bases are recognized and cut out. It is therefore also known as base excision repair (Figure 3.2). In alkylated DNA of *Micrococcus luteus* it has been shown that the 3-methyladenine formed is recognized by an N-glycosylase, which splits the $N_9C'1$ bond between the base and deoxyribose, so that a purine-free site is formed in the DNA. 7-Methylguanine is tolerated by the cells so that formation of this alkylated base is not mutagenic. The lesion is recognized by a special AP endonuclease (AP = apurinic or apyrimidinic site), which splits the DNA strand. The sequence following is similar to nucleotide excision repair, but it has not yet been determined whether identical enzymes are involved. Uracil DNA glycosylase and hypoxanthine DNA glycosylase, both involved in the repair of deaminated DNA, have been found in various bacteria.

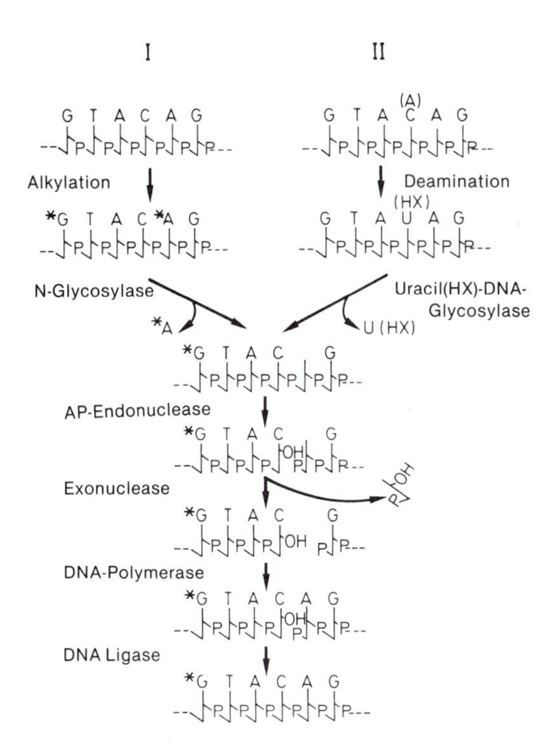

Figure 3.2 Base excision as a mechanism for the repair of alkylated DNA (I) or deaminated DNA (II) with uracil (U) or hypoxanthine (HX). *G = 7-methylguanine; *A = 3-methyladenine

Postreplicative recombination repair If a DNA strand contains a lesion which hinders base pairing, a gap in the daughter strand of approximately 1000 nucleotides is formed during replication. According to Figure 3.3, which is partially hypothetical, these gaps are filled with material from the parent strands through recombination processes (by the action of the *recA* gene product). The repair of the parent strands occurs through repair replication with the daughter strands as a matrix by means of DNA Polymerase I. Several replication and recombination stages are involved in this repair process.

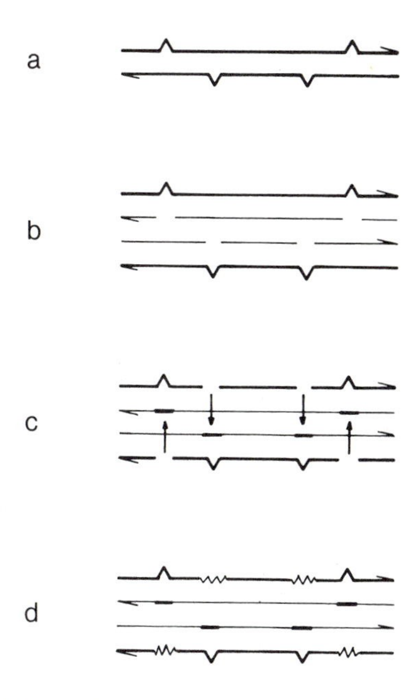

Figure 3.3 Model of post-replicative repair. a. DNA double strand with premutative lesions (__/__); b. Gaps in daughter strands which have resulted during replication; c. and d. Hypothetical exchange of parent strand DNA and replication repair (Smith, 1978)

"SOS" Repair Photoreactivation, excision repair, and postreplicative recombination generally operate as error-free mechanisms. Error-prone—and therefore mutation-inducing—methods also exist, of which the best known is the SOS-repair system of *E. coli*. Overlapping gaps, such as the ones which can result in errors in both complementary strands in the replication of DNA, would be lethal for the cell, but are filled by the SOS repair despite the absence of DNA template (Figure 3.4). Thus the chemical structure of DNA is reconstructed, but the heredity information is defective; as a result, SOS repair very likely results in mutations. In contrast to the constitutive repair systems thus far described, the SOS repair activity is inducible, being repressed in untreated wild type cells. Exposure to ultraviolet radiation or the action of other mutagens which damage DNA or cause an inhibition of replication (such as treatment with the antibiotic mitomycin C) induce the mechanism of SOS repair. In the uninduced condition, the product of the *lexA* gene binds to the identical operator sequence of several unlinked genes: *recA*, *umuDC*, and *uvrA*). After initiation by a regulatory signal, several SOS functions are derepressed (for example, DNA exonuclease V), and DNA replication can restart at the chromosomal origin. The repair is performed by a DNA polymerase which differs from the constitutive DNA polymerases I, II and III in that it continues to carry out DNA synthesis in spite of the lesion in the template, which takes place mainly when "false" bases are incorporated. It is assumed that a new polymerase is not induced but rather a factor is formed which lowers the proof-reading of one or more DNA polymerases.

Adaptive repair In *Escherichia coli*, increased resistance to the mutagenic and lethal effects of high doses of alkylating agents has been found to occur after lengthy treatment with sublethal concentrations of such agents. This effect, which

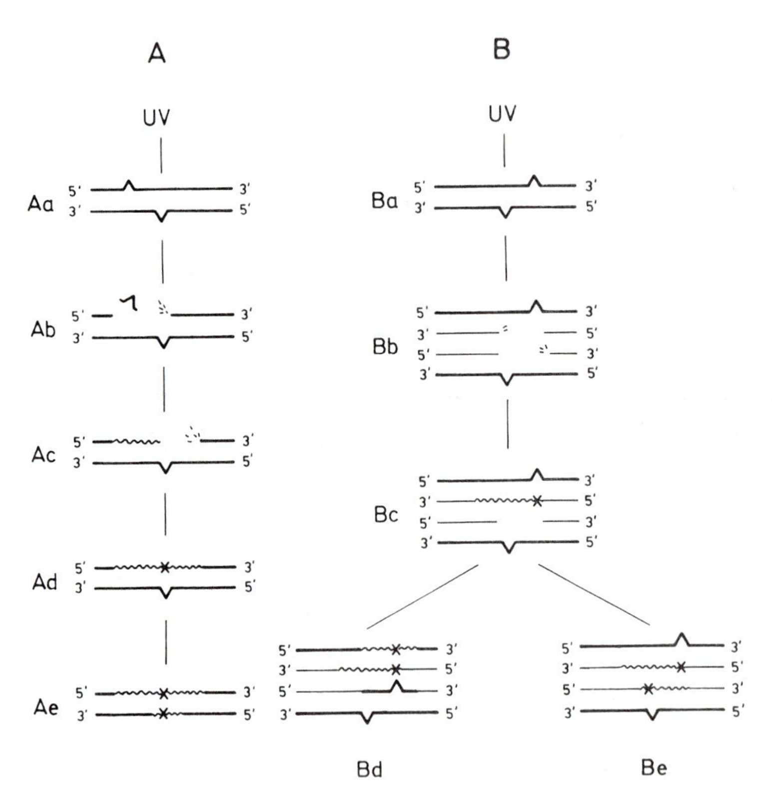

Figure 3.4 SOS repair in *Escherichia coli* (based on Witkin, 1976)
A. Excision repair: a. DNA with pyrimidine dimers (__/__). b. Excision. c. Repair replication stops at the second dimer. Exonuclease activity continues. d. SOS repair is induced. It polymerizes the DNA above and beyond the dimer and causes a mutation through erroneous incorporation (x). e. The second dimer is repaired through nucleotide excision so that both DNA strands carry the error.
B. Post-replication Repair: a. DNA with pyrimidine dimers; b. During DNA replication, gaps result in the daughter strands, which overlap and cannot be closed by recombination repair. c. SOS repair is induced, as in Ad; d. The second gap is closed by recombination. In the following replication, a daughter strand produces overlapping gaps again. See Bb; e. The second gap is also closed through SOS polymerase

cannot be elicited by ultraviolet radiation, can be traced to a further **inducible repair system**. The number of alkylated bases, particularly O^6-methylguanine, in the genome is reduced by this adaptive repair mechanism which works almost without error. In this way, the synthesis of an O^6-methylguanine DNA methyl transferase is induced. In uninduced *E. coli* between 13–60 molecules of this methyl transferase are present whereas induced cells contain more than 3000 molecules. Another enzyme, 3-methyladenine DNA glycosylase II, is involved in the breakdown of 3-methyl adenine, 3-methyl guanine, 7-methyl adenine, and 7-methyl guanine. In the induced cell the glycosylase content is increased by a factor of 20. The mechanism of the induction of these enzymes is not understood. The capacity of this repair system is limited however. O^6-methylguanine accumulates when larger mutagen doses are used, and the mutation frequency is directly proportional to the O^6-methylguanine concentration. The enzymatic mechanism involved in the elimination of the compound is not yet understood.

Reaction mechanisms of mutagens

Many mutagens induce more than one type of potentially mutagenic lesion. Thus, they frequently cause mutation directly as a result of pairing errors and indirectly as a result of errors during the repair process. In the following, the most commonly used mutagens are listed, together with their molecular reaction mechanisms. Detailed descriptions can be found in references cited at the end of this chapter.

Mutagenesis through radiation

Both ultraviolet radiation and ionizing radiation are used in mutagenesis studies. The mechanisms of mutagenesis are quite different for each type of radiation, however.

Short-wavelength ultraviolet One of the more effective mutagenic agents is short-wavelength ultraviolet radiation (UV). The wavelengths effective for mutagenesis are between 200–300 nm with an optimum at 254 nm, which is the absorption maximum of DNA. The most important products of UV action are dimers (thymine-thymine, thymine-cytosine and cytosine-cytosine; Figure 3.5) formed between adjacent pyrimidines or between pyrimidines of complementary strands, which results in crosslinks. Ultraviolet radiation mainly induces transitions of GC $\rightarrow$ AT; transversions, frameshift mutations and deletions are also found.

During the repair of ultraviolet lesions, up to 1000 pyrimidine dimers can be repaired, and with the exception of adaptive repair all repair systems are involved. To increase the frequency of mutation, the error-free mechanisms of photoreactivation and excision repair must be prevented by carrying out all manipulations under long-wavelength visible light ($>$ 600 nm) and/or through the use of caffeine or similar inhibitors of repair. The SOS repair system is primarily responsible for the production of mutations.

Long-wavelength ultraviolet radiation Radiation at wavelengths of 300–400 nm has less lethal and mutagenic effects than short-wavelength UV. However, if the exposure of cells or bacteriophages to long-wavelength UV is carried out in the presence of various dyes which interact with DNA, greater death rates and increased mutation frequency result. Especially effective activators of long-wavelength UV are the psoralen derivatives. 8-Methoxypsoralen (Figure 3.6, Structure

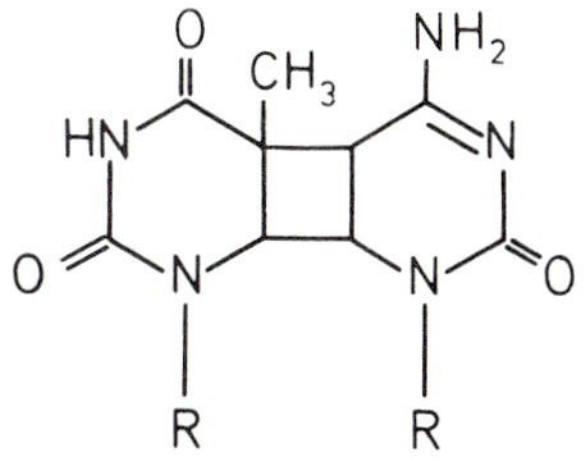

Figure 3.5 Thymine-cytosine-cyclobutane dimer, the photoproduct formed as a result of ultraviolet radiation

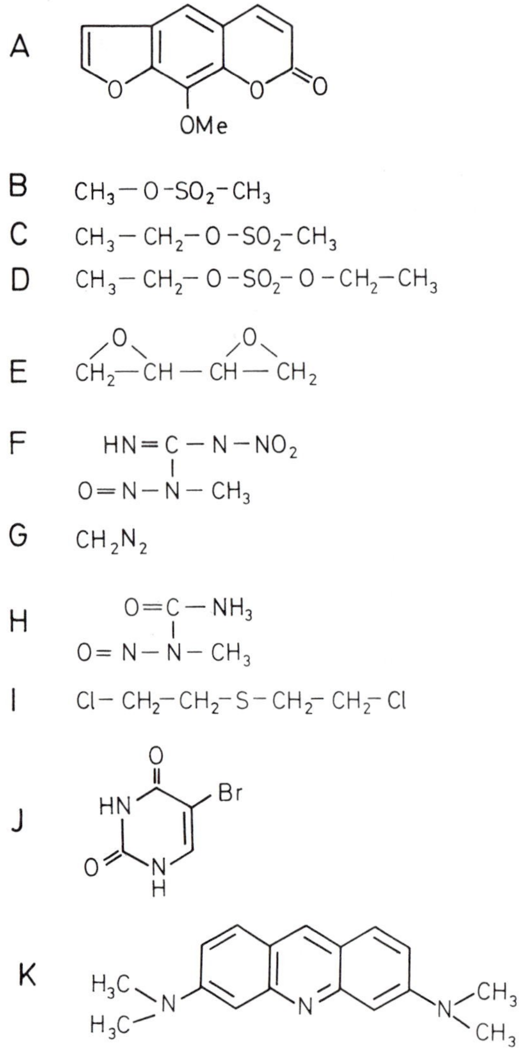

A OMe

B $CH_3-O-SO_2-CH_3$

C $CH_3-CH_2-O-SO_2-CH_3$

D $CH_3-CH_2-O-SO_2-O-CH_2-CH_3$

E $CH_2-CH-CH-CH_2$ (O, O)

F $HN=C-N-NO_2$ / $O=N-N-CH_3$

G CH_2N_2

H $O=C-NH_3$ / $O=N-N-CH_3$

I $Cl-CH_2-CH_2-S-CH_2-CH_2-Cl$

J O, HN, Br, O, N, H

K H_3C, H_3C, N, N, N, CH_3, CH_3

Figure 3.6 Structure of different mutagens
A. 8-Methoxypsoralen; B. Methylmethanesulfonate (MMS); C. Ethylmethanesulfonate (EMS); D. Diethylsulfate (DES); E. Diepoxybutane (DEB); F. N-Methyl-N′-nitro-N-nitrosoguanidine (NTG); G. Diazomethane; H. N-Methyl-N-nitrosourea (NMU); I. Di-(2-chlorethyl)-sulfide (mustard gas); J. 5-Bromouracil (BU); K. Acridine orange (AO)

A) intercalates between the base pairs of double-stranded DNA and after the absorption of long-wavelength UV, an adduct is formed between the 8-methoxypsoralen and a pyrimidine base. Absorption of a second photon causes the coupling of the pyrimidine-psoralen monoadduct with an additional pyrimidine. Biadduct formation between complementary strands of nucleic acid results in crosslinks. These lesions cannot be photoreactivated, although they are eliminated through nucleotide excision repair in conjunction with the mutation-causing SOS repair system.

Ionizing radiation Ionizing radiation includes X-rays, γ-rays, and β-rays, which act by causing ionization of the medium through which they pass. These rays are usually used for mutagenesis only if other mutagens cannot be used (e.g. for cell material impenetrable to ultraviolet rays). Single- and double-strand breaks occur with a significantly higher probability than with all other mutagens. Ninety percent of the single-strand breaks are repaired by nucleotide excision; the *recA*-dependent repair mechanism is also involved. Double-strand breaks result in major structural changes, such as translocation, inversion or similar chromosome mutations. Therefore ultraviolet radiation or chemical agents are normally preferable for mutagenesis in industrial strain development.

Mutagenesis with chemical agents

A variety of chemicals are known which are mutagenic, and these may be classified into three groups according to their modes of action:

- Mutagens which affect **nonreplicating DNA**
- **Base analogs**, which are incorporated into replicating DNA due to their structural similarity with one of the naturally occurring bases.
- **Frameshift mutagens**, which enter into DNA during replication or repair and through this intercalation cause insertion or deletion of one or a few nucleotide pairs.

Chemicals which affect nonreplicating DNA A number of chemicals are known which cause direct damage to nonreplicating DNA. **Nitrous**

acid (HNO_2) deaminates adenine to hypoxanthine and cytosine to uracil. Through the changed pairing properties of the deamination products (hypoxanthine pairs with cytosine, uracil with adenine) AT → GC and/or GC → AT transitions occur. In addition, nitrite induces crosslinks between the complementary strands. Figure 3.7 shows the establishment of the mutation after two generations. Excision and recombination repair are involved in the elimination of deamination products. Besides point mutations, deletions occur relatively frequently with nitrous acid treatment.

Hydroxylamine (NH_2OH) reacts with pyrimidines, but only the reaction with cytosine is mutagenic, whereby the amino group is replaced with a hydroxylamino group. The hydroxylamine derivative from cytosine shows tautomerization and pairs then with adenine, so that through hydroxylamine action GC→AT transitions are caused.

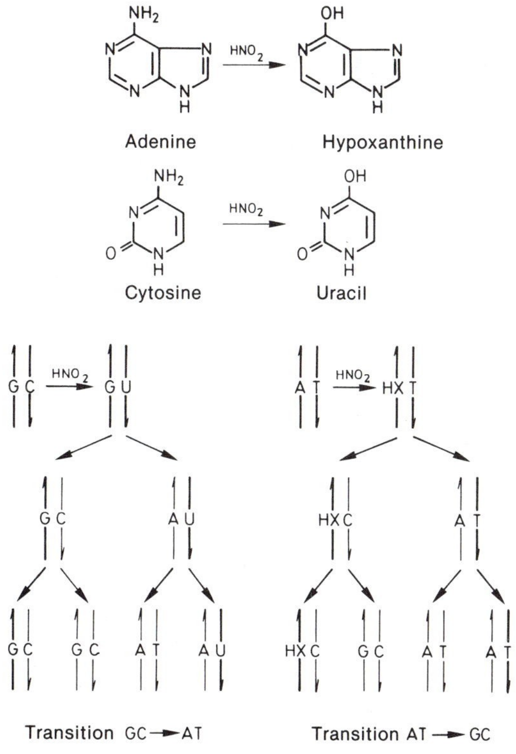

Figure 3.7 Mutation caused by nitrous acid

Another group of chemicals affecting non-replicating DNA are the **alkylating agents.** Except for ultraviolet radiation, alkylating agents are the most potent mutagenic system for practical application. Compounds frequently used include ethyl methanesulfonate (EMS), methyl methanesulfonate (MMS), diethylsulfate (DES), diepoxybutane (DEB), N-methyl-N′-nitro-N-nitrosoguanidine (NTG), N-methyl-N-nitroso-urea and mustard gas (see structures B–I, Figure 3.6). Transitions, transversions, deletions and frameshift mutations occur as a result of the action of alkylating agents.

Mutagenesis with alkylating agents occurs via various pathways. These compounds cause the formation of a whole spectrum of alkylated bases in DNA, along with phosphotriester, purine-free sites and single-strand breaks. Although 7-alkylguanine is in all cases the most common alkylation product, it does not result in mutations. O^6-alkylguanine and O^4-alkylthymine are the most important premutational lesions, and as a result of pairing errors, mainly AT → GC transitions are elicited (direct mutagenesis). A second process which also results in mutation is the induction of error-prone SOS repair (see above) when relatively high doses of mutagen are used.

It has been suggested that the occurrence of mutations in *Escherichia coli* is dosage-dependent in relation to its repair system: At a low level of alkylation of DNA the constitutive error-free systems perform the repair and mutations seldom occur. At higher mutagenic doses, on the other hand, the performance of the constitutive repair systems is not sufficient and the adaptive repair enzymes are induced. At even higher doses, the enzymes involved in SOS repair are also induced. Between the error-free adaptive repair and the error-prone SOS system there is competition for the repair of DNA lesions. The frequency of mutation is critically dependent upon which of these repair systems is working.

The use of **N-methyl-N′-nitro-N-nitroso-guanidine** (NTG) in a mutation program is difficult because of its carcinogenic effects, but it is one of the most effective chemical mutagens. A large proportion of mutants is found under optimal conditions with a low killing rate. In *Streptomyces coelicolor*, for instance, 8–10% of the survivors are found to be auxotrophs, and in *Escherichia coli* up to 50% of the surviving population consists of mutants. Ninety percent of the mutations induced by NTG are GC → AT transitions; to a small extent deletions and frameshift mutations are also found as a result of the deletion of GC pairs.

The exact molecular reaction mechanism of NTG is not yet understood. The compound is easily decomposed in vivo, and in acidic solutions nitrous acid is formed. Although nitrous acid is a mutagen, it is not effective in the pH range where NTG is active (pH 6–9); diazomethane (Figure 3.6, Structure G), a strongly methylating agent, is formed under alkaline conditions.

Besides the alkylation of nonreplicating DNA, the main point of action of NTG is at the replication point of DNA, through a change in DNA polymerase III during DNA replication. In this process, there is incorrect duplication in a short segment of the DNA until the defective polymerase is replaced by an intact molecule. This explains the observation that NTG mutations frequently occur in gene clusters.

Base analogs Because of their structural similarity, base analogs such as 5-bromouracil (BU, Figure 3.6 J) or 2-aminopurine (AP) are incorporated into replicating DNA instead of the corresponding bases thymine and adenine. The analogs tautomerize more frequently than the natural bases: BU in keto form pairs with adenine, whereas BU in enol form pairs with guanine. If the keto form of BU is incorporated, there is a AT → GC transition caused by tautomerization; if the incorporation takes place in the enol form a GC → AT transition is caused (Figure 3.8).

Base analogs are of minor importance for practical application, because for industrially important strains which are poorly understood genetically it can be rather costly to set up the optimal conditions for mutagenesis. As an example of the difficulties, the incorporation of BU in DNA takes place only if the organism is growing under thymine deficiency. Thus, thymine auxotrophs are frequently used; however, use of such auxotrophs can lead to complications in strain development, since in many cases auxotrophic mutation results in reduced production of the desired substance.

Frameshift mutagens Frameshift mutagens intercalate into the DNA molecule and cause errors which result in an alteration of the reading frame, resulting in the formation of faulty protein or no protein at all. The most commonly used frameshift mutagens are the acridine dyes, such as acridine orange (structure in Figure 3.6 K), proflavine and acriflavine. The induction of insertions or deletions is dependent on the ability of an acridine dye to become inserted between two neighboring bases of a DNA strand. The size of the acridine molecule, 3.4 Å, is exactly equivalent to that of a mononucleotide. Intercalation occurs most likely in DNA segments with identical base pairs in areas of strand irregularities, such as at the end of a chromosome, near the replication fork, or at a site undergoing recombination. A proposed model for the mode of action is illustrated in Figure 3.9.

Although acridines are useful is research, they are not very suitable for a routine isolation of mutants in strain development: although they are strong mutagens for bacteriophages T2 and T4, they have little or no mutagenic effect in bacteria.

Further methods of mutagenesis

Strains used industrially are usually less well characterized genetically than the organisms commonly used in basic research and there are practically no applicable processes other than

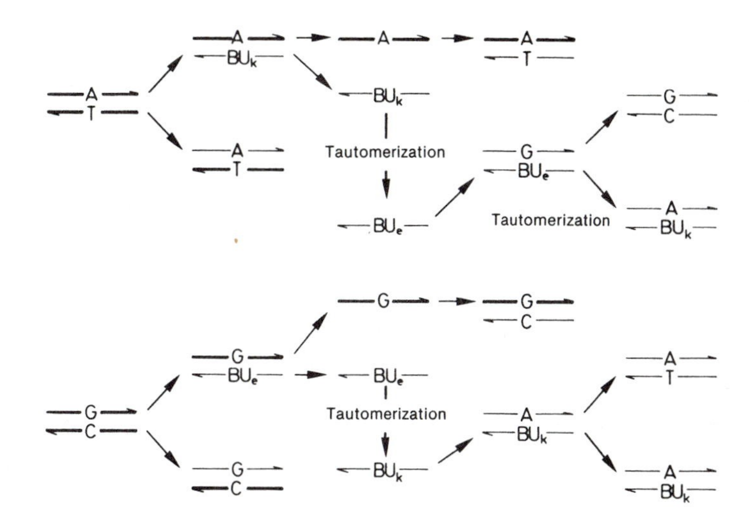

Figure 3.8 Mutation via action of 5-bromouracil (BU). A. BU is incorporated in keto form (BU_k); Tautomerization during replication causes an AT $\rightarrow$ GC-transition. B. BU is incorporated in enol form (BU_e); Tautomerization during replication causes a GC $\rightarrow$ AT-transition

mutagenesis with radiation or chemical agents. For the sake of completeness, several interesting possibilities are mentioned here which have thus far found only limited application.

Mutator genes In *Escherichia coli* the frequency of mutation can be increased by a factor of 100 through the introduction of a mutator gene. The cause of this effect is an error-prone DNA polymerase which frequently makes mistakes copying a template, resulting in either transitions or transversions, depending on the mutator gene. Three such mutator genes have been demonstrated in *Escherichia coli.* Because of the high rate of mutation, the handling of such mutator strains in production may present difficulties.

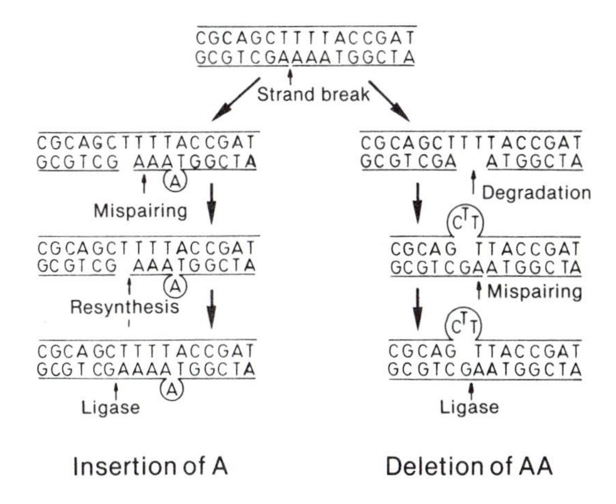

Figure 3.9 Possible mechanism for the production of frameshift mutations

IS-*Elements, transposons, and bacteriophage Mu.* The occurrence of mutations through integration into DNA of prophages or *IS*-elements and transposons is well known in *Escherichia coli* and *Salmonella typhimurium. IS*-elements are DNA sequences of variable length (800–1400 base pairs) which can be incorporated in different sites of the genome and released again. Integration and excision take place in *recA*-independent recombination. This applies also to transposons (genetic elements containing flanking *IS*-elements in inverse orientation, often with antibiotic-resistance genes) and the temperate bacteriophage Mu.

These elements destroy the function of the gene at the site of their integration. The incorporation of *IS1* also has a polar effect on the genes distal with respect to the promotor. The genes are barely or not at all transcribed, probably due to the blocking of mRNA synthesis. *IS2* bears a promotor, which, when incorporated in the appropriate orientation, results in the constitutive expression of genes located downstream. Moreover, *IS*-elements cause chromosome abberations. In particular *IS1* causes deletions, whereas *IS2* causes duplications. Transposon mutagenesis is discussed in a later section.

Comutation and sequential mutagenesis

Nitrosoguanidine (NTG), which causes multiple mutations at the replication point through its effect on DNA polymerase III, can be used to induce mutations in certain sites of the genome by means of a process called comutation or sequential mutagenesis.

Comutation When a mutation is induced in a specific locus, a large number of further mutations, so-called comutations, may be found in closely linked genes. In *E. coli* 40% of comutations are concentrated in a region of about 50,000 base pairs (about 1/60 of the genome). In *Streptomyces coelicolor* the comutation region is about twice as large. By using a **selective marker**, such as the reversion of an auxotrophic mutant, clones can be isolated after NTG mutagenesis which carry mutations in neighboring genes of the selective marker. In some cases this mutagenesis in a specific operon can be used in strain development, provided the genes controlling production have been mapped.

Sequential mutagenesis The replication of circular chromosomes is a sequential process. In *Escherichia coli*, the replication point moves from a fixed starting point O (origin) to an end point T (terminus) in a bidirectional fashion. In an exponentially growing, nonsynchronized bacterial culture, the replication of individual chromosomes is quite varied. In synchronized DNA replication, a specific genome segment is replicated in the vast majority of the chromosomes at a specific time. If mutation is induced by use of NTG pulses and the mutation frequency of certain markers is subsequently plotted against time, maxima are found at certain times during a replication cycle (Figure 3.10). The time between start of replication and the appearance of such a maximum is characteristic of the position of a specific marker on the genetic map. Sequential mutagenesis has already been used in several organisms. By this means, genetic maps can be drawn up; moreover, specific genes can be mutated as desired, provided the time of replication is known.

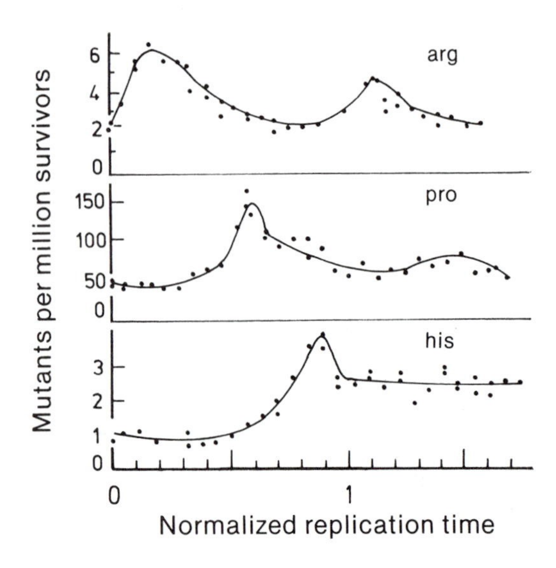

Figure 3.10 Sequential mutagenesis of *Escherichia coli* TAU-bar (Cerda-Olmedo et al., 1968). Samples of a synchronized culture (25°C) were mutagenized at 5 min. intervals with nitrosoguanidine (0.1 mg/ml; pH 5.5) and assayed for revertants to *arg*$^+$, *pro*$^+$ and *his*$^+$.

Directed mutagenesis

The methods of mutagenesis which have been discussed up to this point are completely undirected. The development of gene technology has led to revolutionary new methods which make it possible to isolate mutants of specific genes of interest. Although the details of gene technology are discussed in Section 3.6, some of the approaches to use for mutagenesis purposes are discussed here.

Deletions A circular DNA molecule that has only a single recognition site for a particular restriction endonuclease is linearized when it is treated with this restriction enzyme. The single-strand regions at the site of cutting are then digested with the specific nuclease S1, producing blunt-end fragments. When the linear molecule is introduced into a cell, recircularization can occur, although the shortened ends are generally removed by a polynucleotide ligase (Figure 3.11), thus leading to the formation of a small deletion. Larger deletions are obtained if the DNA contains two recognition sites for the restriction enzyme. For example, after separation of the fragments the larger fragment can be closed into a ring and cloned.

Insertions By use of a **linker-adaptor** approach, an adaptor DNA or a linker DNA can be inserted at the recognition site where the restriction enzyme has acted. An adaptor molecule is a chemically synthesized double-stranded DNA that can be used to connect together the ends of two DNA molecules. A linker molecule is similar but possesses a recognition site for a restriction enzyme.

Transposon mutagenesis is another method for inducing mutation via insertion. Transposons are known in both procaryotes and eucaryotes and can insert at arbitrary sites in the genome.

```
5′-GCCG CCT|CGGC-3′
3′-CGGC|GGA GCCG-5′

        ↓ (1)

5′-GCCGCCT            CGGC-3′
3′-CGGC            GGAGCCG-5′

        ↓ (2)

5′-GCCG        CGGC-3′
3′-CGGC        GCCG-5′

        ↓ (3)

5′-GCCGCGGC-3′
3′-CGGCGCCG-5′
```

Figure 3.11 Directed mutagenesis by site-specific deletion in the region of the recognition site for the restriction enzyme BglI in the genome of virus SV40. 1. Cut with enzyme BglI. 2. Removal of the single-stranded region through the action of the single-strand specific nuclease S1. 3. Joining of the DNA ends by a ligase.

Transposon Tn5 contains the gene for resistance to an aminoglycoside antibiotic which can be expressed in a wide variety of both procaryotes and eucaryotes. Several transposons have been integrated into plasmids (for instance, Tn1 and Tn3), others in either plasmids or chromosomes (for example, Tn5). Thus, transposons are available for a wide variety of purposes in gene technology.

Transposon mutagenesis offers a wide variety of advantages. A mutant phenotype with a very low reversion rate can be obtained. In addition, insertion mutations are relatively easy to isolate, since the transposons contain antibiotic-resistance markers. All one needs to do is plate on a medium containing the antibiotic; only cells containing the transposon will be able to grow and form colonies. The integration of a transposon causes an interruption in transcription, so that transposon mutagenesis exhibits a polar effect. Because of this, the site at which the transposon has been integrated into the operon can be readily determined by assaying for enzyme activity or measuring the accumulation of an intermediate product.

Point mutants at specific locations in the DNA **Bisulfite mutagenesis** can be used to convert cytosine residues in the single-stranded DNA into uracil residues, since treatment of the DNA with sodium bisulfite causes deamination of cytosine. After synthesis of the complementary strand, a transition from GC $\rightarrow$ AT results. The production of single-strand regions can be brought about using certain restriction enzymes which, in the presence of ethidium bromide, split only a single strand of the DNA. The single-stranded site can be extended by treatment of the DNA with an exonuclease enzyme.

Nucleotide analogs can be incorporated in vitro into either RNA or DNA, using an enzyme which will replicate the nucleic acid in a synchronous fashion. Suitable nucleotide analogs include N^4-hydroxycytidine triphosphate (N^4-hydroxy-CTP) or N^4-hydroxydeoxycytidine triphosphate (N^4-hydroxy-dCTP). An example of the use of this procedure is shown in Figure 3.12. N^4-hydroxycytosine is incorporated by means of an in vitro RNA synthesis using the RNA replicase from bacteriophage Qβ. The incorporated N^4-hydroxycytosine pairs in the next round of synthesis with both G and A, leading to the formation of GC ⟶ AT transitions.

Oligonucleotide mutagenesis The availability of synthetic oligonucleotides makes possible a very specific method for mutagenesis, since a nucleotide in any desired position in the DNA sequence can be substituted with any of the other three nucleotides. First the source gene is cloned into a single-stranded vector such as bacteriophage M13. To the system is added a synthetic oligonucleotide of 15 to 100 bases which contains a sequence complementary to the region of interest, but with one base that is mismatched. The added oligonucleotide then serves as a primer for the action of DNA polymerase and a complete second strand is formed. The completed double-stranded heteroduplex DNA, containing a single incorrect nucleotide, is then incorporated into a cloning vector such as *E. coli* by transformation. After replication of the cloning vector, either a mutant or wild type is obtained. If the result is a mutant, the mutation is at the desired location in the gene.

Phenotypic expression of mutations

Many mutations which result in increased formation of metabolites are recessive. When a recessive mutation takes place in a uninuclear, haploid cell (e.g. bacteria and actinomycete spores, asexual conidia of fungi), a heteroduplex results from it (Figure 3.13a); the mutant phenotype can only be expressed after a further growth step. This also applies to exponentially growing bacterial cells, which can contain 2–8 chromosomes (Figure 3.13b); not until several steps in repro-

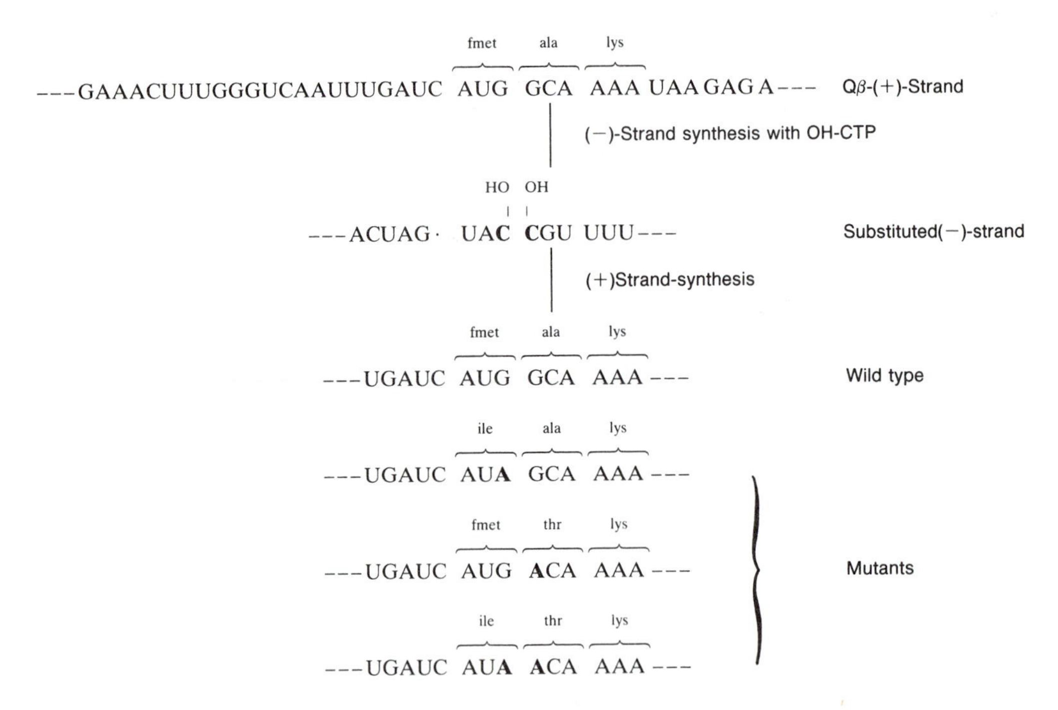

Figure 3.12 Mutagenesis through the incorporation of hydroxy-CTP during the synchronized synthesis of RNA by the RNA replicase of bacteriophage Qβ. Based on Taniguchi and Weissman (1978).

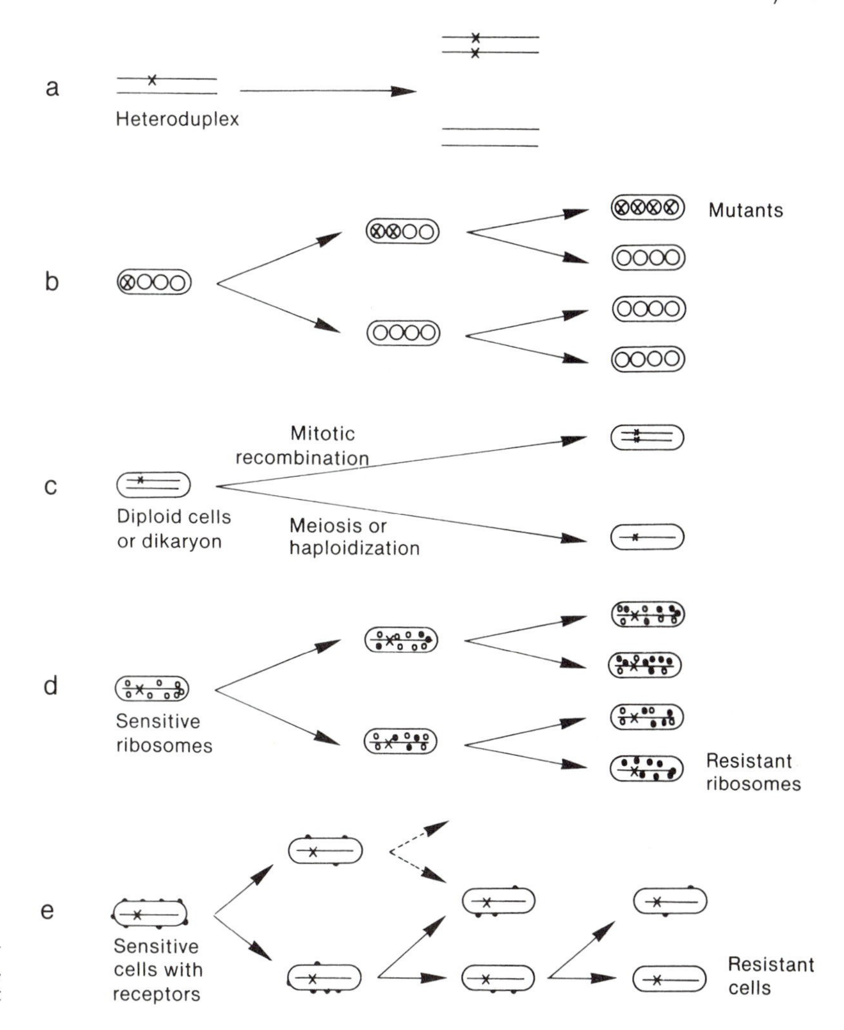

Figure 3.13 Phenotypic expression of mutants (Clarke, 1975). For explanation see text

duction **have** taken place do pure mutant clones appear.

With the filamentous actinomycetes, special procedures for mutant expression must be used. In the course of strain development, actinomycetes can lose their sporulation ability. To obtain cells for plating, the heterokaryotic mycelium which results from mutagenesis is grown and then fragmented by ultrasonic treatment or shaking with glass beads. After filtration through paper, cotton, or an 8 μm membrane filter, mycelium fragments containing only one or a few nuclear bodies are used for plating. Homokaryotic material can be ultimately selected by repeating this segregation process. Another way for attaining segregation is the preparation of protoplasts containing one or few nuclei.

In diploid or heterokaroytic eucaryotes, recessive mutations are allowed to undergo phenotypic expression after meiosis, haploidization, or mitotic recombination (Figure 3.13c).

Delays in expression which are not directly the result of genetic effects are observed, such as mutations which cause changed ribosomes (Figure 3.13d) or mutations resulting in the loss of surface receptors (as in the development of bacteriophage resistance, Figure 3.13e). In both cases, the wild type structures must be diluted out during growth before the mutation is recognizable phenotypically.

Optimizing mutagenesis

Although the molecular mode of action of some mutagens is quite well known, what can never be predicted is the effect of a mutagen on a specific gene or the effect of a mutation on a complex process, such as the biosynthesis of a secondary metabolite. The appearance of mutants, that is, strains in which a mutation has phenotypically resulted in a change, depends on several factors.

- The appearance of mutations is dependent on the **base sequence** of the gene to be mutated. Mutations are not distributed evenly around the genome; there are areas with high mutation frequency, the so-called **hot spots**. Different mutagens cause hot spots at different sites in the genome.
- The **repair systems of the cell** also play a role. In strains with partially defective repair mechanisms, organisms may be killed without having induced mutations, so that specific mutagens can be ineffective.
- A gene activity which has become lost through mutation can be restored at least partially through a second mutation, a **suppressor mutation**.

Suppressor mutations act in several different ways. Suppressors can occur in the same gene that already carries the primary mutation (intragenic suppressors). The primary missense mutation is compensated through the exchange of an amino acid or an additional deletion or insertion which corrects a primary frameshift mutation so that the reading frame remains intact. Suppressor mutations which occur in another gene (extragenic suppressors) compensate the primary mutation particularly at the level of translation, by the formation of mutant transfer RNAs or ribosomes. In strains with suppressor mutations, the function of the enzyme in question is restored up to 10% of the wild type activity. This is enough activity so that the existence of specific mutant types (such as auxotrophy) might be overlooked.

In addition to these strain-specific factors, the **treatment conditions** have a critical effect on mutagenesis. Such factors as the pH, buffer composition, mutagen concentration, exposure time, temperature, and growth phase of the organism may greatly affect the efficiency of the process. By plotting **dose-response curves** (Figure 3.14), all of these factors may be optimized. Action on DNA causes not only mutation, but also killing, due either to irreparable damage to the DNA or to formation of lethal mutations. Therefore mutants are sought from among the few surviving members of the population which had been exposed to a strong mutagenic treatment (death rate >99%), because there is thus a certain probability that each of the surviving cells carries one or several mutations. A high death rate alone,

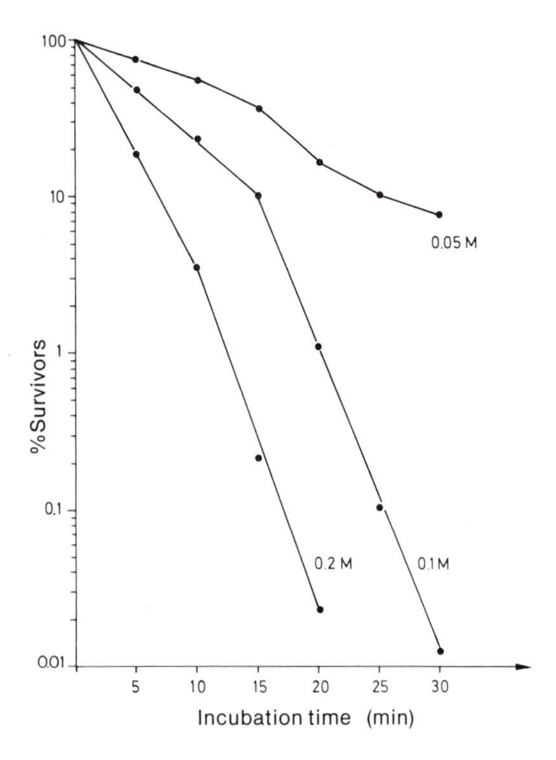

Figure 3.14 Killing of *Micromonospora inyoensis* by nitrous acid

however, is no guarantee of the occurrence of mutations in specific genes. These mutations can only be reliably determined by assessing qualitatively or quantitatively changes in the product of this gene. In the case of an antibiotic, **production** should be considered as a criterion for mutagen influence. To assess this, a random selection of survivors in a population treated with mutagens is assayed for antibiotic formation in laboratory fermentations. The antibiotic titers of mutagen-treated isolates are plotted in a histogram and compared to a control group. Figure 3.15 clearly shows the differences in the variability of the population according to mutagen concentration. Since studies of this type are quite costly, easily detectable changes such as mutations for resistance or reversion to auxotrophy are frequently used to optimize conditions for mutagenesis. However, the results of these latter experiments need not have any bearing on optimal conditions for increased formation of a desired product.

3.3 SELECTION OF MUTANTS

Besides an optimal mutagenesis, the **method of selection** is crucial for the effective screening of mutants. Looking for a desired mutant is analogous to the famous "search for a needle in a haystack". There are basically two ways of

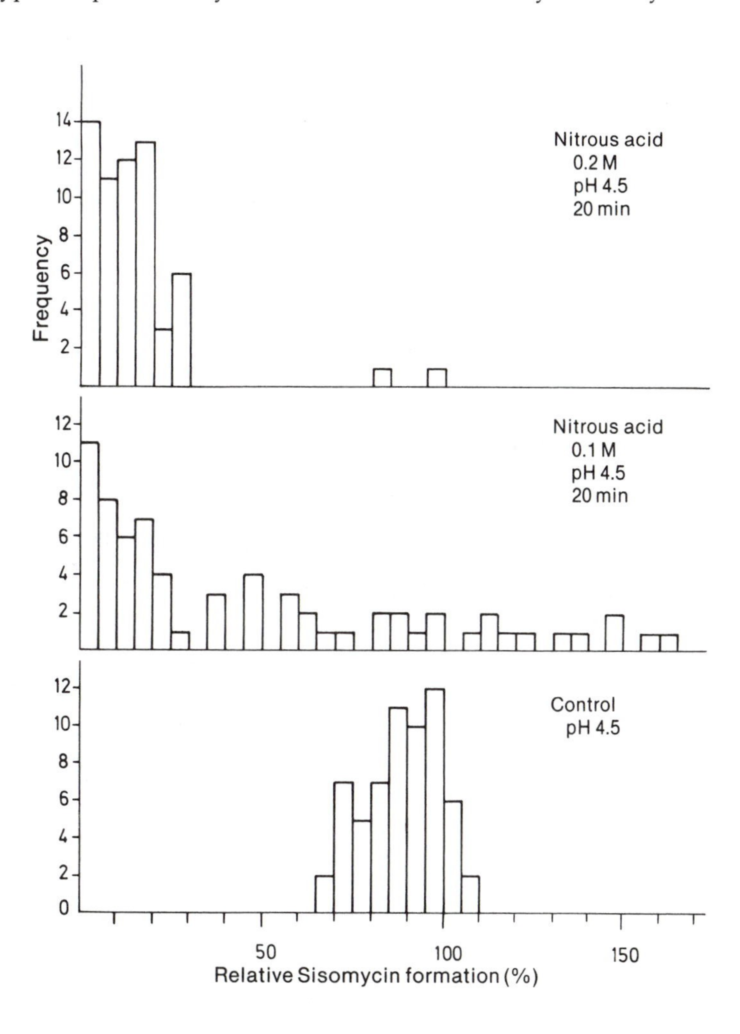

Figure 3.15 Histogram of sisomicin production by *Micromonospora inyoensis* (control strain = 100%) in relation to mutagen treatment with nitrous acid

screening. A random selection of survivors from a mutagenized population can be examined for antibiotic production or other properties in a fermentation process that closely mimics the large-scale process. This procedure is very costly, but is often the only way to find mutants with increased productivity in industrial strains. Wherever possible, a screening method is used in which selective conditions are chosen which promote the growth or early detection of the mutants.

Random screening

The high yields obtained with industrial microorganisms have been possible largely through the process of empirical selection after mutagenesis. After mutagen treatment and expression of induced mutations, a random selection of surviving clones is inspected for ability to produce the product of interest. This is done in model fermentations which are carefully adapted to the medium and fermentation parameters of the large-scale procedure, in order to maximize the likelihood that the strains will be suitable for industrial production. The best strains from such a mutation cycle are repeatedly mutated and selected. A gradual increase in the yield is attained by continuing with these steps. In this **mutation and selection program**, study does not center *only* on the strain exhibiting the best yield. This is because multiple mutations usually occur due to the high mutagen doses and in the course of strain development these unrecognized mutations can cause certain strains to show no rise in productivity. Therefore, depending on the capacity of the screening program, the 5–10 best strains of a mutation-selection cycle should be used as parent strains for future mutagenesis. These strains are normally treated with mutagens different from those used in the initial isolation. Many factors determine how many isolates must be screened to obtain strains with increased productivity. Factors which influence the size of the screening program are: frequency of mutation, extent of yield increases, the amount of time required for a mutation-selection cycle, the available test capacity of the screening program, and the accuracy of the screening test (e.g. antibiotic assay). As a rule, mutants with high yields are much rarer than those with only slight improvements. Moreover, the variability of mutagen treated populations is quite high even when mutagenesis is performed under identical conditions. Thus it is usually more economical to screen a small number of survivors (about 20–50) after many different mutagen treatments and to continue mutating strains having small yield increases as quickly as possible, than it is to test a large number of isolations after a few mutagen treatments and to hope for a one-step large yield increase.

The number of strains which must be screened to obtain mutants with a yield increase depends on the strain, the conditions of mutagenesis, the biosynthesis pathway, and the regulation of the product which is being optimized. Normally, several hundred to several thousand isolates per mutation cycle must be tested. In practice, with nonautomatic methods the number of isolates that can be tested per unit time is usually limited to 1000–2000 per week. Thus the **screening capacity** determines the speed of the progress to be expected. In the first stage of mutant screening, only one fermentation sample per isolation is usually assayed, provided that the test error is smaller than than the yield increase expected. The best isolates of the first series (usually 10–30%) are then tested in a second fermentation. Since the best strains of this second screening are then used in a still further mutation cycle, the yield increase must be statistically significant when compared to the parent strain. The number of replicates from the reference strain and the mutants should be chosen statistically. An optimal increase in yield per test period can then be calculated, if the number of isolates required to attain a specific yield increase can be tested within the time period needed for mutagenesis, colony selection, and assay of the isolates.

Several industrial companies are developing ways to automate mutant screening procedures to increase the screening capacity.

Selective isolation of mutants

Several examples of the many selective methods used in strain development are mentioned here.

Isolation of resistant mutants A high cell density of a mutagenized population can be plated on a selective medium containing a concentration of a toxic substance that prevents the wild type from growing. Only the resistant clones can develop. In this way, mutants may be isolated which are resistant to antibiotics or antimetabolites. The **antibiotic resistance** character can not only be used as a genetic marker, but mutants isolated may also have an increased cell permeability or a protein synthesis with a higher turnover, making them useful for industrial purposes.

Antimetabolite resistance can be used to select mutants which exhibit defective regulation. Altered regulation may occur in such mutants. Antimetabolites, because of their structural similarity to metabolites, may cause feedback inhibition, but are unable to substitute for normal metabolites. Antimetabolites cause death of normal cells, but analog-resistant mutants can form an excess of metabolites, in some cases through changed regulatory mechanisms (elimination of allosteric inhibition; constitutive product formation). Table 3.2 shows some antimetabolites frequently used in screening programs.

Isolation of auxotrophs By using certain blocked mutants, desired products such as amino acids and nucleotides may be formed via branching biosynthetic pathways (see Chapters 9 and 10). Auxotrophic mutations in antibiotic-producing organisms frequently result in reduced product formation. Improved strains can be obtained in some cases (e.g. in tetracycline) by isolation of prototrophic revertants (suppressor mutants) from auxotrophs. In addition, auxotrophic mutations can frequently be used as genetic markers.

The isolation of auxotrophs is done by plating of the mutagenized population on a complete agar medium, on which the biochemically deficient mutants can also grow. By means of Lederberg's well-known replica plating technique, the clones are transferred to minimal medium where the auxotrophic colonies cannot grow. These mutants are picked up from the master plates and their defect is characterized. Since in this method a large number of plates must be observed, various procedures have been devel-

Table 3.2 Frequently used antimetabolites

Natural metabolite	Antimetabolite	Natural metabolite	Antimetabolite
Adenine	Psicofuranine 2,6-Diaminopurine	Leucine	5,5,5-Trifluoroleucine 4-Azaleucine
Guanine	8-Azaxanthine	Methionine	α-Methylmethionine Norleucine, ethionine
Uracil	5-Fluorouracil	Phenylalanine	p-Fluorophenylalanine Thienylalanine
p-Aminobenzoic acid	Sulfonamide	Proline	3,4-Dehydroproline
Nicotinic acid	3-Acetylpyridine	Tryptophan	5-Methyltryptophan 6-Methyltryptophan
Pyridoxine	Isoniazid	Tyrosine	p-Fluorophenylalanine
Thiamine	Pyrithiamine	Threonine	β-Hydroxynorleucine
Arginine	Canavanine	Valine	α-Aminobutyric acid Isoleucine
Histidine	2-Thiazolalanine 1,2,4-Triazol-3-alanine		

oped to enrich for auxotrophic mutants by removing or killing prototrophic organisms. In a process known as **filtration enrichment**, after mutagenesis the spores of filamentous organisms (actinomycetes, fungi) are allowed to develop in a liquid minimal medium. The developing microcolonies of prototrophs are then separated by filtration, leaving behind in the filtrate spores of auxotrophs which have been unable to grow. The filtrate is then plated and the resulting colonies are checked for auxotrophic characteristics.

Another procedure for selection of auxotrophs, which can be used also for unicellular organisms, makes use of the fact that penicillin kills growing cells but not nongrowing cells. In this **penicillin-selection procedure**, growing cells are selectively killed by antibiotic treatment, thus enriching for auxotrophs which cannot grow on minimal medium. Depending on their mode of action several inhibitors other than penicillin can also be used in this procedure: dihydrostreptomycin for *Pseudomonas aeruginosa*, nalidixic acid for *Salmonella typhimurium*, colistin for the penicillin-resistant *Hydrogenomonas* strain H16, and nystatin for *Hansenula polymorpha*, *Penicillium chrysogenum*, *Aspergillus nidulans*, and *Saccharomyces cerevisiae*.

An enrichment procedure with **sodium pentachlorophenolate** makes use of the greater toxicity of this compound against germinating spores than against vegetative cells. The method has been successfully applied with *Penicillium chrysogenum*, *Streptomyces aureofaciens*, *S. olivaceus*, and *Bacillus subtilis*.

By these methods, enrichments for auxotrophs of 10- to 100-fold can be attained, thus increasing the probability of obtaining mutants. However, it should be remembered that the types of mutants present in the original population may be shifted; for instance, an increased proportion of proline auxotrophs has been found in *E. coli* after auxotroph enrichment.

Other procedures The presence or absence of specific enzyme activities can be observed directly in colonies growing on plates by spraying with suitable reagents or by incorporating indicator dyes into culture medium. Antibiotically active substances may be detected by measuring the inhibition of sensitive assay organisms. By using this method, the antibiotic content of a solution can also be determined. A frequently used variant of this method is the "agar plug method", in which agar cylinders with single colonies are transferred to test plates after incubation in a moist chamber (Figure 3.16). The diameter of the resulting inhibition zones serves as a measure of the antibiotic production of each strain. However, one problem with this agar plug method is that frequently there is only a slight correlation between antibiotic formation in plate culture and the antibiotic production in submerged fermen-

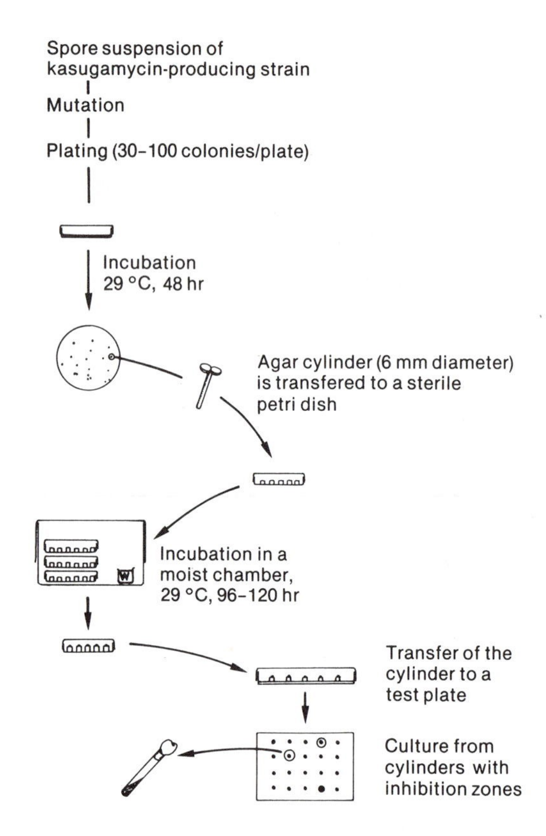

Figure 3.16 Use of the "Agar plug" method in kasugamycin strain development (Ichikawa et al., 1971)

tation. Strains which produce at high yields when grown on plates may produce at only low yields or not at all in liquid culture. Therefore the procedures mentioned are suitable for processes where simply a differentiation between productivity and nonproductivity is sufficient, such as for detecting the formation of constitutive enzymes. If screening is initiated using high-yielding strains, further increases in yield often cannot be detected by this method.

3.4 RECOMBINATION

The genetic information from two genotypes can be brought together into a new genotype through genetic recombination, which is thus another effective means of increasing the genetic variability of a population. As an example of how genetic variability can be increased, consider the following: In mutant screening, each high-yielding mutant is derived ultimately from the wild strain after a series of mutation and selection steps. A cell line with 10 mutations, for example, contains 10 new genotypes. By crossing the last high-yielding mutant with the wild type strain, 2^{10} (or 1024) different genotypes can be elicited. This same reasoning also applies to the crossing of high yielding strains from two different lines.

The advantages of genetic recombination are:

- Different alleles of the parent strains with increased metabolite production can be brought together in one strain, so that the cumulative effect of these mutations can be greater than the effect of the single mutation. However, the original hope of attaining a significant yield increase by merely recombining two high-yielding mutants has only been fulfilled in a few cases. In most cases, the productivity of the recombinants usually is intermediate between the values of the parent strains (see Chapter 13, for examples).
- In the course of strain development, there is frequently a decline in the increase in yield after each stage of mutation. Besides mutants which are selectively enriched because of their increased productivity, there is the development of inapparent mutations which prevent a further increase in the metabolite production through pleiotropic influences. With genetic crosses, these unfavorable mutant alleles may be replaced with alleles of one of the parents in the cross.
- High-yielding strains can actually increase the cost of the fermentation because of changed physiological properties (greater foaming, changed requirements for culture medium, etc.). By crossing back to wild-type strains, high-yielding strains with improved fermentation properties may be formed.

Hence an effective strain development approach should involve the use of sister-strain, divergent strain, and ancestral crosses at specific intervals, besides use of careful mutagenesis to ensure the maintenance of genetic variability.

Sexual and parasexual cycles in fungi

The fungi have two distinct types of genetic recombination processes that can be used in a strain improvement program. These are known as the **sexual** and **parasexual** cycles. We consider each of these processes in turn.

Sexual recombination Some fungi used industrially (e.g. strains of the genera *Aspergillus, Claviceps, Emericellopsis,* and *Saccharomyces*) have a complete sexual cycle. In these organisms, nuclear fusion (karyogamy) results after fusion of hyphae has led to a mingling of nuclei in the heterokaryotic mycelium. After diploid formation, recombination takes place during the subsequent meiosis process. A new genotype results either from the combination of parent chromosomes or through crossing over as a result of segment exchange of paired homologous chromatids.

Parasexual recombination Some of the most economically useful fungi, such as *Penicillium chrysogenum* (producer of penicillin) and *Cephalospo-*

rium acremonium (producer of cephalosporin), do not have a sexual cycle. Fortunately, the discovery of parasexual processes in imperfect fungi has led to the development of suitable breeding techniques. In parasexuality, the fusion of two hyphae of equal or different polarity results in a mycelium with nuclei of both parent strains. This heterokaryon is normally stable with the nuclei mingling but not interacting. However, in rare cases (10^{-7}–10^{-6} in *P. chrysogenum*), nuclear fusion occurs and a diploid nucleus is formed. In such diploid nuclei, mitotic crossing over between chromatids of homologous chromosomes may occur, resulting in genetic recombination (Figure 3.17). To obtain a recombinant, the formation of haploid cells or spores must occur. Spontaneous haploidization is relatively rare (10^{-3}), but can be induced with p-fluorophenylalanine. Haploidy occurs not through meiosis but through random distribution of the chromosomes to the progeny nuclei.

Table 3.3 shows the industrial fungi for which genetic recombination has been established. Heterozygotic diploids from parent strains with different cell lines have been used in attempts to produce strains with increased penicillin titers. The amount of penicillin formed by the diploid strains and their segregants was, however, no greater than that of the parent strains. Only when haploid recombinants were used could stable strains with improved antibiotic production be isolated.

Recombination in bacteria

The parasexual mechanisms established in vivo in bacteria include: **transformation, transduction** and **conjugation**. In each case only a fragment of the genome of the donor cell is transferred into a recipient cell which thus becomes a partial diploid (merozygote). After homologous pairing, recombination occurs, but not every DNA transfer results automatically in recombination. In **transformation**, short pieces of DNA are taken up by competent recipient cells. In **generalized transduction**, temperate phage particles which have lost a piece of their own genomes transfer a DNA fragment of the host bacteria at the optimal rate of 10^5 per phage and per characteristic. Phage P1 is an example of a phage

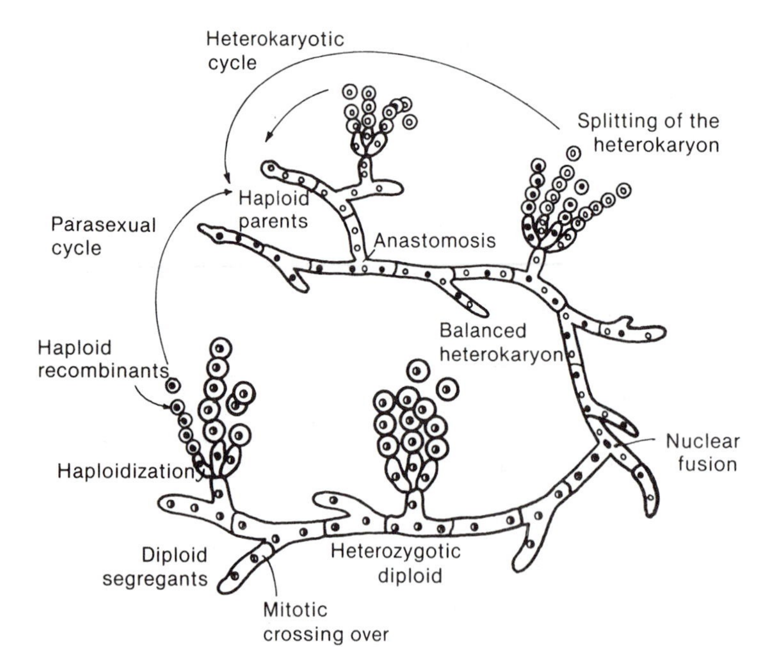

Figure 3.17 Parasexual cycle of *Penicillium chrysogenum* (Sermonti, 1959)

Table 3.3 Antibiotic-producing fungi in which recombination has been discovered

Species	Antibiotic	Recombination type
Aspergillus nidulans	Penicillin G	sexual, parasexual
Cephalosporium acremonium	Cephalosporin C Penicillin N	parasexual
Emericellopsis salmosynnemata	Penicillin N Cephalosporin C	sexual, parasexual
Emericellopsis terricola var. *glabra*	Penicillin N Cephalosporin C	sexual
Penicillium chrysogenum	Penicillin G, O, V	parasexual
Penicillium patulum	Griseofulvin Patulin	parasexual

(Hopwood and Merrick, 1977)

which brings about generalized transduction in *Escherichia coli*. In **specialized transduction**, only the loci which are adjacent to the attachment site of the phage in the bacterial chromosome are transferred. The insertion of the prophage into the chromosome results in the further incorporation of the attached piece of DNA into the genome of the host cell. Phage λ is an example of a phage which brings about specialized transduction. **Conjugation** generally involves the participation of plasmids. In this process, single-stranded DNA is transferred from the donor cell to the recipient cell after the two cells have come into contact. In the *Escherichia coli* system, F^+ cells are donor cells in which plasmid exists in free form. After contact with F^- cells which do not contain the fertility factor, a copy of the F factor is transferred. Thus an F^- population is almost completely transformed into an F^+ type. After the F factor is integrated into the chromosome, the resulting Hfr strains show a significantly higher frequency of recombination (up to 10^{-1} compared to 10^{-5} previously). In conjugation, a fragment of the F factor is first transferred, then the bacterial genome, and finally the remaining fragment of the F plasmid. Since the conjugation process is almost always interrupted before completion, the recipient cell usually does not become an Hfr cell.

Escherichia coli and *Pseudomonas* are Gram-negative bacteria of industrial interest in which conjugation is well developed and in which transduction systems are present. Pseudomonads have sex plasmids similar to F, the factor of *Escherichia coli*. Transformation systems exist for *Bacillus* strains (*B. subtilis, B. licheniformis, B. pumilus*); in addition, transduction systems have been established for *B. subtilis* with phages PBS1 and SP10.

Recombination in actinomycetes

Among industrial microorganisms, actinomycetes (filamentous Gram-positive bacteria) are economically significant as antibiotic producers. In streptomycetes, **transduction** has been sought unsuccessfully in *S. olivaceus* and *S. griseus* and **transformation** has been sought unsuccessfully in *S. aureofaciens* and *S. griseus*. In the thermophile *Thermoactinomyces vulgaris*, however, transformation has been definitely shown.

Conjugation is the most common form of genetic exchange in actinomycetes in vivo (Table 3.4). On the average, one-fifth of the genome is transferred to the recipient cell; the frequency of recombination is between 10^{-6} and 10^{-5}. It has been determined that fertility factors are involved in conjugation in three streptomycetes (*S. reticuli, S. rimosus* and *S. coelicolor*). In *S. coelicolor* A3(2), the best understood actinomycete, two plasmids have been intensively studied: SCP1 with a molecular weight of approximately 100×10^6, which contains the genes for methylenomycin synthesis; and SCP2 (molecular weight 18–20 $\times 10^6$), of which the mutant form SCP2 causes greater recombination frequency.

Recombination has not been of great importance in industrial strain development of actinomycetes. This is because parent strains with selective markers must be used to identify low frequency recombinants and the construction of doubly marked parent strains (such as auxotrophy and antibiotic resistance) is time-consuming. Singly marked strains can not be used in a genetic recombination study because the spontaneous

Table 3.4 Recombination in actinomycetes.

Streptomyces spp.
S. coelicolor (Actinorhodin, Methylenomycin)
S. achromogenes var. *rubradiris* (Rubradirin)
S. acrimycini
S. aureofaciens (Chlortetracycline)
S. bikiniensis (Zorbamycin, Zorbonomycin)
S. erythreus (Erythromycin)
S. fradiae (Neomycin)
S. glaucescens
S. griseoflavus
S. griseus (Streptomycin)
S. olivaceus
S. rimosus (Oxytetracycline)
S. scabies
S. venezuelae (Chloramphenicol)
Nocardia spp.
N. erythropolis
N. mediterranei (Rifamycin)
Micromonospora spp.
M. chalcea
M. echinospora
M. purpurea (Gentamicin)
Mycobacterium smegmatis
Thermoactinomyces vulgaris

The antibiotic produced is indicated within the parenthesis. In *Thermoactinomyces*, transformation is responsible for the genetic exchange; in all other cases, a "conjugation" mechanism has been recognized, but fertility factors have not usually been detected. (Elander et al., 1977)

reversion rate of auxotrophs is in the same range as the frequency of recombination. Moreover, there is the danger that further mutations with a negative effect on antibiotic production may occur unnoticed, since mutations to auxotrophy commonly reduce antibiotic production drastically. In addition, better results occur if the partners used in the cross come from completely different origins. Industrial strain development, which commonly begins with one particular strain, cannot fulfill this requirement. Thus, recombination can only be used at a later point in strain development when distinct cell lines exist. Under these conditions, recombination can result in strains with higher yields; kasugamycin and tetracycline yields have been improved this way. Since the development of genetic engineering methods for the actinomycetes, in vitro recombination has become of increasing importance (see Sections 3.6 and 3.7).

Protoplast fusion

Protoplasts are cells from which the cell wall has been removed by enzyme treatment. For bacteria, the enzyme lysozyme is used, whereas for fungi, chitinase or cellulase is used. The protoplasts must be stabilized against lysis by suspension in a medium containing an osmotic stabilizing agent, such as sucrose. **Recombination by protoplast fusion** or **protoplast transformation** is one of the most important developments in applied genetics in recent years. Protoplast fusion is normally rare because of the strong negative charge of the protoplast surface, but in the presence of polyethyleneglycol (PEG) the protoplasts aggregate and fusion occurs accompanied by DNA exchange. Besides the use of PEG to bring about fusion, the method of electric-field-induced fusion of protoplasts has been developed. This method results in a considerably higher frequency of protoplast fusion. After fusion, the cell wall is allowed to regenerate. In the regenerated progeny there is a significant number of recombinants.

Protoplast fusion can be used for the following:

- **Intraspecific recombination** of strains which lack sexual or parasexual systems or whose frequency of recombination is too low.
- **Interspecific hybridization** to obtain completely new organisms capable of synthesis of modified metabolites.

Engineered genes in plasmids or virus DNA can also be used to transform protoplasts.

An interesting alternative to PEG fusion is **electrofusion,** a technique developed by Zimmerman. When cells are placed in an alternating current electrical field, transient holes develop in the plasma membrane, promoting the process of membrane merging and cell fusion. With electrofusion, two or more protoplasts can be caused to fuse under microscopic control, or several cells can be fused into one giant cell. Further, protoplasts can be induced to fuse artificial phospholipid vesicles called *liposomes*. The fusion rate is around 80–90%, considerable higher than the 60% rate obtained using PEG as a fusogen. In addition to its use in the fusion of plant protoplasts, electrofusion can also be used with yeast and fungal protoplasts. However, the fusion of small bacterial protoplasts is more difficult to accomplish.

Intraspecific recombination Good genetic recombination systems exist for many industrial microorganisms: For various strains of *Bacillus, Lactobacillus, Streptococcus, Corynebacterium,* or *Brevibacterium,* in fungi such as *Aspergillus, Penicillium, Mucor, Claviceps,* and *Cephalosporium* strains, in yeasts such as *Candida, Saccharomyces, Kluyveromyces,* and in actinomycetes such as *Streptomyces, Micromonospora,* and *Nocardia.* A number of advantages arise from the use of these newer methods to achieve intraspecific recombination:

- Protoplast fusion is applicable in bacteria in which other recombination procedures have been successful. It is also applicable in *Bacillus* strains, which do not have a natural conjugation system. Gram-negative bacteria such as *E. coli* and *Providencia alcalifaciens* have also been used although the frequency of protoplast regeneration is rather low.
- The exchange of genetic material does not require the presence of fertility factors, and DNA can be recombined from up to four parental genotypes (Figure 3.18).
- The entire genome can be transferred, rather than only fragments as in conjugation, transduction, and transformation.
- The frequency of recombination is significantly increased.

An apparent increase in the formation of heterokaryons as a result of protoplast fusion has been found in some fungi. This is especially significant in *Cephalosporium acremonium,* in which the hyphal cells are mainly uninucleate, which hinders heterokaryon formation during anastomosis of hyphae. In this fungus, several recombinants with increased cephalosporin formation were isolated after protoplast fusion.

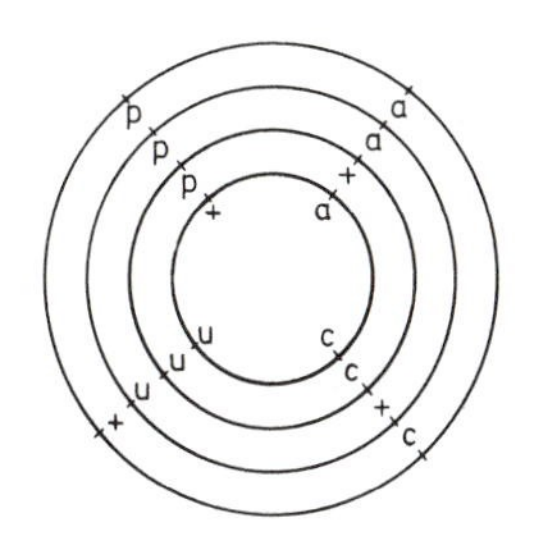

Genotype	Genotype frequency	Number of pairing partners
p a c + p a + u p + c u + a c u	1	1
p a + + p + c + p + + u + a c + + a + u + + c u	3.2×10^{-2}	2
p + + + + a + + + + c + + + + u	3.4×10^{-4}	3
+ + + +	5×10^{-8}	4

Figure 3.18 Protoplast fusion in *S. coelicolor* with 4 mating partners (Hopwood and Wright, 1978)
p = *proA1*; a = *argA1*; c = *cysD18*; u = *uraA1*

Protoplast fusion has resulted in dramatic increases in frequency of recombination in some streptomycetes, as shown in Table 3.5.

After ultraviolet radiation (approximately 99% inactivation), a further enrichment of recombinants can be attained in *S. coelicolor,* because only viable cells remain when lethal lesions are eliminated through recombination. Thus the original frequency of recombination is increased from 10^{-2} to 10^{-1}.

At these high recombination rates, a population can be assayed for recombinants after protoplast fusion without the cumbersome use of selective markers. Hence, because of the development of high-efficiency protoplast fusion tech-

Table 3.5 Frequency of recombination after conjugation and protoplast fusion in streptomycetes

Strain	Mechanism of recombination	Recombination frequency
S. coelicolor A3(2)	Conjugation	1.7×10^{-6}
(SCP1⁻SCP2⁺ × SCP1⁻SCP2⁺)	Protoplast fusion	4.7×10^{-2}
S. coelicolor A3(2)	Conjugation	1.7×10^{-7}
(SCP1⁻SCP2⁻ × SCP1⁻SCP2⁻	Protoplast fusion	6.2×10^{-2}
S. parvulus ATCC	Conjugation	6.3×10^{-6}
12434	Protoplast fusion	0.6×10^{-2}
S. lividans 66	Conjugation	1×10^{-6}
	Protoplast fusion	6.0×10^{-2}
S. griseus CUB 94	Conjugation	2.3×10^{-6}
	Protoplast fusion	1.0×10^{-2}

(Hopwood et al., 1977)

niques, not only mutation and selection but also recombination and selection may be considered equally useful methods of strain development.

Interspecific hybridization This approach allows genetic information from different species to be combined *in vivo* in order to create new or modified products. The use of protoplast fusion for this objective is being examined by different antibiotic manufacturers.

Among the fungi, interspecific crosses have been attempted between *Aspergillus nidulans* × *A. rugulosus; A. nidulans* × *A. fumigatus; Penicillium chrysogenum* × *P. cyaneofulvum; P. cyaneofulvum* × *P. citrinum; Saccharomyces cerevisiae* × *S. diastaticus; Kluyveromyces lactis* × *K. fragilis.* Correct heterokaryons and recombinants have only been obtained in the cross between *P. cyaneofulvum* × *P. citrinum,* in which a fusion frequency of less than 10^{-5} was obtained as compared to an intraspecific frequency of 40–60%. Interspecific crosses among streptomycetes have given similar results. The low frequency of heterospecific fusion may because of genome inhomologies which prevent recombination from taking place or because of the presence of restriction/modification systems that lead to physiological incompatibility.

The fusion frequency of intergeneric crosses between *Candida tropicalis* × *Saccharomyces fibuligera* is also quite low, around 10^{-5}. Crosses between *Saccharomyces cerevisiae* × *Lipomyces kononkoae* showed several fusion products which exhibited metabolic properties of both parental strains, as well as the formation of nonparental segregants after haploidization.

Transformation and transfection of protoplasts Recombinant DNA technology made possible for the first time the development of methods for transformation of protoplasts with plasmid, chromosomal, or viral DNA. In the basic method, which is simple and generally applicable, protoplasts are treated with DNA in the presence of PEG and Ca^{2+}. In this way, all of the numerous techniques of genetic engineering become available for use with microorganisms of industrial interest (fungi, actinomycetes, other bacteria): amplification of genes, restoration of metabolism in deficient mutants, in vitro mutagenesis with protein engineering. Thus, even microorganisms for which conventional host-vector systems are not available can be handled. For example, excellent transformation systems have been developed for microorganisms of interest to the dairy industry such as *Streptococcus lactis* and various lactobacilli, for *Staphylococcus carnosus* and *Bacillus subtilis,* and vitamin B_{12}-producing *Propionibacterium freudenreichii.* Transformation systems have also been developed for β-lactam-producing strains of *Streptomyces clavuligerus* and *S. wadayamensis,* vancomycin-producing *Nocardia orientalis,* erythromycin-producing *S. erythreus,* and gentamicin-producing *Micromonospora purpurea* subspecies *luridus.* Using virus vectors, transformation (transfection) systems have been described for *Streptomyces, Thermomonospora, Mycobacterium smegmatis,* and *Brevibacterium lactofermentum.*

3.5 REGULATION

Both catabolic and anabolic processes are regulated and metabolism is generally so efficient that excess products are not formed. Strains with less efficient regulation can be selected in a screening process. It is well established that strain development and the optimization of fermentation conditions lead to a relaxation of regulation in the producing strains. A broad understanding of biosynthesis, the enzymes involved in these processes, and their regulation is necessary for developing a rational approach to the alteration of the regulation of a fermentation process.

Microbial metabolism is controlled by the regulation of both enzyme activity and enzyme synthesis. We discuss both types of regulatory phenomena here.

Regulation of enzyme activity

Extensive research over the past several decades has shown that the activity of enzymes can be controlled by different mechanisms. We discuss here those mechanisms which are thought to be of significance for industrial process development.

Feedback inhibition In an **unbranched** biosynthetic pathway, the end product inhibits the activity of the first enzyme of the pathway, a process called *feedback inhibition.* A conformation change and hence inactivation (allosteric effect) occurs when an effector (end product) is attached to a specific site of the enzyme (allosteric site) The end product thus inhibits the activity of the enzyme noncompetitively.

In a **branched** biosynthetic pathway, feedback inhibition of the first common enzyme by means of one of the end products would cause more than one end product to be affected. In branched biosynthetic pathways, different kinds of feedback inhibition are found:

- The end product inhibits the first enzyme in each case *after* the branch point.
- The first step in the common synthesis path is catalyzed by several isoenzymes, each of which can be regulated independently.
- The first common enzyme in a branched biosynthetic pathway is influenced by each end product only slightly or not at all; there must be an excess of all end products for inhibition to occur (a phenomenon called multivalent inhibition).
- Each end product of a branched pathway acts as an inhibitor; cumulative inhibition is the effect of all the inhibitors.

Examples of various feedback-inhibition reaction types are given in Chapters 9 and 10.

Energy charge Feedback inhibition also regulates catabolic pathways in which ATP is the main products. The relative concentrations of AMP, ADP and ATP in the cell are used in the following formula to calculate the energy charge (EC):

$$\text{EC} = \frac{(\text{ATP}) + 0.5\,(\text{ADP})}{(\text{AMP}) + (\text{ADP}) + (\text{ATP})}$$

The values of the EC calculated in this way lie between 0 and 1.0.

If the EC is high, the activities of enzymes involved in ATP synthesis are inhibited (for example, isocitrate dehydrogenase is inhibited when the energy charge is ≥ 0.8). Conversely, the activities of anabolic enzymes (e.g., aspartokinases), which consume ATP, are stimulated by a high energy charge. Thus, an alteration in the rate of catabolism, since it affects ATP level and hence energy charge, may cause an increase or decrease in the activity of a variety of enzymes.

Breakdown of enzymes Enzymes which are no longer needed in metabolism may be broken down through the action of highly specific proteases. One of the best-known examples is the enzyme tryptophan synthetase in *Saccharomyces cerevisiae,* which is broken down specifically when the cells go into the stationary phase.

Modification of enzymes The activity of some enzymes (such as glutamine synthetase in *Escherichia coli*) is controlled by conformational changes, such as phosphorylation or adenylylation.

Regulation of enzyme synthesis

At least three mechanisms have been detected which regulate synthesis of enzymes. Note that although feedback inhibition affects enzyme molecules that have already been made, the following mechanisms control the actual synthesis of enzyme protein.

Induction Some enzymes are formed irrespective of the culture medium; such enzymes are called constitutive. Many catabolic enzymes are induced: they are not formed until the substrate to be metabolized is present in the medium. The product of one enzyme can in turn induce the synthesis of another enzyme (sequential induction).

Repression Anabolic enzymes are generally present only when the end product is absent. The excess end product suppresses enzyme synthesis, acting as a co-repressor.

Attenuation This is a further mechanism for the control of gene expression which is involved in the biosynthesis of amino acids in bacteria. Attenuation can be the only regulatory step as in the case of histidine biosynthesis in *Salmonella typhimurium*, or it can work in addition to a repressor-operator mechanism as with tryptophan in *Escherichia coli*. According to the attenuator model, the transcription rate of an operon is regulated by the secondary structure of the leader sequence of the newly transcribed mRNA. The structure of this leader sequence determines whether the transcription of the operon is continued by the RNA polymerase or a termination occurs. If termination occurs, the mRNA transcription ceases and the enzyme or enzymes coded for by that mRNA are not made. In the tryptophan situation, repression has a large effect on enzyme synthesis whereas attenuation has a more subtle, although still important, effect.

Excess production of primary metabolites

The growing understanding of the biochemistry and genetics of microorganisms has led to the production of strains which excrete excess primary metabolites (amino acids, vitamins, purine nucleotides). This has been accomplished primarily by eliminating feedback inhibition.

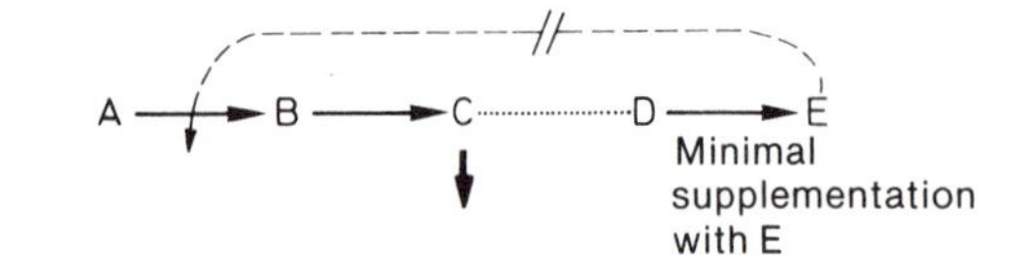

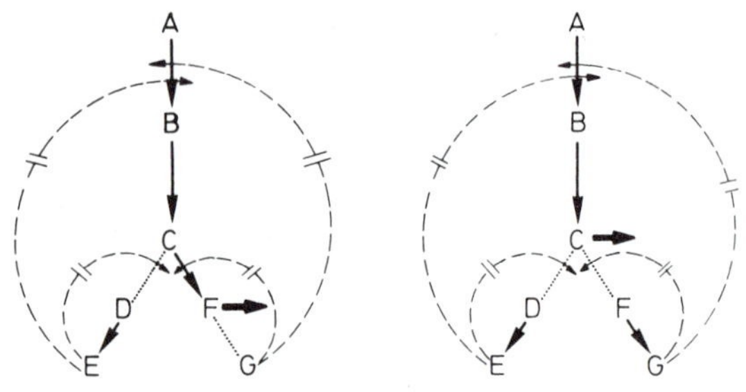

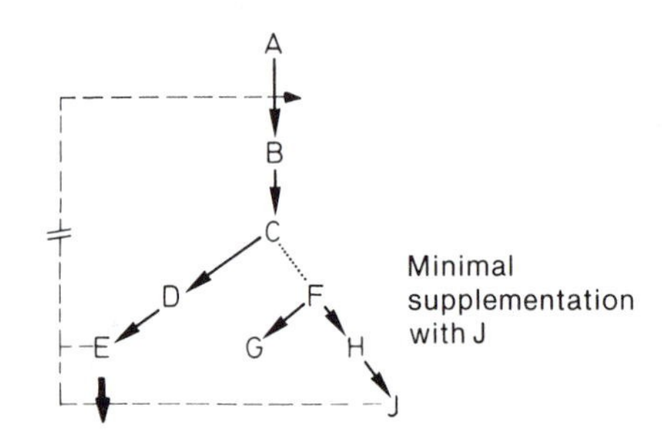

Figure 3.19 Overproduction of primary metabolites by auxotrophic mutants. ···· Reaction step which is blocked due to auxotrophic mutation; – – – Feedback regulation. ➡ Primary metabolite produced in excess (Demain, 1972)

- The elimination of end product inhibition or repression is achieved by using auxotrophic mutants that can no longer produce the desired end product due to a block in one of the steps in the pathway. By adding the required end product in low amounts, growth occurs but feedback inhibition is avoided. Excretion of the desired intermediate product thus occurs. Both branched and unbranched pathways can be manipulated in this way (Figure 3.19).

- A second method is the selection of mutants that are resistant to antimetabolites (see Section 3.3). In this case either the enzyme structure is changed so that the corresponding enzyme lacks the allosteric control site, or mutations in the operator or regulator gene (O^c-, R^--mutants) result in constitutive enzyme production and thus overproduction.
- In mutants with a block in an allosterically regulatable enzyme, suppressor mutations can lead to restoration of enzyme activity; however, these enzymes are not allosterically controllable.

Regulation and overproduction of secondary metabolites

The methods described above, which were used first for primary metabolites, can be successfully applied to secondary metabolites as well (Table 3.6). When a branched biosynthetic pathway simultaneously leads to primary and secondary metabolites (Figure 3.20), an auxotrophic mutation in the biosynthesis of the primary metabolite can lead to an increased production of the secondary metabolite. Production of secondary metabolites is controlled by 5 different classes of genes:

Table 3.6 Overproduction of antibiotics by feedback-resistant mutants

Microorganism	Antibiotic	Characteristic
Penicillium chrysogenum	Penicillin	Reduced sensitivity to feedback inhibition by valine
S. griseus	Candicidin	Resistance to tryptophan antimetabolites
S. lipmanii	Cephamycin	Resistance to leucine antimetabolites
S. viridofaciens	Chlortetracycline	Suppression of a *met*⁻ mutation
S. antibioticus	Actinomycin	Suppression of a *ilv*⁻ mutation
S. lipmanii	Cephamycin	Suppression of a *cys*⁻ mutation
S. viridofaciens	Chlortetracycline	Reversion of a nonproducer

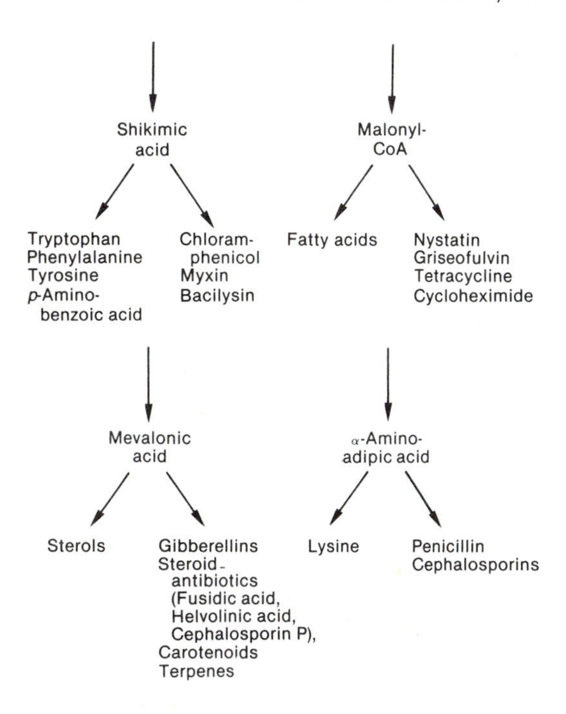

Figure 3.20 Branched biosynthetic pathways which simultaneously lead to primary and secondary metabolites (Demain, 1972)

1. **Structural genes**, which code for enzymes involved in secondary metabolite biosynthesis.
2. **Regulatory genes**, which control secondary metabolite synthesis.
3. **Resistance genes**, which keep antibiotic-producing strains immune to their own products.
4. **Permeability genes**, which control the uptake and excretion of substances.
5. **Regulatory genes**, which control primary metabolism and thus indirectly affect the biosynthesis of secondary metabolites.

Many genes are involved in the synthesis of secondary metabolites. It has been estimated that 300 genes are involved in chlortetracycline biosynthesis and approximately 2000 genes are directly or indirectly involved in neomycin biosynthesis. In such a complex system, a rational approach to increased yield is possible only in rare cases because there is insufficient data on biosynthesis and regulation of secondary metabolite production, on the supply of intermediates of primary metabolism for secondary metabolite synthesis, and on the energy links between catabolic, anabolic and amphibolic pathways.

Regulatory mechanisms that affect the products of secondary metabolism are outlined below.

Induction In batch fermentations with readily metabolizable carbon and nitrogen sources, secondary metabolites are formed primarily after growth has ceased. The logarithmic growth phase is called the **trophophase**, and the subsequent phase, in which the secondary metabolite may be produced, is called the **idiophase**. Secondary metabolites are therefore also sometimes referred to as **idiolites**. In all cases studied thus far, the synthesis of enzymes involved in secondary metabolism is repressed during the trophophase (Table 3.7). The composition of the culture medium can however also be so arranged so that a significant fraction of a slowly metabolizable substrate is used, the organism thus growing under suboptimal conditions, leading to a situation where growth and secondary metabolite formation occur in parallel.

There is only scanty information on the nature of the induction of enzymes of secondary metabolism: Methionine induces cephalosporin and fosfomycin synthesis, factor A induces streptomycin production (see Section 13.4) and tryptophan is a precursor and regulator of ergot alkaloid biosynthesis. (see Section 14.6).

Table 3.7 Key enzymes of secondary metabolism which are induced at the end of the trophophase

Enzyme	Secondary metabolite
Amidinotransferase	Streptomycin
Acyltransferase Phenylacetate-activating enzyme	Penicillin
Oxidoreductase Transmethylase	Tylosin
Synthetase I and II	Gramicidin S
Phenoxazinone synthetase	Actinomycin
Dimethylallyl tryptophan synthetase Chanoclavin-I-cyclase	Ergot alkaloids

Endproduct regulation It is known that antibiotics inhibit their own biosynthesis (e.g. penicillin, chloramphenicol, virginiamycin, ristomycin, cycloheximide, puromycin, fungicidin, candihexin, streptomycin). The mechanism of feedback regulation has only been explained in a few cases: chloramphenicol represses arylamine synthetase, which is the first enzyme in the biosynthetic pathway which branches off from aromatic biosynthesis to chloramphenicol. With chloramphenicol and penicillin, it has been shown that the concentration of the end product which inhibits corresponds to the production level. Thus, if strains could be isolated which were less sensitive to endproduct inhibition by these antibiotics, they might produce higher yields.

Catabolite regulation Catabolite regulation is a general regulatory mechanism in which a key enzyme involved in a catabolic pathway is repressed, inhibited, or inactivated when a commonly used substrate is added. Substrates which have been found to bring about catabolite repression include both carbon and nitrogen sources.

Carbon sources: Biosynthesis of different secondary metabolites (antibiotics, gibberellins, ergot alkaloids) is inhibited by rapidly fermentable carbon sources, particularly glucose (Table 3.8). Depending on the organism and metabolite, the basic mechanism of this carbon catabolite regulation is different. One is the well-known **carbon catabolite repression** found in many bacteria, yeasts and molds which involves a catabolite activator protein (CAP) that must combine at the promoter site before RNA polymerase can attach. The CAP will only bind if it is first complexed with cyclic adenosine monophosphate, *cyclic AMP*. Readily utilizable carbon sources such as glucose stimulate an enzyme which causes the breakdown of cyclic AMP, thus rendering CAP inactive. Thus, glucose inhibits the synthesis of the mRNA for any enzyme requiring CAP for its biosynthesis.

If readily utilizable carbon sources such as glucose elicit catabolite regulation, how is it pos-

Table 3.8 Catabolite regulation in antibiotic biosynthesis

Antibiotic	Carbon Source	
	Inhibitory	Not inhibitory
Penicillin	Glucose	Lactose Glucose-feeding
Cephamycin	Glycerol	Asparagine
Cephalosporin C	Glucose	Sucrose
Actinomycin	Glucose	Galactose
Streptomycin	Glucose	Mannan and L-Rhamnose
Siomycin	Glucose	Maltose
Indolmycin	Glucose	Fructose
Bacitracin	Glucose	Citrate
Chloram-phenicol	Glucose	Glycerol
Mitomycin	Glucose	Glucose-feeding
Neomycin	Glucose	Maltose
Kanamycin	Glucose	Galactose
Butirosin	Glucose	Glycerol
Puromycin	Glucose	Glycerol
Novobiocin	Citrate	Glucose
Candicidin	Glucose	Glucose-feeding
Candihexin	Glucose	Glucose feeding

Listed are carbon sources which cause catabolite regulation and carbon sources which do not inhibit product formation.

sible to use these substrates in an industrial fermentation? In the manufacture of antibiotics, carbon sources other than glucose may be used or glucose can be fed at low rates, to minimize catabolite repression.

Nitrogen sources: In several antibiotic fermentations it has been observed that ammonia or other rapidly utilizable nitrogen sources act as inhibitors. The fundamentals of this regulation have not yet been completely understood, although glutamine synthetase and glutamic dehydrogenase are considered key enzymes. In enteric bacteria it has been established that glutamine synthetase has a regulatory function in the synthesis of additional enzymes which are involved in nitrogen assimilation.

Phosphate regulation In a culture medium **inorganic phosphate (P_i)** is required within a range of 0.3–300 mM for the growth of procaryotes and eucaryotes. However, a much lower phosphate concentration inhibits the production of many secondary metabolites. In a number of systems studied, the highest P_i concentration which allows unimpeded production of secondary metabolites is about 1 mM; complete inhibition of production occurs at about 10 mM P_i. Phosphate regulation has been observed in the production of alkaloids (Section 14.6), gibberellins and particularly in several antibiotics (Table 3.9). The phosphate regulation mechanism is not yet fully understood. P_i controls the metabolic pathways which precede the first stage of secondary metabolite formation, but also affects the biosynthesis of secondary metabolites themselves.

Several varying phosphate effects have been described. In one type of mechanism, phosphate stimulates primary metabolism. The consumption of P_i in the culture medium seems to be one of the main causes for the metabolic change from the trophophase to the idiophase. By adding P_i at the end of the trophophase, the idiophase can be delayed; after addition of P_i in the idiophase, growth resumes and antibiotic synthesis stops. In a second mechanism, phosphate inhibits or represses phosphatases, enzymes which are involved in the biosynthesis of secondary metabolites. This mechanism has been demonstrated for the biosynthesis of the aminoglycoside antibiotics. A third mechanism leads to a shift in the

Table 3.9 Some antibiotics whose formation is inhibited by organic phosphate

Antibiotic	Producer	P_i range (mM) permitting antibiotic formation
Streptomycin	*Streptomyces griseus*	1.5 – 15
Kanamycin	*S. kanamyceticus*	2.2 – 5.7
Bacitracin	*Bacillus licheniformis*	0.1 – 1.0
Gramicidin S	*B. brevis*	10 – 60
Amphotericin B	*S. nodosus*	1.5 – 2.2
Candicidin	*S. griseus*	0.5 – 5.0
Chlortetracycline	*S. aureofaciens*	1 – 5
Vancomycin	*S. orientalis*	1 – 7
Actinomycin	*S. antibioticus*	1.4 – 17
Tetracycline	*S. aureofaciens*	0.14 – 0.2
Cycloheximide	*S. griseus*	0.05 – 0.5

(from Martin, 1977)

carbohydrate metabolism from the pentose phosphate cycle or hexose monophosphate pathway to glycolysis when excess phosphate is present. $NADPH_2$ thus becomes a limiting factor in the synthesis of antibiotics.

Finally, it has been shown that phosphate restricts the induction of secondary metabolite production. For instance, dimethyl allyltryptophan synthetase, the first specific enzyme of ergot alkaloid biosynthesis, is not produced in the presence of high P_i concentrations (see Section 14.6).

Some data indicate that phosphate regulation occurs via a control system at the transcription level. A model of negative control shown in Figure 3.21 has been proposed for the induction of candicidin synthesis. The repressor concentration decreases at the end of the trophophase, so that the transcription of genes that code for antibiotic synthesis is possible. It has not yet been determined whether phosphate itself functions as a corepressor or whether it regulates the content of intracellular effectors (cAMP; ATP; phosphorylated nucleotides such as ppGpp or pppGpp). In the case of candicidin synthesis, an effector may be ATP. The intracellular ATP content decreases sharply before the production of the antibiotic, but increases rapidly after adding P_i.

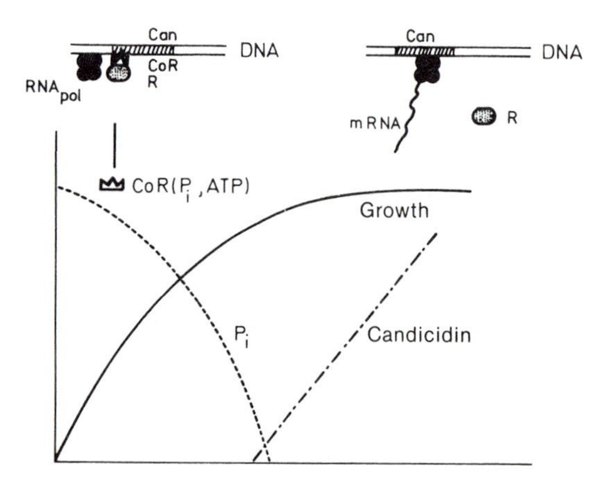

Figure 3.21 Model for control of candicidin synthesis through P_i or ATP (Martin et al., 1979).
RNA_{pol} RNA polymerase; CoR corepressor; R repressor

The industrial production of P_i-regulated secondary metabolites is carried out, as far as possible, under conditions of P_i limitation. The standard culture media used for production often contain relatively high proportions of P_i (around 2.5 mM) and at these levels, phosphate can decrease yields. The isolation of phosphate-deregulated (PD) mutants (as has been done for candicidin producing strains) seems to be a worthwhile possibility for strain development. PD mutants, all of which are less sensitive to phosphate regulation, frequently produce candicidin up to 70% better than the wild type in a normal medium.

Autoregulation In some actinomycetes it has been possible to show that differentiation and secondary metabolism are subjected to a type of "self-regulation" from low-molecular weight substances. For instance, in *Streptomyces griseus* and *S. bikiniensis* the formation of streptomycin, the development of streptomycin resistance, and spore-formation are all affected by factor A, a substance produced by the streptomyces themselves (see Figure 13.24 for the structure). It has been shown that the streptomycin resistance property is due to the increased transcription of the gene for the enzyme, streptomycin phosphotransferase, induced by the factor A. The effect on streptomycin formation is thought to be due to a shift in the metabolism of the carbohydrate source: although the activity of the enzyme glucose-6-phosphate dehydrogenase is high in factor A-deficient mutants, this enzyme cannot be demonstrated in high-yielding strains. Addition of factor A to mutants leads to a strong decrease in enzyme activity. It is assumed that when the pentose phosphate cycle is blocked through the absence of glucose-6-phosphate dehydrogenase, glucose is channeled into pathways involved in the formation of streptomycin units.

In a sense, factor A can be considered analogous to a hormone. Autoregulatory mechanisms similar to that of factor A have been found in other actinomycetes. For instance, a factor is hypothesized in *S. virginiae* which stimulates the formation of the antibiotic virginiamycin. In rifamycin-producing *Nocardia mediterranei* butyryl phosphoadenosine has been characterized as a regulatory factor. Two γ-lactones (L factors) have been shown to be autoregulatory agents in leukaemomycin-producing *S. griseus.*

3.6 GENE TECHNOLOGY

By gene technology is meant those techniques that permit the manipulation of genes as biochemical entities. Gene technology includes in vitro recombination, gene cloning, gene manipulation, and genetic engineering. These techniques permit the introduction of specific DNA sequences into procaryotic or eucaryotic organisms and the replication of these sequences; that is, to clone them. To carry out these procedures, the following steps are necessary:

- The DNA sequence to be cloned must be available.
- The sequence must be incorporated into a vector.
- The vector with the DNA insert must be introduced by transformation into a host cell, where the vector must replicate the insert in a stable manner.
- The clone which contains the foreign DNA must be selectable in some manner.

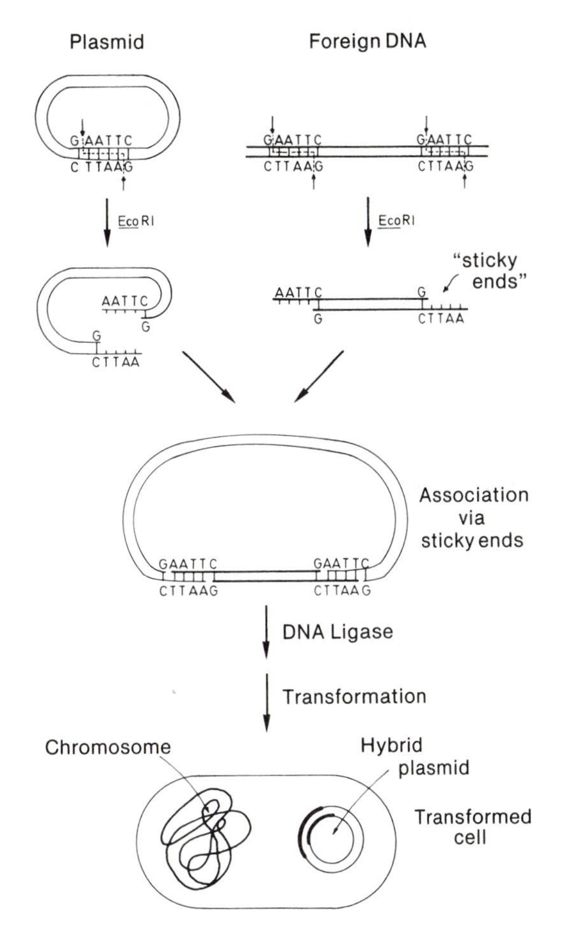

Figure 3.22 Method for the production of recombinant DNA

Methods of gene technology

The basic procedures for the first steps in gene cloning are outlined in Figure 3.22. Details can be found in specialized books on gene cloning (see References).

Isolation of DNA sequences for cloning

Genome fragments **Restriction endonucleases** are used to cut DNA. These enzymes belong to specific restriction and modification systems and are used by the cell to protect itself from foreign DNA. Restriction enzymes split double-stranded DNA at specific sites, usually at palindrome structures 4–11 nucleotides in length. Either blunt double-stranded ends or short, cohesive, single-stranded "sticky ends" develop, depending on the enzyme involved:

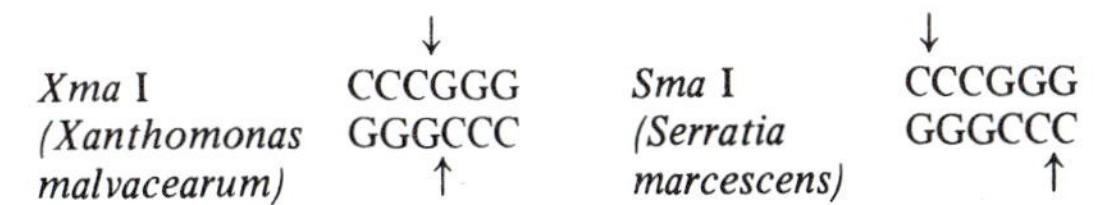

More than 600 restriction endonucleases are known in bacteria.

If the sequence of the DNA to be cloned is unknown, it is possible to use a so-called "shotgun" approach. With this procedure, a *gene bank* is produced by using suitable restriction enzymes to fragment the total genome of the organism into pieces of about 20 kilobases in length. Each of these fragments is linked to a vector (generally a phage or cosmid) and cloned into a suitable host. With an appropriate screening method the cultures containing the clones are selected, among which the target gene should be present. There will be a 99% probability, assuming 20 kb fragments, that in *E. coli* (total genome length of about 4000 kb) all genes will be cloned if 900 separate clones are selected (in yeast, which has a total genome size of 13,500 kb, 3100 clones would be required).

It is preferable if the initial cloning is carried out with enriched fragments. Enrichment can be done either by use of sucrose-gradient centrifugation, agarose-gel electrophoresis, column chromatography, or by use of specific gene probes. **Gene probes** are short DNA or RNA fragments that have been "tagged" in some way which are able to hybridize with the specific genome sequence. Gene probes can be tagged with ^{32}P or ^{35}S labeled nucleoside triphosphates, permitting the probes to be detected by autoradiography. An alternative procedure is the use of biotin-labeled DNA. Because of the strong affinity of biotin for avidin (a glycoprotein obtained from egg white) or streptavidin (a protein obtained from cultures of *Streptomyces avidini*) the presence of the biotin-labeled DNA can be detected using avidin or streptavidin which has been marked with a fluorescence dye, an enzyme, or an antibody.

To obtain expression of the cloned gene, it is necessary to understand the genetic organization of procaryotes and eucaryotes. In eucaryotes, the coding regions of genes (exons) are generally broken with noncoding regions (introns). From the primary transcript from the nuclear gene, the introns must be removed (RNA processing) before the mature translatable mRNA is present. The RNA processing steps cannot be carried out by procaryotes so that a correct translation of cloned genes of eucaryotes is not possible. Therefore, when bacteria are used as cloning hosts it is preferable to use synthetic DNA or complementary DNA (a DNA copy of the mature mRNA) (see below).

Synthetic DNA In order to produce a specific DNA fragment containing the coding region of a protein, the DNA sequence is deduced by "reverse translation" from the amino acid sequence of this protein. Using an automated DNA synthesis machine, it is possible to produce DNA fragments of 20 to 100 bases, which can be connected together to make longer sequences. Examples of the use of this technique are the artificial synthesis of the gene for somatostatin, a peptide hormone with 14 amino acid residues and the synthesis of the A and B chains of insulin, which were cloned and expressed in *E. coli*. Since the DNA synthesis procedure is completely under the control of the investigator, it is also possible to produce sequences in which one or more bases have been changed, making possible the production of highly specific mutations.

Production of complementary DNA (cDNA) Specific mRNA molecules, isolated from mRNA-rich cells or tissues, are used as templates in vitro with the enzyme *reverse transcriptase*, to produce complementary DNA. An example of the use of this method for cloning the insulin gene from rats in *E. coli* is shown in Figure 3.23.

Vectors As vectors for the transfer of DNA, a variety of genetic elements are available. Most commonly used are plasmids, temperate phages, and cosmids.

Plasmids are defined as circular, extrachromosomal, self-replicating DNA molecules. They have molecular weights between 1.5×10^6 and 200×10^6. Large plasmids and small plasmids have different properties. Large plasmids have an average molecular weight of 65×10^6; there are usually 1–2 copies per bacterial chromosome in the cell and they can be transferred during conjugation. Small plasmids (about 5 million molecular weight), of which there are more than 10 copies per bacterial chromosome, are not con-

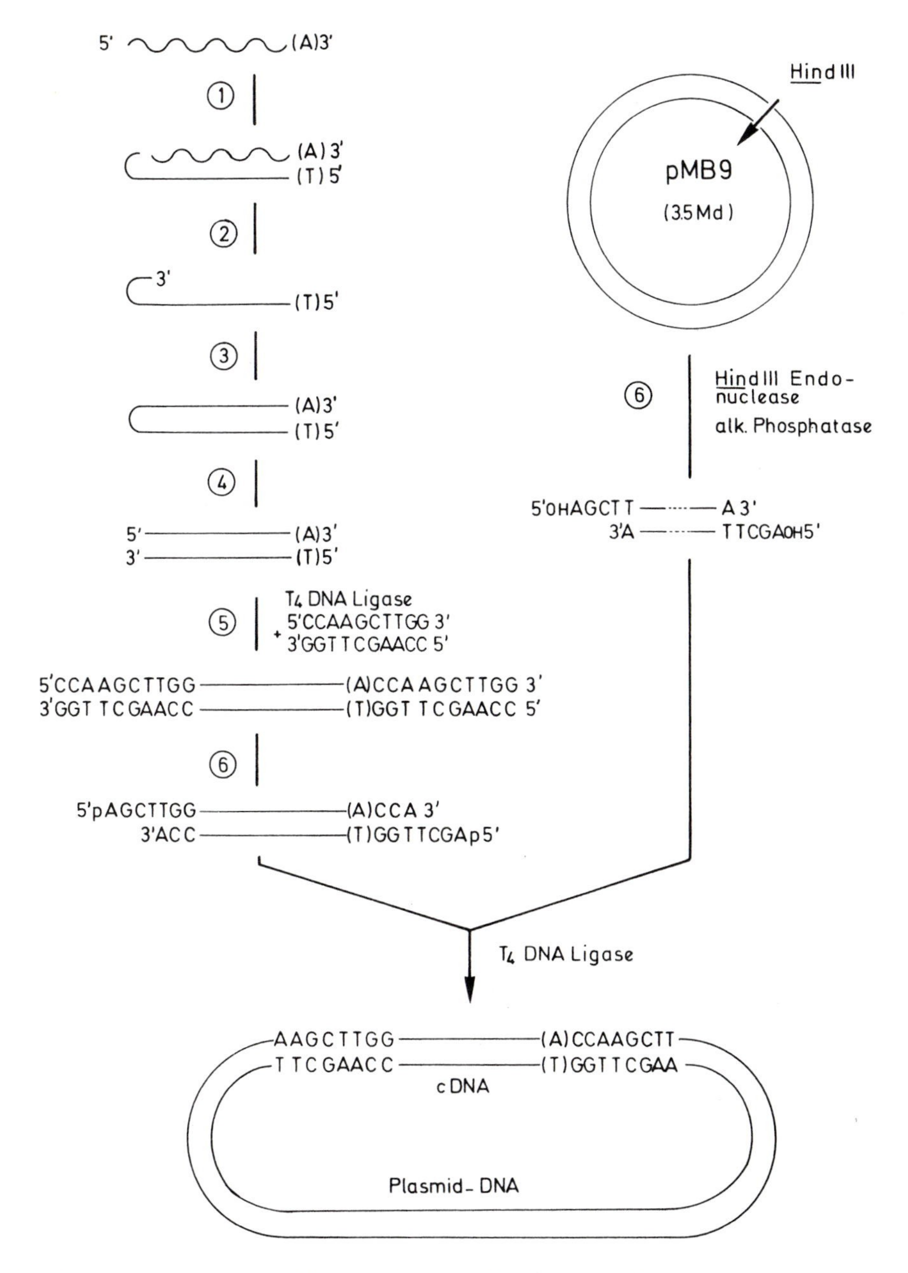

Figure 3.23 Cloning rat insulin genes (Ulrich et al., 1977). 1: mRNA is isolated from an insulin-producing tumor of the rat and transcribed with reverse transcriptase into cDNA. Simultaneously, the enzyme reverse transcriptase produces a hairpin structure (fold-back region) which serves as a primer for the completion of the double strand. 2: RNA is removed from the mRNA-cDNA-hybrid molecule by treatment with alkali. 3: The single-stranded cDNA becomes double-stranded through reverse transcriptase. 4: The fold-back region of the hairpin structure is digested by S1-nuclease. 5: Through T4 ligase, chemically synthesized nucleotide sequences containing the recognition sequence for the restriction endonuclease *Hind*III are attached at both ends of the cDNA. 6: Through the action of *Hind*III, sticky ends are created which permit the attachment to a similarly-treated plasmid DNA

jugative. Plasmids code for numerous and varied cell functions:

- Fertility: ability to transfer genetic material through conjugation.
- Antibiotic resistance: resistance to one or more antibiotics; R-plasmids are known in more than 50 species of bacteria.
- Resistance to heavy metals: Cd^{2+}, Hg^{2+}.
- Resistance to ultraviolet radiation.
- Production of bacteriocins, substances which inhibit or kill cells of the same species.
- Production of antibiotics: a plasmid has been shown to code for the antibiotic methylenomycin.
- Utilization of unusual carbon sources: breakdown of camphor, octane, and octanol by *Pseudomonas*, for example (see Table 3.10).
- Formation of toxins and surface antigens, such as enterotoxin and hemolysin.
- Tumor induction in plants: formation of crown gall tumors by the Ti plasmid of *Agrobacterium*.
- Involvement in sporulation in streptomycetes.

Plasmids are capable of adding foreign genes to the genetic material of the cell. If there is no homologous area in the bacterial chromosome for these foreign genes, they cannot be exchanged during crossing over, but may be maintained through plasmid replication. Some plasmids exhibit the phenomenon of *incompatibility*, in which two related plasmids are unable to be maintained together in the same cell.

Table 3.10 Catabolic plasmids from *Pseudomonas*

Plasmid	Substrate broken down	Transmissibility
NAH	Naphthalene	Conjugative
SAL	Salicylic acid	Conjugative
CAM	Camphor	Conjugative
OCT	Octane, hexane, decane	Nonconjugative
TOL	p- or m-xylene, toluene	Conjugative
pJP1	2,4-dichlorophenoxyacetic acid	Conjugative
pAC25	3-chlorobenzoate	Conjugative
pAC27	3- and 4-chlorobenzoate	Conjugative

Cosmids are plasmids which contain the *cos* region of phage λ, which makes it possible for the plasmid to be packaged within bacteriophage particles and hence transfered by infection to *E. coli.* Cosmids are capable of incorporating extremely long sequences of DNA, 32–47 kb, and hence are especially useful for the production gene banks.

The foreign DNA is either incorporated into a specific restriction site of the cosmid (insertion vector) or replaces a fragment of equivalent length of the vector (substitution vector).

Depending on the purpose, the following kinds of vectors can be constructed:

Cloning vectors are used primarily for the amplification of foreign DNA in the host cell. With certain plasmids, the copy number per cell can be increased by growth in the presence of the antibiotic chloramphenicol, which inhibits the replication of the host genome without affecting the replication of the plasmid. In this way, recombinants plasmids can often be produced to as many as 1000 copies per cell.

Sequence vectors are those which contain an increased number of recognitions sites for restriction enzymes (so-called multi-purpose cloning sites). When the foreign DNA is integrated into such a plasmid, DNA fragments of various sizes can be produced and sequenced.

Expression vectors contain not only the foreign gene but also promoter, operator, and terminator sequences so that the gene can be efficiently transcribed into mRNA. In addition, expression vectors contain the ribosome-binding site so that efficient translation can take place.

The expression of foreign DNA in the host organism is largely dependent on the system used. In *Escherichia coli*, procaryote DNA is transcribed and expressed into protein. Under favorable conditions as much as 30% of the dry weight of the cell can consist of the heterologous protein.

The expression of genes of lower eucaryotes, such as *Saccharomyces cerevisiae* or *Neurospora crassa*, has been successfully achieved in *Escherichia coli*, but in more complex eucaryotes, DNA transcription and translation have not occurred correctly. These difficulties may be overcome by incorporating eucaryote DNA into a procaryote gene containing appropriate regulatory signals, thus permitting expression of the eucaryote DNA. This was first accomplished with somatostatin, which was incorporated into the lactose operon located on plasmid pBR322 (Figure 3.24). The product synthesized by the *Escherichia coli* cell (a fusion protein) was β-galactosido-somatostatin. Active somatostatin was cleaved from the chimeric protein *in vitro* by treatment with cyanogen bromide, which splits peptides at methionine residues. The yield of somatostatin obtained was one-tenth of the value calculated from the plasmid copy number but was still relatively favorable (10 mg somatostatin protein per 100 g *Escherichia coli* wet weight) considering the simplicity of the *Escherichia coli* fermentation.

In the same way, synthetic genes of the A and B chain of human insulin were incorporated into the β-galactosidase region of the pBR322 plasmid and cloned separately in *Escherichia coli.* The transformed bacteria synthesized the A and B chains separately; by mixing both compounds, an active insulin preparation was developed and has now been marketed by the Eli Lilly Company.

Another insulin production method involved use of *Escherichia coli* χ1776. Complementary DNA copies of preproinsulin-mRNA of the rat were cloned into the penicillinase gene of plasmid pBR322. A penicillinase-proinsulin complex, penicillinase(24-182)–$(Gly)_6$-proinsulin(4-86), was formed by the bacteria. The synthesis rate was about 100 molecules per cell. Since *Escherichia coli* penicillinase is an extracellular protein, the complex was excreted from the cell and could be transformed into active insulin by splitting the penicillinase residue from the insulin C chain.

In order for a vector to replicate in the host cell, it must contain the appropriate origin-of-replication site (*ori*). Vectors containing *ori* of two separate host systems are able to replicate in both. Such vectors, called **shuttle vectors**, are known for *E. coli* and *Bacillus subtilis*, and for *E. coli* and yeast.

Incorporation of foreign DNA This is done differently with blunt end or cohesive end fragments. With blunt-end fragments, the 5′ ends are digested by λ-exonucleases and the 3′ ends are extended through action of a terminal transferase enzyme with ATP or TTP. In this process, a terminal poly-A or poly-T sequence results. One strand with a poly-A sequence can combine with another strand containing a poly-T sequence to form a circle as a hybrid plasmid. When there are sticky ends, DNA fragments of any origin combine as long as the same restriction endonuclease has been used in the creation of both fragments.

Insertion of foreign DNA into the host The efficient incorporation of foreign DNA into the target host cell is a critical step in gene manipulation. It is essential that the host is able to take up the engineered DNA. Host organisms include not only microorganisms but also plant, animal, and human cells. In *E. coli* transformation of naked plasmid or phage DNA can be brought about in the presence of $CaCl_2$, which renders the cell wall and cell membrane permeable to free DNA. With plasmid DNA, 10^7–10^8 transformants can be obtained per μg DNA. An alternative procedure is to use a transduction system consisting of phage or cosmid DNA packaged into empty phage heads. For gram-positive bacteria, actinomycetes, yeasts, and filamentous fungi, effective transformation of plasmid or phage DNA can be accomplished using the PEG system in protoplasts (see Section 3.4).

Analysis of recombinant clones There are several possible procedures that can be used to screen transformed cells and determine that the cloning and expression processes have been effective:

- Identification of host cells containing the foreign-DNA-containing vector;

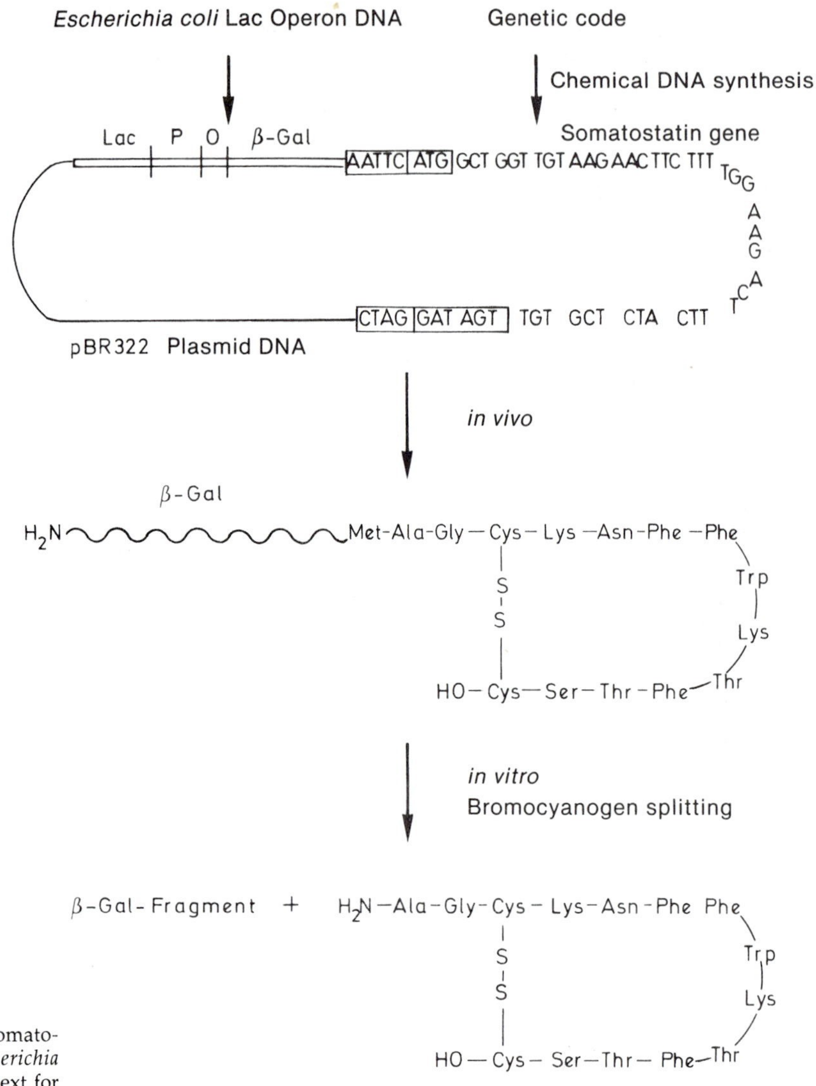

Figure 3.24 Expression of somatostatin gene sequence in *Escherichia coli* (Itakura et al., 1977). See text for explanation

- Detection of the foreign DNA in the host cells;
- Detection of the foreign DNA indirectly by assaying for the expression product (foreign protein).

To select transformed cells, the **marker inactivation technique** can be used. Vectors are used containing two selectable markers (for instance, antibiotic resistance), one of which contains the recognition site for the restriction enzyme used in the cloning process. If the foreign DNA becomes integrated into this antibiotic resistance gene, the activity of that gene is lost (insertional inactivation). Host cells that lack the vector are sensitive to both antibiotics, host cells containing a vector lacking the foreign DNA are resistant to both antibiotics, whereas vectors with inserted foreign DNA are sensitive to the one antibiotic into whose resistance gene the foreign DNA has been inserted.

To assay a complete gene bank for the presence of the gene, it is anticipated that around 1000 recombinant clones must be tested. Further tests must then be used to determine that the appropriate clone has been obtained. To demonstrate the presence of cloned DNA in the cell, two types of procedures are used, colony hybridization and Southern blotting.

Colony hybridization is done by using the replica-plating technique to transfer around 200 colonies from agar plates to a nitrocellulose filter. The colonies on the filter are then lysed and the released DNA denatured with 0.5 N NaOH. After removing excess protein with the enzyme proteinase K, the single-stranded DNA is fixed to the filter by heating to 80°C. After hybridization with a ^{32}P-labeled probe (either DNA or RNA), the filter is subjected to autoradiography. The colonies corresponding to the active spots can then be isolated from the original plate.

After isolation of hybrid DNA molecules, the specific DNA sequence can be identified by means of the **Southern blotting technique**. After treatment of the DNA with a restriction enzyme, the resulting DNA fragments are separated by gel electrophoresis. The DNA bands are then transferred to paper or nylon filter by blotting. The filter provides a stable carrier for the fragments, and the fragment containing the target sequence is identified using a gene probe. (The corresponding technique for RNA is called Northern blotting, that for proteins is called Western blotting.)

A different kind of procedure for detecting the desired clones involves seeking clones in which the gene product (protein) has been expressed. Since the expression efficiency is often quite low, a very sensitive method for detecting the gene product is necessary. One of the most widely used methods is immunological, in which an antibody (marked by radioactivity or enzyme) is used as a probe.

Cloning and expression systems for various microorganisms

The selection of a suitable cloning system depends to a great extent on the protein to be synthesized. Low-molecular weight proteins with few disulfide bridges and which do not require glycosylation for activity can be cloned and expressed in bacteria. Bacteria are unable, however, to glycosylate proteins, whereas yeasts preferentially glycosylate with mannose rather than glucose (the preferred sugar in human proteins). To produce complex proteins, cell cultures are most suitable. In such cultures, not only is the protein properly glycosylated, but it is also secreted in its native configuration.

As shown in Table 3.11, when foreign proteins are expressed, much of the synthesized protein may be retained in the cells rather than being excreted into the medium. It may therefore be preferable to begin with a system in which total activity is lower in order to save costs of isolation and purification.

Gram-negative bacteria Because of its well-known biology and genetics, *E. coli* is the preferred organism for experiments in genetic engineering. A large number of heterologous genes have been cloned and expressed in this organism. However, wild type plasmids such as ColE1, pSC101, and RSF2124 presented a number of disadvantages, so that suitable vectors have been constructed in vitro. A widely used vector is pBR322, a ColE1-like plasmid, which contains

Table 3.11 Final yield of heterologous proteins in relation to protein secretion in various cloning systems

Cloning system	Amount of heterologous protein (mg/l)	Protein excretion	Final yield (mg/l)[a]
E. coli	5000	No	250
B. subtilis	100	Yes	30
Saccharomyces	10	Yes	3
cerevisiae	2000	No	50
Animal cell culture	100	Yes	30

[a]Final yield after isolation and purification. Data of Davies (1986).

the ampicillin-resistance gene (Ap^R) of RSF2124, the tetracycline-resistance gene (Tc^R) of pSC101, and the origin of replication of pMP1. In addition, pBR322 contains recognitions sites for 20 restriction enzymes. Other suitable vectors for *E. coli* include phage λ and single-stranded DNA phages M13, f1 and fd, which are especially suitable for DNA sequencing. Details of cloning systems for *E. coli* are widely available in manuals and reference books (see References).

Large-scale production of foreign proteins in *E. coli* presents some difficulties, especially because foreign proteins are often subjected to extensive proteolysis or are produced in the cell in the form of highly insoluble protein bodies (inclusion bodies). Purification of these atypically folded and frequently denatured proteins leads in many cases to only a very tiny amount of the correct product.

To clone in other gram-negative bacteria, for instance *Pseudomonas*, derivatives have been developed of the wild type conjugative plasmids IncP and IncW as well as the nonconjugative plasmid IncQ. These plasmids possess an extremely wide host range and can be used in almost all species of gram-negative bacterila.

Gram-positive bacteria Species of the genus *Bacillus* have been used for many years in industry for the production of peptide antibiotics and exoenzymes. Their use in large-scale cultivation is therefore well understood. An additional advantage of the bacilli for gene technology is that excellent excretion of foreign proteins takes place. However, the native plasmids and phages of *Bacillus subtilis* are not suitable as cloning vectors. In vitro construction of suitable plasmids was based on the utilization of antibiotic resistance plasmids of *Staphylococcus aureus*. An artificial plasmid pHV11, containing tetracycline-resistance and chloramphenicol-resistance genes (Tc^RCm^R) was developed for *B. subtilis* which was analogous to pBR322 of *E. coli*. Addition of the origin of replication of *E. coli* led to the development of a shuttle vector for *E. coli* and *B. subtilis*. Vectors suitable for various experimental purposes could also be derived from temperate phages p11, ϕ105, and NP02. For the transformation of bacilli, it is necessary to begin with protoplasts and to use single-stranded DNA. One problem with *B. subtilis* is that plasmids are often unstable, leading to rearrangments of the genome or deletions.

Other gram-positive microorganisms of industrial interest for which cloning systems have been developed include streptococci (for the dairy industry), clostridia (for acetone, butanol, isopropanol, butyric acid, and acetic acid fermentations), *Corynebacterium* and *Brevibacterium*, for amino acid and nucleotide production.

A number of plasmid and phage vectors have been developed for the industrially important group of streptomycetes. Among others, these are pIJ61, a derivative of plasmid SLP1.2 of *Streptomyces lividans*, SCP2, a low-copy-number plasmid of *S. coelicolor* A3(2), bacteriophage ϕC31 of *S. coelicolor*, and plasmid pIJ101, a conjugative multicopy plasmid of *S. violaceoruber*. Detailed procedures for cloning in *Streptomyces* have been developed (see References). With these methods, recombinant DNA techniques can be used in the study of biosynthesis and regulation of secondary metabolites in streptomycetes. With such basic information available, intraspecific or interspecific cloning can be used for strain improvement. Some successes in this area have already been obtained. The O-methyltransferase gene of *S. coelicolor* A3(2), involved in the biosynthesis of undecylprodigiosin, has been cloned. Studies on glucose repression of glycerol catabolism in *S. coelicolor* have led to clarification of the mechanism of glucose catabolite regulation.

Genes controlling the production of three antibiotically active substances of *S. coelicolor* have been cloned and expressed: actinorhodin (7 chromosomal genes), undecylprodigiosin (5–6 chromosomal genes), and methylenomycin (5–6 plasmid-controlled genes). In addition, genes involved in antibiotic synthesis and antibiotic resistance have been cloned for a variety of other streptomycetes (see Section 3.7). Because of the extreme complexity in the biosynthetic pathways of antibiotics, much further basic research will be needed before these newer methods will lead readily to strain improvement.

Because streptomycetes readily excrete proteins, *S. lividans* has been used as a model system for the expression of bovine somatotropin, human serum albumin, and human interleukin-2, although the use of streptomycetes as expression systems for heterologous proteins is still in its infancy.

Eucaryotes The yeast *Saccharomyces cerevisiae* has replaced *E. coli* in a number of projects for the expression of eucaryotic proteins. Large-scale production of yeast is well-established. Genes for foreign proteins have been expressed and the proteins secreted into the medium. A further advantage for the production of genes from mammals is that yeast is able to glycosylate foreign proteins. A whole series of mammalian proteins have been successfully expressed and correct splicing of mRNA and correct glycosylation do not always occur.

A range of vectors have been developed for *Saccharomyces cerevisiae*, mostly hybrids between yeast DNA and *E. coli* plasmids which can serve as shuttle vectors. The principle vectors are summarized here:

- **YIp vectors** (yeast integrating plasmids) contain the ColE1 origin of replication of *E. coli* and chromosomal sequences from yeast. They replicate as plasmids in *E. coli*. Although they do not replicate in yeast they are able to integrate into the yeast genome, although the integrated sequences are relatively unstable.
- **YEp vectors** (yeast episomal plasmids) are circular plasmids (2 μm in length, about 6 kb) that are present in yeast cells in around 50–100 copies. When connected with selectable yeast sequences and corresponding sequences from *E. coli* plasmids, they can be used as vectors. Vectors based on the 2 μm DNA are, however, relatively unstable.
- **YRp vectors** (yeast replicating plasmids) are yeast sequences which have a chromosomal origin and contain an origin of replication. After incorporation into bacterial plasmids, they can be used as vectors. They exhibit a relatively high frequency of transformation (10^3–10^4 transformants per μg DNA).
- **YCp vectors** (yeast centromere plasmids) are YRp vectors into which the centromere of chromosome III has been incorporated. These plasmids behave as circular minichromosomes, are stably inherited, and can be used as vectors.

Filamentous fungi Although a variety of antibiotics and enzymes are produced commercially with filamentous fungi, cloning systems for these organisms are not nearly so well developed as for yeast. Plasmids resembling the YIp type have been constructed by combining genes from *Neurospora* with the origin of replication of pBR322 of *E. coli*. Hybrids between pBR322 and mitochondrial DNA (mtDNA) of the ascomycete *Podospora anserina* replicate in both *E. coli* and *Podospora*. Plasmids from *N. crassa* and *P. anserina* can perhaps be produced by insertion of the origin of replication from mtDNA sequences.

Various commercially grown aspergilli, such as *Aspergillus niger*, *A. awamori*, and *A. oryzae*, excrete large amounts of exoenzymes. However, the species *A. nidulans* is the best known genetically and has served as a model for the development of effective transformation and expression systems. Biologically active bovin chymosin, human interferon α2, and human tissue plasma activator (tPA) have been expressed and secreted in *A. nidulans*. However, it is not certain whether the correct glycosylation has been obtained.

The procedures that have been developed in yeast and aspergilli have also been adapted to the important β-lactam-producing fungi *Penicillium chrysogenum* and *Cephalosporium acremonium*. In *C. acremonium* a cosmid gene bank has been been constructed which has permitted the isolation of the *leu*B gene which codes for β-isopropylmalate dehydrogenase. A vector has been constructed, pIT221, which contains, among others sequences from pBR322, a mtDNA fragment of *C. acremonium*, and an ARS region (autonomously replicating sequence) which permits replication in *S. cerevisiae*. Using the gene coding for hygromycin B phosphotransferase (HPT) of *E. coli*, it has been possible to transform protoplasts of *C. acremonium* and demonstrate the presence of the HPT coding sequence. The principal goal of this work is to increase the yield of the enzyme.

Acrasiomycetes In the continuing search for suitable expression systems for mammalian proteins, research has turned to the organism *Dictyostelium discoideum*, a eucaryote which has a life cycle in which a free-living amoeboid form alternates with a fruiting body in which resting cells are produced. A functioning expression system for this organism has been created; for a particular foreign protein, yields 300 times greater than that of an optimized *E. coli* system have been obtained. Since *Dictyostelium* grows in the vegetative phase as an amoeba, diffusion is not a limiting factor in nutrient uptake so that continuous culture can be used to produce the desired protein in an unlimited fashion. However, whether correct protein processing and glycosylation occur has not been determined, and scale-up of this organism to a large-scale industrial process has not yet been accomplished.

Risks in genetic engineering

The creation of new types of viruses and the possible cloning of the DNA of tumor viruses has been perceived to be potentially dangerous. Since *Escherichia coli*, which exists in the human intestinal tract, has been used almost exclusively for the cloning of recombined DNA, the possibility of human infection cannot be ruled out. Thus in 1974 numerous renowned American scientists called for a voluntary moratorium on certain cloning experiments until potential risks were resolved. The first guidelines for work involving recombined DNA were established at an international conference at Asilomar, California in 1975. They became the basis for the rules of the U.S. National Institute of Health (NIH). Since that time, however, the risks have been considered so slight that the NIH guidelines have been suspended. This has been because of:

- The development of suitable safety measures.
- The use of weakened *Escherichia coli* strains (e.g. χ1776) with a probability of survival of 10^{-8} outside of the laboratory.
- Experimental evidence that there is only very slight danger of the transfer of cancer genes to humans.

Certain experiments are still prohibited or require authorization. Each country has its own specific guidelines for recombinant DNA research, and the operative regulations can be obtained by applying to the proper authorities.

3.7 USE OF GENETIC METHODS

Strain optimization

The primary goal of an industrial strain development program is the increase in the yield of the desired product. The genetic techniques for accomplishing this goal have been discussed in Sections 3.2–3.6. If the desired product, for example, an enzyme, is controlled by a single gene or a small group of genes, then high-yielding strains can be produced by:

- Isolation of mutants resistant to inhibitors of protein synthesis, which often overproduced proteins;
- Manipulation of regulatory signals to increase transcription or translation by cloning the gene on an expression vector or inserting the gene into a transposon which has a strong promoter;
- Modification of the gene by use of site-directed mutagenesis.

By increasing the gene dosage (**gene amplification**), the yield may be increased, provided the expression of the duplicated genes is not restricted by regulation. Gene amplification can occur in the following ways:

- Increasing the number of DNA replication sites in growing bacterial cells causes amplification of the genes situated near the origin of replication.
- Diploidization of fungi increases gene dosage, although the strains are usually unstable.
- Isolation of hyperinduced strains, which have been cultivated under selective conditions over a long period. These strains are extremely unstable, however, and are usually not suitable for commercial processes.
- If the gene has been characterized, the greatest success is likely by use of genetic engineering methods, for example, cloning and amplification of the gene by means of a multicopy plasmid or a phage vector. For instance, by use of a cosmid system the formation of the enzyme penicillin acylase in *E. coli* has been markedly increased when compared to the wild type. A whole series of industrial enzymes have been optimized in this way.

Difficulties abound when the goal is the increase in yield of a multi-gene product such as a primary or secondary metabolite, although some successes have been achieved. Amino acid production has been increased by cloning the whole genome, first in *E. coli,* later in production strains such as *Corynebacterium, Brevibacterium,* or *Serratia* (see Chapter 9).

For secondary metabolites such as antibiotics, cloning and amplification of the rate-limiting enzyme of the biosynthetic pathway can be done. As a first step in this direction, the genes for a number of antibiotics have been isolated, cloned, and in a few cases expressed. These include actinorhodin, methylenomycin, and undecylprodigiosin (*Streptomyces coelicolor*), cephalosporin (*Cephalosporium acremonium*), erythromycin (*S. erythreus*), oxytetracylcine (*S. glaucescens*) and tylosin (*S. fradiae*).

In addition to increase in yield, genetic methods have been used to improve the fermentation characteristics of the process. For instance, if the yeast *Saccharomyces cerevisiae* contains an α-amylase from *Bacillus subtilis,* ethanol fermentation can occur at 40°C and a stronger starch hydrolysis can take place. After transfer of the killer factor from a wild type to an industrial mutant of *S. cerevisiae,* wine, beer, sake, and alcohol production was protected from undesirable wild yeast contamination.

Qualitative change in the spectrum of antibiotics

Many producers of secondary metabolites also produce a range of chemically related byproducts. Mutants which produce only the desired antibiotic or only a small percentage of other substances can simplify the product recovery considerably and thus lower its cost. Table 3.12 shows examples which are industrially relevant.

Production of modified secondary metabolites

One way to produce therapeutically effective antibiotics is to use mutants which synthesize derivatives of known antibiotics. Such modified antibiotics may be synthesized either directly or through transformation of added precursors. With the use of the protoplast fusion or recombinant DNA techniques, hybrid antibiotics can be produced by bringing together in one strain the biosynthetic genes from several organisms.

Direct biosynthesis of modified antibiotics Strains with mutations in antibiotic synthesis produce substances whose structure is similar to the antibiotic of the parent strain (e.g. demethyl compounds from tetracycline, chlortetracycline, rifamycin). One of the best examples is the production of adriamycin (14-hydroxydaunomycin) by construction of a mutant from *S. peucetius* var. *caesius*. This derivative of daunomycin is more effective as an antitumor agent than daunomycin itself and is the most commonly used antitumor antibiotic today. Intermediate prod-

Table 3.12 Changes in byproducts of antibiotic-producing organisms through strain development

Fermentation product	By-products eliminated by strain development	Production strain
Polymyxin B	Other polymyxins	*Bacillus polymyxa*
Cephalosporin C	Cephalosporin N and P	*Cephalosporium acremonium*
n-Butanol	Acetone (variation in amount produced)	*Clostridium* sp.
Griseofulvin	Mycelianamide	*Penicillium griseofulvum*
Actinomycin D	Other actinomycins	*S. antibioticus*
Oleandomycin A	Oleandomycin B	*S. antibioticus*
Tetracycline	Chlortetracycline	*S. aureofaciens*
Neomycin B	Neomycin C, Fradicin	*S. fradiae*
Streptomycin	Mannosidostreptomycin, Cycloheximide, Vitamin B_{12}	*S. griseus*
Cycloheximide	Streptomycin	*S. griseus*
Antimycin A_1	Other antimycins	*Streptomyces* sp.
Tobramycin (Nebramycin factor 6) Apramycin (= Factor 7)	Remaining factors of Nebramycin complex (1,1′, 2,3,4,5,5′, 6 and 7)	*S. tenebrarius*
Levorin	Levoristatin (reduced proportion)	*Actinomyces levoris* No. 28
Leucomycin A_4,A_5	Leucomycin A_1–A_3	*Streptoverticillium kitasataensis*

(Perlman, 1973; with additions)

ucts of biosynthesis, which in wild strains may be produced at concentrations too low to detect, may accumulate in blocked mutants, thus leading to the detection of new products. For instance, a new tylosin analog was found in this way in a tylosin-producing culture of *S. fradiae.*

Mutasynthesis In mutasynthesis (mutational biosynthesis), mutants are used which cannot synthesize a part of the antibiotic molecule. If analogs of this missing component are added, modified antibiotics may be produced, provided the analog penetrates the cell and is recognized by the respective enzymes. Table 3.13 shows several examples. Attempts have been made to use mutasynthesis to produce modified macrolide and β-lactam antibiotics, as well as novobiocin analogs. Although several new compounds have been found, commercial use has not yet been possible due to the poor rate of precursor incorporation and the difficulty of obtaining proper mutants.

Cosynthesis In a mixed culture containing two mutants, the first mutant may accumulate an intermediate (C in the diagram below) which is transformed by the second mutant (blocked earlier in the synthetic pathway) into the end product.

A ------ B ------ C --//-- D ------ E
↓
A ------ B --//-- C ------ D ------ E

This method, which has been primarily used in the study of biosynthetic pathways, has also been successful in the isolation of modified antibiotics. For example, cosynthesis between two mutants of the 6-demethyl tetracycline producer *S. psammoticus* resulted in a new end product of tetracycline biosynthesis having a different spectrum of activity.

Hybrid antibiotics Genes for biosynthetic steps in different organisms can be combined in the same organism, thus leading to the production of new metabolites. By protoplast fusion, a blocked mutant of the streptomycin-producing *S. griseus* was combined with the istamycin-producing *S. tenjimariensis,* leading to the production of the hybrid antibiotic indolizomycin. In another example, the genes for actinorhodin biosynthesis from *S. coelicolor* were cloned into a producing strain of the same antibiotic class and various new compounds were detected. Also, the gene for oleandomycin-producing *S. antibioticus* was transferred to a blocked mutant of eryth-

Table 3.13 Some examples of mutasynthesis

Production strain (antibiotic produced)	Block in the biosynthesis of	Analog compound added	Mutasynthesis product
S. fradiae (Neomycin)	2-Deoxystreptamine	Streptamine	Hybrimycin A_1, A_2
		2-Epistreptamine	Hybrimycin B_1, B_2
Bacillus circulans (Butirosin)	2-Deoxystreptamine	2,5-Dideoxystreptamine	5-Deoxybutirosamine
		Streptamine	2-Hydroxybutirosin
	Neamine	6′-N-Methylneamine	6′-N-Methylbutirosin A,B
Micromonospora inyoensis (Sisomicin)	2-Deoxystreptamine	Streptamine	Mutamicin 1
		2,5-Dideoxystreptamine	Mutamicin 2
		2-Epistreptamine	Mutamicin 4
Micromonospora sagamiensis (sagamicin)	2-Deoxystreptamine	Streptamine	2-Hydroxysagamicin
S. griseus (Streptomycin)	Streptidine	2-Deoxystreptidine	Streptomutin A

romycin-producing *S. erythreus*. Although new compounds can undoubtedly be produced in this way, it appears unlikely that these more or less random approaches will lead to the formation of useful new compounds.

Another approach to the detection of new substances is the use of DNA sequences as gene probes for the detection of new antibiotics of the same chemical type.

Recombinant DNA products and specific applications of gene technology

Advances in the use of recombinant DNA technology to produce pharmacologically interesting proteins, enzymes, or amino acids is proceeding at a rapid pace. A whole series of recombinant DNA products are on the market, and more can be anticipated in the near future. Applications are in medical diagnostics, agriculture, and environmental protection as well as in basic research. The production of heterologous proteins by recombinant DNA technology has become widespread. However, the high development cost of these products is often not supported by a sufficiently large market. For some products, only kilograms are required on a world-wide basis. In general, the cost for developing a recombinant product may be as much as 100-fold higher than that for development of a conventional pharmaceutical product. A further handicap to rapid development is the competitive situation, since many companies are often pursuing research on the same compound.

In the following, a brief summary of the current developments in use of recombinant DNA products is given.

Gene technology in basic research Genetic engineering has revolutionized research in basic biology. The clarification of control methods in both procaryotes and eucaryotes, embryological development and differentiation, sequence determination of proteins has all made enormous advances. The first completely sequenced eucaryotic gene was the β-globin gene of rabbit. This led immediately to the discovery that eucaryotic genes were often split into coding regions (exons) and noncoding intervening regions (introns). Also discovered was the fact that the genome of an organism is not static but that genes exist whose role is in the regulation and expression of other genes (dynamic and hierarchical aspects of genetics).

As further examples from microbiology: the identification and characterization of antibiotic resistance genes, as well as regulatory and structural genes of biosynthetic pathways, the study of genetic instability and strain degeneration (primarily in streptomycetes), and genome analysis (for example, the demonstration that the archaebacteria constitute a distinct group of organisms).

In medicine, the methods of gene technology have made important contributions in cancer research, immunology, circulatory system, as well as in the nerve and endocrine systems.

Protein engineering The use of X-ray crystallography has permitted the determination of the three-dimensional structure of proteins. By the use of computer-aided protein design, molecular models can be readily produced and their structure/function relationships determined. Changes in amino acid sequence lead to changes in protein structure, not always in a predictable way. By use of modelling, the intelligent selection of desirable mutations can be made and these mutations can then be created by recombinant DNA technology. The technique offers the possibility of modifying enzyme sensitivity to inhibitors, turnover rate, substrate specificity, stability, etc. As an example of the potentialities: changes in various domains of the protein tPA (tissue plasminogen activator) led to increase in the half-life of the enzyme in animal studies. In another example, changing a single amino acid (aspartic acid to serine at position 99) in subtilisin (*Bacillus amyloliquefaciens*) led to an alteration in the pH optimum.

Human therapy By the use of recombinant DNA technology, substantial quantitities of proteins can now be produced that previously were only available in extremely small quantitities, if they were available at all. Such proteins can be used therapeutically in those patients deficient in these proteins (for example, insulin for diabetics, human growth hormone for dwarfs, factor VIII for hemophiliacs). Another use is in the increase in an immune response, as for instance interferon. Over 100 proteins of human origin have been cloned and expressed in experimental cell systems. Several such products are now being marketed commercially (Table 3.14) or are in clinical trials (Table 3.15).

Recombinant DNA technology opens up further possibilities through:

- The development of **vaccines**, especially so-called "subunit vaccines" that represent the antibody-recognition sites of pathogens; for instance, surface proteins of viruses or pathogenic bacteria. Hepatitis B vaccine is already on the market in this area.
- The development of **gene probes for use in diagnosis** for virus, bacteria, and fungal diseases. Such probes are of special use for those disease agents for which culture methods are not possible, such as many of the viruses.
- **Monoclonal antibodies** are immunoglobulins that are very specific for their antigens. They are produced by the hybridoma technique, in which an antibody-producing lymphocyte is fused with a tumor cell (myeloma) to produce the so-called hybridoma. The value of the hybridoma is that it can be cultured indefinitely, permiting unlimited production of the particular immunoglobulin. Monoclonal antibodies have already been perfected for analytical and diagnostic purposes as well as for use in protein purification (for example, interferon). It can be anticipated that monoclonal antibodies will also play a role in cancer diagnosis, as well as in virus therapy.

Agriculture Advances in the production of cultivated plants, such as the production of disease-resistant plants, can be expected from genetic engineering. Genetic manipulation of plants has conventionally been a slow and difficult task, but has been greatly facilitated by the new genetic methods. It is possible to use plant cell culture procedures to select clones of plant cells that are genetically changed and then, with proper treatments, induce these cell cultures to make whole plants which can be propagated vegetatively or by seeds. One approach to the introduction of foreign genes into plants is the use of the bacterium *Agrobacterium tumefaciens,* which contains a large plasmid (the Ti plasmid) which can become integrated into the plant genome. The Ti plasmid can thus be used as a vehicle for the introduction of foreign genes into plant cells. In

Table 3.14 Recombinant DNA products on the commercial market

Substance	Trade name	Company	Application
Human insulin	Humulin	Eli Lilly/Genentech	Diabetes
Interferon Alpha 2	Berofor	Boehringer Ingelheim	Antiviral/Antitumor
	Intron A	Schering Plough/Biogen	
Interferon Alpha (Namalva)	Wellferon	Wellcome Laboratories	
Interferon Beta	Frone	Serono/Interpharm	
Hepatitis B vaccine (HBsAG)	Recombivax HB	Merck/Chiron	Vaccine
Human growth hormone	Protropin	Genentech	Dwarfism
Tissue plasma activator (tPA)	Actilyse	Boehringer Ingelheim/ Genentech	Thromobolytic activity (myocardial infarction)

Table 3.15 Recombinant DNA products in clinical trial

Substance	Application
Antithrombin III	Inhibition of blood coagulation
Atrial Natriuretic Factor	Diuretic
Blood coagulation factors VIII and IX	Hemophilia
Epidermal growth factor	Healing of wounds
Erythropoietin (EPO)	Anemia in dialysis patients
Human gonadotropin (hCG, hMG)	Sterility
Human serum albumin	Blood substitute
Interferon Gamma	Antitumor therapy/ Arthritis
Lymphokines (especially interleukin 2)	Stimulation of immune defenses (bacteria/virus infections; antitumor therapy)
Proteins as antigens (herpes, malaria, influenza, AIDS)	Vaccine production
Superoxide dismutase	Heart attacks
Tumor necrosis factor	Antitumor therapy

one example, the gene for the insect toxin produced by the bacterium *Bacillus thuringiensis* has been introduced into the tomato plant, rendering these plants resistant to insect infestation.

In animal agriculture, genetic engineering is being used for the production of proteins that modify the growth or productivity of farm animals. For instance, swine growth hormone and bovine growth hormone produced by recombinant DNA technology can be used to promote more rapid animal growth. Bovine growth hormone can also be used to increase milk production in dairy cattle by as much as 20%.

Environmental protection Although still far in the future, there is the possibility of the use of engineered microorganisms for the elimination of toxic wastes. Many compounds produced industrially are highly persistent in the environment and, because of their real or potential toxicity, should be eliminated. Genetic engineering can be used to develop new strains of microorganisms capable of breaking down these persistent chemicals, opening up the possibility that these new microorganisms could play a role in environmental protection. However, much more information on the microbial ecology of these new microorganisms will be needed before they can be effectively and safely released into the environment.

REFERENCES

Alikhanian, S.I. 1962. Induced mutagenesis in the selection of microorganisms. Adv. Apl. Microbiol. 4:1–50.

Atkinson, D.E. and G.M. Walton. 1967. Adenosine triphosphate conservation in metabolic regulation. Rat liver citrate cleavage enzyme. J. Biol. Chem. 242:3239–3241.

Baltz, R.H., and P. Matsushima. 1980. Applications of protoplast fusion, site directed mutagenesis and gene amplification to antibiotic yield improvement in *Streptomyces*. Actinomycetes 15:18–34.

Beppu, T. 1986. Application of recombinant DNA technology of breeding of amino acid-producing strains, pp.24–35. In: Aida, K., I. Chibata, K. Nakayam, K. Takinami, and H. Yamada (eds.). Biotechnology of amino acid production. Progress in industrial microbiology, Vol. 24. Kodansha Ltd., Tokyo.

Betz, J. 1985 Automatisierte Stammverbesserung mittels durchflusscytometrischer Einzelzellklonierung. (Automatic strain improvement by use of the flow cytometer for cloning single cells.) BTF-BiotechForum 2:74–80.

Budowsky, E.I. 1976. The mechanism of the mutagenic action of hydroxylamines. Progr. Nucleic Acid Res. Molec. Biol. 16:125–188.

Castro, J.M., P. Liras, J. Cortes, and J.F. Martin. 1985. Regulation of α-aminoadipyl-cysteinyl-valine, isopenicillin N sythetase, isopenicillin N isomerase and deacetoxycephalosporin C synthetase by nitrogen sources in *Streptomyces lactamdurans*. Appl. Microbio. Biotechnol. 22:32–40.

Catenhusen, W.M. and H. Neumeister (editors). 1987. Gentechnologie. Chancen and Risiken. Bd. 12. Chancen and Risiken der Gentechnologie: Dokumentation d. Berichts an d. Dt. Bundestag/Enquete Kommission des Dt. Bundestages. (Gene technology: opportunities and risks. Volume 12. Opportunities and risks in gene technology: report of the German Senate.) J. Schweitzer Verlag, München.

Cerda-Olmedo, E., P.C. Hanawalt and N. Guerola. 1968. Mutagenesis of the replication point by nitrosoguanidine: Map and pattern of replication of the *Escherichia coli* chromosome. J. Mol. Biol. 33:705–719.

Chakrabarty, A.M. 1986. Genetic engineering and problems of environmental pollution, pp. 515–530. In: Rehm, H.J., and G. Reed (eds.). Biotechnology, Vol. 8a VCH Verlagsgesellschaft, Weinheim.

Chater, K.F. 1987. Technical advances in Streptomyces genetics, pp. 271–281. In: Alacevic, M., D. Hranueli, and Z. Toman (eds.). Proc. of the Fifth Int. Symp. on the Genetics of Industrial Microorganisms, Zagreb.

Churchward, G., and M. Chandler. 1985. *In vitro* recombinant DNA technology, pp. 77–111. In: Moo-Young, M. (ed.). Comprehensive Biotechnology, Vol. 1. Pergamon Press, Oxford.

Clarke, P.H. 1975. Mutagenesis and repair in microorganisms. Sci. Prof. Oxf. 62:559–577.

Clarke, P.H. 1976. Mutant isolation, pp.15–28. In: Macdonald, K.D. (ed.). Second Int. Symp. Genet. Indust. Microorg. Academic Press, London.

Coetzee, J.N., F.A. Sirgel and G. Lecatsas. 1979. Genetic recombination in fused spheroplasts of *Providence alcalifaciens*. J. Gen. Microbiol. 114:313–322.

Cooney, C.L. 1979. Conversion yields in penicillin production: Theory vs. practice. Process Biochem. 14:31–33.

Cooper, P. 1982. Characterization of long patch excision repair of DNA in ultraviolet-irradiated *Escherichia coli:* an inducible function under *rec-lex* control. Mol. Gen. Genet. 185:189–197.

Crueger, A. 1988. Microbielle Protoplasten—ihre Anwendung in Biochemie und Genetik. (Uses of microbial protoplasts in biochemistry and genetics.) In: P. Präve, M. Schlingmann, Crueger, W., K. Esser, R. Thauer, and F. Wagner (eds.). Jahrbuch Biotechnologie, Vol. 2. 1988/1989. Carl Hanser Verlag, München.

Daum, S.J. and J.R. Lemke. 1979. Mutational biosynthesis of new antibiotics. Ann. Rev. Microbiol. 33:241–265.

Davies, J. 1986. Retrospect on heterologous gene expression. Vortrag; Fifth Int. Symp. on the Genetics of Industrial Microorganisms, Split.

Demain, A.L. 1972. Cellular and environmental factors affecting the synthesis and excretion of metabolites. J. Appl. Chem. Biotechnol. 22:345–362.

Drake, J.W. 1970. The molecular basis of mutation. Holden-Day, San Francisco.

Dubey, A.K., S.N. Mukhopadhyay, V.S. Bisaria and T.K. Ghose. 1987. Sources, production, and purification of restriction enzymes. Proc. Biochem., Febr. 1987, pp. 25–34

Elander, R.P., L.T. Chang, and R.W. Vaughan. 1977. Genetics of industrial microorganisms. Ann. Rep. Ferment. Proc. 1:1–40.

Esser, K., U. Kück, C. Lang-Hinrichs, P. Lemke, H.D. Osiewacz, U. Stahl, and P. Tudzynski. 1986. Plasmids of Eukaryontes. Fundamentals and Applications. Springer Verlag, Berlin.

Finkelstein, D.B., J.A. Rambosek, J. Leach, R.E. Wilson, A.E. Larson, P.C. McAda, C.L. Soliday, and C. Ball. 1987. Genetic transformation and protein secretion in industrial filamentous fungi, pp. 101–110. In: Alacevic, M., D. Hranueli, and Z. Toman (eds.). Proc. of the Fifth Int. Symp. on the Genetics of Industrial Microorganisms, Zagreb.

Friedberg, E.C. 1985. DNA repair. W.H. Freeman, San Francisco.

Goeddel, D.V., D.G. Kleid, F. Bolivar, H.L. Heyneker, D.G. Yansura, R. Crea, T. Hirose, A. Kraszewski, K. Itakura, and A.D. Riggs. 1979. Expression in *Escherichia coli* of chemically synthesized genes for human insulin. Proc. Natl. Acad. Sci. USA 76:106–110.

Heim, J., Y.Q. Shen, S. Wolfe, and A.L. Demain. 1984. Regulation of isopenicillin N synthetase and deacetoxy-cephalosporin C synthetase by codon source during the fermentation of *Cephalosporium acremonium*. Appl. Microbiol. Biotech. 19:232–236.

Hopwood, D.A., and M.J. Merrick. 1977. Genetics of antibiotic production. Bacteriol. Rev. 41:595–635.

Hopwood, D.A. and H.M. Wright. 1978. Bacterial protoplast fusion: recombination in fused protoplasts of *Streptomyces coelicolor*. Mol. Gen. Genet. 162:307–317.

Hopwood, D.A., H.M. Wright, M.J. Bibb, and S.N. Cohen. 1977. Genetic recombination through protoplast fusion in *Streptomyces*. Nature 268:171–174.

Hopwood, D.A., M.J. Bibb, K.F. Chater, T. Kieser, C.J. Bruton, H.M. Kieser, D.J. Lydiate, C.P. Smith, J.M. Ward, and H. Schrempf. 1985. Genetic manipulation of *Streptomyces*. A Laboratory Manual. The John Innes Foundation, Norwich.

Hopwood, D.A., F. Malpartida, H.M. Kieser, H. Ikeda, J. Duncan, I. Fujii, B.A.M. Rudd, H.G. Floss, and S. Omura. 1985. Production of "hybrid" antibiotics by genetic engineering. Nature 314:642–644.

Horinouchi, S., and T. Beppu. 1987. A-Factor and regulatory network that links secondary metabolism with cell differentiation in *Streptomyces*. In:Alacevic, M., D. Hranueli, and Z. Toman (eds.). Proc. of the Fifth Int. Symp. on the Genetics of Industrial Microorganisms, Zagreb.

Hurst, A., and A. Nasim. 1985. Repairable lesions in microorganisms. Academic Press, Orlando.

Hütter, R. 1986. Overproduction of microbial metabolites. pp. 3–17. In: Rehm, H.J. and Reed, G. (eds.). Biotechnology, Vol. 4, VCH Publishers, New York.

Ichikawa, T., M. Date, T. Ishikura, and A. Ozaki. 1971. Improvement of kasugamycin-producing strain by the agar piece method and the prototroph method. Folia Microbiol. 16:218–224.

Imanaka, T. 1986. Application of recombinant DNA technology in the production of useful biomaterials, pp. 1–27. In: Fiechter, A. (ed.). Advances in Biochemical Engineering/Biotechnology, Vol. 33. Springer Verlag, Berlin.

Itakura, K., T. Hirose, R. Crea, A.D. Riggs, H.L. Heyneker, F. Bolivar, and H.W. Boyer. 1977. Expression in *Escherichia coli* of a chemically sythesized gene for the hormone somatostatin. Science 198:1056–1063.

Kaplan, B.E. 1985. The automated synthesis of oligodeoxy-ribonucleotides. Trends Biotechnol. 3:253–257.

Kiyoshima, K., K. Takada, M. Yamamoto, K. Kubo, R. Okamoto, Y. Fukagawa, and T. Ishikura. 1987. New tylosin analogs produced by mutants of *Streptomyces fradiae*. J. Antibiot. 40:1123–1130.

Kollek, R., B. Tappesser, and G. Altner (eds.). 1986. Gentechnologie. Chancen and Risiken. Bd. 10. Die ungeklarten Gefahrenpotentiale der Gentechnolgie. (Opportunities and risks in gene technology. Volume 10. The uncertain dangers of gene technology.) J. Schweitzer Verlag, München.

Laval, J. 1978. Recent progress in excision repair of DNA. Biochimie 60:1123–1134.

Lebrihi, A., G. Lefebvre and P. Germain. 1988. Carbon catabolite regulation of cephamycin C and expandase biosynthesis in *Streptomyces clavuligerus*. Appl. Microbiol. Biotechnol. 28:44–51.

Lemoine, Y., M. Courtney and J.P. Lecoq. 1987. Biotechnology: Vectors for gene expression in microorganisms, pp. 39–49. In: Alcevic, M., D. Hranueli and Z. Toman (eds.). Proc. of the Fifth Int. Symp. on Genetics of Industrial Microorganisms, Zagreb.

Malik, V.S. 1979. Genetics of applied microbiology. Adv. Genet. 20:37–126.

Martin, J.F. 1977. Control of antibiotic synthesis by phosphate. Adv. Biochem. Eng. 6:105–127.

Martin, J.F., J.A. Gil, G. Naharro, P. Liras and J.R. Villanueva. 1979. Industrial microorganisms tailor made by removal of regulatory mechanisms, pp. 205–209. In: Sebek, O.K. and A.I. Laskin (eds.). Genetics of industrial microorganisms (GIM 78). American Soc. Microbiol. Washington, D.C.

McAlpine, J.B., J.S. Tuan, D.P. Brown, K.D. Grebnes, D.N. Whittern, A. Buko, and L. Katz. 1987. New antibiotics from genetically engineered actinomycetes. I.2 Norerythromycins, isolation and structural determinations. J. Antibiot. 40:1115–1122.

Müntz, K. 1987. Engineering pflanzlicher Speicherproteine. (Engineering of plant storage proteins.) Bio-Engineering 2: 36–43.

Nellen, W., S. Datta, T. Crowley, C. Reymond, A. Sivertsen, S. Mann, and R.A. Firtel. 1987. Molecular biology in *Dictyostelium*: tools and applications. Methods Cell Biol. 28:67–100.

Old, R.W., and S,B. Primrose. 1985. Principles of gene manipulation. An introduction to genetic engineering. 3rd edition. Blackwell Scientific Publications. Oxford.

Perlman, D. 1973. Directed fermentations. Proc. Biochem. July 1973, 18–20.

Pontecorvo, G., and G. Sermonti. 1954. Parasexual recombination in *Penicillium chrysogenum*. J. Gen. Microbiol. 11:94–104.

Roth, J.R. 1974. Frameshift mutations. Ann. Rev. Genet. 8:319–343.

Saunders, V.A., and J.R. Saunders. 1987. Microbial genetics applied to biotechnology. Principles and techniques of gene transfer and manipulation. Croom Helm, London.

Sengbusch, V.P. 1979. Molecular and Cell Biology. Springer Verlag, Berlin.

Sermonti, G. 1959. Genetics of penicillin production. Ann. N.Y. Acad. Sci. 81:950–973.

Shortle, D., D. DiMaio, and D. Nathans. 1981. Directed mutagenesis. Ann. Rev. Genet. 15:265–294.

Skatrud, P.L., D.L. Fisher, T.D. Ignolia, and S.W. Queener. 1987. Improved transformation of *Cephalosporium acremonium*, pp. 111–119. In: Alacevic, M., D. Hranueli, and Z. Toman (eds.). Proc. of the Fifth Int. Symp. on the Genetics of Industrial Microorganisms, Zagreb.

Smith, K.C. 1978. Multiple pathways of DNA repair in bacteria and their roles in mutagenesis. Photochem. Photobiol. 28:121–129.

Smith, C.P., and K.F. Chater. 1987. Physiology, genetics and molecular biology of glycerol utilization in *Streptomyces coelicolor*, pp. 7–15. In: Alacevic, M., D. Hranueli, and Z. Toman (eds.). Proc. of the Fifth Symp. on the Genetics of Industrial Microorganisms, Zagreb.

Southern, E. 1975. Detection of specific sequences among DNA fragments seperated by gel electrophoresis. J. Mol. Biol. 98:503–517.

Starlinger, P. 1980. *IS* elements and transposons. Plasmid 3:241–259.

Taniguchi, T., and C. Weissmann. 1978. Site-directed mutations in the initiator region of the bacteriophage Qβ coat cistron and their effect on ribosome binding. J. Mol. Biol. 118:533–565.

Thoma, R.W. 1971. Use of mutagens in the improvement of production strains of microorganisms. Folia Microbiol. 16:197–204.

Ullrich, R., J. Shine, J. Chirgwin, R. Pictet, E. Tischer, W.J. Rutter, and H.M. Goodman. 1977. Rat insulin genes: Construction of plasmids containing the coding sequences. Science 196:1313–1319.

Vanek, Z., J. Cudlin, M. Blumauerova, and Z. Hostalek. 1971. How many genes are required for the synthesis of chlortetracycline? Folia Microbiol. 16:225–240.

Villa-Komaroff, L., A. Efstratiadis, S. Broome, P. Lomedico, R. Tizard, S.N. Naber, W.L. Chick, and W. Gilbert. 1978. A bacterial clone synthesizing proinsulin. Proc. Natl. Acad. Sci. USA 75:3727–3731.

VuTrong, K., and P.P. Gray. 1987. Influence of ammonium on the biosynthesis of the macrolide antibiotic tylosin. Enzyme Microb. Technol. 9:590–593.

Walker, G.C. 1984. Mutagenesis and inducible responses to deoxyribonucleic acid damage in *Escherichia coli*. Microbiol.Rev. 48:60–93.

Weinberg, E.D. 1974. Secondary metabolism: Control by temperature and inorganic phosphate. Dev. Ind. Microbiol. 15:70–81.

Werner, R.G., and W. Feurer. 1987. Neue Arzneimittel Generation: Therapie mit korpereigenen Wirkstoffen. (New-generation pharmaceuticals.) BioEngineering 4:12–20.

Winnacker, E.L. 1985. Gene and Klone. Eine Einführung in die Gentechnologie. (Gene and clone: An introduction to gene technology.) VCH Verlagsgesellschaft, Weinheim.

Witkin, E.M. 1976. Ultraviolet mutagenesis and inducible DNA repair in *Escherichia coli*. Bacteriol. Rev. 40, 868–907.

Yamashita, F., K. Hotta, S. Kurasawa, Y. Okami, H. Umezawa. 1985. New antibiotic-producing *Streptomycetes*, selected by antibiotic resistance as a marker. I. New antibiotic production generated by protoplast fusion treatment between *Streptomyces griseus* and *S. tenjimariensis*. J. Antibiot. 38:58–63.

Yanofsky, C. 1981. Attenuation in the control of expression of bacterial operons. Nature 289:751–758.

Zimmerman, U., J. Vienken, and P. Scheurich. 1980. Electric field induced fusion of biological cells. Biophys. Struct. Mechanisms 6:86.

4

Substrates for industrial fermentation

Media used in the cultivation of microorganisms must contain all elements in a form suitable for the synthesis of cell substances and for the production of metabolic products. In laboratory research with microorganisms, pure defined chemicals may be used in the production of culture media, but in industrial fermentations, complex, almost undefinable substrates are frequently used for economic reasons. Depending on the particular process, from 25 to 70% of the total cost of the fermentation may be due to the carbohydrate source. In many cases, media ingredients are byproducts of other industries and are extremely varied in composition. For strain development and fermentation control, the consequences are as follows:

- An **optimally balanced culture medium** is mandatory for maximal production. A supplement of critical elements must be used if necessary. Using statistical methods in computer-based programs, culture media may be optimized.
- The composition of culture media must constantly be adapted to the fermentation process. New batches of substrate have to be carefully evaluated in trial fermentations before they can be used in production.
- In addition to product yield, product recovery must be examined in trial fermentations.
- If **catabolite repression** or **phosphate repression** cannot be eliminated by optimization of the nutrient medium or suitable fermentation management (e.g. feeding), **deregulated mutants** must be used as production strains.

Besides **material cost** and **product yield**, it must be considered whether materials used are readily available in sufficient supply without high transportation costs, and whether impurities will hinder product recovery or increase cost of product recovery.

We discuss below some of the frequently used substrates in industrial fermentation. More details can be found in references cited at the end of the chapter and in chapters on the individual fermentations.

4.1 SUBSTRATES USED AS CARBON SOURCES

Carbohydrates are traditional energy sources in the fermentation industry. For economic reasons, pure **glucose** or **sucrose** can seldom be used as the sole carbon source, except in processes which demand exact fermentation control. **Molasses**, a byproduct of sugar production, is one of the cheapest sources of carbohydrate. Besides a large amount of sugar, molasses contains nitrogenous substances, vitamins, and trace elements. However, the composition of molasses varies depending on the raw material used for sugar production. Table 4.1 shows a comparison of the analysis of sugar beet and sugar cane molasses. Considerable variation in the quality of molasses occurs, depending on the location, the climatic conditions, and the production process of each individual sugar factory. In addition to conventional molasses, the residue from starch saccharification which accumulates after the crystalization of glucose is also widely used as a fermentation substrate. For example, "hydrol" molasses is a byproduct of glucose production from corn.

Table 4.1 Composition of sugar beet and sugar cane molasses

Composition	Sugar beet molasses	Sugar cane molasses
Dry matter %	78 –85	77 –84
Sucrose	48.5	33.4
Raffinose	1.0	–
Invert sugar	1.0	21.2
Miscellaneous organic materials	20.7	19.6
N	0.2 – 2.8	0.4 – 1.5
P_2O_5	0.02 – 0.07	0.6 – 2.0
CaO	0.15 – 0.7	0.1 – 1.1
MgO	0.01 – 0.1	0.03– 0.1
K_2O	2.2 – 4.5	2.6 – 5.0
SiO_2	0.1 – 0.5	–
Al_2O_3	0.005– 0.06	–
Fe_2O_3	0.001– 0.02	–
Ash	4 – 8	7 –11
Thiamine μg/100 g dry weight	130	830
Riboflavin	41	250
Pyridoxine	540	650
Niacinamide	5100	2100
Pantothenic acid	130	2140
Folic acid	21	3.8
Biotin	5.3	120

(Rhodes and Fletcher, 1966; Imrie, 1969)

Malt extract, an aqueous extract of malted barley, is an excellent substrate for many fungi, yeasts, and actinomycetes. Dry malt extract consists of about 90–92% carbohydrates, and is composed of hexoses (glucose, fructose), disaccharides (maltose, sucrose), trisaccharides (maltotriose), and dextrins, as shown in Table 4.2. Nitrogenous substances present in malt extract include proteins, peptides, amino acids, purines, pyrimidines, and vitamins. The amino acid composition of different malt extracts varies according to the grain used, but proline always makes up about 50% of the total amino acids present. Culture media containing malt extract must be carefully sterilized. When overheating occurs, the Maillard reaction shown in Figure 4.1 results, due to the low pH value and the high proportion of reducing sugar. In this conversion, the amino groups of amines, amino acids (especially lysine), or proteins react with the carbonyl groups of reducing sugars, aldehydes, or ketones, which results in the formation of brown condensation products. These reaction products are not suitable substrates for microorganisms. The Maillard reaction is one of the main causes of damage to culture media during heat sterilization, resulting in considerably reduced yields.

Table 4.2 Typical composition of malt extract

Component	% of Dry weight
Maltose	52.2
Hexoses (glucose, fructose)	19.1
Sucrose	1.8
Dextrin	15.0
Other carbohydrates	3.8
Nitrogenous materials	4.6
Ash	1.5
Water content	2.0

pH (10% solution) = 5.5

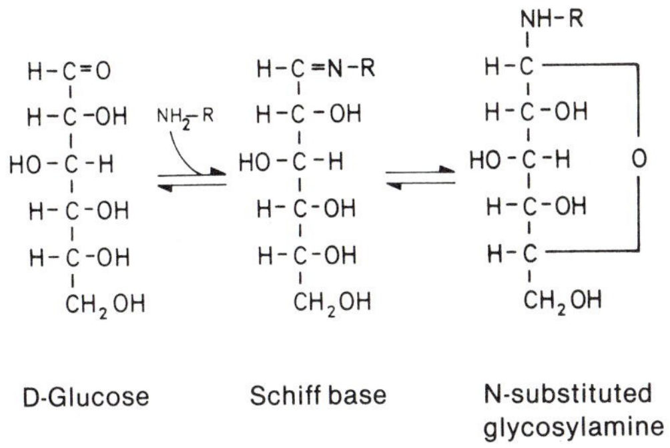

D-Glucose

Schiff base

N-substituted glycosylamine

Figure 4.1 Maillard reaction

Starch and **dextrins** can be directly metabolized as carbon sources by amylase-producing organisms. In addition to glucose syrup, which is frequently used as a fermentation substrate, starch has become more important as a substrate for ethanol fermentation. The largest use of carbohydrate for ethanol fermentation has been the National Alcohol Program begun in 1975 by the Brazilian government. This Program, which is based on starch from the cassava plant, produced about 9 million metric tons of ethanol in 1984.

Sulfite waste liquors, sugar-containing waste products of the paper industry, which have a dry weight of 9–13%, are primarily used in the cultivation of yeasts. Sulfite liquors from coniferous trees have a total sugar content of 2–3%, and 80% of the sugars are hexoses (glucose, mannose, galactose), the others being pentoses (xylose, arabinose). Sulfite liquors from deciduous trees contain mainly pentose sugars.

Because of its wide availability and low cost, **cellulose** is being extensively studied as a substrate for conversion to sugar or alcohol. The world's annual cellulose production is estimated at 10^{11} metric tons; much of it exists as waste, in such forms as straw, corn cobs, wood wastes, peat, bagasse, and waste paper. It is usually not possible to use cellulose directly as a carbon source, so it must first be hydrolyzed chemically or enzymatically. The sugar syrup formed from cellulose hydrolysis has been used for ethanol fermentation, and the fermentative production of butanol, acetone, and isopropanol is also being considered. Work is in progress to develop one-step processes for direct conversion of cellulose to ethanol, using fermentative organisms which produce cellulases.

Whey, a byproduct of the dairy industry, is produced annually on a world-wide basis to the amount of 74 million tons (containing 1.2 million tons of lactose and 0.2 million tons of milk protein). Only about 56% of this product is used for human or animal feed. The lactose is used primarily for the production of ethanol or single-cell protein, but also in the production of xanthan gum, vitamin B_{12}, 2,3-butandiol, lactic acid, and gibberellic acid. Because of storage and transportation costs, whey is often not economical as a substrate.

Animal **fats** such as lard and animal and plant **oils** are readily utilized by some microorganisms, but are generally added as supplemental substrates rather than as the sole fermentable carbon source. For instance, in certain antibiotic fermentations, soy, palm, and olive oils are used.

With respect to its carbon content, **methanol** is the cheapest fermentation substrate, but it can be metabolized by only a few bacteria and yeasts. Methanol has commonly been used as a substrate for single cell protein production (see Chapter 16). Research has been carried out on processes for the production of glutamic acid, serine, and vitamin B_{12} using methanol as the sole carbon source or as a cosubstrate.

Ethanol is available in ample supply from the fermentation of either saccharified starch or cellulose, and can be metabolized by many microorganisms as the sole carbon source or as a cosubstrate. Acetic acid, for instance, is presently made by the oxidation of ethanol. At the present time, the cost of ethanol is too high to make it utilizable as a general industrial carbon source.

Alkanes with a chain length of C_{12} to C_{18} are readily metabolized by many microorganisms. The use of alkanes as an alternative to carbohydrates depends on the price of petroleum (see Section 16.2).

4.2 SUBSTRATES USED AS NITROGEN SOURCES

Many large-scale processes utilize **ammonium salts**, **urea**, or **gaseous ammonia** as nitrogen source. A nitrogen source which is efficiently metabolized is **corn steep liquor**, which is formed during starch production from corn. The concentrated extract (about 4% nitrogen) contains numerous amino acids, such as alanine, arginine, glutamic acid, isoleucine, threonine, valine, phenylalanine, methionine, and cystine. The sugar present in corn steep liquor becomes largely converted to lactic acid (9–20%) by lactic acid bacteria.

Yeast extracts are excellent substrates for many microorganisms. They are produced from baker's yeast through autolysis at 50–55°C or through plasmolysis in the presence of high concentrations of NaCl. Yeast extract contains amino acids and peptides, water-soluble vitamins, and carbohydrates. The glycogen and trehalose of yeast cells are hydrolyzed to glucose during yeast extract production—Table 4.3 shows some typical analyses. The composition of yeast extract varies, partly because the substrates used for yeast cultivation affect the quality of the yeast extract.

Peptones (protein hydrolysates) can be utilized by many microorganisms but they are relatively expensive for industrial application.

Table 4.3 Composition of yeast extract

	Yeast extract produced by means of	
	Autolysis	Plasmolysis with NaCl
Composition (%)		
Dry matter	70	80
Total nitrogen	8.8	7.4
Protein (N × 6.25)	55	46
NaCl	< 1	18
Amino acids (% of total)		
Alanine	3.4	2.3
Aminobutyric acid	0.1	0.1
Arginine	2.1	1.1
Asparagine	3.8	3.1
Cystine	0.3	0.2
Glutamic acid	7.2	5.1
Glycine	1.6	1.6
Histidine	0.9	0.8
Isoleucine	2.0	1.6
Leucine	2.9	2.3
Lysine	3.2	2.9
Methionine	0.5	0.5
Ornithine	0.3	0.9
Phenylalanine	1.6	1.6
Proline	1.6	1.5
Serine	1.9	1.5
Threonine	1.9	1.4
Tyrosine	0.8	0.5
Valine	2.3	1.9
Vitamin content (ppm)		
Thiamine	20–30	10–15
Riboflavin	50–70	50–70
Pyridoxine	25–35	20–30
Niacinamide	600	100
Pantothenic acid	200	350

(Data from Ohly Inc., Hamburg)

Table 4.4 Compositions of some typical peptones of different origins

	N-Z-Amine®A	HY-Case® Amine	Soy Peptone
Starting material	Casein	Casein	Soy meal
Hydrolytic method	Enzymatic	Acid	Enzymatic
Composition (%)			
Water content	4.8	3.5	4.0
Ash	5.8	39.7	14.0
Total nitrogen	12.8	8.3	9.2
Amino nitrogen	6.7	6.4	1.8
NaCl	3.0	38.0	5.9
Carbohydrate	–	–	29.0
Free Amino acids (mg/g)			
Alanine	16.8	14.1	3.3
Arginine	30.2	5.0	1.1
Aspartic acid	8.7	33.2	2.4
Cystine	–	0.7	–
Glutamic acid	38.6	79.6	5.6
Glycine	4.4	9.3	2.3
Histidine	13.3	4.6	1.1
Isoleucine	29.9	10.7	2.8
Leucine	71.9	29.7	9.4
Lysine	61.1	12.8	7.9
Methionine	22.7	11.2	3.3
Phenylalanine	33.0	9.7	3.7
Proline	7.4	34.5	0.8
Serine	28.7	19.8	4.8
Threonine	21.5	13.1	1.9
Tryptophan	8.6	0.05	1.6
Tyrosine	14.0	4.9	0.7
Valine	36.4	16.7	3.8

(Data from Sheffield Chem. Co.)

Sources of peptones include meat, casein, gelatin, keratin, peanut seeds, soy meal, cotton seeds, and sunflower seeds. Peptone composition varies depending upon its origin. For instance, peptone from gelatin is rich in proline and hydroxyproline, but has almost no sulfur-containing amino acids. On the other hand, peptone from keratin has a large proportion of proline and cystine, but lacks lysine. Peptones of plant origin (soy peptone, cottonseed peptone) have large proportions of carbohydrates. The end product is also influenced by the type of hydrolysis, whether acid or enzymatic, especially in regard to its tryptophan content, as shown in Table 4.4.

Soy meal, the residue from soybeans after the extraction of soybean oil, is a complex substrate. Analysis shows a protein content of 50%, a carbohydrate content of 30% (sucrose, stachyose, raffinose, arabinoglucan, arabinan, and acidic polysaccharides), 1% residual fat, and 1.8% lecithin. Soy meal is frequently used in antibiotic fermentations; catabolite regulation does not occur because of the slow catabolism of this complex mixture.

REFERENCES

Bridson, E.Y. and A. Brecker. 1970. Design and formulation of microbial culture media. Methods in Microbiol., vol. 3A, pp. 229–295.

Dale, B.E. and J.C. Linden. 1984. Fermentation substrates and economics. In: Tsao, G.T. (editor). Annual Reports on Fermentation Processes, Vol. 7. Academic Press, Orlando.

Greenshields, R.N. and A.W. Macgillivray. 1972. Caramel—Part 1. The browning reactions. Proc. Biochem. December:11–13,16.

Imrie, F.K.E. 1969. Fermentation media. Sugar and molasses. Proc. Biochem. January:34–35.

Klyosov, A.A. 1986. Enzymic conversion of cellulosic materials to sugars and alcohol. Appl. Biochem. Biotechnol. 12: 249–300.

Kosarik, N. and Y.J. Asher. 1985. The utilization of cheese whey and its compounds. pp. 25–60 In: Fiechter, A. (editor). Advances in Biochemical Engineering Biotechnology, Vol. 32. Springer-Verlag, Berlin.

Minoda, Y. 1986. Raw materials for amino acid fermentation. pp. 51–66 In: Aida, K., I. Chibata, K. Nakayama, K. Takinami, and H. Yamada (editors). Progress in industrial microbiology, Vol. 24. Biotechnology of amino acid production. Kodansha Ltd., Tokyo.

Ratledge, C. 1977. Fermentation substrates. Ann. Rep. Ferm. Proc. 1:49–71.

Rhodes, A. and D.L. Fletcher. 1966. Principles of industrial microbiology. Pergamon Press, Oxford.

Rosillo-Calle, F. and H. Rothmann. 1984. The Brazilian National Biotechnology Program. Bio/Technology 5: 421–431.

Saguy, I. 1982. Utilization of the "complex method" to optimize fermentation processes. Biotechnol. Bioeng. 24: 1519–1525.

Schliephake, D. (editor). 1986. Nachwachsende Rohstoffe. Holz und Stroh, Natürlich Öle und Fette, Alkohole für Kraftstoffe. (Recyclable raw materials. Wood and straw, natural oils and fats, alcohol as energy source.) Verlag J. Kordt, Bochum.

5

Methods of fermentation

5.1 INTRODUCTION

The goal of biotechnology (bioengineering) is to obtain useful metabolic products from biological material. Biotechnology encompasses two distinct phases: fermentation and product recovery. **Fermentation procedures** must be developed for the cultivation of microorganisms under optimal conditions and for the production of desired metabolites or enzymes by the microorganisms. **Product recovery** or **downstream processing** involves the extraction and purification of biological products. Biochemical process recovery differs from chemical recovery primarily in that biological materials are frequently much more labile. Although many of the techniques used in biological or biochemical product recovery overlap with those used for strictly chemical processes (for instance, separation, distillation, heating, cooling, and drying), there is increasing use of methods specifically designed for biological products, such as chromatography and electrophoresis.

In fermentation processes, engineering is only an aid in the development and regulation of biological processes and the microorganism is the center of attention. All methods, such as genetic manipulation, regulation of metabolism by optimizing culture media, and provision of adequate oxygen supply under sterile conditions, are only ways of directing the process toward the desired product. The more precisely metabolic processes during growth and product formation are understood, the greater is the possibility of operating fermentation processes rationally rather than empirically. However, a fully automatic, computer-run fermentation process to obtain a metabolite such as an antibiotic has not yet been realized at the industrial scale.

5.2 GROWTH KINETICS OF MICROORGANISMS

A clear understanding of microbial growth kinetics is necessary if a large-scale process is to be properly managed. Growth kinetics are treated differently for conventional batch processes than for continuous processes.

Batch fermentation

A batch fermentation can be considered to be a "closed system." At the time T = 0 the sterilized nutrient solution in the fermenter is inoculated with microorganisms and incubation is allowed to proceed under optimal physiological conditions. In the course of the entire fermentation, nothing is added except oxygen (in the form of air), an antifoam agent, and acid or base to control the pH. The composition of the culture medium, the biomass concentration, and the metabolite concentration generally change constantly as a result of the metabolism of the cells. After the inoculation of a sterile nutrient solution with microorganisms and cultivation under physiological conditions, four typical phases of growth are observed: lag phase, log phase, stationary phase and death phase (Figure 5.1).

Lag phase When cells are transferred from one medium to another, there is initially no increase in the number of cells, although cell weight may change. During the lag phase, the microorganisms adapt to their new environment. Because of this transfer to a new medium, several parameters will probably be altered for the inoculum cells: change in pH value, increase in supply of nutrients, decrease of growth inhibitors. New transport systems for nutrients must be induced within the cells. Essential cofactors may diffuse out of a cell and enzymes of primary metabolism must be adjusted to the new conditions.

Moreover, the physiological condition of the inoculum is crucial to the length of the lag phase. If a culture for use as inoculum is still in the log phase (see below), a lag may not occur and growth may begin immediately. However, inoculum taken from a culture in which growth has stopped due to substrate limitation needs more time to adapt to the new nutrient solution. The concentration of the inoculum also has an influence on the lag phase.

The proper cultivation and transfer of inoculum are essential for the later production of both primary and secondary metabolites. Although the regulation of product formation is not yet fully understood, other preculture media and culture conditions often have to be designed for optimal yields. However, the kinetics of product formation are not necessarily correlated with the length of the lag phase.

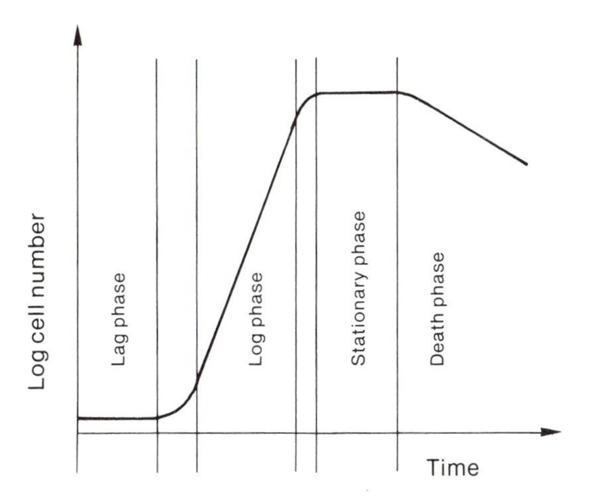

Figure 5.1 Growth curve of a bacterial culture

Log phase By the end of the lag phase, the cells have adapted to the new conditions of growth. Growth of the cell mass can now be described quantitatively as a doubling of cell number per unit time (yeasts and bacteria) or a doubling of biomass per unit time (filamentous organisms such as streptomycetes and fungi). By plotting the number of cells or biomass against time on a semilogarithmic graph, a straight line results, hence the term "log phase".

Although the cells alter the medium through uptake of substrates and excretion of metabolic products, the growth rate remains constant during the log phase. Growth rate is independent of substrate concentration as long as excess substrate is present. The rate of increase in biomass is correlated with the specific growth rate μ and the biomass concentration X (g/l), whereas the rate of increase in cell number is correlated with μ and cell density N (1/l)

$$\frac{dX}{dt} = \mu \cdot X \qquad [1]$$

or

$$\frac{dN}{dt} = \mu \cdot N \qquad [2]$$

The specific growth rate, μ, is generally found to be a function of three parameters: the concentration of limiting substrate, S, the maximum growth rate μ_m, and a substrate-specific constant K_S.

$$\mu = \mu_m \frac{S}{K_S + S} \qquad [3]$$

This equation is generally called the Monod equation since the relationship expressed was first described by Jacques Monod. The constant K_S is the substrate concentration at which half the maximum specific growth rate is obtained ($\mu = 0.5\ \mu_m$). K_S is equivalent to the Michaelis constant in enzyme kinetics. If several limiting substrates are present, eq. (3) can be expanded.

$$\mu = \mu_m \cdot \left(\frac{k_1 S_1}{K_1 + S_1} + \frac{k_2 S_2}{K_2 + S_2} + \cdots + \frac{k_i S_i}{K_i + S_i}\right)\left(\frac{1}{\sum_{j=1}^{i} K_j}\right) \qquad [4]$$

If there is an excess of all substrates, then $\mu = \mu_m$, and the culture is in the log phase at its maximal growth rate. If one substrate has been exhausted, and another substrate is still present, there may be a second log phase during the metabolism of this second substrate with a second specific growth rate μ_1.

The value for K_S is generally very low; in *Escherichia coli*, K_S values of 1.0 mg/l have been measured for glucose and 1.1 mg/l for tryptophan. Such low values also account for the fact that substrate levels equivalent to the K_S value are not obtained in normal batch fermentations during the log phase. Because of the large biomass which is present by the end of the log phase, the substrate is quickly exhausted so that the period of time during which the substrate concentration is near that of the K_S is very short and the stationary phase is approached abruptly.

Maximum specific growth rates are of considerable industrial importance. The μ_m is dependent on the organism and on the conditions of fermentation. The values for several fungi are listed in Table 5.1. The specific growth rates vary between 0.090–0.61 h^{-1}.

Since an organism needs extra energy to split long-chain substrates, the specific growth rate for simple substrates is greater than for long-chain molecules. One example in Table 5.2 shows the

Table 5.1 Maximal specific growth rates (μ_m) of some fungi on glucose

Organism	T°C	μ_m(h^{-1})	Doubling time (h)
Aspergillus niger	30	0.20	3.46
Aspergillus nidulans	20	0.090	7.72
	25	0.148	4.68
	30	0.215	3.23
	37	0.360	1.96
Penicillium chrysogenum	25	0.123	5.65
Mucor hiemalis	25	0.17	4.1
Fusarium avanaceum	25	0.18	3.8
Fusarium graminearum	30	0.28	2.48
Verticillium agaricinum	25	0.24	2.9
Geotrichum candidum	25	0.41	1.7
Geotrichum candidum	30	0.61	1.1
Neurospora sitophila	30	0.40	1.73

(Anderson et al., 1975)

Table 5.2 Effect of glucose chain length on the maximal specific growth rate (μ_m) in *Fusarium graminearum* at 30°C

Substrate	Number of glucose units	μ_m	Doubling time (h)
Glucose	1	0.28	2.48
Maltose	2	0.22	3.15
Maltotriose	3	0.18	3.85

(Anderson et al., 1975)

growth of *Fusarium graminearum* with glucose, maltose (2 glucose units) and maltotriose (3 glucose units).

The Monod equation for the correlation of specific growth rate and substrate concentration does not hold true when the intracellular substrate concentration is reduced during fast growth, even though adequate substrate is still available in the medium. Additional mathematical models have been developed for handling such situations, in which the concentrations used are not those of the medium but of the intracellular milieu.

When complex nutrient solutions are used, two log phases frequently occur separated by a second lag phase. This process is called **diauxy** and arises because one of the substrates (usually glucose) is catabolized preferentially (see the discussion of catabolite regulation in Section 3.5). The presence of this one substrate represses the break down of the other substrates. The catabolic enzymes for the other substrates are induced (during the second lag phase) only after the first substrate has been completely metabolized.

Stationary phase As soon as the substrate is metabolized or toxic substances have been formed, growth slows down or is completely stopped. The biomass increases only gradually or remains constant during this stationary phase, although the composition of the cells may change. Due to lysis, new substrates (carbohydrates, proteins) are released, which then may serve as energy sources for the slow growth of the survivors. The various metabolites formed in the stationary phase are often of great biotechnological interest.

Death phase In this phase the energy reserves of the cells are exhausted. A straight line may be obtained when a semilogarithmic plot is made of survivors vs. time, indicating that the cells are dying at an exponential rate.

The length of time between the stationary phase and the death phase is dependent on the organism and the process used. In commercial processes, the fermentation is usually interrupted at the end of the log phase or before the death phase begins.

Fed-batch fermentation processes

In the conventional batch process just described, all of the substrate is added at the beginning of the fermentation. An enhancement of the closed batch process is the fed-batch fermentation, which is used in the production of substances such as penicillin. In the fed-batch process, substrate is added in increments as the fermentation progresses. The formation of many secondary metabolites is subject to catabolite repression by high concentrations of glucose, other carbohydrates, or nitrogen compounds (see Section 3.5). For this reason, in the fed-batch method the critical elements of the nutrient solution are added in small concentrations at the beginning of the fermentation and these substrates continue to be added in small doses during the production phase.

Since it is usually not possible to measure the substrate concentration directly and continuously during the fermentation, indirect parameters which are correlated with the metabolism of the critical substrate have to be measured in order to control the feeding process. For instance, in the production of organic acids, the pH value may be used to determine the rate of glucose feeding. In fermentations with critical osmotic values, feeding can be regulated by monitoring the pO_2-value or the CO_2-content in the exhaust air.

Continuous fermentation

In continuous fermentation, an open system is set up. Sterile nutrient solution is added to the bioreactor continuously and an equivalent amount of converted nutrient solution with microorganisms is simultaneously taken out of the system. Among the diverse kinds of continuous fermentation, two basic types can be distinguished:

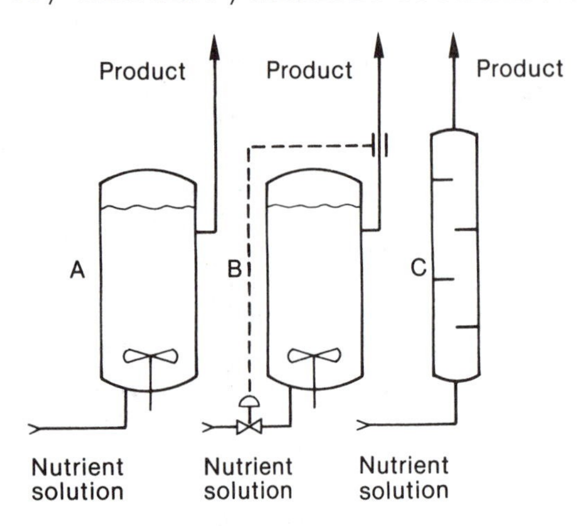

Figure 5.2 Continuous fermentation in the chemostat (A), turbidostat (B), and plug-flow reactor (C)

1. **Homogenously mixed bioreactor.** This is run as either a **chemostat** or a **turbidostat**. In the chemostat in the steady state, cell growth is controlled by adjusting the concentration of one substrate (Figure 5.2 A). Any required substrate (carbohydrate, nitrogen compound, salts, O_2) can be used as a limiting factor. In the turbidostat, cell growth is kept constant by using turbidity to monitor the biomass concentration and the rate of feed of nutrient solution is appropriately adjusted (Figure 5.2 B).
2. **Plug Flow Reactor.** In this type of continuous fermentation, the culture solution flows through a tubular reactor without back mixing. The composition of the nutrient solution, the number of cells, mass transfer (O_2-supply) and productivity vary at different locations within the system (Figure 5.2 C). At the entrance to the reactor, cells must continuously be added along with nutrient solution (usually as a back flow in a bypass from the fermenter outlet or from a second continuous fermentation).

In a continuous process under **steady state conditions**, cell loss as a result of outflow must be balanced by growth of the organism.

$$D = \mu \qquad [5]$$

$$\mu = 1/X \cdot \frac{dX}{dt} \qquad \frac{\text{Growth rate}}{\text{Cell concentration in bioreactor}} \qquad [6]$$

The rate of flow D is defined as the volumetric flow rate (in and out) through the fermenter volume.

By using the Monod equation describing dependence of the specific growth rate on the limiting substrate concentration (equation 3) and the yield constant (equation 7), the cell density X (equation 8) and substrate concentration S (equation 9) in the bioreactor can be determined.

$$Y^X_S = \frac{\text{Biomass (g)}}{\text{Substrate consumption (g)}} \qquad [7]$$

$$X = Y^X_S \left(S_o - \frac{D \cdot K_S}{\mu_m - D}\right) \qquad [8]$$

$$S = \frac{D \cdot K_S}{\mu_m - D} \qquad [9]$$

The substrate concentration in the incoming medium is designated as S_o. Equations 8 and 9 show the dependence on flow rate. At a lower flow rate, the substrate is almost completely used up by the cells ($S \rightarrow 0$). The cell concentration is then $X = S_o/Y$. If D increases, X decreases slowly at first in a linear fashion and then at $D = \mu_m$ it drops sharply to 0. S increases slowly at first as D increases and approaches S_o at $D = \mu_m$. When X is zero in equation 8, the wash-out point D_m has been reached.

$$D_m = \frac{\mu_m \cdot S_o}{K_S + S_o} \qquad [10]$$

$$\frac{S_o}{K_S + S_o} \approx 1 \quad \text{therefore} \quad D_m = \mu_m \qquad [11]$$

Above D_m, a steady state is not possible. If the flow rate is only slightly lower than D_m, the system is very sensitive to external influences. Extreme changes in biomass may result from small deviations in feeding or cell removal. Figure 5.3 shows the interdependence of rate of flow, substrate concentration, cell concentration and rate

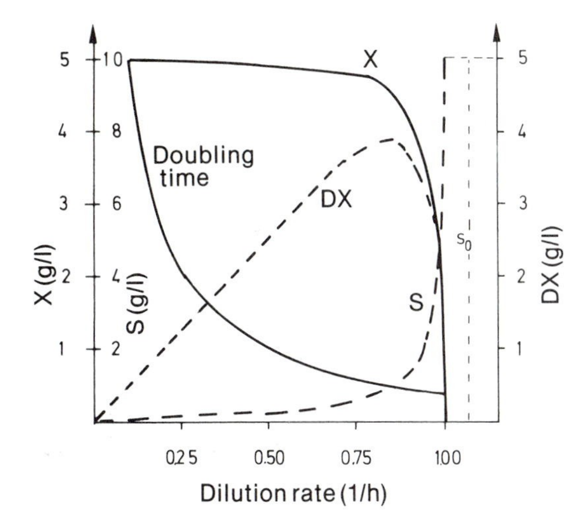

Figure 5.3 Effect of flow rate on substrate concencentration (S), cell concentration (X), Doubling time t_d and cell formation rate (D·X).

of cell formation ($\mu_m = 1\ h^{-1}$, $K_S = 0.2$ g/l, Y = 0.5).

The maximum specific growth rate is chosen in industrial processes so as to obtain the largest yield in the smallest fermenter volume and the shortest time.

The Monod model is not usually applicable to continuous fermentation at very low and very high dilution rates. At mean dilution rates the molar relationship between biomass produced and CO_2 given off is constant. At low dilution rates, the proportion of CO_2 becomes higher. This is because more of the energy source is used for maintenance of cell structure, for osmotic regulation, or for motility. On the other hand, at high flow rates, a part of the added carbon is incompletely oxidized (to products such as pyruvate, acetate or tricarboxylic acids) and productivity is thus reduced. With nitrogen as the limiting substrate, reserve materials such as polysaccharides are formed at low flow rates, which results in an increase in cell size and hence in biomass.

The **production of metabolites** is also not well described by the Monod model at very low or high flow rates. For example, with *Penicillium chrysogenum* the production of penicillin ceases below a minimal growth rate, and conidiospores are formed.

In continuous fermentation at a constant flow rate, steady state conditions do exist and the substrate content, the biochemical reactions within the cells, and the rate of product formation do not change. However, there are differences in metabolism at a constant flow rate depending on the substrate limitation. Table 5.3 shows the influence of substrate limitation on the amino acid pool and glutamic acid dehydrogenase in *Saccharomyces cerevisiae*.

Furthermore the **rate of product formation** is dependent on the flow rate. This can be applied not only to individual enzymes, but to the excretion of primary and secondary metabolites as well (Figure 5.4). When the idiophase (product formation phase) is separate from the trophophase (growth phase), product formation cannot be carried out continuously in an optimally mixed bioreactor. Fermentation must then be done using a series of fermenters as a cascade or in a tower reactor using single reaction chambers separated by sieve plates.

Continuous fermentation processes have been developed for the production of single-cell protein, antibiotics, organic solvents, starter cultures and for cellulose decomposition (in laboratory studies). Pilot plants or production plants have been installed for continuous production of single-cell protein based on n-alkanes, C_1-compounds and starches (see Chapter 16).

Table 5.3 Effect of growth conditions on the amino acid metabolism of *Saccharomyces cerevisiae*
Dilution rate = 0.1 h^{-1}, 30°C

Limiting substrate	Amino acids (mM)		Glutamic acid dehydrogenase (mMol/min/mg protein)
	Total	Glutamic acid	
Ammonium ion	144	35.6	1730
Glutamic acid	180	49.2	510
Glucose	381	135.4	620
Phosphate	386	121.4	465

(Brown and Stanley, 1972)

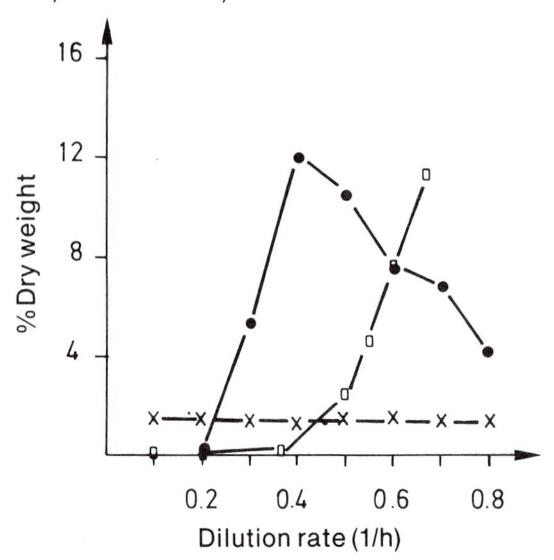

Figure 5.4 Effect of flow rate on the production of poly-hydroxybutyric acid (PHB •——•) and glycogen (X —— X) with limited carbon, and PHB with excess acetate (□——□). (Wilkinson and Munro, 1967)

Industrial processes using continuous fermentation include waste-water treatment (see Chapter 17) as well as the production of beer, glucose isomerase, and ethanol.

Although many fermentations for metabolite production work well as continuous processes at a laboratory scale, only a few processes have proved useful for practical application for several reasons:

- Many laboratory methods operate continuously for only 20 to 200 hours; for industrial use the system must be stable for at least 500 to 1000 hours.
- Maintaining sterile conditions on an industrial scale over a long period of time is difficult.
- The composition of substrates must be constant in order to obtain maximal production. The composition of industrial nutrient solutions varies (e.g. corn steep liquor, peptone, starch).
- When high-yielding strains are used, reverse mutants arise, which can overgrow the production strains in continuous culture.

Continuous cultures also find considerable use in research on ecological processes. Growth at low-substrate levels in a continuous culture is analogous to growth at low-substrate levels in nature, so that the continuous cultures provide excellent models for analysis of ecological processes.

Classification of fermentation processes

We have previously subdivided fermentation processes into three types: batch, fed-batch, and continuous. Another means of classification has been developed in which fermentations are classified according to the dependence of product formation on energy metabolism.

Type I In a type I process, the product is derived directly from the primary metabolism used for energy production. The product can also be biomass itself. Growth, carbohydrate catabolism, and product formation run almost in parallel (Figure 5.5 A) and trophophase and idiophase are not separated from each other. The reaction can be expressed as follows:

Substrate A → Product
or Substrate A → B → C → Product

Processes for the production of single-cell protein, ethanol, and gluconic acid all belong to this category. Continuous fermentations in reactors without a cascade system are classified here as well.

Type II Here the product is also derived from the substrate used for primary energy metabolism, but production takes place in a secondary pathway which is separate from primary metabolism.

Substrate A → B → C → D → Primary metabolism
↓
E → F → Product

In batch fermentations of this type there are two maxima. At first, good growth occurs accom-

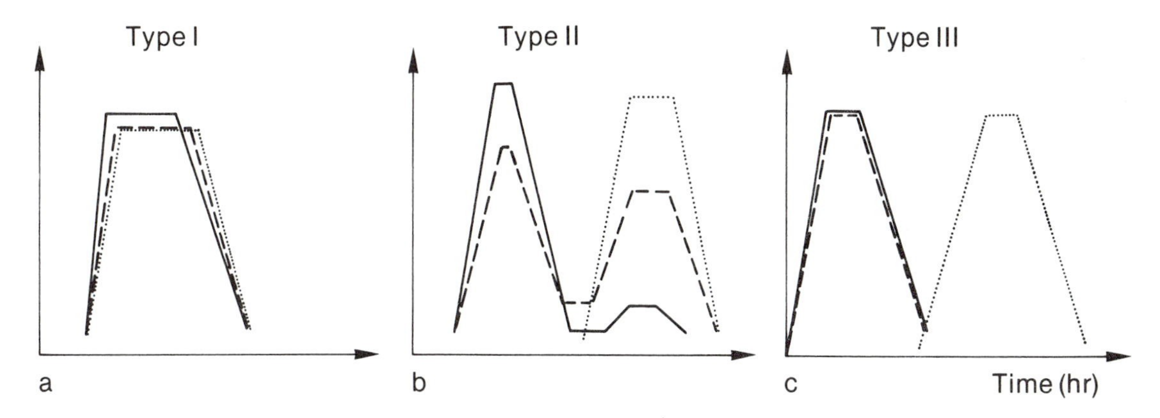

Figure 5.5 Fermentation categories defined by Gaden (1959). Specific growth rate (——), specific carbohydrate consumption (– – –), specific product formation rate (····)

panied by high substrate consumption and little or no product formation. Then growth slows down and product formation begins, accompanied by a high substrate consumption rate (Figure 5.5 B). In type II processes, the trophophase and idiophase are separated in time. Citric acid, itaconic acid, and some amino acids are produced by this type of process.

Type III In the type III processes, primary metabolism and product formation occur at completely separate times. The product is not derived from catabolism, but from amphibolic pathways. In this type of fermentation, primary metabolism functions first, accompanied by substrate consumption and growth. Afterwards, the product is formed by the reactions of intermediary metabolism (Figure 5.5 C). Many antibiotics and vitamins are produced by this type of fermentation.

This classification cannot be applied to all fermentation processes. Depending on metabolite production, composition of culture medium, and regulation in the strain used, there may be intermediate forms. For example, the production of lactic acid falls between types I and II, and aminoglycoside antibiotic production falls between types II and III.

Production involving mycelium-producing microorganisms can not be classified among these simple fermentation types, since the correlation of mycelium production and changes in mycelium structure with antibiotic production is complex. The stages in oxytetracycline formation with the mycelium-forming organism *Streptomyces rimosus* are shown in Table 5.4.

Table 5.4 Stages in the oxytetracycline fermentation with *Streptomyces rimosus*

Stage	Fermentation process
1 Lag phase	Duration 90 minutes for low inoculum. No metabolism evident.
2 Growth of primary mycelium	Duration 10–25 h, depending on inoculum. Hyphae dense. Respiration, nucleic acid synthesis and some other enzymatic conversions are high. Pyruvate content reaches the maximum. No antibiotic production.
3 Fragmentation of primary mycelium	Duration 10 h, no growth during this time. Respiration and nucleic acid synthesis decline. Pyruvate content falls to low value.
4 Growth of secondary mycelium	Duration 25 h, mycelium volume is 2 to 4 times that of Stage 2, hyphae thin, good antibiotic production. Nucleic acid production resumes, but respiration is low. Carbohydrate and ammonium content fall to zero. Pyruvate content can begin to increase again.
5 Stationary phase	Growth ceases, metabolism is low, antibiotic production continues but at a low rate.

(Bailey and Ollis, 1977)

Productivity and specific rate of production

The productivity of a fermentation is defined as

$$\text{Productivity (P)} = \frac{\text{Product concentration /l}}{\text{Fermentation time}} = \text{Units/hr}^{-1}$$

In evaluating the cost-effectiveness of a process, several factors must be considered: the production time of the fermentation, the time required to clean and set up the fermenter, the sterilization time, and the duration of the lag phase. In Figure 5.6, the slope of the tangent of the product formation curve is a measurement of maximal productivity and the slope of the straight line to the end of the product formation curve indicates total productivity.

Whether the fermentation process is terminated at the time of maximal productivity or later depends on the operating costs, which include energy, overhead, labor, and the capacity of the system. By using a graph such as that of Figure 5.6, an estimate can be made of how to optimize the process. In short-term fermentations (8–70 h) the set-up time is significant, whereas in long-term fermentations (> 3 days) the set-up time is less crucial.

The **fermentation conditions** directly affect productivity. Figure 5.7 shows the productivity of a single-cell protein process in relation to the substrate (hydrocarbon) content. There is a linear increase in productivity as the substrate content rises until the concentration is reached at which the petroleum hydrocarbon becomes the continuous phase (oil and water are not miscible). Above this concentration, productivity then decreases.

In continuous fermentations, productivity can be expressed as:

$$P = D \cdot X \ (g/\ell \cdot h) \qquad [12]$$

When the relationship of equation 8 is used, the following equation results:

$$P = D \cdot Y_S^X \left(S_o - \frac{D \cdot K_S}{\mu_m - D}\right) \qquad [13]$$

In continuous fermentations, maximal productivity is equal to total productivity since set-up time and lag phase can be disregarded. For instance, if the initial set-up time in a continuous fermentation requires 24 hours, and the continuous fermentation itself is run for 24 hours with a rate of flow of 0.1 h^{-1}, 24 hours with 0.15 h^{-1} and at least 72 hours with 0.2 h^{-1}, the total productivity for product X is calculated as shown in Table 5.5.

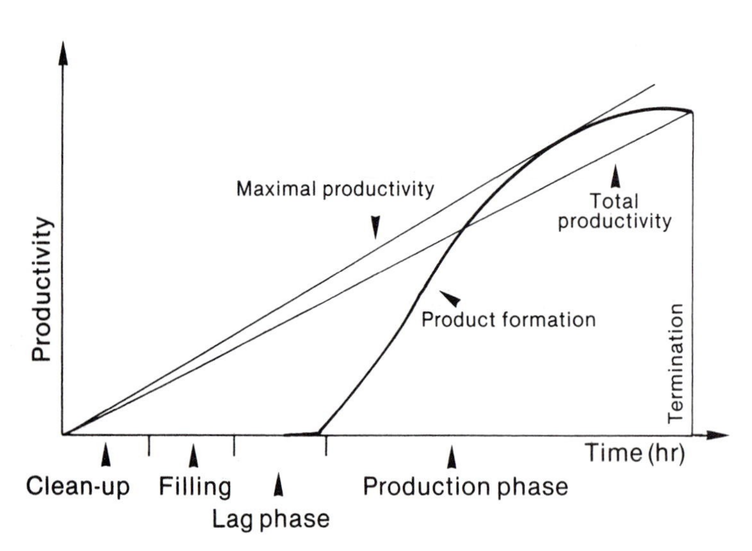

Figure 5.6 Total productivity and maximal productivity

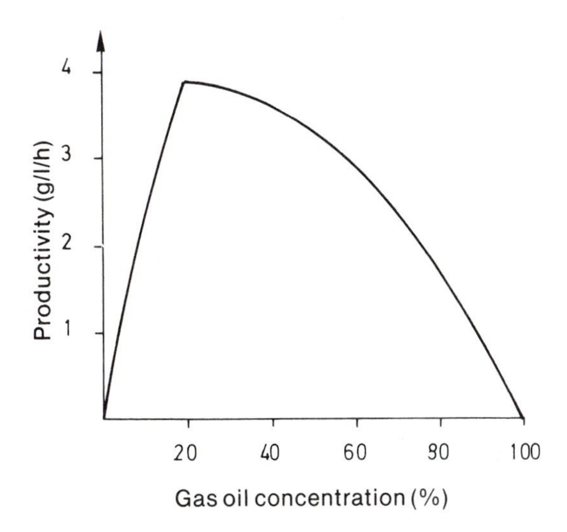

Figure 5.7 Productivity in relation to gas oil concentration in a single-cell-protein process

The specific rate of production q_p is derived from equation 1 in the following way:

$$\frac{dP_1}{dt} = q_p \cdot X \qquad [14]$$

The value P_1 is the concentration of product and the specific rate of production q_p is equivalent to the specific growth rate μ in equation 1. In type II or III processes, however, there is no direct correlation between μ and q_p.

Yield coefficients

Monod originally defined the yield coefficient, Y, as the ratio of cells produced to substrate consumed:

$$\frac{dX}{dt} = -Y\frac{dS}{dt} \qquad [15]$$

or

$$Y_S = \frac{\text{Biomass (g)}}{\text{Substrate consumption (g)}} \qquad [16]$$

Today general yield coefficients are used to characterize fermentation processes, i.e. the relationship of cells produced to individual substrates

Table 5.5 Total productivity of a continuous fermentation

Time period h	Flow rate h	Total productivity $X(g/l)\cdot D(h^{-1})$
0–24	Batch	–
24–48	0.1	X·0.05
48–72	0.15	X·0.08
> 72	0.2	
300	0.2	X·0.171
1000	0.2	X·0.191
5000	0.2	X·0.198

converted or to energy released or energy consumed. However, yield coefficients are not constants, since they are dependent on biological parameters (X, μ) and chemical parameters (pO_2, C/N ratio and P content of the medium).

Yield coefficients have been classified in three groups. The first group of coefficients (Y_S, Y_{O_2}, Y_{kcal}) gives information about the technical process and thus about cost-effectiveness. The second group (Y_C, Y_P, Y_N, Y_{ave^-}) describes catabolism, amphibolism, and anabolism. The third group includes Y_{ATP}, a coefficient which describes the energy relations of the cell. Some authors use other yield coefficients, which must be designated properly, as shown:

$$Y_S^{CO_2} = \frac{\text{Mol } CO_2 \text{ formed}}{\text{Mol substrate consumed}} \qquad [17]$$

Heat production

In order to obtain optimal yields, fermentations must be carried out at **constant temperature**. We now discuss the parameters affecting the heat balance of a fermentation process.

The rate of heat production due to stirring and due to the metabolic activity of the microorganisms must be balanced by the heat loss resulting from evaporation and radiation plus heat removal by the cooling system (the jacket of the fermenter or cooling coils).

The heat production during stirring can be calculated from the energy consumption of the motor minus the losses through the seals, bearings, and drive shaft.

Heat production of the microorganism can be determined using the yield coefficient Y_{kcal}:

$$Y_{kcal} = \frac{Y_S}{H_S - Y_S \cdot H_C} \quad [18]$$

The yield coefficient Y_{kcal} is given in grams of cells/kcal energy released; Y_S indicates the ratio of grams of cells/gram of substrate utilized; the heats of combustion of the substrate H_S and of the biomass H_C are measured in kcal/g. Although there are tables giving the heats of combustion of different substrates, the heat of combustion of the cells (H_C) must be determined experimentally or calculated by using various assumptions.

One possibility for calculating H_C makes use of the electron theory of valence. In the transfer of an electron equivalent, 26.05 kcal are released. Since four valence electrons are shared by oxygen, 104.20 kcal/M O_2 are produced. One calculation using *Pseudomonas fluorescens* assumed that when the biomass is burned, CO_2, N_2 and H_2O are produced.

$$\underset{\textit{Pseudomonas}}{C_{4.41}H_{7.3}N_{0.86}O_{1.19}} + 5.64\ O_2 \rightarrow 4.41\ CO_2 + 0.43\ N_2 + 3.65\ H_2O \quad [19]$$

The heat of combustion is calculated as:

$$H_C = \frac{5.64 \cdot 104.2}{4.41 \cdot 12 + 7.3 \cdot 1.0 + 0.86 \cdot 14 + 1.19 \cdot 16}$$

$$= 6.44\ \text{kcal/g} \quad [20]$$

The value obtained must be corrected by about 10% due to the ash content of the cells.

$$H_C = 6.44 \cdot 0.90 = 5.8\ \text{kcal/g} \quad [21]$$

The **heat production coefficients** for some bacteria on different substrates are listed in Table 5.6. For the more highly oxidized substrates (e.g. glucose and acetate), less heat is produced per biomass than with the more reduced substrates (methane and methanol). In general, Y_{kcal} values increase linearly in relation to the specific growth rate μ of the cultures.

Table 5.6 Yield coefficients for bacteria with different substrates

Substrate	Y_{kcal}(g cells/kcal)
Malate	0.30
Acetate	0.21
Glucose	0.42
Methanol	0.12
Ethanol	0.18
Isopropanol	0.074
n-Paraffin	0.16
Methane	0.061

The rate of microbial heat production per unit time, which along with the heat produced by stirring must be removed from the bioreactor, is calculated from the rate of cell production (equation 1), the reactor volume and the heat coefficient:

Growth rate $$\frac{dX}{dt} = \mu \cdot X \quad [1]$$

Heat production rate $$Q_W = V \cdot \mu X \frac{1}{Y_{kcal}} \quad [22]$$

5.3 FERMENTER SYSTEMS

In a bioreactor, production of metabolites must be accomplished with maximum emphasis on reliability for the process and a minimum of capital investment and operating cost. Reliability is more difficult to achieve in a microbiological than in a chemical process so that bioreactors are more expensive to design and construct than chemical reaction vessels.

Microbiology must be the focus of all considerations concerning construction of a fermenter system. It must be decided in the planning stages whether the fermenter is to be used for a special process with one organism or for a variety of processes with different microorganisms. A number of different bioreactors for aerobic fermentation are diagrammed in Figure 5.8. They may be classified in four groups with respect to gas distribution.

1. **Gas distribution by stirring**. The stirred vessel (Figure 5.8, part 1) is the best understood

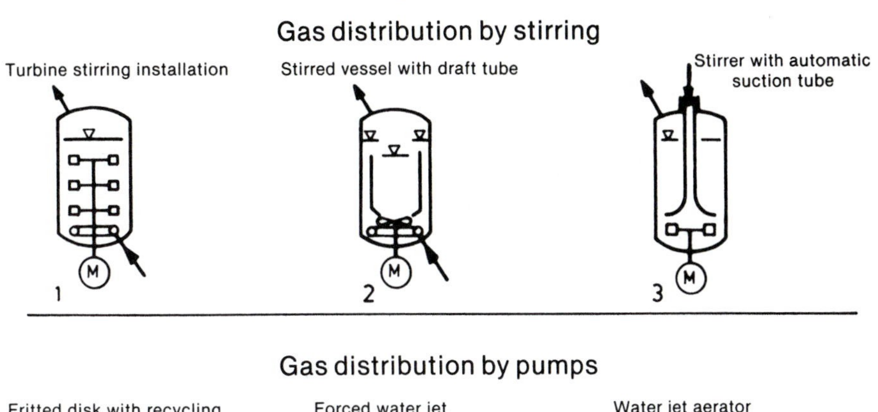

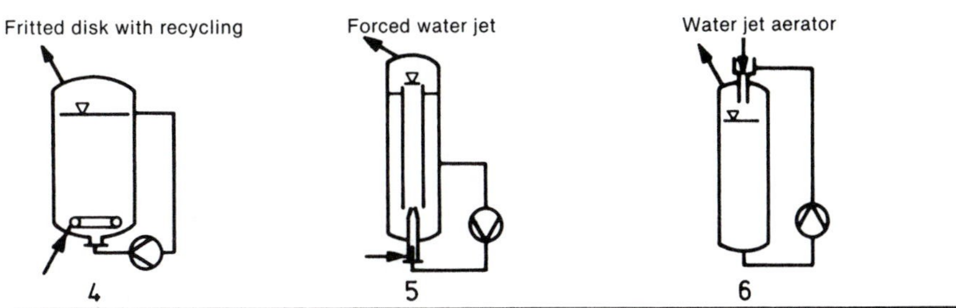

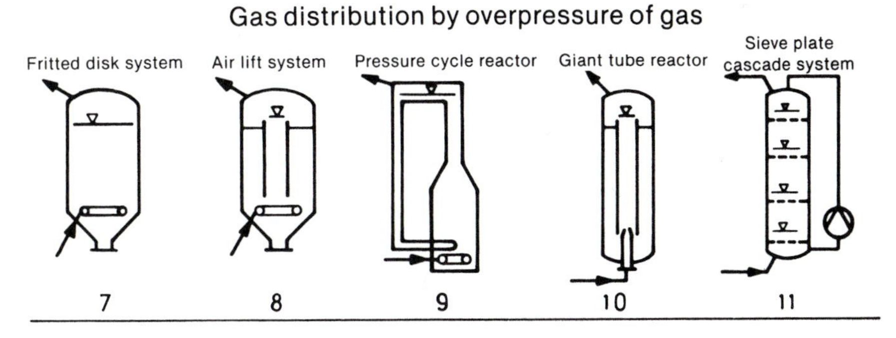

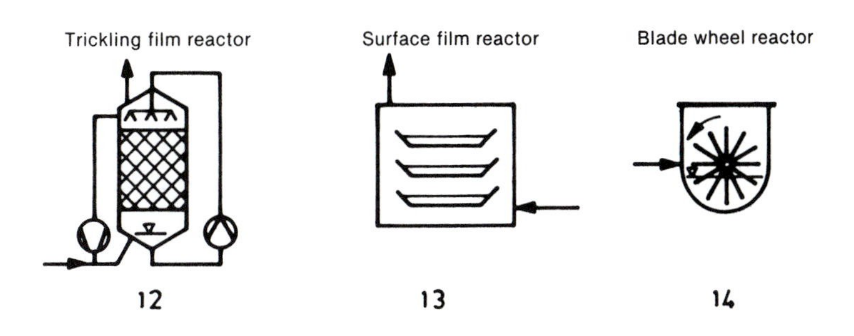

Figure 5.8 Basic construction types for bioreactors

and most widely used bioreactor and is quite flexible. Loop reactors with draft tubes (2) are also suitable for mass production. A system in which air is sucked through the tube by the low pressure on the lee side of the stirrer blades is used in vinegar production (3). The various surface aerators used in waste-water treatment also belong to this group.

2. **Gas distribution through pumps**. In a modified airlift system (4), the pump primarily transports liquid and the air is distributed more or less independently through the sparger. In the more efficient systems (5, 6), there is a direct mixing of air and fluid by means of a water jet pump.
3. **Gas distribution by means of pressurized air**. In most of these systems, no moving parts are present in the sterile area. The "pressure cycle fermenter" (9) was the first model of this kind used in biotechnology for single-cell protein production on an industrial scale.
4. **Continuous gas phase**. In these processes, air circulates over a film of microorganisms. In the trickle-film reactor (12), the organism grows on a solid, inert carrier material. In the surface reactor (13), it floats on or in the nutrient solution or grows in a semi-solid or solid culture medium. In the blade-wheel reactor (14), it grows on movable blades or drums and is dipped alternately in nutrient solution and in gas phase as the wheel turns. Reactors used in waste-water treatment, for the leaching of ores, and for vinegar or citric acid production, as well as the culture flasks on laboratory shaking machines, are included in this group.

Stirred bioreactors

For industrial use, especially in the pharmaceutical industry, the most versatile bioreactor is the simple stirred, aerated fermenter. However, no single system which adequately meets the needs of *all* biological systems can be constructed. Fermenters for laboratory experiments requiring volumes of up to 20 l are made of glass or stainless steel (Figure 5.9); for larger volumes construction is of stainless steel. Figure 5.10 illustrates the design of a large scale fermenter.

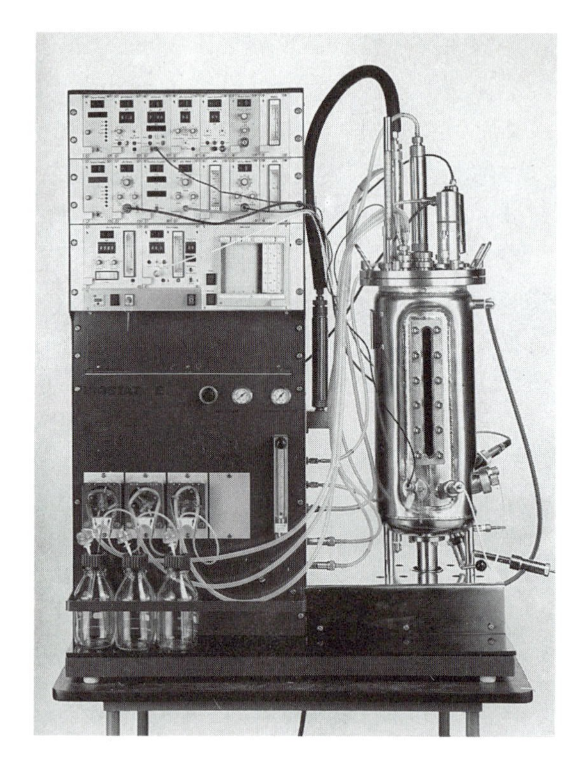

Figure 5.9 Laboratory fermenter

Up to the 3000 l scale, fermenters can be standardized. However, the geometry of production fermenters changes as the volume and the process requirements change. The height and width relationship may vary from 2:1 to 6:1 and the stirrer may be top or bottom driven.

Baffles Baffles bring about the transfer of turbulence to the fermenter wall. Four baffles are commonly installed, with a width 1:10 or 1:12 of the reactor diameter. If heat dissipation is a problem, as it often is in large production fermenters ($> 100\ m^3$), up to 12 baffles can be used as heat exchangers.

Foam separators Foaming is frequently a problem in large-scale aerated systems. Antifoam chemical agents cannot always be used for the

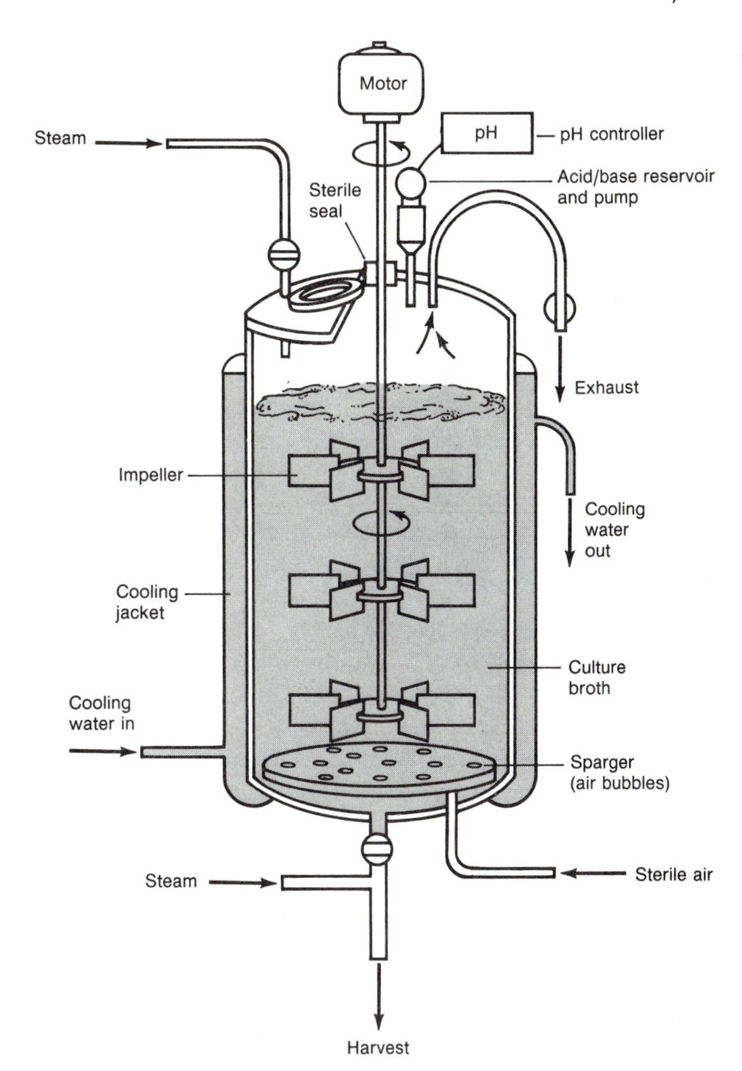

Figure 5.10 An industrial fermenter, illustrating the construction and facilities for aeration and process control.

reduction of foam, since they may have inhibitory effects on the fermentation. There are several methods by which the foam in the sterile area can be destroyed by mechanical methods. The simplest devices have rakes mounted on the stirrer. In the Frings system for vinegar production and the Fundafom system (Figure 5.11), both of which are currently used industrially, foam is destroyed by centrifugal force. The nutrient solution held in the foam flows back into the bioreactor and the air released from the foam leaves the sterile system.

Types of stirrers Only specific types of stirrers developed for chemical technology can actually be used for microbiological processes. Figure 5.12 shows the most important ones. The disc stirrer is the most common type; 4–8 radial blades project out beyond the edge of a disc of appropriate diameter. In the turbine-type stirrer, blades are curved. Compared with the disc stirrer, the turbine type requires 50% less air for the same yield and energy consumption. Two other types consist of stirring arms with blades attached at an angle. Compared with both of the above-men-

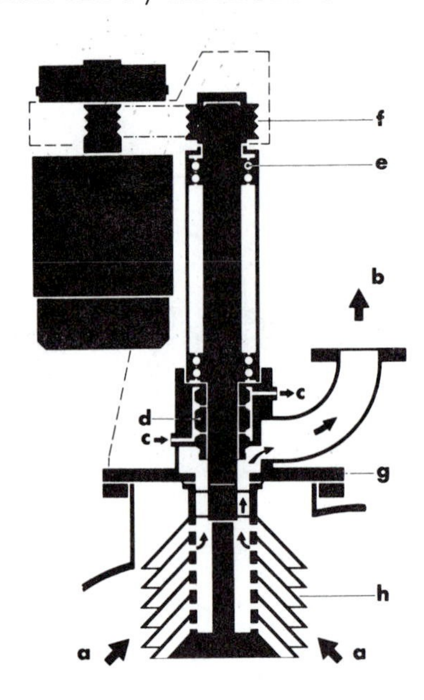

Figure 5.11 Mechanical foam separator (Fundafom, Chemap). a) Foam entrance, b) Gas exit, c) Lubrication, d) Double seal, e) Packing, f) Drive, g) Intermediate flange, h) Rotating plate

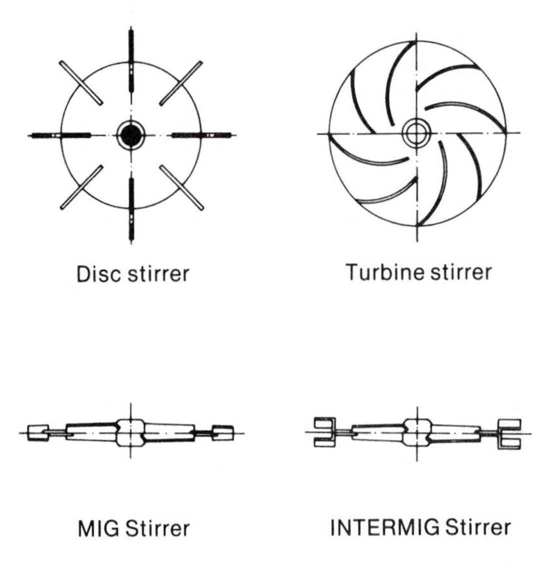

Figure 5.12 Impeller systems for fermenters

tioned stirrers, the MIG and INTERMIG stirrers (Figure 5.12) require 25% and 40% less energy respectively for equivalent yields.

Maintenance of sterility There should be a minimum number of openings in the fermenter to favor maintenance of sterility. Small openings must be made leakproof with O-rings, larger openings with flat gaskets. Wherever a movable shaft penetrates the fermenter wall, special problems of sterility maintenance arise. Double mechanical seals on the agitator shaft are currently used and present noticeable advantages in comparison to the more conventional stuffing-box seal. If possible, the joints of all parts connected within the sterile area as well as all of the pipes both inside and outside the fermenter should be welded. There should not be any direct connection between the nonsterile and sterile area; that is, sampling devices and injection ports must be covered with steam-sterilizable closures. Sterile pipes must be slanted to collect the condensate and to drain it.

Although the fermenter vessel is the main focus of interest in the overall fermenter system, in an actual installation a large number of additional devices must be provided in order to obtain a functioning fermenter system. Figure 5.13 shows the layout for the installation of a fermenter system.

Reactors for immobilized enzymes or cells

An important type of large-scale biological reactions involves the attachment of the biocatalyst, enzyme or cell, to an inert support through which the substrate is passed. The product is then present in the effluent and can be directly purified. Bioreactors for use with immobilized enzymes or cells must be so constructed that that rate of movement between substrate and the biocatalyst is not rate-limiting. Internal mass transfer can only be influenced by the method by which the catalyst is immobilized. Figure 5.14 summarizes the various types of bioreactors for immobilized systems. Two basic types are used:

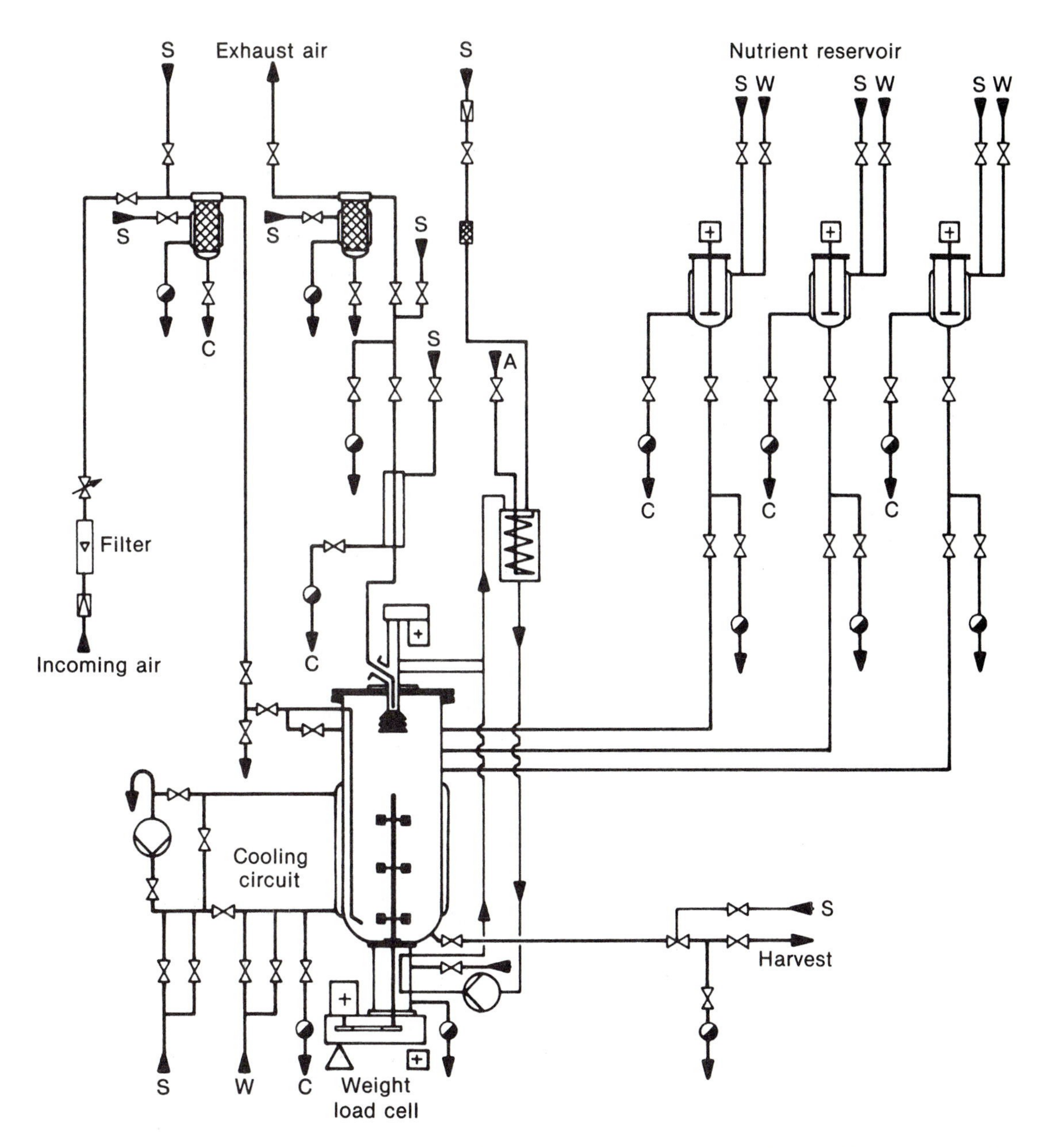

Figure 5.13 Installation of a fermenter. S, steam; C, condensate; W, water; A, air. The steam lines permit in-place sterilization of valves, pipes, and seals. The input air is sterilized by both incineration and filtration.

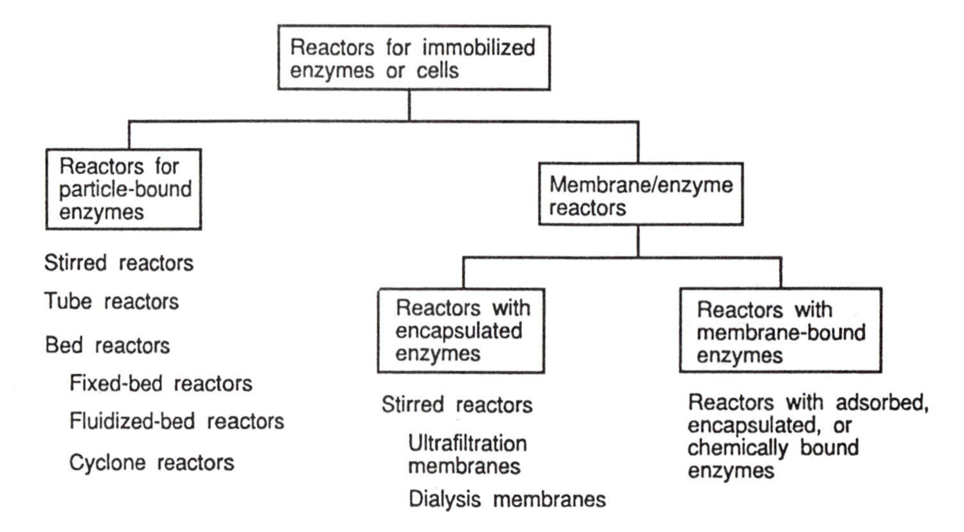

Figure 5.14 Types of systems for immobilized enzymes/cells

those in which the enzyme is incorporated into a membrane and those in which the enzyme is bound to an inert particle.

Particle-bound systems For systems that used particle-bound enzymes, stirred reactors or those mixed by pumps are used (see Figure 5.8, parts 1,2, and 4) if high aeration or careful pH control is necessary, or if the fluid has a high viscosity. A disadvantage of the use of a stirred reactor is that the shearing force of the stirrer or pump is high, thus raising the possibility that the catalyst may be damaged.

Another approach is the use of a system in which the catalyst is immobilized in a bed, such as a fixed-bed, fluidized bed, or cyclonic reactor (Figure 5.15). The fixed-bed reactor adapts itself well to large-scale operation; because of the dense packing of the immobilized material a high rate of substrate conversion per unit time is obtained. Since the flow is from the bottom to the top, there are gradients of concentrations of substrate and product. Disadvantages of the up-flow fixed-bed reactor are: gas formation may reduce the contact surface between substrate and enzyme; channels in the fixed bed may develop, leading to the appearance in the product of un-

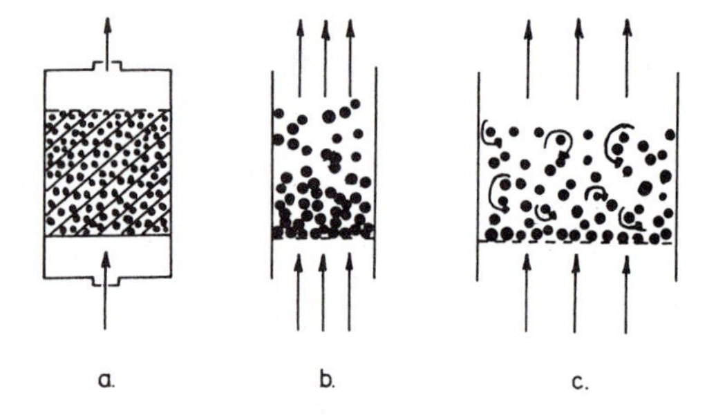

Figure 5.15 Bed reactions for immobilized enzyme reactions: a) Fixed bed; b) Fluidized bed; c) Cyclone

reacted substrate; regulation of temperature and pH is difficult. In the fluidized-bed reactor the height of the catalyst must be somewhat larger than that of the substrate solution. Because of the plug-flow nature of fluid movement, the catalyst is kept in suspension. Recycling of the fluid is not possible, the flow-rate cannot be varied, and scale-up is quite difficult.

Cyclone reactors can be well mixed, either by installation of a stirrer or by use of a gas under pressure. Because of the broad distribution of residence times of the substrate, complete conversion of substrate to product is not possible.

Membrane/enzyme reactors In membrane/enzyme reactors the soluble enzyme and substrate are introduced on one side of an ultrafilter membrane. By means of a pump, the product is forced through the membrane under 0.5 to 5 atmospheres pressure; the membrane holds back the enzyme (Figure 5.16). Enzyme/membrane reactors have several advantages:

- The expense of immobilization is eliminated and the loss of enzyme is reduced.
- By periodic addition of enzyme to make up that lost by denaturation, the rate of reaction can be maintained at a constant rate.
- Multi-enzyme systems can be used.
- By recycling of the reaction mixture internal and external mass transfer can be optimized.
- Substrate and enzyme can be easily replaced.
- Scale-up presents no particular problems.

The main disadvantages of enzyme/membrane systems are the shearing stress from the fluid flow, which tends to denature the enzyme, and the relatively large surface area of membrane required.

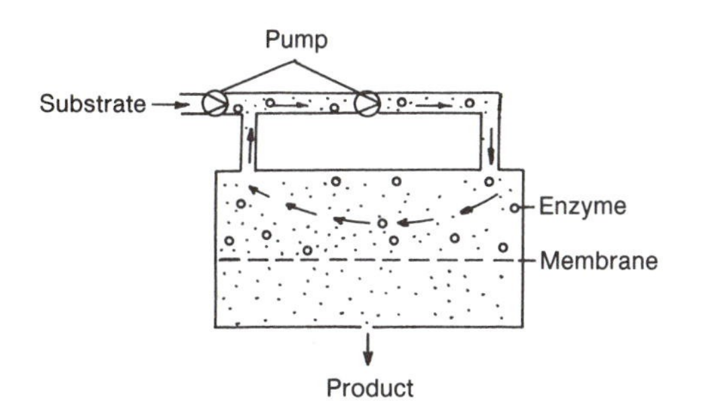

Figure 5.16 Use of ultrafilter membrane to separate enzyme and substrate from product

Modern ultrafiltration membranes made of polyamides (nylon) or polysulfones offer excellent molecular-weight separation in the range of 500–300,000 Daltons). They are also quite stable to temperature, pressure, and chemical attack and are also resistant to microbial degradation. In order to reduce the tendency for clogging, asymmetrical membranes can be used with smaller pores on the high-pressure side. Such membranes are also readily cleanable by backwashing. For industrial-scale operation, the separation is not carried out in the bioreactor itself but in a bypass in which the membrane is installed. Instead of ultrafiltration, it is also possible to separate product from enzyme and substrate by dialysis.

Bioreactors have also been developed in which the enzyme is adsorbed onto or incorporated within the membrane, either through physical or chemical means. In such systems the substrate is converted to product as it passes through the membrane. For instance, a cross-linked sulfonic acid cellulose has been developed which can be fabricated into paper sheets and wrapped into cartridges. In a radial reactor of this type, the enzyme mutase was used to quantitatively convert a solution of sucrose to isomaltulose. By passage through the cartridge the substrate was completely converted to product.

5.4 STIRRING AND MIXING

A microbial fermentation can be viewed as a **three-phase system**, involving liquid-solid, gas-solid, and gas-liquid reactions.

1. The **liquid phase** contains dissolved salts, substrates, and metabolites. A second liquid phase may occur in some cases if there is an insoluble substrate, e.g. alkane fermentations.
2. The **solid phase** consists of the individual cells, pellets, insoluble substrates, or precipitating metabolic products.
3. The **gaseous phase** provides a reservoir for oxygen supply, for CO_2 removal, or for the adjustment of pH with gaseous ammonia.

The transfer of energy, substrate, and metabolite within the bioreactor must be brought about by a suitable mixing device. The efficiency of the transport of any one substrate may be crucial to the efficiency of the whole fermentation.

For the three phases, the stirring of a bioreactor brings about the following:

- Dispersion of air in the nutrient solution.
- Homogenization to equalize the temperature and the concentration of nutrients throughout the fermenter.
- Suspension of microorganisms and solid nutrients.
- Dispersion of immiscible liquids.

Reynold's number

Stirring ensures the transport of nutrients within the culture liquid. In turbulent flow, two molecules of liquid move in relation to each other. The relative velocity between the nutrient solution and the individual cell should be about 0.5 m/sec. Turbulent flow can be characterized by the dimensionless Reynold's number, well discussed in texts on fluid mechanics. The Reynold's number is calculated as follows

$$\text{Reynold's number} \quad R_e = N_{Re} = \frac{D_i^2 \cdot N \cdot \rho}{\eta} \qquad [23]$$

D_i = Stirrer diameter (cm)
N = Stirrer speed (sec^{-1})
ρ = Density (g/cm^3)
η = Dynamic viscosity ($g/cm \cdot sec$)

The following Reynold's numbers have been calculated for the penicillin fermentation with *Penicillium chrysogenum*. Note how much lower the values are than for pure water:

	Small fermenter	Production fermenter
Water ($\eta = 10^{-2}$ g/cm · sec = 1 centi Poise)	Re $4.0 \cdot 10^5$	$6.9 \cdot 10^6$
Culture medium ($\eta = 5$ g/cm · sec = 500 centi Poise)	Re $8.0 \cdot 10^2$	$1.4 \cdot 10^4$

The Reynold's number describes the flow only at the periphery of the stirrer. To distribute the turbulence homogeneously within the entire reactor, an impeller of appropriate shape and diameter must be used. The flow rate is crucial to the distribution of turbulence. However, as turbulence may damage filamentous organisms, there are frequently limitations as to how fast a system can be stirred. The mixing time, θ_δ is the time needed for homogenization to the required degree of homogeneity in the entire reactor.

Power number

The power number (N_p = Newton's number = Ne) has been defined as a dimensionless parameter relating to the energy required by stirred reactors. It is calculated as follows:

$$\text{Power number} = N_p = \frac{\text{Imposed force}}{\text{Inertial force}} = \frac{P_o}{N^3 \cdot D_i^5 \cdot \rho} \qquad [24]$$

P_o = Stirring power (kW)
N = Stirring speed (sec^{-1})
D_i = Stirrer diameter (cm)
ρ = Density of the medium (g/cm^3)

The power number has been correlated with the Reynold's number for several types of stirrers. Figure 5.17 shows this relationship with six types of stirrer blades.

In the **laminar flow** range of mixing speed ($N_{Re} < 10$) the Reynold's number is correlated with the power number as follows:

$$N_p = K_1 \cdot (N_{Re})^{-m} \qquad [25]$$

K_1 = Constant dependent upon the container geometery and the shape of the stirrer, but not dependent upon the reactor size
m = 1

With laminar flow, the power required for stirring is not dependent on the density, but is correlated with fermentation parameters as follows:

$$P_o \approx N^2 \cdot D_i^3 \cdot \eta \qquad [26]$$

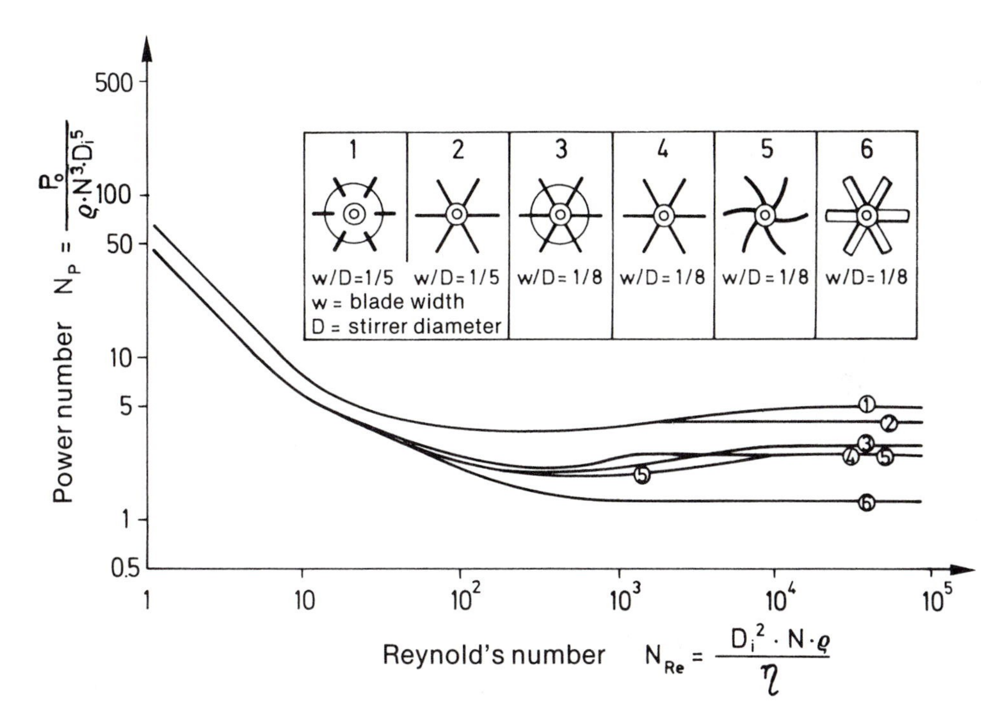

Figure 5.17 Correlation between power number and Reynold's number in Newtonian solutions for various impellers (Bates et al., 1963)

In the **turbulent flow** range of mixing speed ($N_{Re} > 10^4$), the power number is independent of the Reynold's number and is constant

$$N_p = K_2 = \text{Constant} \quad m = 0$$

Under these conditions the power number is independent of the viscosity

$$P_o = K_1 \cdot N^3 \cdot D_i^5 \cdot \rho \qquad [27]$$

In the **transient** range of mixing speed ($N_{Re} = 10\text{–}10^4$), there is no simple correlation between the power number and the Reynold's number. Disc impellers have a higher power number than slanted blade or propeller stirrers.

In bioreactors where a vortex is formed, another dimensionless number, Froude's number, is useful. It is described as follows

$$N_{Fr} = \frac{N^2 \cdot D_i}{g} \qquad [28]$$

g = Acceleration due to gravity

Power number, Reynold's number and Froude's number are correlated with each other

$$N_p = K_1 \cdot (N_{Re})^m \cdot (N_{Fr})^y \qquad [29]$$

No vortex is formed in bioreactors with baffles, so that Froude's number is not relevant in these cases.

Effect of viscosity

Nutrient solutions can be subdivided into two groups according to the way they behave when stirred: **viscous** solutions with Newtonian and non-Newtonian properties; and **viscoelastic** solutions, in which normal liquid-state properties are not observed in stirred vessels.

There are only a few examples which fall into the second group (polysaccharides and certain antibiotic fermentations). During agitation of such solutions, the liquid does not flow tangentially, but rises up along the impeller, a behavior called the Weissenberg effect. With such solutions, scale-up calculations are difficult to make.

Most fermentation solutions fall into the first category. Uninoculated solutions and bacterial cultures often behave as simple Newtonian liquids. In such liquids, the dynamic viscosity (η) of a nutrient solution is a constant at a constant temperature, and it is dependent on the ratio of the shearing stress (T) to the rate of shear (γ).

Newton's law of friction

$$T = \eta \cdot \dot{\gamma} \qquad [30]$$

Newtonian fluid

$$\eta = \frac{T}{\dot{\gamma}} = \text{Constant} \qquad [31]$$

T = Shear stress (kp/m^2)
$\dot{\gamma}$ = Shear rate (sec^{-1})

With many mycelial organisms, changes occur during the fermentation not only in the amount of mycelium, but in the characteristics of the nutrient solution. Substrates are taken up during metabolism and the proportion of undissolved substrates is reduced. At the same time, metabolites are excreted, thus affecting the viscosity of the solution.

Figure 5.18 presents data on a fermentation system with non-Newtonian properties. In this figure, the dynamic viscosity of a *Micromonospora* culture is plotted against time (ν = dynamic viscosity/density; 1 Stokes = 1 St = $10^{-4} m^2/sec$). In this case, the viscosity increases by a factor of 100 during mycelium formation.

In non-Newtonian nutrient solutions, the dynamic viscosity is dependent not only on the temperature, but also on the rate of shear. The

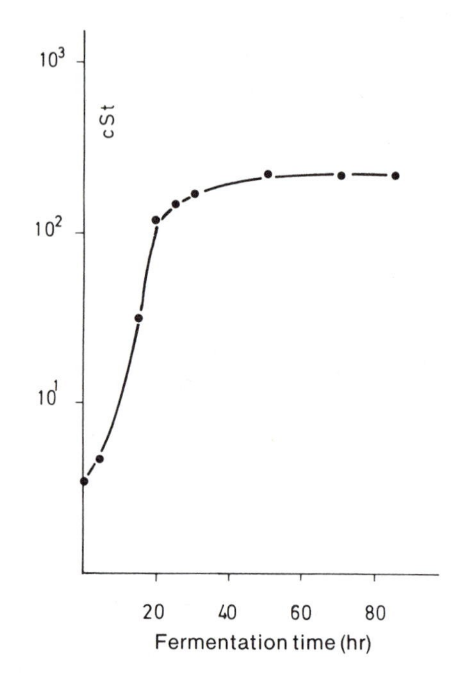

Figure 5.18 Progress of kinematic viscosity in *Micromonospora* fermentations (1cSt = $10^{-6} m^2/sec$, D = $10 s^{-1}$)

various types of non-Newtonian solutions are shown in Figure 5.19, which illustrates the dependence of shearing stress on rate of shear.

For solutions which exhibit **pseudoplastic** behavior, the apparent viscosity decreases as the rate of shear increases. With **dilatant** nutrient solutions, the apparent viscosity increases as the rate of shear increases. **Bingham-plastic** behavior is exhibited by nutrient solutions which will not flow unless a stress T_0 is imposed. The law is stated as follows

$$\frac{T - T_0}{\dot{\gamma}} = \eta = \text{Constant}$$

Table 5.7 lists the type of viscosity of some microbial mycelia.

It is more difficult to determine the amount of energy required for non-Newtonian solutions than for Newtonian solutions. The rate of shear changes with changes in the impeller speed, temperature, and time. The curves in Figure 5.19 are

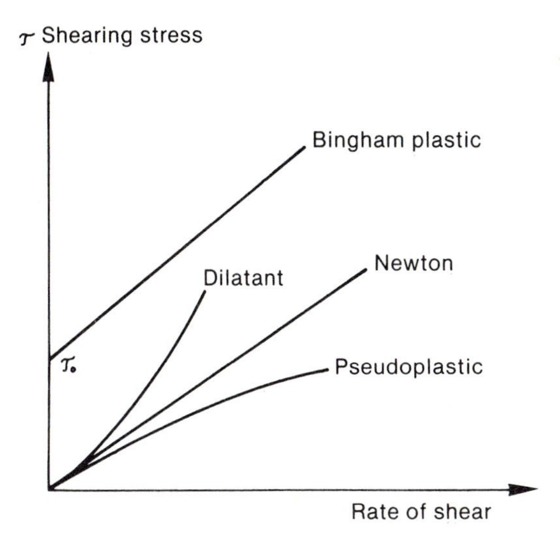

Figure 5.19 Correlation between shear rate and shear stress in nutrient solutions with Newtonian and non-Newtonian properties

Table 5.7 Viscosity of mycelium-producing microorganisms

Microorganism	Use	Viscosity
Penicillium chrysogenum	Penicillin	Pseudoplastic
Coniothyrium hellbori	Steroid hydroxylation	Bingham
Streptomyces noursei	Nystatin	Newton
Aspergillus niger		Bingham
Streptomyces niveus	Novobiocin	Bingham
Streptomyces griseus	Streptomycin	Bingham
Streptomyces sp.		Newton and pseudoplastic
Endomyces sp.	Glucoamylase	Pseudoplastic

not linear and the Bingham-plastic line does not cross zero. Mycelial suspensions which are taken from a bioreactor to determine viscosity may change their behavior within minutes without constant gaseous conditions, stirring, and temperature.

Power requirement in aerated bioreactors

During aeration, as the effective density of the gas-liquid mixture is reduced, the power requirement decreases. The correlation between power requirements of aerated and unaerated fermenters is given below:

$$P_g = K \cdot \left(\frac{P_1^2 \cdot N \cdot D_i^3}{Q^{0.56}} \right)^{0.45} \qquad [32]$$

P_g = Power requirement of the aerated bioreactor (horsepower)
K = Constant (function of fermenter geometry)
P_1 = Power requirement of the unaerated fermenter (horsepower)
N = Stirring rate (rpm)
D_i = Stirrer diameter (cm)
Q = Aeration rate (vvm)

This calculation describes both Newtonian and non-Newtonian fermentation systems. This equation has been shown to hold for a series of

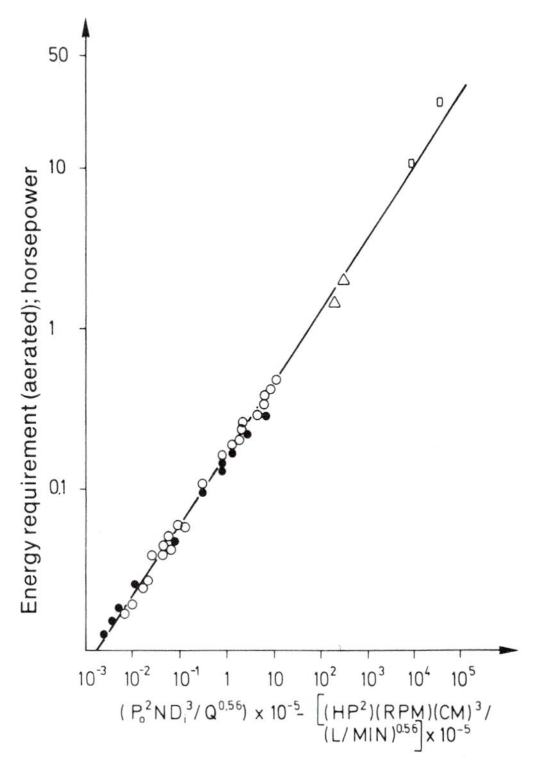

Figure 5.20 Michel and Miller correlation of energy required in geometrically similar aerated and unaerated bioreactors (Taguchi et al., 1968). • 30 l, ○ 100 l, △ 5000 l, and □ 50,000 l fermenters

fermenters from 20 l to 30,000 l for the glucoamylase fermentation by *Endomyces* sp. (Figure 5.20).

Table 5.8 shows data for actual power requirements for some industrial fermenters in both pilot-plant and production-sized systems. Small fermenters are usually constructed so as to be oversized in power, in order to accommodate future needs. The power requirements of large fermenters are similar to those in the chemical industry.

Table 5.8 Energy demand in pilot plant and production fermenters

Fermenter size (m^3)	Power kWh/m^3
0.1	11–13
1 – 5	4– 7
15 –120	1– 3

(Einsele, 1978)

5.5 GAS EXCHANGE AND MASS TRANSFER

One of the most critical factors in the operation of a large-scale fermenter is the provision of adequate gas exchange. **Oxygen** is the most important gaseous substrate for microbial metabolism, and **carbon dioxide** is the most important gaseous metabolic product. When oxygen is required as a microbial substrate, it is frequently a limiting factor in fermentation. Because of its low solubility, only 0.3 mM O_2, equivalent to 9 ppm, dissolves in one liter of water at 20°C in an air/water mixture. Due to the influence of the culture ingredients, the maximal oxygen content is actually lower than it would be in pure water. The solubility of gases follows Henry's Law in the gas pressure range over which fermenters are operated.

Henry's law

Henry's law describes the solubility of O_2 in nutrient solution in relation to the O_2 partial pressure in the gas phase.

$$C^* = \frac{p_O}{H} \qquad [33]$$

In this equation, C* is the oxygen-saturation concentration of the nutrient solution, p_o is the partial pressure of the gas in the gas phase and H is Henry's constant, which is specific for the gas and the liquid phase. As the oxygen concentration increases in the gas phase, the O_2 proportion of the nutrient solution increases. Consequently, the highest O_2 partial pressures are attained during aeration with pure oxygen. Compared to the value in air (9 mg O_2/l), 43 mg O_2/l dissolves in water when pure oxygen is considered. Henry's constants for various gases are summarized in chemical handbooks for standard conditions.

As temperature rises, the O_2 solubility decreases

$$C^* = \frac{468}{31.6 + T} \qquad [34]$$

T = Temperature (°C)

Oxygen transfer

For oxygen to be transfered from a gas bubble to an individual cell, several independent partial resistances must be overcome (Figure 5.21): 1) resistance within the gas film to the phase boundary; 2) penetration of the phase boundary between gas bubble and liquid; 3) transfer from the phase boundary to the liquid; 4) movement within the nutrient solution; 5) transfer to the surface of the cell.

For fermentations carried out with single-celled organisms such as bacteria or yeasts, the resistance in the phase boundary between gas bubble and liquid is the most important factor controlling the rate of transfer.

Microbial cells near gas bubbles may absorb oxygen directly through the phase boundary and the rate of gas transfer to such cells is increased. In cell agglomerates or pellets, the O_2 transfer

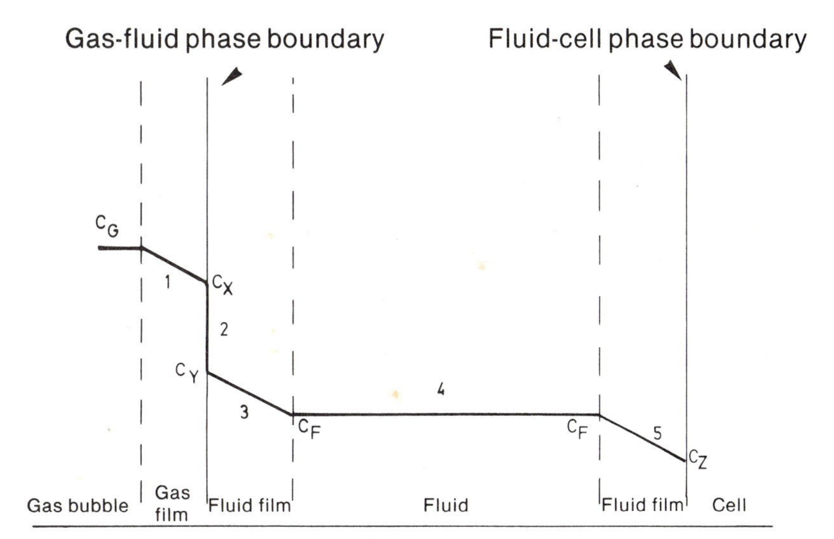

Figure 5.21 Resistances for oxygen transfer from the air bubble to the microbial cell

within the agglomerate can become the limiting factor.

Mass transfer of oxygen into the liquid can be characterized as follows:

$$N_A = k_L \cdot a \cdot (C^* - C_L) = OTR \qquad [35]$$

N_A = Volume-dependent mass transfer ($mM\ O_2/\ell \cdot h$)
k_L = Transfer coefficient at the phase boundary
a = Specific exchange surface
$k_L \cdot a$ = Volumetric oxygen transfer coefficient (h^{-1})
C^* = Saturation value of the dissolved gas in the phase boundary
C_L = Concentration of the dissolved gas (mM/ℓ)
OTR = Oxygen transfer rate ($mM\ O_2/\ell \cdot h$)

Table 5.9 summarizes several oxygen transfer rates for three impeller systems.

The volumetric oxygen transfer coefficient has been thoroughly examined as a critical parameter for bioreactor function. The coefficient is dependent on the diameter, capacity, power, aeration system, and aeration rate of the bioreactor and on the density, viscosity, and composition of the nutrient solution, the structure of the microorganism, the antifoam agent used, and the temperature.

Table 5.9 Oxygen transfer rates (OTR) in bioreactors

Reactor volume m^3	Impeller	Method of assay	OTR $mM\ O_2/l{\cdot}h$
0.1	Turbine	Sulfite	100–223
0.8	Turbine	Sulfite	94
1.2	Turbine	Sulfite	64
5.0	Turbine	Sulfite	45– 72
47.7	Turbine	Sulfite	42
34.2	Waldhof	Yeast	16– 22
58.5	Vogelbusch	Yeast	26– 43

(Hatch, 1975)

The volumetric oxygen transfer coefficient is dependent on the following fermentation conditions:

$$k_La = k \cdot (P_g/V)^{0.4} \cdot (V_s)^{0.5} \cdot (N)^{0.5} \qquad [36]$$

k = Constant
V = Bioreactor volume
V_s = Gas exit speed (cm/min)

The k_d value is frequently substituted for the k_La value in multi-level systems, as described in the following relationship:

$$k_d = (2.0 + 2.8\ N_i) \cdot (P_g/V)^{0.56} \cdot V_s^{0.7} \cdot N^{0.7} \cdot 10^{-3} \qquad [37]$$

N_i = Stirrer number

The value of the combined parameters k_La can be directly calculated, but k_L and a are difficult to calculate individually. The interfacial area a is usually unknown since it depends on bubble size. Table 5.10 shows some values for the interfacial area a.

Surface-active substances such as antifoam agents reduce the value of k_La. In pure water, the bubble surface is constantly renewed through vibration and oscillation. As soon as surface-active substances are added, the renewal of the bubble surface by bubble movement ceases. Microorganisms themselves have an effect on the oxygen transfer by acting as a barrier, thus inhibiting the O_2 transfer. With filamentous organisms, there are variations depending on whether the mycelium is in loose form or in pellets. While the k_La value decreases gradually as the pellet increases in size, there is a much steeper decline with loose forms (Figure 5.22).

Table 5.10 Values of the specific exchange coefficient *a* in some bioreactors

Bioreactor	Aeration type	Aeration rate vvm	Specific energy demand kW/m^3	Exchange coefficient a (m^{-1})
Stirred vessel			1	120
			1	300
			1	400
			2	600
Bubble column	Fritted disk	0.63	0.6	650
Frings bioreactor			1.2	1000
Stirred vessel			4	1000
			10	1100
Tubular loop reactor	Ejector		0.9	1300
			0.9	1300
			1.8	1700
			1.8	1800
			7.2	2000
Bubble column	Fritted disk	0.56	0.9	2000
Stirred vessel			10	2200
Frings bioreactor		0.48	2.77	2315
Tubular loop reactor	Ejector		7.2	2500
Bubble column	Ejector	0.42	1.5	6000
Bubble column	Ejector	1.4	2.2	8000

(Schügerl and Lücke, 1977)

The gas bubbles are replenished in locations of the bioreactor where there is negative pressure, such as behind the agitator blades. As the aeration rate increases, various conditions can be characterized. At low aeration rates, large gas bubbles form behind individual turbine blades and smaller bubbles are spun off centrifugally into the nutrient solution. As the aeration rate is increased, gas bubbles collect behind all turbine blades and continue to accumulate. The energy input is one-third less than that used in unaerated systems. In this intermediate stirring range, gas dispersion is the best. At very high aeration rates, many large gas bubbles adhere to each other and the impeller is flooded with gas, resulting in sharply lowered gas dispersion.

Oxygen as a substrate

As previously mentioned, growth in a microbial culture with limited substrate is calculated as follows:

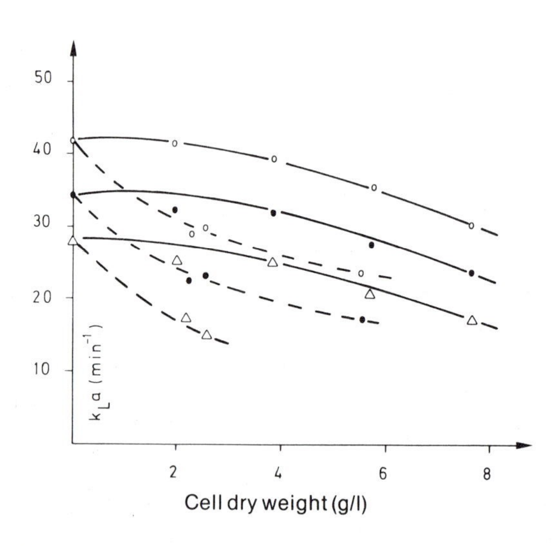

Figure 5.22 Relation of the k_La value to the mycelium concentration (Miura, 1976). Impeller rotations per min are given as follows: ○ — ○ 480, • — • 390, and △ — △ 290; —— mycelium in pellet form, – – – mycelium without pellet formation

$$\mu = \mu_m \frac{S}{K_S + S} \quad [3]$$

With oxygen as substrate, $S = C_L$, which is the current oxygen concentration. When the specific growth rate in the steady state is correlated with the specific oxygen absorption rate, the equation can be stated

$$Q_{O_2} = Q_m \cdot \frac{C_L}{K_O + C_L} \quad [38]$$

K_O = Michaelis-Menten constant for O_2
Q_{O_2} = Specific O_2 uptake rate/cell dry weight ($mM\ O_2/g \cdot h$)
Q_m = Maximum specific uptake rate

The specific maximal O_2 uptake is designated as the specific O_2 requirement; some examples are compiled in Table 5.11.

When the Q_m value is correlated with biomass/liter, one obtains the gas absorption rate or oxygen uptake rate which must be achieved during a fermentation:

$$X \cdot Q_m = k_L a \cdot (C^* - C_L) \qquad (mM\ O_2/\ell \cdot h) \quad [39]$$

X = Cell concentration (dry weight g/l)

The **gas absorption rate** is not constant during the fermentation, since if a substrate other than O_2 becomes limiting, such as at the end of the log phase, the value can be reduced.

During the fermentation of single-celled and mycelium-producing organisms, there is a characteristic difference in the oxygen absorption rate (Figure 5.23). During log growth, the O_2 absorption rate increases and the O_2 content in the broth decreases until it becomes limiting. Thereafter, with unicellular bacteria the O_2 absorption rate is constant until another substrate becomes limiting. In mycelial (streptomycete and fungal) fermentations, however, the O_2 absorption rate decreases when O_2 becomes limiting, due to the increase in mycelium volume and the related viscosity increase.

Table 5.11 Specific O_2 requirements of some microorganisms

Organism	Q_m(mM O_2/g cells·h)
Aspergillus niger	3.0
Streptomyces griseus	3.0
Penicillium chrysogenum	3.9
Klebsiella aerogenes	4.0
Saccharomyces cerevisiae	8.0
Escherichia coli	10.8

(Brown, 1970)

Critical oxygen concentration

Critical oxygen concentration is the term used to indicate the value of the specific oxygen absorption rate which permits respiration without hindrance. Table 5.12 shows several examples of critical oxygen concentrations. They are 5–25% of the oxygen saturation value in cultures.

At oxygen absorption rates which are lower than the critical concentrations, respiration rate is correlated with the O_2 concentration in the solution. Above this value, no dependence between respiration rate and dissolved oxygen has been observed. In Newtonian fluids, such as those occurring in yeast and bacterial fermentations, the critical oxygen concentration is constant and is not affected by fermentation conditions. In non-Newtonian solutions, such as those occurring with filamentous organisms (e.g., during novobiocin production with *Streptomyces niveus*), the critical oxygen concentration has been shown to be dependent on fermentation conditions. Table 5.13 shows an experimental result in which the critical oxygen concentration in a strain varied (within the error of measurement) between 0 and 55% of the saturation value.

Determining the oxygen uptake rate

There are four acceptable methods of determining the O_2 uptake rate: the dynamic method, the sulfite method, the direct measurement of the volumetric O_2 transfer rate, and calculation from measurements of the growth of the microorganisms.

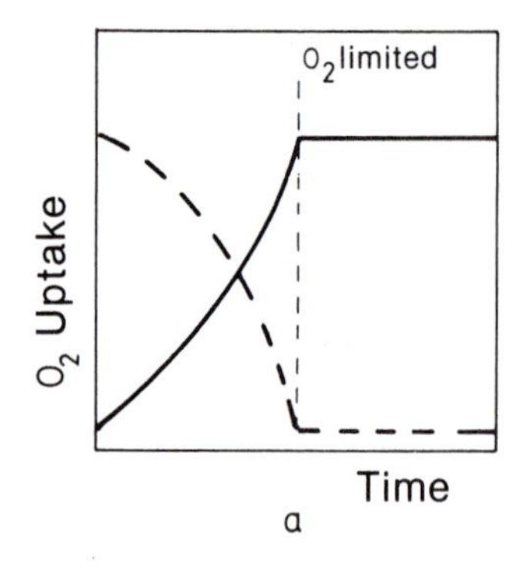

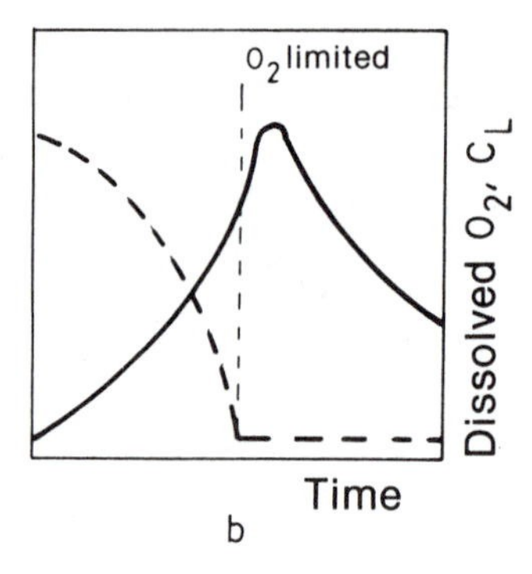

Figure 5.23 O_2 uptake rate and concentration of dissolved oxygen under O_2 limitation. (a) Bacterial culture, (b) Fungal fermentation; —— O_2 absorption rate, - - - C_L, dissolved O_2 concentration

Table 5.12 Critical oxygen concentrations of some organisms

Organism	$C_{krit.}$ (mg/l)
Escherichia coli	0.26
Penicillium chrysogenum	0.40
Saccharomyces cerevisiae	0.60
Pseudomonas ovalis	1.10
Torulopsis utilis	2.00

(Brown, 1970)

Table 5.13 Dependence of the critical oxygen concentration of *Streptomyces niveus* on fermentation parameters

Fermenter Volume (l)	Impeller D_i (cm)	Impeller speed (rpm)	Critical O_2 Concentration (% saturation)
20	7.0	400	0
20	12.2	275	0
20	21.0	180	50
250	15.8	175	5
250	15.8	220	18
250	23.7	175	22
15,000	76.8	164	55
15,000	90.5	124	55
15,000	118	82	55

(Wang and Fewkes, 1977)

Dynamic method In batch fermentations, the actual dissolved O_2 concentration C_L can be determined by measuring the speed of the C_L decrease after interrupting the aeration, thus determining the volumetric oxygen transfer coefficient. The following expresses the change of the current O_2 content:

$$\frac{dC_L}{dt} = k_L a \cdot (C^* - C_L)_m - Q_{O_2} \cdot X \qquad [40]$$

or

$$C_L = \frac{1}{k_L a} \cdot \left(\frac{dC_L}{dt} + Q_{O_2} \cdot X\right)_m + C^* \qquad [41]$$

m = average value over the bioreactor

When the decline in O_2 concentration in the broth is plotted against time after interruption of aeration, a straight line results with a slope of $-Q_{O_2} \cdot X$. When the change in the O_2 concentration is plotted against time according to equation 1 (Figure 5.24), the slope of the straight line equals the reciprocal of the volumetric transfer coefficient $k_L a$.

The advantage of this procedure is that only one parameter must be calculated. The disadvantages are that the results are inaccurate due to the long response time of the oxygen electrodes and due to oxygen transfer from the head space of the fermenter, thus requiring a gas exchange with nitrogen. In some cases the value of C_L is so low that an exact measurement in the fermenter is not possible. Also there may be a production deficit whenever the slightest interruption of aeration occurs (as is found in acetic acid production).

Sulfite method The sulfite method is frequently used to obtain a quantitative measure of the O_2

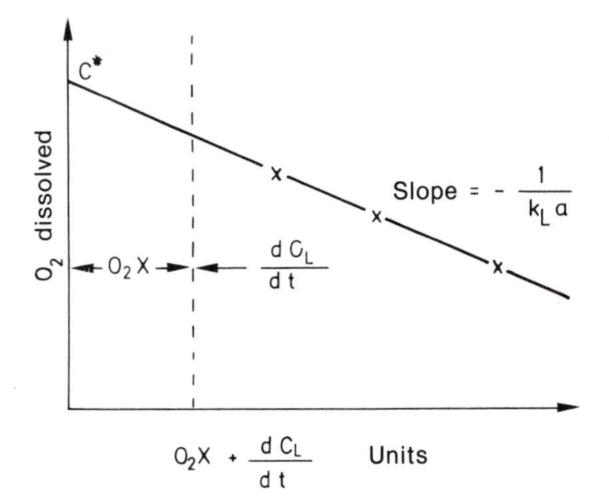

Figure 5.24 Determination of mass transfer coefficient using the dynamic method

transfer in bioreactors. In the presence of 10^{-3}M Cu^{2+} or Co^{2+} ions, sodium sulfite is oxidized with dissolved oxygen according to the formula

$$Na_2SO_3 + \frac{1}{2}O_2 \xrightarrow{Cu^{2+} \text{ or } Co^{2+}} Na_2SO_4 \qquad [42]$$

Since this reaction takes place in a very short time, if there is excess sodium sulfite, the O_2 concentration in the solution equals zero and the O_2 transfer from the gas phase into the solution is the limiting factor. Nonoxidized sulfite is determined by adding a sample to an excess iodine solution and by titrating the residual iodine with sodium thiosulfate.

$$Na_2SO_3 + I_2 + H_2O \rightarrow 2\ NaI + H_2SO_4 \qquad [43]$$

$$I_2 + 2\ Na_2S_2O_3 \rightarrow 2\ NaI + Na_2S_4O_6 \qquad [44]$$

$$1\ mM\ Na_2S_2O_3 = 0.25\ mM\ O_2 \qquad [45]$$

If two samples are taken at intervals, the O_2 absorption per unit time can be calculated from the rate of sulfite decrease.

The sulfite procedure has a disadvantage: The process is performed as a chemical reaction in an aqueous solution without further additives. The sulfite value (oxygen transfer rate, OTR) is a characteristic of the bioreactor and gives some information about the aeration device but gives no indication of the current oxygen energy supply in a fermentation broth. In addition, the cost of analysis materials needed for use of this method in a large bioreactor is considerable.

Direct measurement of the volumetric O_2 transfer rate A direct measurement of oxygen transfer is the most exact method, but it is time-consuming and requires precise analytical devices. In this method, the O_2 content in incoming and outgoing air is measured to determine the O_2 transfer rate. The exact oxygen absorption can be determined by multiplying the aeration rate by the difference in O_2 content (in and out) and taking the absolute temperature and pressure into consideration. The oxygen content in the air is measured specifically and very accurately by its paramagnetic properties or with a mass spectrometer, so that no other gases interfere.

$$Q(Y_{in} - Y_{out}) = N_A \qquad [46]$$

Q = Aeration rate
Y_{in} = Oxygen content of the input
Y_{out} = Oxygen content of the exit

Determining the O_2 transfer rate via measurement of microbial growth The amount of oxygen required can be calculated from thermodynamic considerations, if it is assumed that a single carbon substrate is oxidized to carbon dioxide, water and ammonia by an organism. Under aerobic conditions the oxygen requirement C is calculated per gram of dry biomass produced:

$$C = \frac{A}{Y_S} - B \qquad [47]$$

C = Oxygen demand/dry weight biomass ($mM\ O_2$/g cell dry weight)
A = Oxygen required for oxidation of 1 g of substrate to CO_2, H_2O, NH_3
B = Oxygen required for oxidation of 1 g biomass to CO_2, H_2O, NH_3
Y_S = Cell yield/substrate (g/g)

Table 5.14 Cell yields obtained with various substrates

Substrate	Y_S Bacteria	Y_S Yeast
Acetate	0.36	0.37
Glycerol		0.51
Glucose	0.51	0.50
Methanol	0.40	
Ethanol	0.68	
Alkanes	1.03	1.00
Methane	0.62	

The bacterial data are from Abbott and Clamen (1973) and the yeast data from Bronn (1966).

The value of A can be calculated from the theoretical oxidation of the substrate. For B the biomass must be analyzed (in yeast B = 934 ml O_2/g dry weight).

Some examples of cell yields (Y_S) in yeast and bacteria are given for several substrates in Table 5.14.

Experimental values for the amount of O_2 needed are listed in Table 5.15. Since these are the total amounts needed, they do not provide any information about when during a fermentation the O_2 is needed. In order to determine the amount of O_2 required in a culture per hour, the specific growth rate μ (h^{-1}), the biomass X (g/l), and the Y_O value as grams of cells per gram of oxygen (g/g) must be known.

Table 5.15 Oxygen requirements of some microorganisms

Microorganism	Substrate	O_2 required (ml O_2/g dry weight)
Aerobacter aerogenes	Glucose-NH_3	515
Escherichia coli	Glucose-NH_3	400
Pen. chrysogenum	Glucose-NH_3	410
Sacch. cerevisiae	Molasses	500
Sacch. cerevisiae	Molasses	420–590
Sacch. cerevisiae	Molasses	430–550
Rhodotorula glutinis	Glucose-peptone	510
Candida utilis	Glucose-peptone	550
Candida utilis	Glucose-urea	460
Candida utilis	Glycerol-peptone	480

(Bronn, 1966)

Hence

$$N_A = \mu \cdot X \cdot \frac{k}{Y_O} \qquad [48]$$

k = Correction factor $\quad 31.3 \frac{\text{mM } O_2}{\text{g } O_2}$

For the cell yield per substrate, Y_O = g cells/g oxygen, the following equation is used:

$$\frac{1}{Y_O} = 16\left[\frac{2C + H/2 - O}{Y_S \cdot M} + \frac{O'}{1600} - \frac{C'}{600} + \frac{N'}{933} - \frac{H'}{200}\right] \qquad [49]$$

C, H, O = Number of carbon, hydrogen, and oxygen atoms in substrate
C′, H′, O′, N′ = Percent of carbon, hydrogen, oxygen, and nitrogen in biomass
M = Substrate molecular weight

Effects of carbon dioxide on the fermentation As mentioned earlier, carbon dioxide is the most important gaseous metabolite produced in fermentation. Although much study has been devoted to the effects of oxygen, too little attention has been given to the correlation between carbon dioxide and microbial metabolism. The negative influence of CO_2 on penicillin formation has been studied; erythromycin and rifamycin B formation rates are also reduced by excess CO_2.

Figure 5.25 and Table 5.16 show the effect of CO_2 on production of the aminoglycoside antibiotic sisomicin. When 1% CO_2 is added to the incoming air in a 300 l fermenter, substrate is metabolized more gradually, mycelium is formed slowly, and the sisomicin yields are 33% less than in the control.

5.6 SCALE UP

Significance of scale up

The conversion of a laboratory procedure to an industrial process is termed **scale up**. It is well established in the field of industrial microbiology that a process which works well at the laboratory scale may work poorly or not at all when first attempted at large scale. It is generally not possible to take fermentation conditions that have

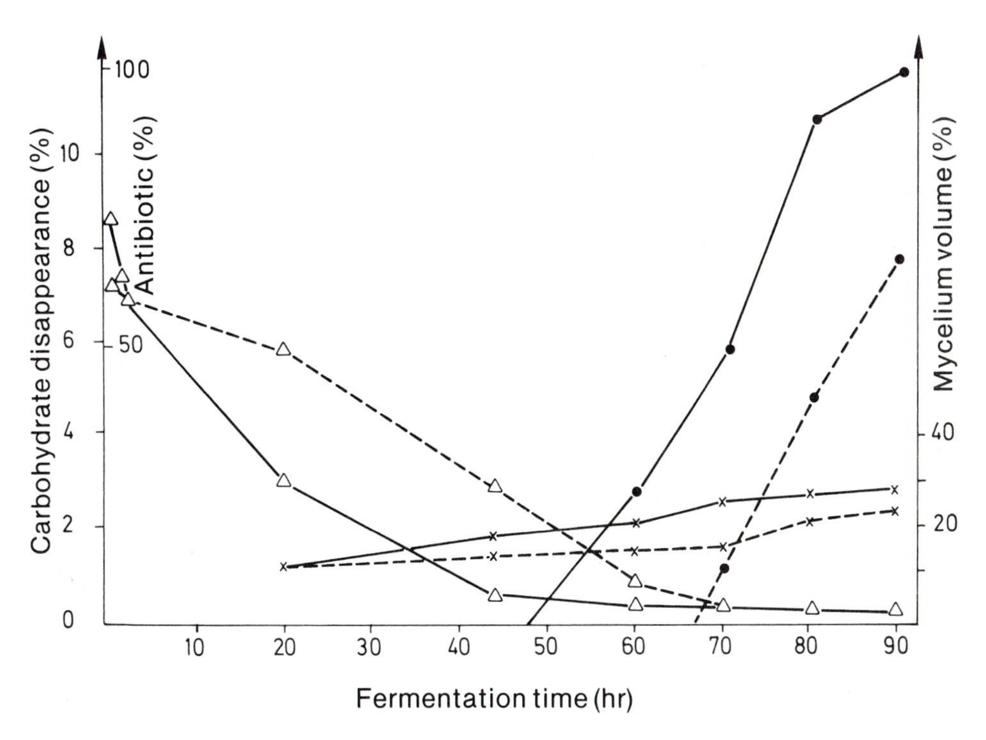

Figure 5.25 Sisomicin formation during aeration with a 1% CO_2 air mixture. --- CO_2/air mixture, — control, • antibiotic, △ carbohydrate, × mycelium wet volume

Table 5.16 Influence of CO_2 on yield of the antibiotic sisomicin

CO_2 content of incoming air (%)	Relative sisomicin yield
0	100
1	66
2	15
3	0
4	0

worked in the laboratory and blindly apply them to industrial-scale equipment. All of the skills of the biotechnologist must be brought into play in order to develop a successful large-scale process.

Scale up is necessary in the following circumstances:

- A new process is implemented in the plant
- Mutants with 10–20% greater yield are to be introduced into large-scale production as soon as possible
- Construction of a completely new fermentation plant, a rare occurrence.

While laboratory microbiologists are primarily interested in yield based on weight of biomass, units per ml broth, or maximal yield per unit time, success in scale up is evaluated on the basis of maximal yield for minimal operating cost and time.

Comparing fermenters with similar geometries, Table 5.17 shows that at different fermenter sizes not all parameters can be kept constant. If the impeller increases in diameter by a factor of five, the fermenter volume increases by a factor of 125, from 80 liters to 10,000 liters. If one of several criteria is kept constant in scale up, e.g. energy consumed/volume or Reynold's number, the other parameters are quite different from the values obtained with the small fermenter.

Although many parameters have been tested for use as scale up criteria, there is no general

Table 5.17 Interdependence of scale-up parameters

Scale up criterion	Designation	Small fermenter 80 l	Production fermenter 10,000 l			
Energy input	P_o	1.0	125	3125	25	0.2
Energy input/volume	P_o/V	1.0	1.0	25	0.2	0.0016
Impeller rotation number	N	1.0	0.34	1.0	0.2	0.04
Impeller diameter	D_i	1.0	5.0	5.0	5.0	5.0
Pump rate of impeller	F	1.0	42.5	125	25	5.0
Pump rate of impeller/volume	F/V	1.0	0.34	1.0	0.2	0.04
Maximum impeller speed (max. shearing rate)	N/D_i	1.0	1.7	5.0	1.0	0.2
Reynolds number	$ND_i^2\rho/\eta$	1.0	8.5	25.0	5.0	1.0

(Oldshue, 1966)

formula because of the variation in fermentation processes. The most important methods are:

- Constant power consumption per unit of broth
- Constant volumetric oxygen transfer rate

Some metabolites (for example, phenylalanine, valine, leucine, and capreomycin) are formed best at oxygen concentrations below the critical level. Figure 5.26 shows the use of the critical O_2 concentration to scale up the formation of nikkomycin, an acracide, from shake flasks to 40 m^3 fermenters.

An additional criterion may be the impeller tip velocity, which is in the range of 5–7 m/sec

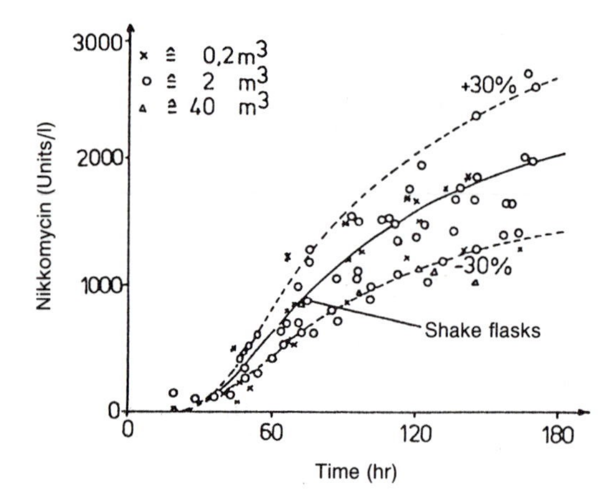

Figure 5.26 Scale up of nikkomycin production using the critical O_2 concentration as the scale-up criterion. (C_{crit} = 1.5 mg O_2/l). (Crueger et al., 1982).

in many production fermenters. Mixing time is not suitable as a scale-up parameter because it increases as reactor size increases. It can be seen in Table 5.18 that mixing time is approximately related to the cube root of the fermenter volume.

In reality, scale up is not usually done with geometrically similar fermenters in laboratory, pilot plant and production scale. There is usually no means nor any necessity of scaling up while keeping one parameter constant.

Scale up with constant power consumption per volume

In this simple method, one of the factors in the determination of energy input in aerated bioreactors is the previously mentioned correlation (see Section 5.4) between energy consumption (P_g) and fermenter parameters:

$$P_g = K(P_i^2 \cdot N \cdot D_i^3/Q^{0.56})^{0.45}$$

Table 5.18 Mixing time in relation to fermenter volume

Fermenter volume l	Impeller speed rpm	Mixing time sec
3	750	5
9	2,000	3
100	230	6.6
300	350	5
1,000	200	25
3,000	180	20
24,000	30	66

Studies have been done on the power input in a penicillin fermenter with an energy input of 1.5–3.0 horsepower/m^3 and in a streptomycin fermenter with 2 horsepower/m^3.

In Figure 5.27, novobiocin titers are plotted against energy input for three impeller types. The curves are parallel and maximal yields are obtained above 1.5 horsepower/m^3. Below 1 horsepower/m^3 considerably lower yields are observed. This graph also shows that at equivalent power inputs, antibiotic biosynthesis is dependent on the diameter of the impeller.

In another example, the flavomycin process was transferred from a 4 m^3 to a 40 m^3 fermenter by keeping constant the P/V relationship.

In a comparative study of many production plants, no constant P/V ratio was found. Rather, the following relationship was observed:

$$P_1/V \sim (V)^{-0.37} \quad [50]$$

Scale up with constant oxygen transfer rate

Scale up is most commonly performed on the basis of k_La data, which are calculated from sulfite values, as described earlier. But this calculation does not give any information on oxygen transfer in highly viscous non-Newtonian fermentation solutions (such as found in many antibiotic fermentations) or under conditions of high-speed stirring.

Figure 5.28 shows the production of baker's yeast in vessels ranging in volume from a small shake flask up to 114 m^3, plotted as the sulfite value. In this study, neither the fermenter type nor the volume were crucial in determining yeast growth. Above 125 mM O_2/l·h, optimal yields were attained. Scale up of penicillin and streptomycin processes has been successful in 15 l through 100 l and 3 m^3 up to 63 m^3 fermenters. Instead of the k_La value, a modified k_d value could be used as the constant parameter.

Figures 5.29 shows scale-up tests by Jarai (1972) on the value of k_La for the secondary metabolites (nystatin and fumagillin). For both these antibiotics and several industrial enzymes, there was a good agreement between k_La values and the yields in 6 l and 3000 l fermenters.

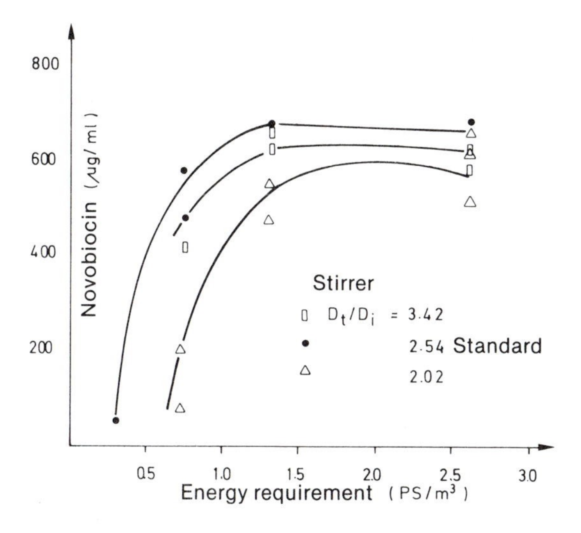

Figure 5.27 Effect of impeller geometry and energy input on novobiocin biosynthesis (Maxon, 1959)

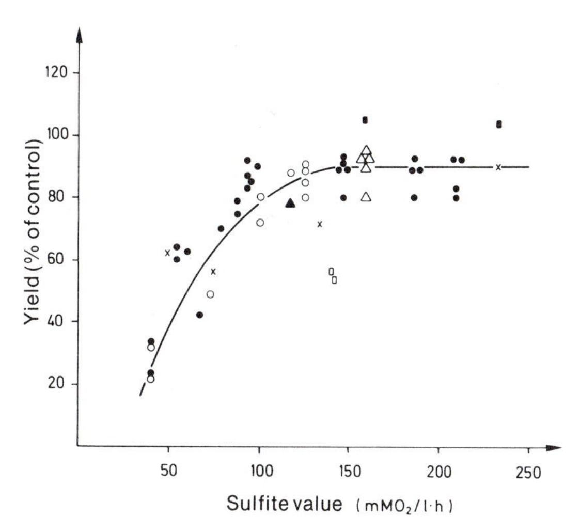

Figure 5.28 Results of the scale up of baker's yeast production (Strohm et al., 1959). Reactor volume: × 190 ml shaker ; □ 19 l, 600 rpm, impeller flooded; • 19 l, 800 rpm; ■ 265 l, 550 rpm; O 265 l, unstirred, fritted disk with small holes; △ 265 l, unstirred, fritted disk with large holes; ▲ 114 m^3, unstirred

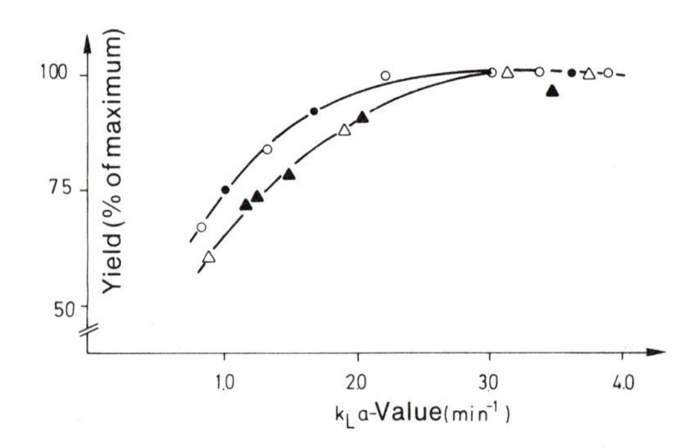

Figure 5.29 Nystatin and fumagillin titers in relation to K_La values during scale-up (Jarai, 1972). —○—●—○— nystatin, —△—▲—△— fumagillin; ○, △ 6 l, ●, ▲ 3000 l

5.7 STERILIZATION OF GASES AND NUTRIENT SOLUTIONS

In virtually all fermentation processes, it is mandatory for a cost-effective operation to have contamination-free seed cultures at all stages, from the preliminary culture to the production fermenter. A bioreactor can be sterilized either by destroying the organisms with some lethal agent such as heat, radiation, or a chemical, or by removing the viable organisms by a physical procedure such as filtration.

The process of destroying a population generally follows first-order kinetics. Using the initial number of cells N_o/ml, the number of destroyed organisms N' at time t (min), and the surviving cells N, the death rate can be calculated as follows

$$-\frac{dN}{dt} = k(N_o - N') \quad = kN \qquad [51]$$

(k is the specific death constant/min)

When integrated between N_o at time t = 0 and N at time t = t, the following equation is obtained

$$kt = \ln\frac{N_o}{N} \quad \text{or} \qquad [52]$$

$$\ln\frac{N}{N_o} = -kt \qquad [53]$$

The ratio of N_o/N is the inactivation factor, the ratio of N/N_o is the survival factor and the ln of $N_o/N = \bar{V}$ is the **design criterion**, a parameter which encompasses the contamination level of the medium to be sterilized, N_o, and the desired sterility level, N.

The ideal curve expressing the exponential decline in survivors with time is plotted in Figure 5.30. This line has a negative slope and in Figure 5.31 part II corresponds to the death curve for vegetative cells of *Escherichia coli* at 54–60°C. This type of killing kinetics is referred to as a logarithmic death curve. In actuality, this curve is not always linear, as shown for the death curve of *Bacillus stearothermophilus* spores (Figure 5.31, part I). Mathematical models for such nonlogarithmic destruction curves have been developed.

In the above equations, k is a constant which expresses the specific death rate. It increases sharply with temperature and can be experimentally determined for an organism using equation 53. According to the Arrhenius equation the following connection between temperature and k value exists:

$$\frac{d\ln k}{dt} = \frac{E}{R \cdot T^2} \qquad [54]$$

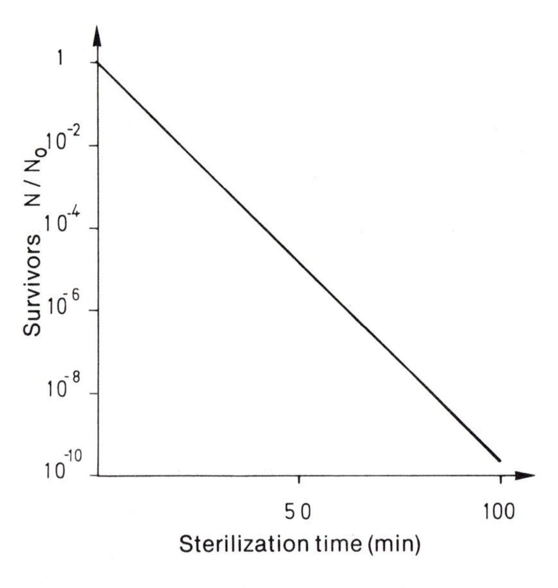

Figure 5.30 Theoretical death curve of a bacterial culture

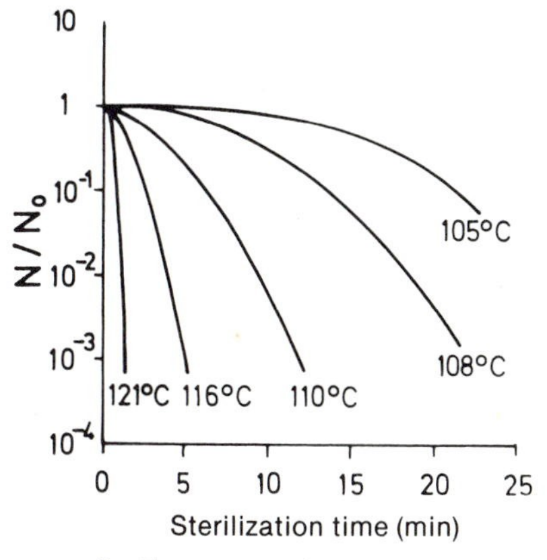

I *B. stearothermophilus*

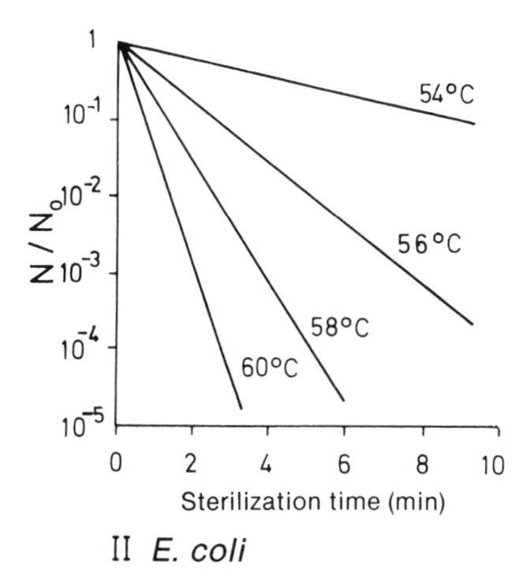

II *E. coli*

Figure 5.31 Nonlogarithmic and logarithmic death rates

R is the gas constant (J/Mol°C), T the absolute temperature (°K) and E a specific constant for the population (activation energy, J/Mol).

If the experimentally determined ln k value from this equation is plotted against the reciprocal temperature value, a straight line should be obtained from which the k value can be calculated for a desired temperature.

The Arrhenius relationship has been observed only for pure cultures. The populations which normally exist in unsterile solutions are generally nonhomogeneous **mixed cultures** containing organisms of varying heat resistance. The destruction rate is thus not a straight line, but a curve, as shown in Figure 5.32 for a culture consisting of two strains.

During fermentation the following points must be observed to ensure sterility:

- Sterility of the culture media
- Sterility of incoming and outgoing air
- Appropriate construction of the bioreactor for sterilization and for prevention of contamination during fermentation.

Sterilization of culture media

Nutrient media as initially prepared contain a variety of different vegetative cells and spores, derived from the constituents of the culture medium, the water, and the vessel. These must be eliminated by a suitable means before inoculation. A number of means are available for sterilization, but in practice for large-scale installations, heat is the main mechanism used.

Heat sterilization This is the most useful method for the sterilization of nutrient media. A number of factors influence the success of heat sterilization: the number and type of microorganisms

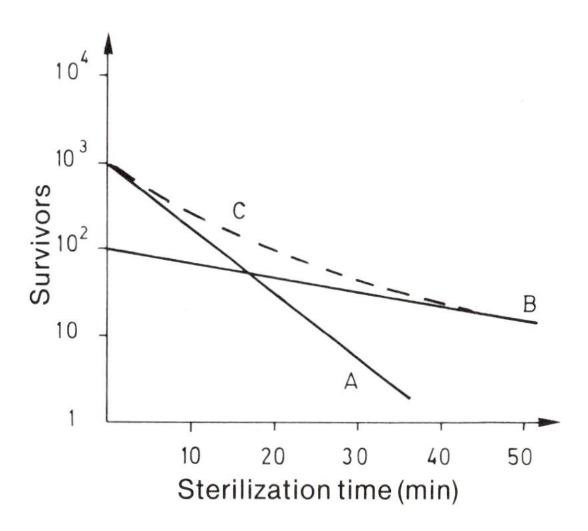

Figure 5.32 Death curve of a mixed culture (C). The straight lines (A) and (B) indicate the death rates of both pure cultures

present, the composition of the culture medium, the pH value, and the size of the suspended particles. Vegetative cells are rapidly eliminated at relatively low temperatures, as shown in Figure 5.31, but for destruction of spores, temperatures of 121°C are needed.

Spores of *Bacillus stearothermophilus* are the most heat resistant. Therefore they are used as assay organisms for testing the various procedures used to sterilize equipment. Table 5.19 provides data on the relationship between temperature, k value, and design criterion for this organism. A list of sterilization times and temperatures for various organisms is given in Table 5.20.

Radiation (UV, X Rays, or γ rays) Although occasionally used in the food industry, these agents are not used in industrial fermentation.

Chemical methods Although a number of chemical disinfectants are known, they cannot be used to sterilize nutrient media because there is a risk that inhibition of the fermentation organism could occur from the residual chemical.

Table 5.19 Relationship of temperature, k-value and the design criterion in *Bacillus stearothermophilus*

T°C	k(min^{-1})	$\bar{V}$
100	0.019	–
115	0.666	3.154
118	1.307	6.341
121	2.538	12.549
130	17.524	90.591
140	135.9	
150	956.1	

Table 5.20 Sterilization time and sterilization temperature of various groups of organisms

Cells	Sterilization time min	Sterilization temperature °C
Vegetative cells	5–10	60
Fungus spores/yeast spores	15	80
Streptomyces spores	5–10	60–80
Bacterial spores, general	5	121
Spores of *Bacillus stearothermophilus*	15	121

Mechanical removal of organisms Alternatives such as centrifugation, adsorption to ion exchangers, adsorption to activated carbon, or filtration are possible. Filtration is the only method in practical use. Filter sterilization is often used for all components of nutrient solutions which are heat sensitive and would thus be denatured through the steam sterilization process normally used in industrial fermentation. Vitamins, antibiotics or blood components are examples of heat-labile compounds which must be sterilized by filtration. Such materials must be completely dissolved before filtration, otherwise they would be filtered out of the mixture with the microorganisms. Deep filters (plate filters) are sometimes used to filter complex nutrient solutions. Two disadvantages of filtration are: 1) certain components of the nutrient solution may be adsorbed on the filter material, and 2) high pressures must be used (up to 5 bar), which are undesirable in industrial practice.

One approach which is cost-effective is the filtration of just the water which is to be used in the preparation of the culture medium. For instance, in steroid bioconversion processes, a concentrated nutrient solution is sterilized by heat in the fermenter and is then diluted to the normal concentration with water which has been filter-sterilized.

Batch sterilization

Most nutrient media are presently sterilized in batch volumes in the bioreactor at 121°C. Approximate sterilization times can be calculated from the nature of the medium and the size of the fermenter. Not only the nutrient media, but also the fittings, valves and electrodes of the fermenter itself must be sterilized. Therefore, actual sterilization times are significantly longer than calculated ones and must be empirically determined for the specific nutrient solutions in the fermenter. One method of sterilization is to inject steam into the fermenter mantle or interior coils (indirect sterilization). Another method is to inject steam into the nutrient solution itself (direct procedure), in which case pure steam (free of chemical additives) is a prerequisite. Many industrial steam supplies contain potentially toxic chemicals derived from anti-corrosive additives used in the steam-manufacturing process. With direct steam injection, condensate accumulates within the fermenter and the volume of liquid thus increases during the sterilization process.

The drawbacks of the heat-sterilization process are shown for a typical sterilization of a 3000 l batch fermenter in Figure 5.33. It takes 2–3 hours to reach the sterilization temperature of 121°C, depending on the steam conduction and fermenter size. Once the proper temperature has been reached, another 20–60 minutes are required for the actual killing process, followed by cooling for about one hour. The energy required for heating must subsequently be removed in order to cool the fermenter and if the hot water obtained during the cooling cannot be put to some use, batch sterilization becomes very costly.

Another disadvantage of heat sterilization (and from the standpoint of microbiology the most significant shortcoming) can be seen in Figure 5.33. The heating, sterilization and cooling phases not only kill microorganisms but also severely alter nutrient solutions. Discoloration and changes in the pH value result from caramelization and Maillard reactions (see Chapter 4). Vitamins are destroyed and the quality of the culture medium deteriorates. The extent to which the subsequent fermentation is affected depends on the organism and the process.

Continuous sterilization

The two main disadvantages of batch sterilization just mentioned, culture medium damage and

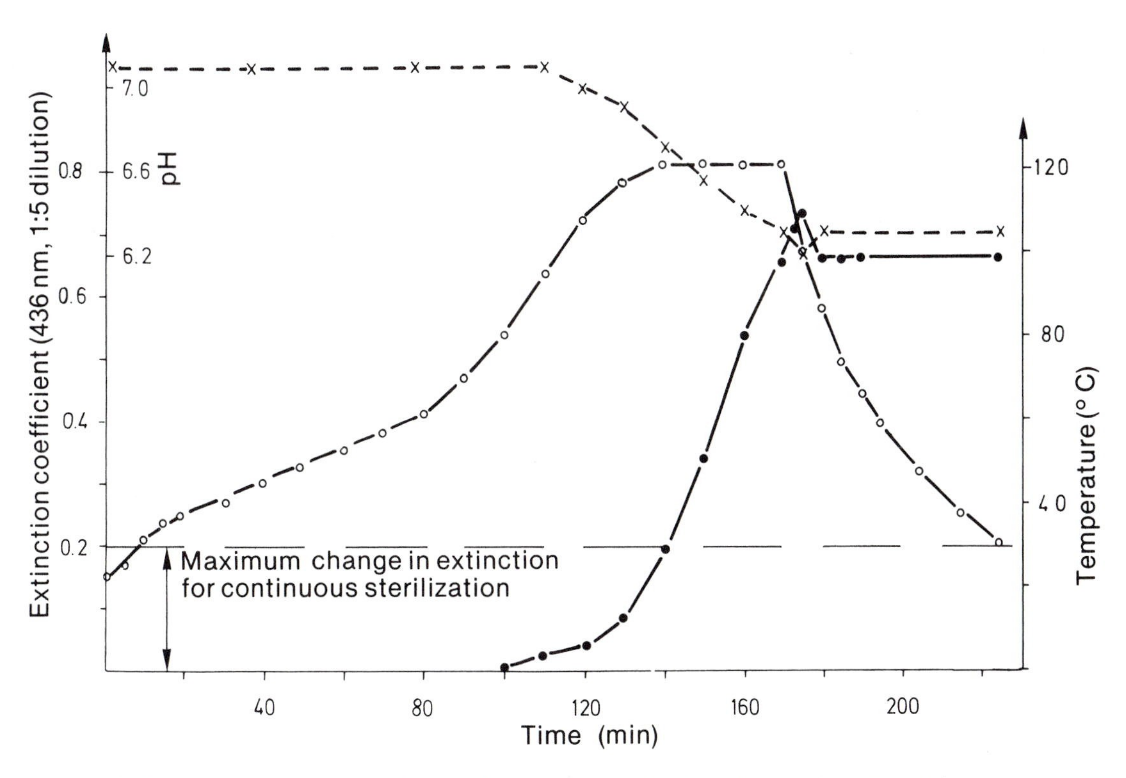

Figure 5.33 Temperature profile and culture medium change in a batch sterilization for a 3000 l. fermenter. o — o °C, ● — ● extinction (1:5 diluted, 436 nm), × — × pH value

high energy consumption, can be largely avoided by use of a continuous sterilization procedure. Although continuous sterilization is the logical preliminary step for continuous fermentations in industrial scale, it is also of value in batch fermentations, making greater yields possible for the time and space allotted. The reason for this is because of the exponential relationship between death rate and temperature, making the time required for the complete elimination of life shorter if higher temperatures are used. While batch sterilization is carried out in 30–60 minutes at 121°C, continuous sterilization is normally accomplished in 30–120 seconds at 140°C.

The heating of culture media for continuous sterilization can be done either by injection of steam or by means of heat exchangers. Sterilization with steam injection is done by injecting steam into the nutrient solution. The temperature is raised quickly to 140°C and is maintained for 30–120 seconds. Due to the formation of condensate, the nutrient solution becomes diluted; to correct this, the hot solution is pumped through an expansion valve into a vaporizer and the condensate is removed via vacuum pumps so that the sterilized nutrient solution has the same concentration after the cooling process as before. The disadvantage of this process is the sensitivity it exhibits to changes in the viscosity of the medium and to pressure variations.

In the continuous process using heat exchangers (Figure 5.34), the nutrient solution in the first heat exchanger is preheated to 90–120°C within 20–30 seconds by the exiting previously sterilized nutrient solution. Then in the second heat exchanger, it is heated indirectly with steam to 140°C. This temperature is maintained for 30–120 seconds in a holding pipe before it is placed in the first exchanger for preliminary cooling and then in a third exchanger for cooling to the temperature of the fermenter. The cooling phase is only 20–30 seconds. Figure 5.35 shows the temperature profile of the nutrient solution during sterilization.

In the process using heat exchangers, 90% of the energy input is recovered. The disadvantage of this method is that with some nutrient solutions, insoluble salts (e.g., calcium phosphate or calcium oxalate) are formed and crusts appear in the first heat exchanger, due to the extreme temperature differences between the sterilized nutrient solution and the cold incoming solution. The heat transfer coefficient is calculated as follows:

$$K = \frac{\dot{Q}}{A \cdot \Delta T_m}$$

K = Heat transfer coefficient
$\dot{Q}$ = Heat consumed
A = Transfer surface
ΔT_m = Average temperature gradient

If precipitation occurs, the heat transfer coefficient decreases and the system must then be stopped, treated with cleaning agents (acid or base), and resterilized. By sterilizing the critical components of the nutrient solution separately, the value of k can be kept constant and the useful period can be extended for weeks.

Starch-containing solutions which become viscous when heated are difficult to use in continuous sterilization processes. Before the actual sterilization, a liquefaction and partial hydrolysis through acids or amylases must be carried out. Moreover, if there are suspended particles in the nutrient solutions, the short sterilization times in the continuous process may not be sufficient for the heat to permeate them thoroughly. The heating time for 1 mm particles is 1 second; for 1 cm particles it is 100 seconds. Therefore the particle size should be restricted to 1–2 mm in continuous sterilization processes.

Sterilization of fermentation air

Most industrial fermentations are operated under vigorous aeration and the air supplied to the fermenter must be sterilized. The number of particles and microorganisms in air varies greatly depending on location of the plant, air movement, and previous treatment of the air. On the

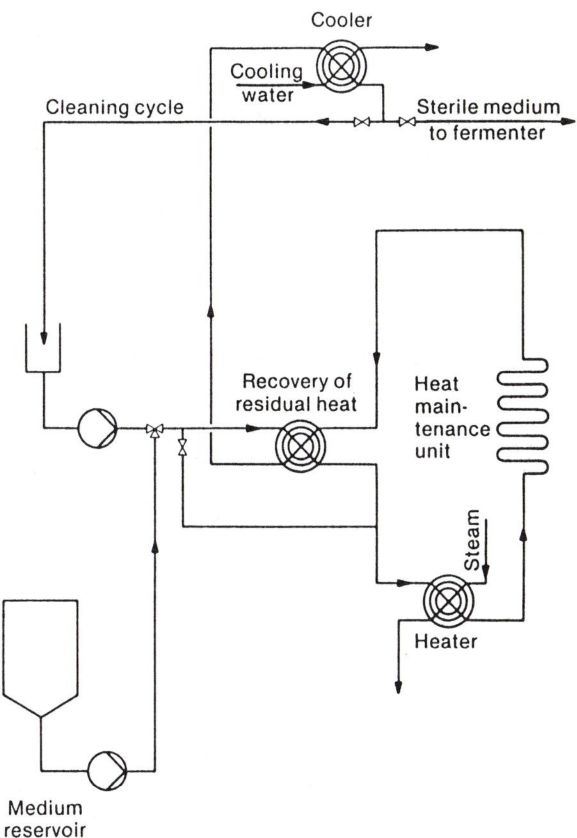

Figure 5.34 Diagram of continuous sterilization via spiral heat exchangers

average, outdoor air has 10–100,000 particles per m^3 and 5–2,000 microorganisms/m^3. Of these, 50% are fungus spores and 40% are Gram-negative bacteria.

Fermenters generally work with aeration rates of 0.5–1.0 vvm (air volume/liquid volume·minute. A fermenter having a working volume of 50 m^3 with an aeration rate of 1 vvm needs 3000 m^3 sterile air per hour. The critical importance of air sterilization in industrial microbiology can be seen from these values.

The methods available for **sterilizing gases** include filtration, gas injection (ozone), gas scrubbing, radiation (UV), and heat. Of these, only filtration and heat are practical at an industrial scale. For many years, air was sterilized

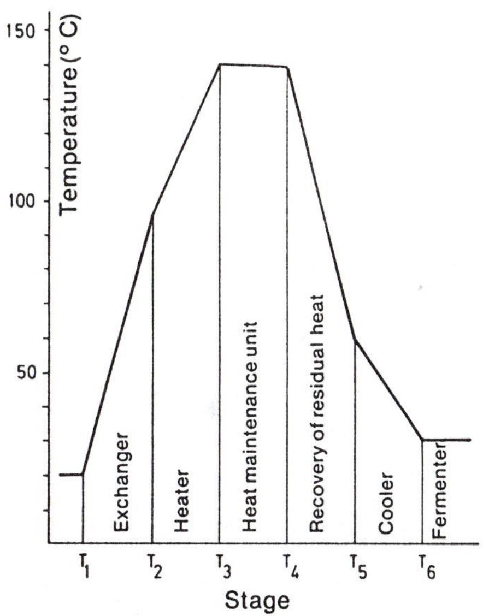

Figure 5.35 Temperature profile of nutrient solution during a continuous sterilization process

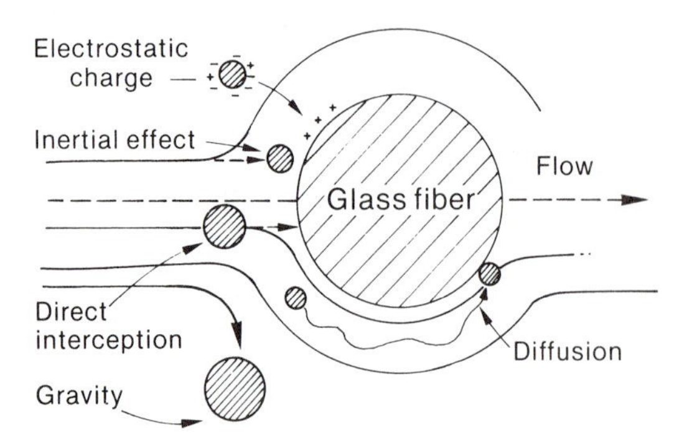

Figure 5.36 Mechanism by which particles are removed by a depth filter

by passing it over electrically heated elements, but due to the high cost of electricity, this process has been replaced by filtration.

In **industrial systems**, the air is sterile filtered. In older systems pure **depth filters** such as glass wool filters were installed, in which particles would be trapped by a combination of physical effects. As shown in Figure 5.36, particle filtration involves inertial effects, blocking effects, diffusion, gravity separation, and electrostatic attraction. The last two mechanisms have a minimal effect on the removal of particles. The disadvantages of glass wool filters are shrinkage and solidification during steam sterilization. **Glass fiber filter cartridges**, which do not have these shortcomings, have replaced glass wool filters.

New cartridge filter systems using pleated membranes are now widely available. The advantage of these filters is that they are substantially smaller than glass wool or activated carbon towers. Operating the systems has become much simpler; because of the removable cartridge construction, replacing used filter elements is easy. Constructed of cellulose ester, polysulfone, or nylon, these membrane filters have the same structure as depth filters, but because they have a membranous structure they have an absolute filter effect. Figure 5.37 shows an example of the manner by which a filter cartridge is constructed.

Figure 5.38 shows a fermenter installation with incoming and outgoing air filters. The dis-

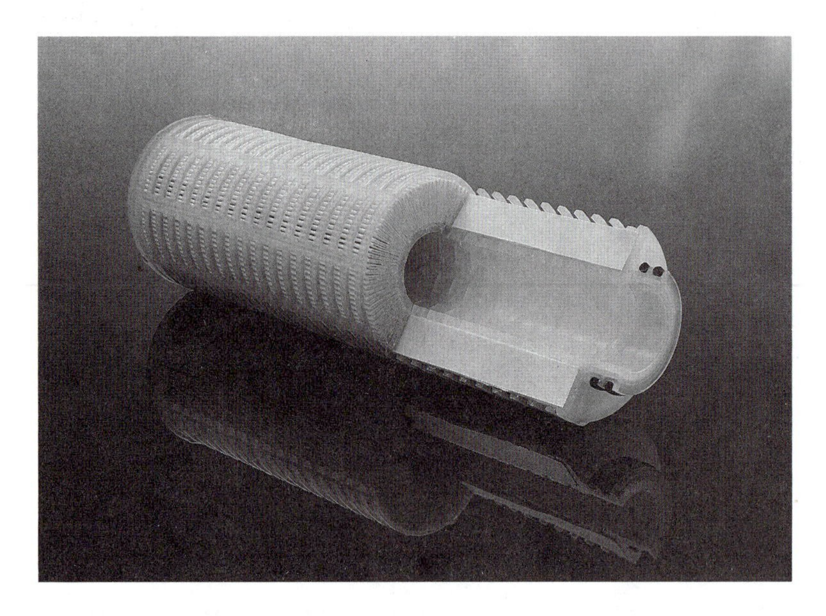

Figure 5.37 Structure of a filter cartridge

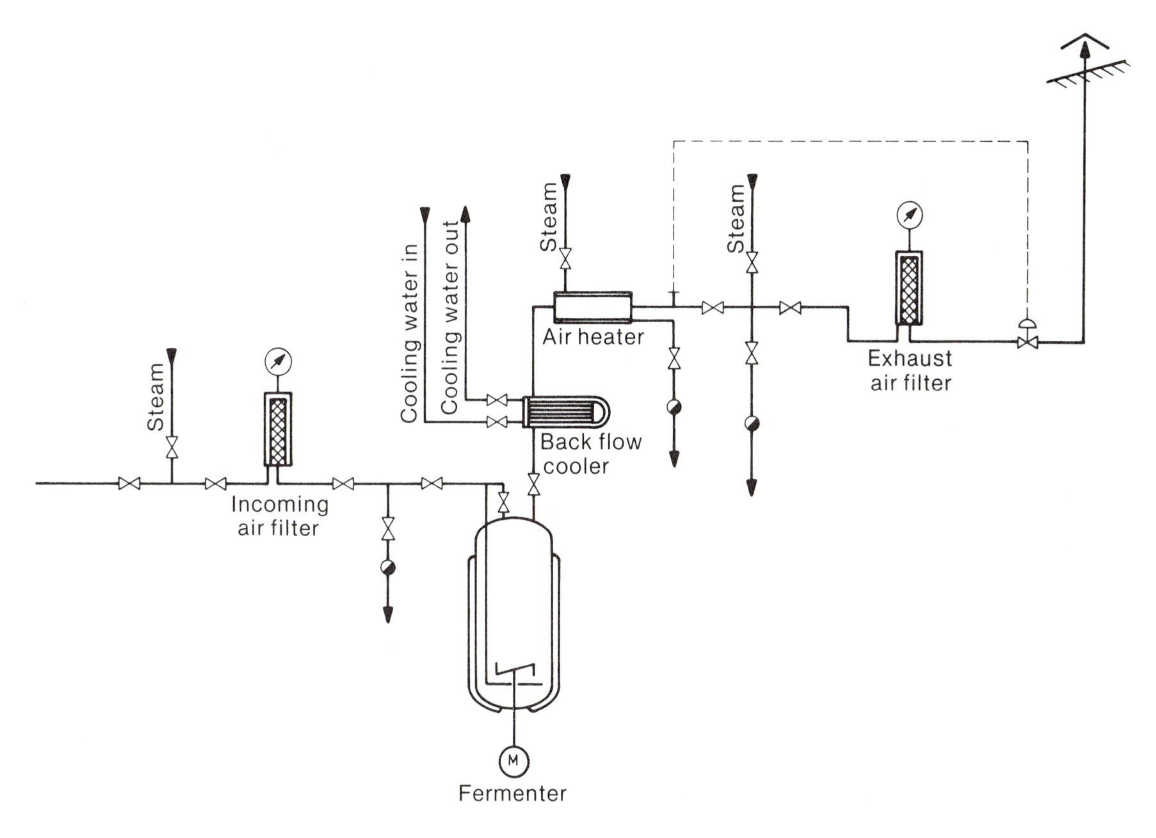

Figure 5.38 Installation of an air filter system (incoming and outcoming) in a fermenter

advantage of most systems installed today is that there is not yet any **absolute filter for bacteriophages** in industrial use. Bacteriophages can cause total failure in a system: for example, in the production of glutamic acid with *Corynebacterium glutamicum* or in the production of penicillin acylase with *Escherichia coli.*

5.8 FERMENTATION PROCESSES

An overall scheme of a fermentation process for the production of primary or secondary metabolites is given in Figure 5.39. The various stages of this scheme are discussed below.

Stage I: Inoculum preservation

The preservation of production strains over a long period is a basic requirement for a practical

Stage	Course of the fermentation		
I	Inoculum preservation		
II	Inoculum build-up	a.	1–2 Shake flask cultures
		b.	Spore formation of solid medium
III	Prefermenter culture		1–3 Preculture fermentations
IV	Production fermenter	a.	Batch fermentation
		b.	Continuous fermentation

Figure 5.39 General course of a fermentation in the production of primary and secondary metabolites

fermentation. The mere survival of strains is not the main objective. Microorganisms can easily be kept viable through periodic transfer, but it is their capability for product formation which must be preserved. High-yielding strains have often become damaged in primary metabolism during the strain selection process, and such strains frequently degenerate during successive transfers, probably as a result of spontaneous mutation.

The objective of preservation is thus to maintain strains as long as possible without cell division. "Master strains" should not be cultivated more than once in two years and activity levels must be checked with each usage. Depending on the strain, selection procedures must be undertaken periodically. "Working strains" are derived from "master strains". Working strains should be inspected for sterility and capability of product formation, and then stored until used.

The optimal method of preservation must be worked out for each process, i.e., each strain. The following three techniques are most commonly used:

Storage at low temperatures (2–6°C) This method is the easiest, but also the least secure. Microorganisms are kept as stab cultures on agar or in liquid culture in the refrigerator. There is a relatively high risk of contamination and reverse mutation through frequent transfer (normally every 8–16 weeks, at least once annually).

Frozen storage The most common method is freezing at −18°C or −80°C in freezers or at −196°C under liquid nitrogen. Freezing down to −196°C must be done gradually (1°C/min), although rapid freezing can be used if protective substances are added to prevent the formation of crystals. Frozen cultures may be kept for several years. The proportion of survivors is critical because up to 95% of the microorganisms are generally killed during freezing and subsequent thawing.

Lyophilization The best method of strain preservation is freeze-drying (lyophilization) after cultivation in special media. The addition of protective agents (such as skim milk or sucrose) reduces the lethality during the lyophilization process. Special equipment can be purchased which reduces the difficulties involved in freezing, drying, and aseptically sealing the ampules. Lyophilization is the method of choice in large culture collections because the cultures can be satisfactorily maintained for an essentially unlimited amount of time.

Stage II: Growth of the Inoculum

The preserved culture is initially revived by growth in shaken liquid culture or on solid medium (if spore formation is needed). The conditions used in the initial culture (medium, temperature of incubation, etc.) will depend upon the specific process. Standard growth times can be expected as follows:

Lyophilized cultures	4–10 days
Frozen cultures	
Bacteria	4–48 hours
Actinomycetes	1– 5 days
Fungi	1– 7 days
Refrigerated cultures	
Bacteria	4–24 hours
Actinomycetes	1– 3 days
Fungi	1– 5 days

In order to obtain sufficient inoculum for small fermenters, a second series of shake cultures is usually made in more flasks. In some fermentations, the large-scale inoculum must consist of spores. To obtain a spore crop, the preserved culture is cultivated on a solid substrate in 2–10 liter glass vessels under conditions of constant temperature and sterile aeration for 8–24 days. The substrate for the production of large amounts of spores is generally a granular material such as bran, peat, rice, or barley. In order to ensure continued aeration, the substrate must be shaken daily, which makes maintenance of aseptic conditions difficult. Further, many of the spores produced may be incapable of germinating.

For inoculation, the total culture (spores plus culture medium) is suspended with the aid of a surface-active agent (e.g., Tween 80) and transferred into the fermenter.

The proper cultivation of inoculum is vital for optimal titers in the later production-scale process. For optimal yields, not only the number of cells and spores have an influence but also the nutrient medium used for the inoculum, the temperature of growth, and the inoculum age. Induction or repression phenomena in the culture used for inoculum may affect the rate of production.

Stage III: Fermenter preculture

Fermenter precultures must be made in order to have enough inoculum for a large fermenter. If a production fermenter is started with too little inoculum, growth is delayed and the product formation rate can be unsatisfactory (see Section 5.2). The optimal inoculum concentration for the production fermenter determines the number of stages of fermenter preculture that are needed. In general the following inoculum concentrations are required:

Bacteria	0.1– 3.0%
Actinomycetes	5 –10 %
Fungi	5 –10 %
Spore suspension	$1–5\cdot10^5$/l culture solution

At times, the production culture medium is used for the last stage of inoculum build-up in order to induce product formation.

Stage IV: Production fermentation

Depending on the fermentation, reactors of various sizes are used and no general scheme for the inoculation of a production fermenter can be given. Table 5.21 gives the sizes of various production fermenters in actual use.

Nutrient media for production must be optimized not only in the ingredients used but in how the nutrient medium is prepared. Several parameters which must be optimized are:

- Composition of ingredients, quality, carbon/nitrogen relationship, impurities, variability from batch to batch.
- Order of solution or suspension of ingredients, pH value before and after sterilization, effect of sterilization on the entire nutrient solution or on individual components.
- Changes in the sterilized nutrient solution before inoculation due to increase in temperature and aeration.

The most important parameters during the fermentation are:

Temperature Fermentations are run either in the mesophile range (temperature optimum 20–45°C) or thermophile (>45°C) range. The appropriate temperature must be chosen to achieve maximum growth on the one hand and optimal product formation on the other hand. In some fermentations, higher temperatures are used to obtain increased growth of the culture and then the temperature is decreased at the onset of the idiophase. As an example of how important temperature control is, an increase of 1°C above the optimum produces a 20% lower yield in the penicillin acylase fermentation!

Aeration The aeration rate is 0.25–1.0 vvm (air volume/liquid volume·minute). The aeration rate must be adjusted to the amount of O_2 required.

Table 5.21 Fermenter sizes for various processes

Size of fermenter m^3	Product
1– 20	Diagnostic enzymes, substances for molecular biology, recombinant organisms
40– 80	Some enzymes, antibiotics
100–150	Penicillin, aminoglycoside antibiotics, proteases, amylases, steroid transformations, amino acids
–450	Amino acids (glutamic acid), single-cell protein

Pressure In order to minimize the risk of contamination, an overpressure of 0.2–0.5 bar is used. The hydrostatic pressure also has to be taken into account in large fermenters, since this influences the O_2 and CO_2 solubility in the nutrient solution.

Stirring Depending on construction, the following stirring rates are used with disc impellers.

Fermenter size (m^3)	Impeller speed (rpm)
0.02	250–450
0.2	250–350
1 - 20	120–180
40 -150	120–150
450	60–120

The installation of a continuous drive system is desirable in industrial fermenters in order to be able to precisely adjust the stirring rate to the process.

Table 5.22 Parameters that can be measured in fermentation processes

Physical parameters	Chemical parameters	Biological parameters
Temperature	pH	Biologically active product
Pressure	Dissolved O_2	Enzyme activity
Power consumption	O_2 and CO_2 in waste gas	DNA and RNA content
Viscosity	Redox potential	$NADH_2$ and ATP content
Flow rates (air and liquid)	Substrate concentration	Protein content
Turbidity	Product concentration	
Weight of fermenter	Ionic strength	

5.9 INSTRUMENTATION

To carry out measurements during fermentation for data analysis and control of the process, **special sensors**, which differ somewhat from those in the chemical industry, have been developed for bioreactors: 1. All sensors located in the sterile area must be sterilizable. 2. Some sensors must be specifically adapted to biochemical needs. The physical and chemical parameters listed in Table 5.22 can either be measured directly at many pilot plant or production fermenters or can be measured off-line in the laboratory.

The biological parameters listed must all be measured outside of the fermenter, with the exception of the $NADH_2$ measurement, which can be done on-line using a fluorescent method. There are interesting developments in the field of enzymatic electrodes, so-called **biosensors** (see Section 11.11). Such sensors, however, cannot be sterilized.

A standard procedure is to determine O_2 and CO_2 in the ingoing and outgoing air separately through the paramagnetic property of O_2 and the infrared absorption of CO_2. Sensors for measuring these gases are well-developed and function with few interruptions. But the **mass spectrometer** is more versatile, since it can also measure N_2, NH_3, methanol, and ethanol simultaneously as well as give qualitative and quantitative information on exchange of O_2 and CO_2. By the use of gas-permeable membranes, it is also possible to measure dissolved gases in nutrient media. Devices have been developed which analyze up to eight gases in the fermentation simultaneously.

Equipment for making accurate pH measurements is readily available. Combination electrodes (glass electrode, reference electrode, and temperature compensator in a single unit) are available which are able to withstand sterilization temperatures, pressures, and mechanical stresses. Response time and sensitivity of these electrodes is satisfactory for the usual fermentation requirements. However, although electrodes are available for measuring many other inorganic ions, they are not as sensitive as those for the hydrogen ion (see the paper by Fiechter et al., 1987).

CO_2 electrodes and **oxygen electrodes** have been used in commercial operations with varying success. The electrodes are of the amperometric type (galvanic or polarographic).

The current and not the voltage is altered by the oxygen concentration. In the polarographic electrode, oxygen is reduced at the cathode and silver is oxidized at the anode:

Cathode (Pt)	$O_2 + 2\,H_2O + 4\,e^- \rightarrow 4\,OH^-$	
Anode (Ag)	$4\,Ag + 4\,Cl^-$	$\rightarrow 4\,AgCl + 4\,e^-$

Overall reaction $4\,Ag + O_2 + 2\,H_2O + 4\,Cl^- \rightarrow 4\,AgCl + 4\,OH^-$

The current produced in this reaction is proportional to the oxygen partial pressure.

Commercial oxygen electrodes suitable for fermentation processes are widely available. However, the durability of these electrodes varies. Failure rates after sterilization lie above 50%, depending on the nutrient medium.

5.10 USE OF COMPUTERS

Computers can serve a variety of functions in fermentation process control and analysis:

1. **Optimization via computer**. Computers are used in scale up to store and evaluate fermentation parameters and to measure the effects of individual parameters on the metabolic behavior of cultures.
2. **Control via computer**. Computers can control fermentation processes. On-line fermentation control is widely used in the production scale in many companies.

Computer applications in biotechnology are not yet as widespread as in the chemical industry for several reasons: sensors suitable for use in sterile systems are not yet reliable enough to take advantage of computer capacity; biosynthesis and regulation of metabolite formation are not yet fully understood; fermentation cost reduction by using computers is difficult to calculate.

Thus, in biotechnology, computers are used primarily for data acquisition, data analysis, and development of fermentation models.

Data acquisition Data can be acquired directly at the fermenter with on-line sensors. The information acquired can be data such as pH, pO_2, temperature, pressure, viscosity, fermenter weight, power uptake, aeration rate, and O_2 and CO_2 content in the gas stream. Other data can be obtained from laboratory measurements and fed into the computer off-line, e.g. biomass concentration, nutrient content, metabolite formation. This information can be entered as raw data and can be converted by the computer to standard units; for example, to adjust volumes for a standard temperature, temperature-correction data can be used to calculate the true aeration rate for a production system.

An alarm system can be hooked up to the data-acquisition system to inform the attendant when deviations from standard values occur. Data about the course of fermentation can be stored, retrieved, and printed out and product calculations can be documented.

Data analysis The data entered or measured is used in calculations such as CO_2 formation rate, O_2 uptake rate, respiratory quotient, specific substrate uptake rate, yield coefficient, heat balance, productivity, volume-specific energy uptake, and Reynold's number. When biomass is not contin-

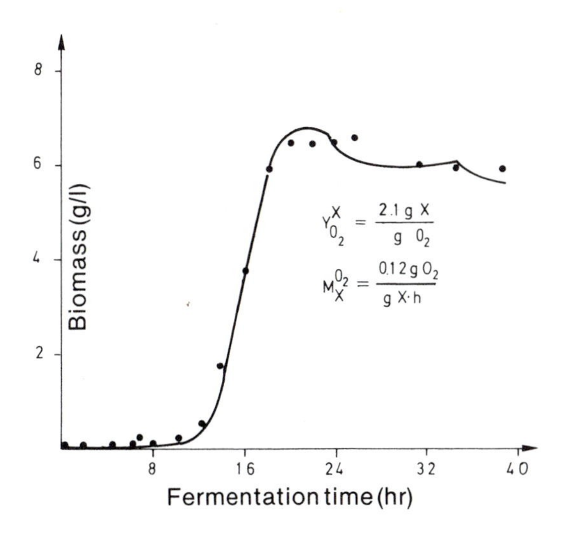

Figure 5.40 Comparison of experimentally determined *Thermoactinomyces* sp. concentrations (•) with growth curve generated from model calculations (Zabrieskie, 1976)

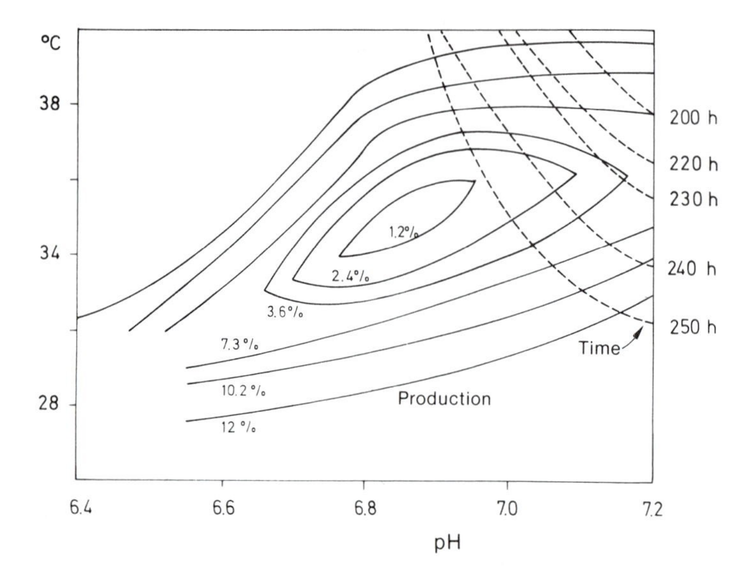

Figure 5.41 Isoproduction and isotime curves for the erythromycin production process in relation to temperature and pH. The percentages are based on the lowest yields obtained (Cherny and Durand, 1979)

uously measured, the biomass concentration can be calculated through the O_2 uptake rate. It is assumed that the yield constant and the proportion of O_2 needed for maintenance metabolism are known. The calculations must be adjusted if, for example, secondary metabolites are formed or the yield constant changes during the fermentation. Figure 5.40 shows the agreement between the measured biomass of *Thermoactinomyces* and the calculated growth curve. The value M is the amount for maintenance metabolism.

As a further example of data analysis, the optimization of erythromycin biosynthesis is illustrated in Figure 5.41. After fermentation at different pH and temperature levels, "isoproduction" and "isotime curves" are computed. The yields are given as percentage of the minor production compared with the maximum erythromycin titer. Optimal productivity (production/fermentation time) for a given set of operating conditions can be ascertained from this graph.

Development of fermentation models By using mathematical models, it is possible to better understand the fermentation process and to calculate the effects of process variables on the fermentation results, and thus optimize a process faster and more cost-effectively. The use of models can also aid the development of better control strategies for fermentations. A large number of models exists for batch and continuous fermentations. However, each model is only applicable to a specific process and cannot be used for other processes.

REFERENCES

Abbott, B.J. and A. Clamen. 1973. The relationship of substrate, growth rate, and maintenance coefficient to single-cell protein production. Biotechnol. Bioeng. 15: 117–127.

Anderson, C., J. Longton, C. Maddix, G.W. Scammell and G.L. Solomons. 1975. The growth of microfungi on carbohydrates, pp. 314–329. In: Tannenbaum, S.R. and D.I.C. Wang (eds.), Single cell protein, vol. II. MIT Press, Cambridge, MA.

Bailey, J.E. and D.F. Ollis. 1977. Biochemical engineering fundamentals. McGraw-Hill Book Comp., New York.

Bartholomew, W.H. 1960. Scale-up of submerged fermentation. Adv. Appl. Microbiol. 2:289–300.

Bates, R.L., P.L. Fondy, and R.R. Corpstein. 1963. An examination of some geometric parameters of impeller power. I.A.E.C. Process design and development. 2:310–314.

Bohnet, M. (editor). 1987. Fouling of heat transfer surfaces. Chem. Eng. Technol. 10:113–125.

Brauer, H. (editor). 1985. Biotechnology, Volume 2. Fundamentals of Biochemical Engineering. VCH Publishers, Deerfield Park, Florida.

Bronn, W.K. . 1966. Sauerstoffbedarf aerober Mikroorganismen und Technik der Sauerstoff-Versorgung (Oxygen requirement of aerobic microorganisms and measurement of oxygen demand), pp. 45–64. In: Windisch, S. (ed.), Zentralblatt für Bakteriologie, Parasitenkunde, Infektionskrankheiten und Hygiene. 1st Abt. Supplementheft 2.

Brown, C.M. and S.O. Stanley. 1972. Environment-mediated changes in the cellular content of the "pool" constituents and their associated changes in cell physiology. J. Appl. Chem. Biotechnol. 22:363–389.

Brown, D.E. 1970. Aeration in the submerged culture of microorganisms, pp. 125–174. In: Norris, J.R. and D.W. Ribbons (eds.), Methods in microbiology, vol. 2. Academic Press, New York.

Charles, M. 1978. Technical aspects of the rheological properties of microbial cultures. Adv. Biochem. Eng. 8:1–62.

Cherny, A. and A. Durand. 1979. Optimization of erythromycin biosynthesis by controlling pH and temperature: Theoretical aspects and practical application. Biotechnol. Bioeng. Symp. 9:303–320.

Cooper, C.M., G.A. Fernstrom, and S.A. Miller. 1944. Performance of agitated gas-liquid contactors. Ind. Eng. Chem. 36:504–509.

Crueger, W. 1973. Belüftungsfilter in der Fermentation, Fragen zur Bakteriophagen-Dichtigkeit (Air filters in fermentation; resistance to passage of bacteriophage particles), pp. 181–186. In: Dellweg, H. (ed.), 3rd Symp. Technische Mikrobiologie. Verlag Versuchs- und Lehranstalt für Spiritusfabrikation. Berlin.

Crueger, W., H.-J. Henzler, and M. Schedel. 1982. Scale up of nikkomycin. 13th International Congress of Microbiology, Boston.

European Federation of Biochemistry. 1983. European Federation of Biochemistry Working Party of Immobilized Biocatalysts. "Guidelines for the characterization of immobilized biocatalysts." Enzyme Microb. Technol. 5:304–307.

Egerer, P., W. Crueger, and G. Schmidt-Kastner. 1985. Continuous technique for the enzymatic production of isomaltulose. German Patent 3528752.

Einsele, A. 1978. Scaling-up of bioreactors. Proc. Biochem. July: 1–14.

Fiechter, A., M. Meiners, and D.A. Sukatsch. 1987. Biologische Regulation und Prozessführung (Biological regulation and process control). pp. 189–228 In: Präve, P., U. Faust, W. Sittig, and D.A. Sukatsch (editors). Handbuch der Biotechnologie. Oldenbourg Publishers, Munich.

Fox, R.I. 1984. Computers and microprocessors in industrial fermentation. pp. 125–174 In: Wiseman, A. (editor). Topics in enzyme and fermentation biotechnology, Volume 8. Ellis Horwood Publishers, Chichester.

Fukuda, H., Y. Sumino, and T. Kanzaki. 1968. Scale-up of fermentors. I. Modified equations for volumetric oxygen transfer coefficient. J. Ferment. Technol. 46:829–837.

Gaden, E.L. 1955. Fermentation kinetics and productivity. Chem. Ind. 154–159.

Gaden, E.L. 1959. Fermentation process kinetics. J. Biochem. Microbiol. Techn. Eng. 1:413–429.

Gekas, V.C. 1986. Artificial membranes as carriers for the immobilization of biocatalysts. Enzyme Microb. Technol. 8:450–460.

Hartmeier, W. 1986. Immobilisierte Biokatalysatoren (Immobilized Biocatalysts). Springer-Verlag, Berlin.

Hatch, R.T. 1975. Fermenter design, pp. 46–68. In: Tannenbaum, S.R. and D.I.C. Wang (eds.), Single cell protein, vol. II. MIT Press, Cambridge, MA.

Hempel, D.C. 1986. Grundlagen des Scale up für biotechnologische Prozesse in Rührfermentern (Basic concepts of scale up for stirred fermenters). pp. 77–123 In: Crueger, W., K. Esser, P. Präve, M. Schlingmann, R. Thauer, and F. Wagner (editors). Jahrbuch Biotechnologie 1986/87. Hanser Publishers, Munich.

Jarai, M. 1972. Oxygen transfer in the fermentations of primary and secondary metabolites, pp. 97–103. In: Terui, G. (ed.), Fermentation technology today. Yamada-Kami, Osaka.

Kargi, F. and M. Moo-Young. 1985. Transport phenomena in bioprocesses. pp. 5–56 In: Moo-Young, M. (editor). Comprehensive Biotechnology, Volume 2. Pergamon Press, Oxford.

Kim, J.H., J.M. Lebeault, and M. Reuss. 1983. Comparative studies on rheological properties of mycelial broth in filamentous and pelleted forms. Appl. Microbiol. Biotechnol. 18:11-16.

Laine, B.M. and J. du Chaffaut. 1975. Gas-oil as a substrate for single-cell protein production, pp. 424–437. In: Tannenbaum, S.R. and D.I.C. Wang (eds.), Single cell protein, vol. II. MIT Press, Cambridge, MA.

Lengyel, Z.L. and L. Nyiri. 1966. Studies on automatically aerated biosynthetic processes II. Occurrence and elimination of CO_2 during penicillin biosynthesis. Biotechnol. Bioeng. 14:337–352.

Lin, S.H. 1979. Residence time distribution of flow in a continuous sterilisation process. Proc. Biochem. July: 23–27.

Luong, J.H.T. and B. Volesky. 1983. Heat evolution during the microbial process—estimation, measurement, and applications. Advances in Biochem. Engineering/Biotechnology 28:1–40.

Malek, I., J. Ricica, and J. Votruba. 1984. Continuous cultivation of microorganisms—ecological significance of physiological state studies. In: Dean, A.C.R., D.C. Ellwood, and C.G.T. Evans (editors). Continuous culture 8: Biotechnology, Medicine, and Environment. Ellis Horwood Publishers, Chichester.

Mateles, R.J. 1971. Calculation of the oxygen required for cell production. Biotechnol. Bioeng. 13:581–582.

Maxon, W.D. 1959. Aeration-agitation studies on the novobiocin fermentation. J. Biochem. Microbiol. Techn. Eng. 1:311–324.

Michel, B.J. and S.A. Miller. 1962. Power requirements of gas-liquid agitated systems. A.I.Ch.E. (Amer. Inst. Chem. Eng.) Journal. 8:262–266.

Miura, Y. 1976. Transfer of oxygen and scale-up in submerged aerobic fermentation. Adv. Biochem. Eng. 4:3–40.

Monod, J. 1942. Recherches sur la croissance des cultures bacteriennes. Herrman and Cie, Paris.

Moo-Young, M. and H.W. Blanch. 1981. Design of biochemical reactors. Mass transfer criteria for simple and complex systems. Adv. Biochem. Engineering 19: 1–69.

Oldshue, J.Y. 1966. Fermentation mixing scale-up techniques. Biotechnol. Bioeng. 8:3–24.

Perkowski, C.A. 1983. Fermentation process air filtration via cartridge filters. Biotechnol. Bioeng. 25:1215-1221.

Prokop, A. and A.E. Humphrey. 1970. Kinetics of disinfection, pp. 61–84. In: Bernado, M.A. (ed.), Disinfection. Marcel Dekker, New York.

Radovich, J.M. 1985. Mass transfer effects in fermentations using immobilized whole cells. Enzyme Microb. Technol. 7:2–10.

Reuss, M. 1986. Messtechnik am Bioreaktor (Measurement techniques in bioreactors). pp. 189–227 In: Crueger, W., K. Esser, P. Präve, M. Schlingmann, R. Thauer, and F. Wagner (editors). Jahrbuch Biotechnologie 1986/87. Hanser Publishers, Munich.

Rushton, J.H., E.W. Costich, and H.J. Everett. 1950. Power characteristics of mixing impellers. Chem. Eng. Progr. 46:395–404 and 467–476.

Schügerl, K. and J. Lücke. 1977. Begasung von Blasensäulen, pp. 59–84. In: Rehm, H.J. (ed.), Dechema Monographien, vol. 81, Biotechnologie. Verlag Chemie, Weinheim.

Strohm, J., R.F. Dale, and H.J. Peppler. 1959. Polarographic measurement of dissolved oxygen in yeast fermentations. Appl. Microbiol. 7:235–238.

Taguchi, H., T. Imanaka, S. Teramoto, M. Takatsu, and M. Sato. 1968. *Endomyces* sp. glucoamylase. Scale-up of glucoamylase fermentation by *Endomyces* sp. J. Ferment. Technol. 46:823–828.

Tempest, D.W. and O.M. Neijessel. 1984. The status of Y_{ATP} and maintenance energy as biologically interpretable phenomena. Ann. Rev. Microbiol. 38:459–486.

Tsao, G.T. 1979. Elementary principles of microbial reaction engineering, pp. 223–241. In: Peppler, H.J. and D. Perlman (eds.), Microbial technology, vol. II. Academic Press, New York.

Tsao, G.T., A. Mukerjee, and Y.Y. Lee. 1972. Gas-liquid-cell oxygen transfer in fermentation, pp. 65–71. In: Terui, G. (ed.), Fermentation technology today. Yamada-Kami, Osaka.

Wallhäuser, K.H. (editor). 1984. Praxis der Sterilisation, Desinfektion—Konservierung, Keimidentifizierung—Betriebshygiene (Practical aspects of sterilization, disinfection—preservation, species identification—factory hygiene). Georg Thieme Publishers, Stuttgart.

Wang, D.I.C. and R.C.J. Fewkes. 1977. Effect of operating and geometry parameters on the behavior of non-Newtonian, mycelial antibiotic fermentations, pp. 39–56. In: Underkofler, L.A. (ed.), Developments in industrial microbiology. Amer. Inst. of Biological Science, Washington.

Wang, D.I.C., C.L. Cooney, A.L. Demain, P. Dunhill, A.E. Humphrey, and M.D. Lilly. 1979. Fermentation and enzyme technology. John Wiley and Sons, New York.

Wilkinson, J.F. and A.L.S. Munro. 1967. The influence of growth-limiting conditions on the synthesis of possible carbon and energy storage polymers in *Bacillus megaterium*, pp. 173–184. In: Powell, E.O., C.G.T. Evans, R.E. Strange, and D.W Tempest (eds.), Microbial physiology and continuous culture. HMSO, London.

Wittler, R., R. Matthes, and K. Schügerl. 1983. Rheology of *Penicillium chrysogenum* pellet suspensions. Appl. Microbiol. Biotechnol. 18:17-23.

Yamane, T. and S. Shimizu. 1984. Fed-batch techniques in microbial processes. Advances in Biochem. Eng./Biotechnology 30:147–184.

Zabrieskie, D.W. 1976. Real-time estimation of aerobic batch fermentation biomass concentration by component balancing and culture fluorescence. In: Armiger, W.B. and A.E. Humphrey (eds.), Microbial technology, vol. II, Computer applications in fermentation technology. Academic Press, New York.

6

Product recovery

6.1 INTRODUCTION

One of the most critical aspects of an industrial fermentation process is the recovery and purification of the product. The selection of the appropriate purification steps depends on the nature of the end product, its concentration, the side products present, the stability of the biological material, and the necessary degree of purification.

When discussing product recovery, one must distinguish between the concepts of **purification** and **concentration**. A recovery step changes either the purity or the concentration of a metabolite. In the ideal recovery process, both of these parameters are optimized. In the early days of industrial microbiology, the techniques used in product recovery (for instance, extraction, distillation, dialysis, crystallization, precipitation, drying) were taken over more or less directly from chemical engineering without more than an approximate attempt to adapt them to biological materials. Gradually it came to be realized that specific purification processes had to be perfected for biological materials. The bioreactor is no longer considered in an isolated manner from the apparatus used in purification. Today, the bioreactor and the purification techniques are considered as an integrated system, all components of which must be optimized together.

Further, the work of the biochemical engineer has been considerably expanded by the development of processes for expressing foreign genes of eucaryotes in procaryotes. As discussed in Chapter 3, eucaryotic gene products are often formed in procaryotes in the form of highly insoluble inclusion bodies, frequently in inactive granules held together by disulfide bridges. Such inclusion bodies must be released and purified and eventually converted into active proteins.

6.2 UNIT OPERATIONS IN PRODUCT RECOVERY

The microorganism itself can be the desired end product, as, for example, the ICI Pruteen process for the production of biomass as single-cell protein (SCP). In this case, if the cells are heated they aggregate into large clumps which can be readily separated from the fermentation broth by sedimentation. However, in most large-scale processes, the desired product is a metabolite, which is present either intracellularly or extracellularly. Examples of intracellular metabolitles include nucleic acids, vitamins, enzymes, and certain antibiotics, such as sisomicin and griseofulvin. Examples of extracellular metabolites include amino acids, citric acid, alcohol, some enzymes (for example, amylases and proteases) and most antibiotics (for example, penicillin and streptomycin). In a few cases, metabolites are found both in the cells and the culture filtrate (for example, flavomycin, vitamin B_{12}).

The first step in product recovery is therefore the separation of cell biomass and insoluble nutrient ingredients from the supernatant. For this purpose, several methods are available, including flocculation, flotation, filtration, or centrifugation. If an intracellular metabolite is to be isolated, it must be liberated from the cells.

Once the metabolite has been separated from the cells, the selection of further purification steps will depend upon the desired product. Figure 6.1 summarizes various purification processes, arranged according to the separation principle for particles of various sizes. The last stages in process recovery involve **precipitation, crystallization,** and/or **drying**.

Fermentation products are present in culture solutions primarily at quite low concentrations (Table 6.1). In order to keep costs down, it is desirable in the very first stages of purification to reduce the volume to the smallest possible.

Table 6.1 Concentrations of some metabolites at the end of fermentation

Metabolite	Yield (g/l)
Vitamin B_{12}	0.06
Riboflavin	0.1 - 7
Antibiotics	0.2 - 60
Amino acids	2 -100

Flocculation and flotation

Single cells in the size range of 1 to 10 μm settle only very slowly and are difficult to bring down even with the centrifuge. In some cases (for example, SCP production and sewage treatment), **flocculation** can be used to produce large aggregates which will settle more readily. In most cases, a flocculating agent is added, such as an inorganic salt, an organic polyelectrolyte, or a mineral hydrocolloid. Depending on the agent used, the flocculation process can be either reversible or irreversible. The flocculation process is also influenced by the nature of the cells and the ionic constituents and their concentrations.

The reverse of flocculation is **flotation**, which is most readily accomplished by introducing gas into the liquid. The cells become adsorbed to the gas bubbles and rise to the foam layer at the top of the vessel, where they can be collected and removed from the bioreactor.

Filter systems

Most commonly, the first step in process recovery, separation of the biomass and the culture filtrate, is carried out by some sort of **filtration** process. Two major types of filters are so-called **depth filters** and **absolute filters.** A depth filter is constructed of a filamentous matrix, such as glass wool, filter paper, or asbestos, the filtration process occurring because the particles to be removed are trapped within the matrix. The particles being filtered are often smaller than the spaces in the filter, but are removed anyway as they pass through the contorted interstices of the filter. Absolute filters, on the other hand, are membranes, in which the pore sizes are smaller than the particles being filtered. The particles are therefore removed on the surfaces of the filters. Filamentous fungi are most commonly filtered through depth filters such as cloth, sometimes in the presence of a filter aid. Bacterial cultures, on the other hand, must be filtered with absolute filters. In either case, the efficiency of filtration is influenced by numerous factors, such as the

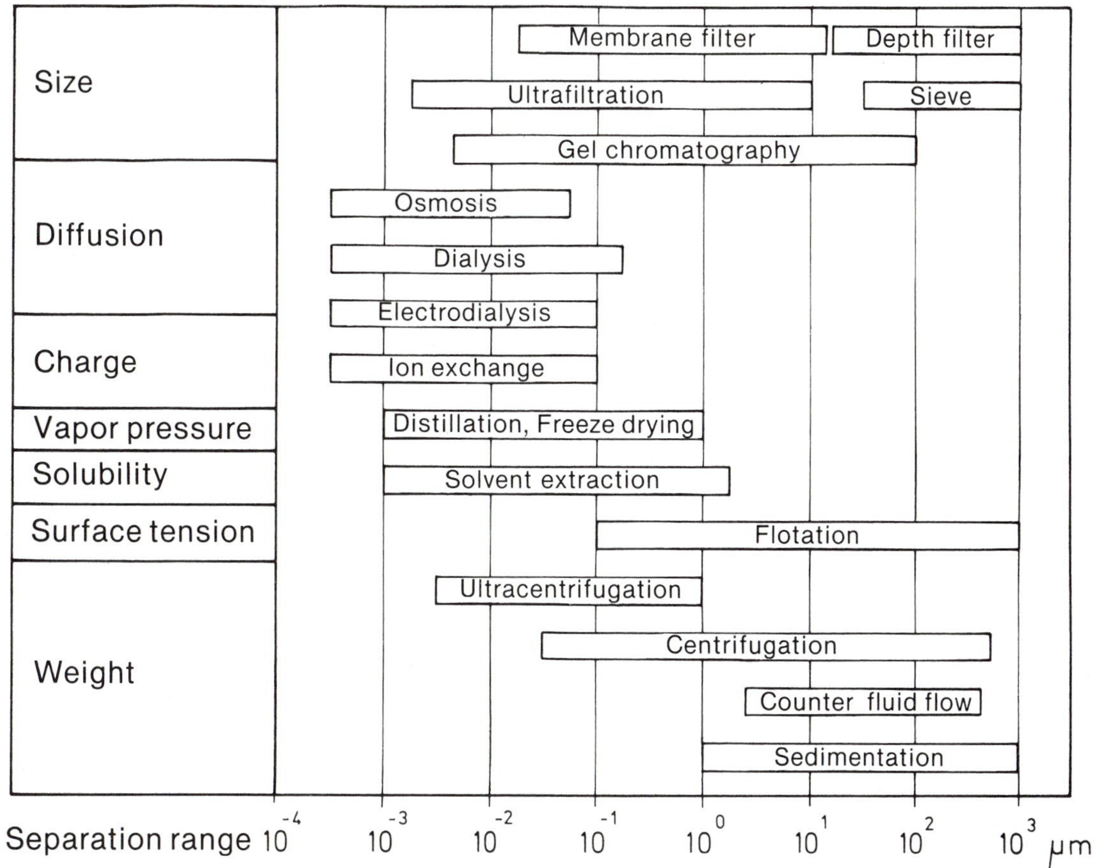

Figure 6.1 Concentration processes using different chemical and physical properties of particles and molecules

size of the organism, its morphology, the pH, viscosity, presence of slimes, temperature, and presence of other organisms as possible contaminants. As the cell material piles up on the filter, a filter cake is formed which reduces the filtration rate.

In the classical purification process used in the antibiotic industry, biomass is separated from the culture filtrate on a vacuum rotating drum filter. A cloth or rope is wound around the outside of the apparatus and a vacuum is pulled from the inside of the drum. To increase the efficiency of the filtration process, a filter aid such as diatomaceous earth is generally used. To maintain the efficiency of filtration, an automatic knife is used to continuously scrape the biomass plus a thin layer of the filter aid from the filter. By arranging the drum in segments, the filter aid can be washed as the drum turns. Typical filtration efficiencies for arrangements of this type are shown in Table 6.2. Vacuum drum filters are especially good for suspensions which contain a high concentration of solids (20–60% mycelial volume).

Another type of filter, better suited for solutions in which the solid concentration is lower, is the **filter press,** in which a stack of flat porous plates are used as supports for a filter, either cloth

Table 6.2 Typical filtration rates for antibiotic fermentations

Antibiotic	Organism	Filtration rate l/h·ft²
Penicillin G	*P. chrysogenum*	130–170
Kanamycin	*S. kanamyceticus*	8
Lincomycin	*S. lincolnensis*	35
Neomycin	*S. fradiae*	12

(Belter, 1985)

or membrane. The volume of the chamber within the plates determines the biomass capacity of this type of filter.

Although mycelium (from either fungi or actinomycetes) can be removed with a drum filter, single-celled bacteria are generally separated from the medium by use of membrane filters. Two types of membrane filtration processes, **static** and **cross-flow,** are compared in Figure 6.2. Because the filtration process is strictly dependent on the size of the pores and the size of the particles, clogging of filters is a problem. The cross-flow method was developed in order to reduce the tendency for clogging. In this method, the solution is pumped in a crosswise fashion across the membrane. The filtrate passes through the membrane and the biomass is washed off the filter and carried out with the retentate. With the cross-flow method, an increase in filtration rate of 100-fold can be obtained in comparison to static filtration.

Depending on the sizes of particles being filtered, three major types of filtration process are recognized: reverse osmosis, for particles of 0.0001–0.001 μm; ultrafiltration, for particles of 0.001–0.1 μm; and microfiltration, for particles of 0.1–10 μm. The capabilities of the ultrafiltration and reverse osmosis processes are generally given in terms of the nominal molecular weights of substances separated (molecular-weight cutoff). Reverse osmosis generally involves substances of molecular weight less than 1000, ultrafiltration molecular weights greater than 1000.

The microfiltration process concerns mainly the separation of cells or cell fractions. Membranes made of cellulose esters, polyvinylfluorides, polycarbonates, polysulfones, and cellulose are used. Membranes are arranged either in cassettes, as spiral-wound modules, as bundles of tubes of 1–2 cm diameter, or as capillary bundles (Figure 6.3). Numerous parameters must be considered when selecting a filtration system. Among these are the pore size and particle selectivity, the rate at which fluid can be passed through when viscous solutions are being filtered, the ease with which the filter can be cleaned, and the number of times the filter can be reused. Other factors to consider are the cost of the filter system, the membrane surface area presented, the dead volume within the filter (which determines how much of the product will be lost), and the sterilizability of the system. Another factor that often reduces the efficiency of

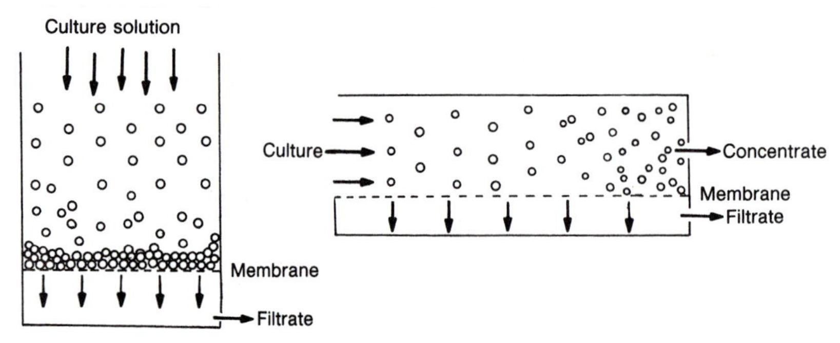

Figure 6.2 Comparison of static and cross-flow filtration

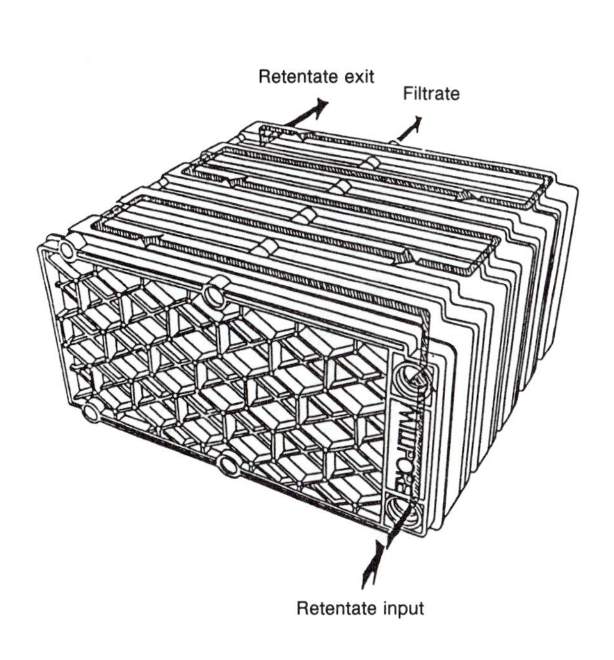

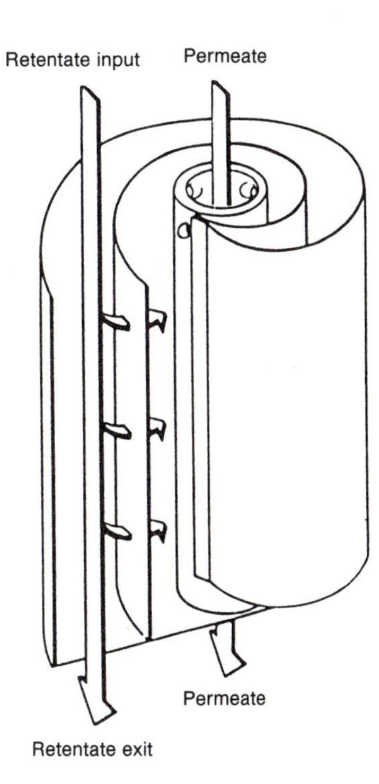

Figure 6.3 Comparison of a casette system and a cross-flow cartridge module

a filter is the presence of antifoam agents, which are often used in large-scale fermentations.

The following data are typical for an industrial process for the concentration of *Escherichia coli* from a culture broth: fermenter volume, 6 m^3; working volume, 4 m^3; filtration rate, 25–80 $l/h \cdot m^2$; process time, 12 h; concentration factor, 1:10. Table 6.3 presents data for filtration processes with several organisms with various types of filters.

The following are typical costs that must be incurred for filtration in a large-scale industrial process: **Production of 1 m^3 retentate**: membranes ($420), capital investment ($1600), labor ($70), energy ($230). **Production of 1 m^3 filtrate:** membranes ($14), capital investment ($60), labor ($3), energy ($8).

In addition to their use in filtration, membranes also play a role in the fermentation process itself. They can function in providing sterile air, and they can be used in other ways in the fermenter (Figure 6.4).

Centrifugation

The sedimentation rate of a particle in a gravitational field can be represented by Stoke's law:

$$v = \frac{d^2 \cdot (\delta_s - \delta_l)}{18\eta} \cdot r \cdot \omega^2$$

where v = the sedimentation rate (m/s), d = the diameter of the particle (m), δ_s = the density of the particle (kg/m^3), δ_l = the density of the liquid (kg/m^3), r = the radius of the centrifuge head (m), ω = the angular velocity (rad/s), and η = the dynamic viscosity (Pa·s).

Centrifugation is used not only for separating solid particles from the liquid phase (fluid/par-

Table 6.3 Concentration of microorganisms by use of cross-flow filtration

Organism	Type of membrane	Concentration % cell wet volume	Average flow rate l/h·m²
Bacillus cereus	Capillary tubes polypropylene 0.3 μm	10 → 30	124
B. cereus	Capillary tubes polysulfone 10^5 dalton	0.8 → 15	47
Brevibacterium sp.	Casette polyacryl 0.2 μm	3.5 → 32	37
Escherichia coli	Capillary tubes polysulfone 10^5 dalton	4.0 → 40	15
E. coli	Casette PVDF 0.45 μm	4.2 → 48	16
Candida boidinii	Capillary tubes polypropylene 0.45 μm	10 → 40	56
Klebsiella pneumoniae	Capillary tubes polycarbonate 2 × 10^6 dalton	2 → 58	73
Lactobacillus casei	Cassette PVDF 0.45 μm	1.5 → 28	28

Kroner et al. (1984)

ticle separation) but also for fluid/fluid and fluid/fluid/particle separation. Fluid/particle separation is of most significance, although fluid/fluid separation is used in penicillin production (for the separation of the antibiotic-extracting solvent from the aqueous phase by means of a two-stage continuous countercurrent extractor; see Section 13.2).

Two distinct types of centrifuges are used, filter- and sieve centrifuges, and baffle centrifuges. In the **filter-** or **sieve centrifuge** the separation occurs as the particles are forced against a filter material. In the **baffle centrifuge** the separation occurs because of density difference betweeen the particles and the liquid. The product can be removed either continuously or batchwise. A wide variety of centrifuges are marketed for large-scale centrifugation processes, and the capabilities of the major types are given in Table 6.4. In the case of yeast, machines capable of handling volumes as large as 300 m^3/h have been developed.

Figure 6.5 shows cross-sectional diagrams of two major types of baffle centrifuges, a *plate separator* and a *decanter*. The major considerations

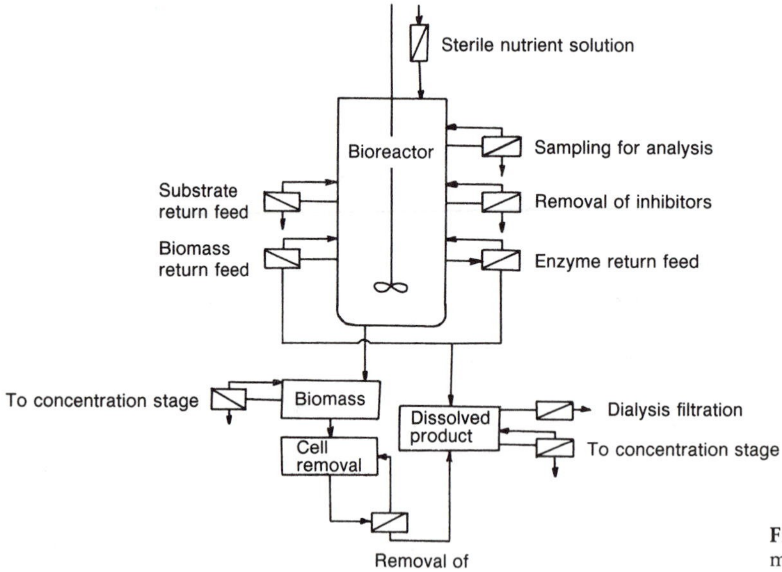

Figure 6.4 Various locations where membrane systems can be used in a bioreactor unit

in the selection of a centrifuge for biotechnological processes depend on the task at hand. For fluid/particle separation, factors to consider are: 1) the needed purity of the fluid phase; 2) the needed recovery of the fluid phase; 3) the needed recovery of the particle phase; 4) the needed permissible moisture content of the particles; 5) the specific density of the particles. For fluid/fluid separation, the major factors to consider are: 1) the needed purity of the lighter or heavier liquid; and 2) the needed recovery of the lighter or heavier phase.

Table 6.4 Comparison of a separator, a decanter, and a tube centrifuge

Parameter	Separator	Decanter	Tube centrifuge
Solids, %	1–30	5–80	1–5
Maximal centrifugal force (g)	5000–15,000	1500–4500	13,000–17,000
Dewatering capacity	Average	Average	Good
Removal of small particles	Good	Moderate	Very good
Cleanability	Good	Good	Very good

For use in the pharmaceutical industry, separators are available that are capable of withstanding temperatures of 121°C and hence can be completely sterilized. An example of such a machine is shown in Figure 6.6.

Disintegration of microorganisms

In some cases, the recovery of product requires that the microorganisms be fragmented, either by chemical, physical, or biological means. A summary of methods for disintegrating microorganisms is given in Figure 6.7. The selection of a method depends principally on the nature of the cells. Although the cell membrane offers no special resistance to breakage, cell walls vary widely in how readily they are broken. Gram-positive bacteria and yeasts are much more difficult to break than gram-negative bacteria or filamentous fungi. Even within a given organism, cells vary significantly in sensitivity to breakage depending upon their physiological state. It is also important, when selecting a disintegration method, that the target biomolecule not be destroyed. For instance, acid can be used to break many orga-

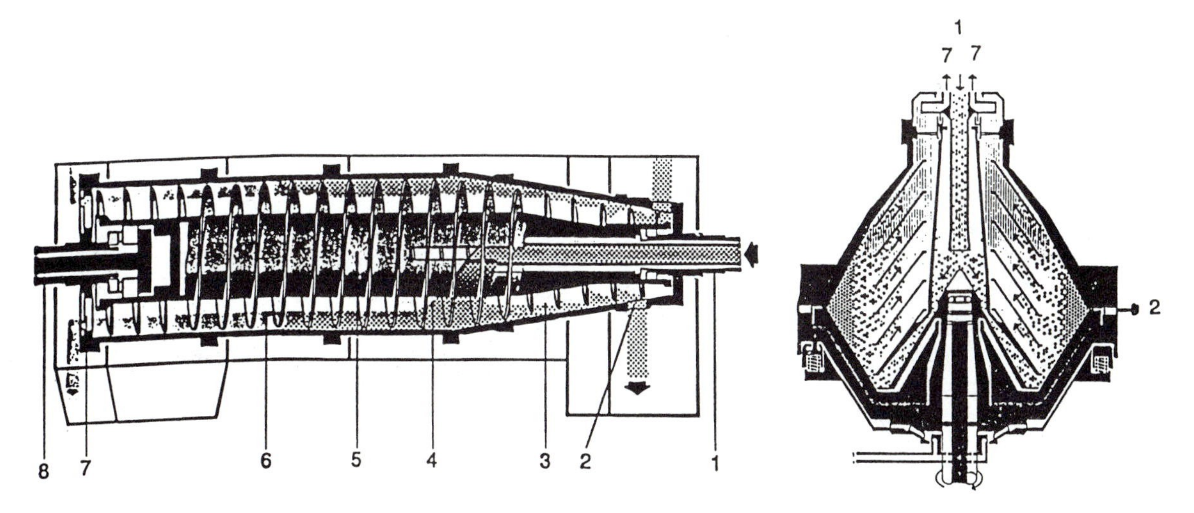

Figure 6.5 Construction details for a decanter (left) and a separator. 1 = inflowing stream; 2 = removal of particles; 3 = zone of low water content; 4 = particle sedimentation and drum cover; 5 = decanter screw; 6 = fluid level; 7 = clear liquid effluent; 8 = drive shaft

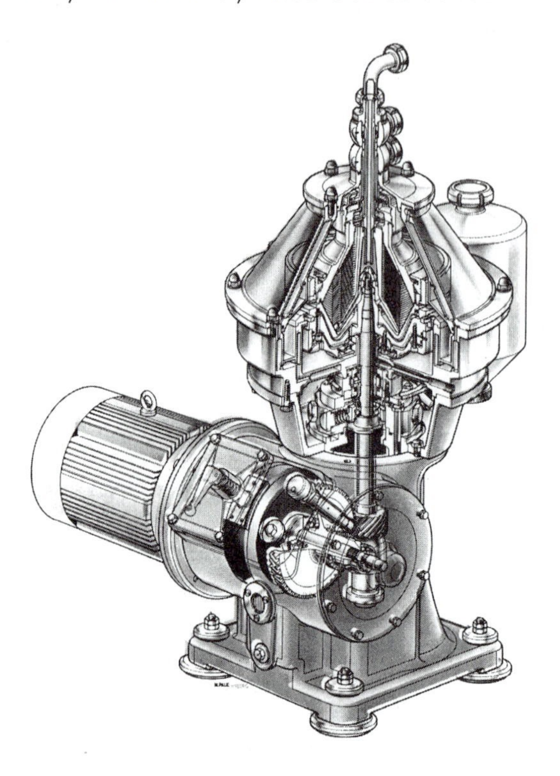

Figure 6.6 Sterilizable separator. This is type BTPX 20s manufactured by Alfa Laval. It contains 80 plates, an interior volume of 3.1 liter, a maximal centrifugal force of 12,800 × gravity, a maximal rotation of 9,650 revolutions per minute, a throughput for biotechnological processes of around 400–500 liters/hour.

nisms but cannot be used if the product is acid labile.

For the extraction of yeast cells, **autolysis** is often used. Endogenous autolytic enzymes of the yeast cells themselves bring about the destruction of the cell wall. Addition of a sodium chloride, ethyl acetate, or chloroform, and incubation for 24 h at 45 °C increases the rate of the autolytic process. Another technique is **plasmolysis**, following the addition of a high concentration of NaCl. **Ultrasonic disintegration** is widely used in the laboratory, but because of the high cost it is not suitable for large-scale industrial processes.

Industrially the most widely used methods of cell disintegration involve mechanical action. **Ball mills** bring about the breakage of cells through the action of glass beads. The cells and the glass beads are mixed together and subjected to high-speed mixing in a reaction vessel. Breakage occurs when a cell is forced against the wall of the vessel by a glass bead. The efficiency of the process depends on the following parameters: 1) the cell concentration (optimal at about 40% cell wet volume); 2) rate of flow through the mill; 3) orientation of the apparatus (horizontal or vertical); 4) the cell/glass bead ratio; 5) the size of the glass beads (generally 0.5–0.8 mm). The procedure must, of course, be optimized for the particular process. One can expect a maximal breakage rate of about 80% of the cells.

Another approach is the use of **homogenizers.** In such devices, the biological material is placed under a high hydrostatic pressure (around 500 atmospheres) and the pressure suddenly released by allowing the liquid to exit through a valve. A cross section of the valve arrangement is shown in Figure 6.8. The operating conditions must be carefully optimized for each situation. Depending on the organism and the desired metabolite, from 10–60% of the cells can be broken. In order to achieve 90% breakage, two or three passes of the material through the homogenizer must be carried out. As seen in Figure 6.9, the efficiency of the process depends on the applied pressure. In general, the best pressure to use is between 450 and 600 atmospheres.

Chromatography

Some of the most expensive stages in product recovery involve the use of chromatographic methods. Such methods are especially value for the purification of biologically sensitive materials and find widespread use in the preparation of pharmaceuticals, diagnostic reagents, and research materials. In many cases, several separate types of chromatographic methods must be used, one following the other (Figure 6.10). Separation generally occurs in a column: a stationary phase (generally a resin) is used to adsorb the product which is then eluted with a mobile (liquid) phase. The elution liquid can be at a single density or

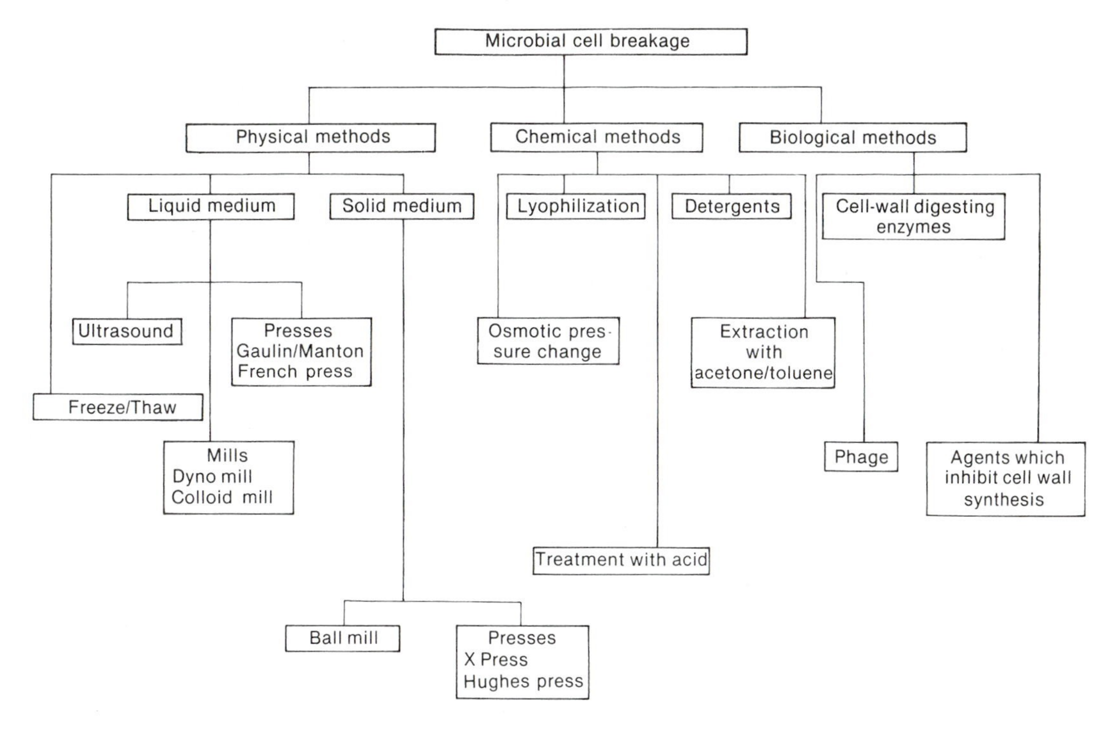

Figure 6.7 Methods of opening cells to isolate microbial metabolites

a gradually changing density (density gradient) can be used. By use of a suitable detector at the exit of the column, the eluted fractions which contain the desired product can be determined. Suitable detectors use ultraviolet or visible light absorption or conductivity. Summarized below are various chromatographic methods that are widely used.

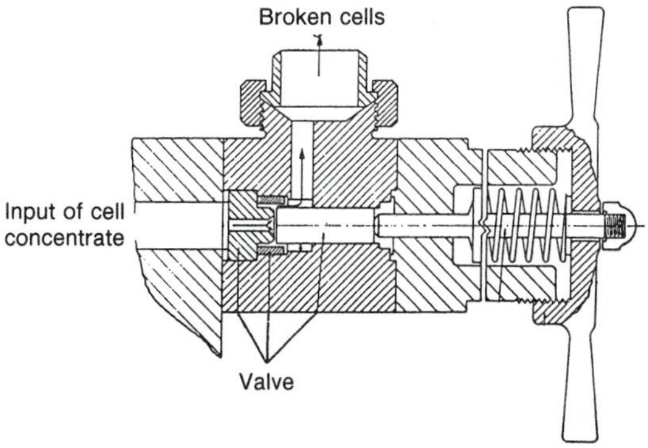

Figure 6.8 Outlet valve for a homogenzier

Gel filtration (molecular sieve chromatography) Molecules of different sizes can be separated by passage through gels with different pore sizes. Large molecules do not penetrate the gel and pass directly through the column, exiting with the elution front. Small molecules penetrate more or less deeply into the gel and their mobility is hence retarded from the solvent front. Standard curves can be developed which relate molecular weight to the elution position. At the industrial scale, gel filtration is used mainly to remove salts and to separate low-molecular-weight impurities. At a smaller scale, gel filtration is used to fractionate and purify protein molecules (for instance, insulin or interferon). The most widely used gels for gel filtration are the Biogel types P and A from BioRad and the Sephadex series S, G, and LH from Pharmacia.

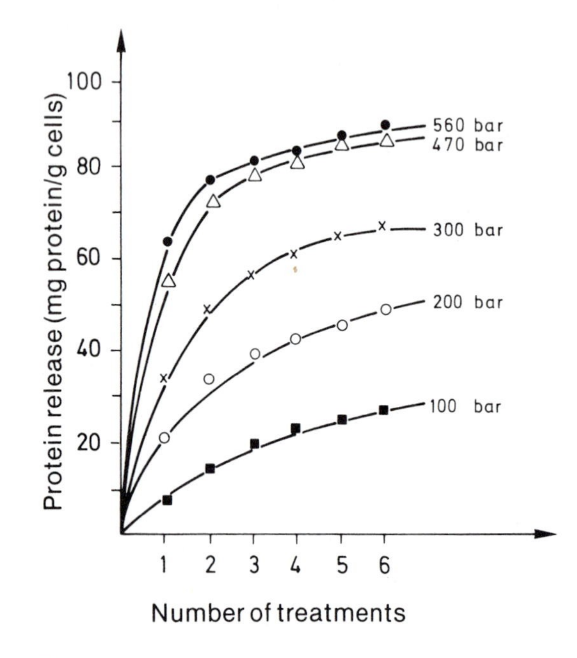

Figure 6.9 Protein release from yeast cells with a pressure homogenizer (Whitworth, 1974)

Adsorption chromatography In adsorption chromatography, separation involves hydrophilic or hydrophobic interactions (van der Waals forces) between the carrier and the biological material. Elution and fractionation are accomplished by means of solutions of higher or lower polarity or ionic strength. Adsorption materials include silicates, alumina, activated carbon, cross-linked dextrans, and hydroxylapatite.

Ion exchange chromatography Macromolecules with ionic groups can be separated by ion exchange chromatography. The macromolecule adsorbs to the carrier and is eluted by a solution of defined ionic strength. Depending on the ionic nature of the molecule and the ionic strength of the eluting solution, the separation can be quite specific. Ion exchange chromatography finds wide use for purification of antibiotics from fermentation broths. It also is widely used in large-scale purification of proteins. The most widely used ion materials for cation exchange are Dowex HCR and OCR, Amberlite IR and IRC, and Lewatit S. Anion exchangers used are Dowex SAR and MSA, Amberlite IRA, and Lewatit M.

Affinity chromatography Affinity chromatography is an absolutely specific method for purification of biological materials. The desired material binds specifically and reversibly to a ligand which has been fixed to an inert carrier. For example, nucleotide adenine dinucleotide (NAD) can be purified by allowing it to bind to a carrier containing a dehydrogenase enzyme. The antibiotic bacitracin can be used as a ligand to isolate Asp–, Ser–, Cys– and metalloproteases from a crude mixture.

Isoelectric focussing By means of isoelectric focussing, proteins can be purified by use of pH gradients in association with ion exchange gels. A linear pH gradient is constructed and the proteins are separated according to their isoelectric

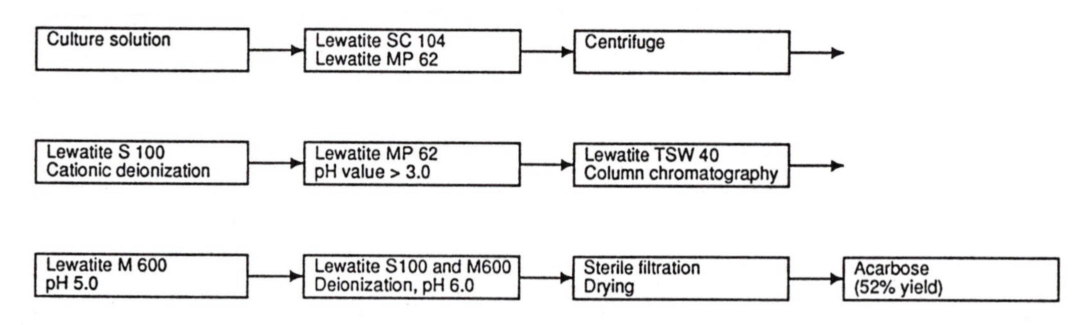

Figure 6.10 Flow diagram for the purification of a particular product, the invertase inhibitor Acarbose® (From Rauenbusch and Schmidt, 1978).

points. The separation of proteins by this method is very good, but only small volumes can be handled.

Extraction

Many products can be purified by use of solvent extraction. A two-phase system is set up, using a solvent which is immiscible with the aqueous fermentation broth. After the desired product is concentrated in the solvent phase, it can be further purified. Ideally, the product will transfer quantitatively to the solvent phase. Solvent extraction has been widely used in the antibiotic industry, for antibiotics such as penicillin, using organic solvents such as amyl or butyl acetate.

For the purification of enzymes in the active state, organic solvents cannot be used, but two-phase aqueous systems can be used, prepared in a salt solution with hydrophilic polymers. In both phases, water is the main component, but the two phases are not miscible. Cells remain in one of the phases and the enzyme is transferred to the other phase without any loss of activity. Human fibroblast interferon can be concentrated 70-fold with 100% yield using this method, at the same time that a four-fold reduction in volume is achieved.

An interesting variant of this procedure is extraction with supercritical solutions. This approach is used in the extraction of caffeine from coffee beans and hops for brewing, but can also be used for the extraction of pigments and flavor ingredients from biological materials.

Crystallization and precipitation

Once the metabolite has been extracted, it can be further concentrated and purified by crystallization, either by evaporation or by transfer to low temperature. Low-temperature crystallization is a very gentle way of purification. In the case of precipitation, a chemical agent is sometimes added to promote the concentration reaction.

Drying

For biological materials, it is essential to dry the final product in such a way that its activity is not lost. Drying essentially involves transfer of heat to the wet product, and removal of the moisture in a stream of gas. Heat transfer can be either by direct contact, by convection, or by radiation. Numerous types of apparatus are commercially available for carrying out the drying process. They include convection dryers (such as pulverizing, rotating, and band dryers) and contact dryers (such as thin layer and chamber dryers).

In some industries, the final product is a living microbial culture (for instance, starter cultures in the dairy industry). To prepare dry cultures, the freeze-dry technique must be used, since in this way the living organism is never subjected to heating. Freeze drying is also used in the preparation of some pharmaceutical products (for instance, penicillin is freeze dried directly in ampules).

6.3 YIELD

Recovery losses depend on the sensivitity of the substance to the process and on the number of purification stages. Table 6.5 shows typical yield losses. Further details will be given in the discussions of particular processes later in this book.

The fraction of the total cost attributed to purification is on the average around 20%, but in extreme cases with certain kinds of intracellular metabolites can be as high as 90%.

In Table 6.6, the data for glutamic acid purification demonstrate the extent to which recovery methods must be combined and optimized.

Table 6.5 Recovery losses in typical fermentation processes

Fermentation product	Recovery loss %
Single cell protein	5
Antibiotics	20–50
Extracellular enzymes	10
Intracellular enzymes	90

Table 6.6 Glutamic acid purification process

Process	Direct crystalization	Yield %	Anion exchange process	Yield %	Cation exchange process	Yield %
Stages of process	Culture solution	100	Culture solution	100	Culture solution	100
	Filtration pH 7.5–8.0	90	Filtration pH 7.5–8.0	90	Filtration pH 7.5–8.0	90
	Filtrate evaporation	98	Adsorbtion to anion exchanger	95	Adsorbtion to cation exchanger	98
	pH adjustment		Elution with acid	98	Elution with base	95
	Crystalline acid	73	Crystalline acid	90		
	Crystalline Na salt	88	Crystalline Na salt	93	Crystalline Na salt	93
Chemicals used	3 Equivalents of acid 1 Equivalent of NaOH		3 Equivalents of acid 3 Equivalents of NaOH		3 Equivalents of acid 2 Equivalents of NaOH	
Total yield		56.6		70.0		78

(Wolf, 1974)

The final yield is only one factor, since the cost of chemicals, equipment, and wages must also be considered in the overall calculation.

REFERENCES

Alt, C. 1972. Filtration, pp. 154–198. In: Ullmanns Encyklopädie der technischen Chemie, Vol. 2. Verlag Chemie, Weinheim.

Atkinson, B., and I.S. Daoud. 1976. Microbial flocs and flocculation in fermentation process engineering. Adv. Biochem. Eng. Biotechnol. 4:41–124.

Belter, P.A. 1985. Filtration of fermentation broths, pp. 347–350. In Moo-Young, M. (ed.). Comprehensive Biotechnology. Vol. 2. Pergamon Press, Oxford.

Cheryan, M. 1986. Ultrafiltration Handbook. Technomic Publ., Lancaster, PA.

Edebo, L. 1983. Disintegration of cells by extrusion under pressure, pp. 93–114. In: Lafferty, R.M. (ed.). Enzyme Technology. Springer Verlag, Berlin.

Egerer, P. 1986. Chromatographische Methoden in der Aufarbeitung von Naturstoffen (Chromatographic methods for the purification of natural products). pp. 125–187. In: Crueger, W., K. Esser, P. Präve, M. Schlingmann, R. Thauer, and F. Wagner (eds.). Jahrbuch Biotechnologie 1986/87. Hanser Verlag, Munich.

Fritz, J.S., D.T. Gjerde, and C. Pohlandt. 1982. Ion Chromatography. Hüthig Verlag, Heidelberg.

Hustedt, H., K.H. Kroner, N. Papamichael, and U. Menge. 1987. Verteilung zwischen wässrigen Phasen unter Mikrogravität. (Use of microgravity in the partitioning between aqueous phases.) BioEngineering, 1/1987:12–29.

Kroner, K.H., H. Schütte, H. Hustedt, and M.R. Kula. 1984. Cross-flow filtration in the down stream processing of enzymes. Process Biocem. April 67–74.

Kula, M.R. 1985. Recovery operations, pp. 725–760. In: Rehm, H.J., and G. Reed (eds.). Biotechnology. Vol. 2. VCH Verlag, Weinheim.

Kula, M.R., K.H. Kroner, and H. Hustedt. 1982. Purification of enzymes by liquid-liquid extraction. Adv. Biochem. Eng. Biotechnol. 24:73–118.

Kula, M.R., K. Schügerl, and Ch. Wandrey (eds.). 1986. Technische Membranen in der Biotechnologie. (Membranes for large-scale biotechnological processes.) GBF Mongraphien 8. VCH Verlag, Weinheim.

Mellor, J.D. 1978. Fundamentals of Freeze Drying. Academic Press, New York.

Moo-Young, M. (ed.) 1985. Comprehensive Biotechnology. Vol. 2. The Principles of Biotechnology: Engineering Considerations. Pergamon Press, Oxford.

Mullin, J.W. 1972. Crystallisation. 2nd Ed. Butterworth, London.

Rauenbusch, E., and D. Schmidt. 1978. Verfahren zur Isolierung von 0-(4,6-Dideoxy-4[[1S-(1,4,6/5)-4,5,6-trihydroxy-3-hydroxymethyl-2-cyclohexen-1-yl] amino]-α-D-glucopyranosyl)-(1-4)-0-α-glucopyranosyl-(1→4)-D-glucopyranose aus Kulturbrühen. (Technique for the isolation of 0-(4,6-Dideoxy-4[[1S-(1,4,6/5)-4,5,6-trihydroxy-3-hydroxymethyl-2-cyclohexen-1-yl] amino]-α-D-glucopyranosyl)-(1-4)-0-α-glucopyranosyl-(1→4)-D-glucopyranose from culture broths.) DE-OS 2719912.

Scouten, W.H. (ed.). 1981. Affinity chromatography. Bioselective adsorption on inert matrices. J. Wiley and Sons, New York.

Trawanski, H. 1972. Zentrifugen und Hydrozyklone (Centrifuges and hydrocyclones), pp. 204–224. In: Ullmans Encyklopädie der technischen Chemie, vol. 2. Verlag Chemie, Weinheim.

Verrall, M.S. (ed.) 1985. Discovery and Isolation of Microbial Products. Ellis Horwood Ltd., Chichester.

Vogelpohl, A. and E.U. Schlünder. 1972. Trocknung fester Stoffe (Drying of solids), pp. 698–721. In: Ullmans Encyklopädie der technischen Chemie, vol. 2. Verlag Chemie, Weinheim.

Weiss, J. 1985. Handbuch der Ionenchromatographie. VCH Verlag, Weinheim.

Whitworth, D.A. 1974. Assessment of an industrial homogenizer for protein and enzyme solubilization from spent brewery yeast. Compt. Rend. Trav. Lab. Carlsberg. 40: 19–32.

Wolf, F.J. 1974. Outline for fermentation technology. MIT, Cambridge, MA.

7

Organic feedstocks produced by fermentation

7.1 GENERAL

The chemical industry makes use of a variety of simple organic compounds as feedstocks, that is, raw materials, for synthetic processes. Due to the low cost of petrochemical products in the post–World War II era, little attention was given to studies involving the microbial production of organic feedstocks from plant substances. Since the mid-1970's, due to political and economic factors, petroleum and natural gas have become scarce and more attention has turned to microbial production of feedstocks. Of the 180×10^9 tons of plant material annually produced on earth, only 1–2% is utilized for animal and human nutrition and 1% for energy and fiber production.

The requirements for a cost-effective fermentation of organic feedstocks from plant carbohydrates are as follows:

- Low transportation costs of raw materials.
- Low costs to convert polymers (wood, cellulose, hemicellulose, starch) to usable mono- and disaccharides.
- Use of mixed cultures in order to catabolize different substrates and to convert them into the desired metabolite.
- Use of thermophilic strains to save costs for cooling, to bring about higher conversion rates, and to reduce contamination risks.
- Due to the high energy demand for aeration, anaerobic processes are preferable.
- Amenable to a continuous process.
- Low recovery and concentration costs.

7.2 ETHANOL

For thousands of years, ethanol has been produced for human consumption, and for at least a thousand years it has been possible to make concentrated alcoholic drinks by means of distillation. Ethanol for use as a chemical feedstock was produced by fermentation in the early days of industrial microbiology; however, for many years it has been obtained by chemical means instead, primarily through the catalytic hydration of ethylene. In recent years, attention has turned again to the production of ethanol for chemical and fuel purposes by fermentation. For example, in three major countries the following amounts of industrial ethanol were produced by fermentation in 1986: United States, 2.5×10^9 liters; Federal Republic of Germany, 1×10^8 liters; Brazil, 1.1×10^{10} liters. In countries with large agricultural areas, such as Brazil, South Africa, and the United States, intensive studies are being conducted on the production of ethanol from carbohydrates such as sucrose and starch. The objective of such research is to use ethanol for automobile fuel (generally mixed with gasoline). In some countries, fermentation ethanol is also used to produce ethylene and other petrochemicals.

The efficiency of energy conversion by ethanol fermentation varies considerably depending on the starting material. For instance, under optimal conditions the efficiency of energy yield (ratio of energy demand to energy produced) is as follows: sugar beet, 86%; potatoes, 59%; corn, 25%; cassava, 50%; sugar cane, 66%. Ethanol is more expensive than petroleum when used as a motor fuel but some countries, for political or internal economic reasons, subsidize ethanol production.

In Brazil, as early as 1982, about 30% of petroleum imports were replaced by ethanol production from sugar cane (5.2×10^6 m^3), involving 60–80 separate fermentation plants. In 1986 in the United States, 65 separate fermentation plants were in operation, with a total capacity of 2.6×10^6 m^3 per year, using corn starch as the starting material. The capacity of these plants varied from 20,000 to 570,000 m^3. However, the total capacity was only about 25% of that which had been planned and some of the plants which had opened were unsuccessful and closed. The main problems were lack of technical and microbiological expertise and the high price of corn starch (which constituted 60–70% of the total production cost).

Biosynthesis of ethanol

Both yeasts and bacteria have been used for the production of ethanol. Among the bacteria, the most widely used organism is *Zymomonas mobilis. Saccharomyces cerevisiae* is the most commonly used yeast but *Kluyveromyces fragilis* has also been employed.

Under aerobic conditions and in the presence of high glucose concentrations, *Saccharomyces cerevisiae* grows well, but produces no alcohol. Under anaerobic conditions, however, growth slows and pyruvate from the glycolytic pathway is split with pyruvate decarboxylase into acetaldehyde and CO_2 (Figure 7.1). Ethanol is then produced from the acetaldehyde by reduction with alcohol dehydrogenase.

Batch systems for ethanol production are started aerobically to obtain maximum biomass, since if anaerobic conditions begin too early, the population density is not high enough to obtain a good conversion rate. Forced aeration may even be necessary for a short time in order to avoid yield losses. In continuous processes, optimal yeast growth and ethanol production are carried out under sugar limitation (< 1 g/l) and in a microaerobic environment (0.2–5 mg O_2/g dry substance·h). Theoretically, from one gram of glucose, 0.511 grams of ethanol can be obtained. When pure substrates are fermented, the yield is 95% and reduces to 91% when industrial-grade starting materials are used. One hundred grams of pure glucose will yield 48.4 grams of ethanol, 46.6 grams of CO_2, 3.3 grams of glycerol, and

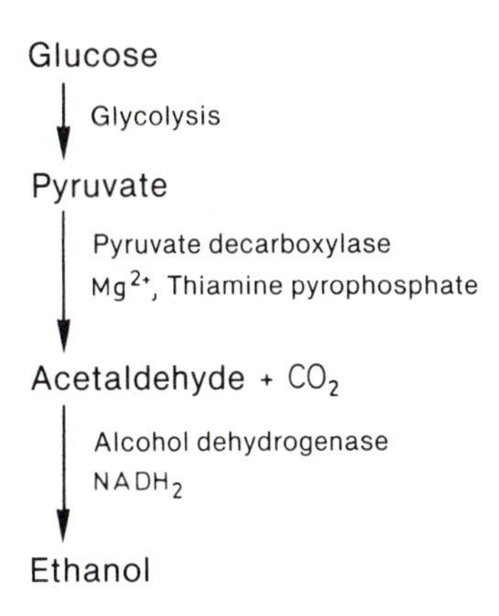

Figure 7.1 Biosynthesis of ethanol

1.2 grams of biomass (yeast cells). If corn starch is used, 100 kg starch (corresponding to 180 kg of corn) ferments to 51.5 grams of ethanol.

Ethanol is inhibitory at high concentrations and the alcohol tolerance of the yeast is critical for high yields. The ethanol tolerance of different strains varies considerably. As the concentration of ethanol increases, the growth rate is first reduced, whereas at higher concentrations the biosynthesis of ethanol itself is inhibited. However, yeast is more sensitive to endogenously produced ethanol than to that added from external sources to the fermentation system. Growth generally ceases at 5% ethanol (volume ethanol in volume water = vv) and the production rate is reduced to zero at 6–10% (vv). Using molasses (12% sucrose, weight per volume water = wv), sucrose is normally converted to 6% (vv) ethanol within 36 hours.

With pure sugar solutions, the ethanol concentration can reach as high as 10%, although ethanol-tolerant mutants have been isolated that can produce 12-13% ethanol and are currently being scaled up to full production. The bacterium *Zymomonas mobilis* has in recent years come under increasing study because it has a number of potential advantages:

- Osmotic tolerance to higher sugar concentrations (up to 400 g/l).
- Relatively higher ethanol tolerance (up to 130 g/l).
- Higher specific growth rate than yeast (growth rate, μ of 0.27 compared to 0.13 for yeast; laboratory culture studies).
- Anaerobic carbohydrate metabolism is carried out through the Entner-Doudoroff pathway, where only one mole of ATP is produced per mole of glucose used, thus reducing the amount of glucose that is converted to biomass rather than ethanol.

The pH optimum of the bacterial ethanol fermentation is considerably broader (pH 5–7) and the temperature optimum is higher (30°C). Even at 37°C the yield of ethanol from glucose is 97% of theoretical. A comparison of the kinetic parameters for *Zymomonas* and *Saccharomyces* shows the following: ethanol formation rate (g/g·h) 2.9 times higher, growth rate (μ) 2.4 times higher, and glucose uptake rate 2.6 times higher.

In order to increase even more the productivity, both the bacteria and yeast processes are being carried out in large scale by continous fermentation. The optimal amount of cell recycling is under study. Maximal productivity with glucose as carbon source that has been reached is about 82 g/l·h for yeast and 120 g/l·h for *Zymomonas*. A further possibility is the use of immobilized cells. In a direct comparison of yeast using molasses as substrate, the following results have been obtained: batch process, 2.0 g/l·h ethanol; continuous process, 3.35 g/l·h; immobilized cells under continuous conditions, 28.6 g/l·h. The immobilized cell process has already been placed in large-scale production in Japan.

Ethanol production process

Ethanol is produced in three steps, each of which must be optimized: 1) Preparation of the nutrient solution; 2) fermentation; 3) distillation of ethanol. Figure 7.2 shows the stages in the production of ethanol from corn meal.

Preparation of the nutrient solution Three types of substrates are used in the ethanol fermentation: 1) starch-containing roots, tubers, or grains; 2) molasses or juice from sugar cane or sugar beet; 3) wood or waste products from processed wood. The production of alcohol from milk whey has been tested but is not yet in use commercially.

The most important root starch is derived from the tropical plant *Manihot esculenta*, from which cassava, manioc or tapioca flour is obtained. In Brazil, 28×10^6 tons of this plant are harvested annually.

Since *Saccharomyces cerevisiae* has no amylases, the starch must be hydrolysed. The roots (containing 20–35% starch and 1–2% protein), are first ground, squeezed, and dried. The starch is liquefied by boiling under pressure, and then

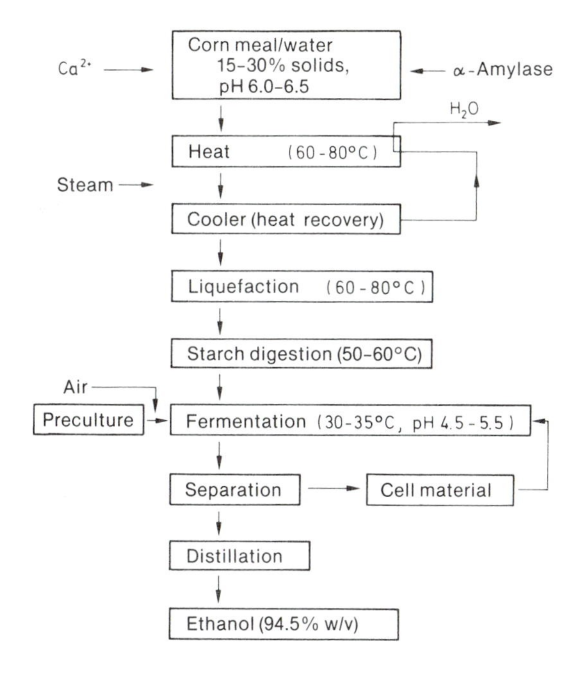

Figure 7.2 Steps in ethanol production from a substrate containing starch

hydrolysed enzymatically. By adding cellulose-splitting enzymes of different microbial origin, the proportion of reducing sugars can be increased in the cassava paste.

The starch-containing grain used in Asia is rice, in the United States mainly corn but also millet in some areas, in Europe potatoes in addition to other grains. Grain can be used whole or, in the case of corn, coarsely ground; the kernels are soaked at 40–50°C for several hours, then steeped and liquefied.

Continuous liquefaction and saccharification processes using first steam injection (3 min at 150°C) and then vacuum cooling are used in modern plants. Some α-amylase is added before heating in order to decrease the viscosity which develops after the steeping process. For glucose production, glucoamylase and α-amylases are added after the heating and cooling process. Approximately 1 l commercial α-amylase and 3.5 l glucoamylase are added for 1 ton of starch.

Molasses, a byproduct of sugar crystallization, is commonly used today as an ethanol source, but may become less important in the future since it has better uses as animal feed or in other fermentations.

When molasses is made from sugar beets, the juice is extracted from the beet chips with hot water. With sugar cane, the cane juice is released by means of presses. The residue from the pressing of sugar from sugar cane stems is called bagasse. Eighty percent of the bagasse formed can be burned as an energy source in the distillation process and 20% more can be fermented after chemical hydrolysis. In the Federal Republic of Germany a hybrid plant which is a cross between sugar beet and turnip is being developed as a source of carbohydrate for ethanol production.

Wood has not yet been used in the commercial production of ethanol. Because of the large amount of waste wood available, the direct fermentation of wood which has been hydrolyzed with cellulases would be of considerable importance.

Sulfite waste liquor is another potential source of sugar. This is formed during production of paper from conifers and contains fermentable hexoses. Sulfite waste liquor from deciduous trees cannot be used for commercial processes because of the high fraction of the sugar which is in the form of pentoses rather than hexoses.

Fermentation Continuous fermentations are only slowly being brought into large-scale operation. In the United States, among 10 large-scale plants using yeast, four are operating continuously, one of which also employs cell recycling. A few continuous plants have also been started in Brazil. A Danish system has been described in detail: The nutrient solution was molasses with diammonium phosphate additive. The pH value was adjusted to 5 with H_2SO_4 and pasteurization was then carried out. The fermentation temperature was 35°C and the yeast production was 10 g/l. After the fermentation, the cells were separated by centrifugation and channeled back into the first fermenter. After 10.5

hours, the yield in the first fermenter was 6.1% (vv) alcohol with 1% residual sugar, whereas in the second fermenter the yield was 8.4% (vv) with 0.1% residual sugar.

In the Federal Republic of Germany, ethanol production with *Zymomonas mobilis* is being carried out in two parallel bioreactors, each of which has a volume of 70 m^3. The fermentation operates continuously but without cell recycling, using as a starting material a starch fraction which was a byproduct of enzymatic glucose production. The ethanol concentration ranged from 7–8%. The process could be operated for up to three weeks before a lactobacillus contaminant became established and the process had to be restarted from scratch.

Batch fermentation is more commonly employed for ethanol production. Production is carried out in batch processes (fermenter volume of 600 m^3) with either starch hydrolysate (up to 20% dry weight) or molasses, using a 3% inoculum (cell density 3×10^6/ml). Within 12 hours, *Saccharomyces* produces 10% (vv) ethanol with 10–20 g cell dry weight/l when the process is carried out at 35–38°C, pH 4.0–4.5; the maximal productivity was 1.9 g/l·h. The short fermentation time was accomplished by use of cell recycling; 80% of the cells were removed in a separator and brought back again into the fermenter. When high quality molasses is used, the maximal yield is 95% of the theoretical. Table 7.1 shows a comparison of the operating parameters of three different types of fermentation systems, batch, continuous, and continuous with cell recycling.

Before distillation, the cell mass is separated by centrifugation or sedimentation. Figure 7.3 shows the layout of a modern system.

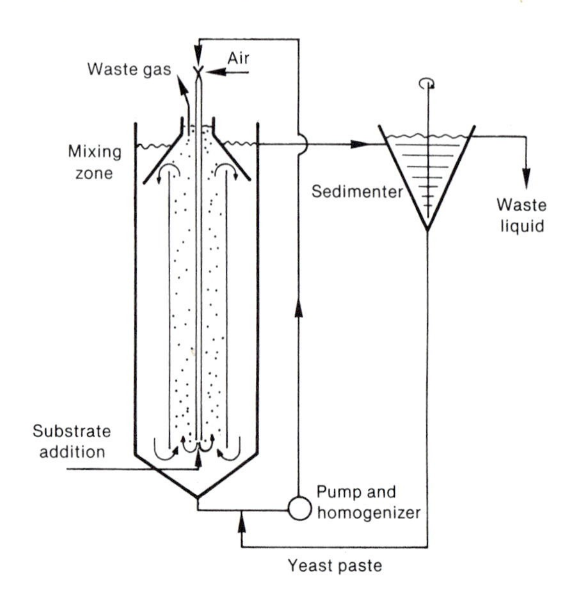

Figure 7.3 ALCO-FLOC system using air suction and cell flocculation (Meyrath, 1979)

Energy balance of the ethanol fermentation Since ethanol is produced primarily as a fuel and energy source, the energy balance of the whole process determines its economic viability. A summary of the energy requirements of the different stages in the process for ethanol production from various substrates is given in Table 7.2.

It can be seen that the product-recovery stage, the distillation of ethanol, is the single most energy-demanding step of the whole process. Because of this, improvements in the distillation process will have more impact on the success of the process than will improvements in the fermentation itself. If the energy yield of the ethanol produced is related to the total energy inputs from the various stages of the process, it is found

Table 7.1 Comparison of several kinds of fermentation systems for ethanol production

Fermentation process	Sugar concentration %	Cell concentration g/l	Ethanol concentration g/l	Specific productivity g/g cells·hr	Volume-specific productivity g/l·h
Batch	16.7	21.3	85.6	0.42	11.8
Continuous	15.6	19.7	79.1	0.72	14.1
Continuous with cell recycling	16.7	100	85.6	0.42	42.5

(Maiorella et al., 1984)

Table 7.2 Energy (MJ/1 of pure ethanol) required to produce absolute alcohol

Process Stage	Substrate		
	Beets	Cane	Starchy raw materials
Digestion/ hydrolysis,			
batch	4 – 5	–	7 – 8
continuous	–	–	2
Cane mill	–	1.1 – 1.5	–
Extraction	0.8 – 1	2 – 3	–
Fermentation,			
batch	0.06	0.06	0.06
continuous	0.1	0.1	0.1
Distillation,			
single-stage	10 –13	10 –13	10 –13
optimized	5 – 7	5 – 7	5 – 7
Process,			
conventional	16	13	19
optimized	7	7	8

(Misselhorn, 1979)

that there is actually either an approximate balance or a net energy loss. This emphasizes the importance in optimizing all steps in the process to the utmost. Because the profit margin is so low, such optimization is economically much more important for the production of an energy source than it is for the production of a fine chemical or antibiotic. Ethanol with a purity of 92.4% is used as a solvent in the cosmetic, pharmaceutical, and chemical industry and at 99.2% purity as a motor fuel. In order to produce the virtually water-free product, better distillation methods would be desirable.

7.3 ACETONE/BUTANOL FERMENTATION

Pasteur first observed the production of butanol by bacteria in the 19th century. Before World War I, microbial processes were developed for the purpose of obtaining butadiene for **synthetic rubber**, at which time Chaim Weizmann performed fundamental research on the fermentation of *Clostridium acetobutylicum* for the production of acetone, butanol, and ethanol. During World War I, the product of interest was acetone used for the production of the explosive trinitrotoluene (TNT), but after the war butanol became more important, used primarily for the production of nitrocellulose lacquers. After World War II, petroleum-based processes replaced biological fermentation processes, and the majority of plants with fermenters less than 200 m^3 in volume were closed. However, some plants of this size are still functioning today in countries where economic, political, geographical, or climatic factors are favorable, such as in Taiwan and South Africa.

Butyric acid, butanol, acetone, and isopropanol are obtained through clostridial fermentation of starch, molasses, sucrose, wood hydrolysates, and pentoses. The relative proportions of each of these products in the fermentation depends on the bacterial strain used and on the fermentation conditions. Three fermentation types can be categorized, according to their fermentation products:

- Acetone-butanol fermentation with *Clostridium acetobutylicum*
- Butanol-isopropanol fermentation with *Clostridium butylicum*
- Butyric acid-acetic acid fermentation with *Clostridium butyricum*.

Biosynthesis

Figure 7.4 shows the biosynthesis of the main fermentation products: acetone, butanol, butyric acid, isopropanol, and acetic acid. The butanol referred to here is entirely n-butanol. Hydrogen and carbon dioxide are also produced. In addition, ethanol can be produced through the reduction of acetaldehyde. The organism forms a net yield of 2 ATP in the steps from glucose to pyruvate and this is the total energy released when no acetic acid is produced.

Only the acetone-butanol fermentation is of current economic interest. Butanol at a concentration of less than 0.5% has no influence on the cells, whereas at higher concentrations it causes

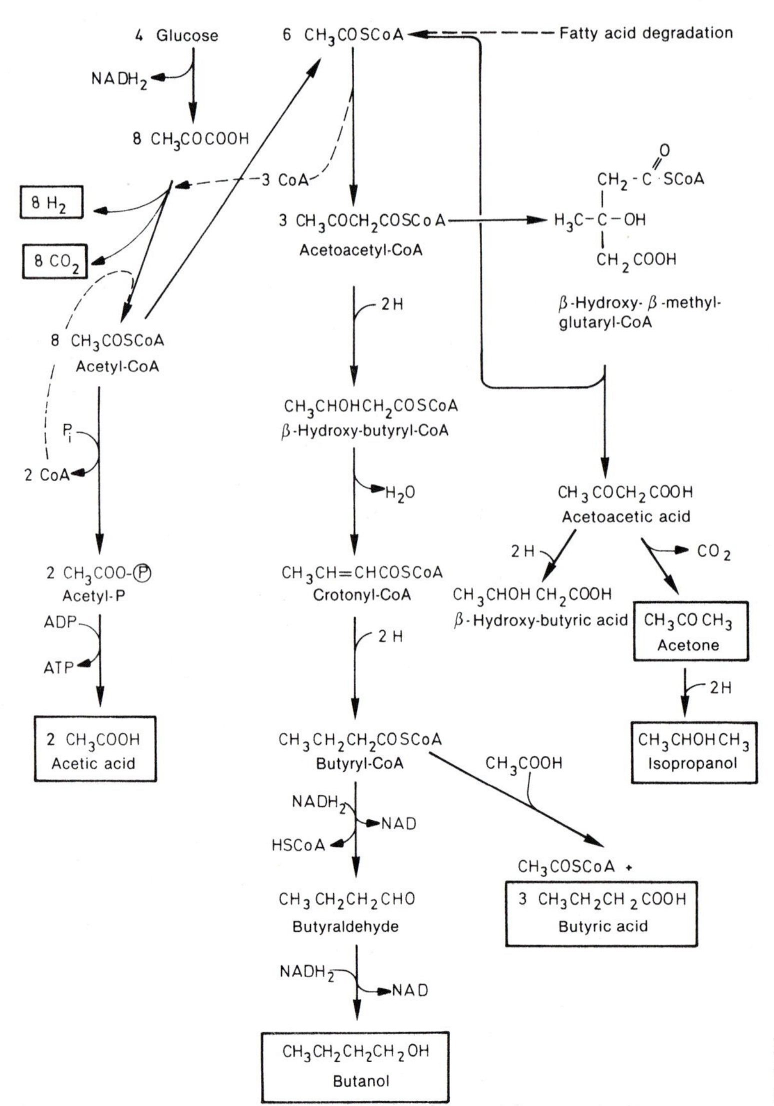

Figure 7.4 Pathway of biosynthesis of butanol, butyric acid, acetic acid, acetone, and isopropanol

damage to the phospholipids of the cell membrane; at concentrations above 1.3% butanol production ceases. In addition, depending on the strain and the fermentation conditions, autolysins may be produced that cause the cells to lyse.

Laboratory studies are underway to develop a continuous fermentation process for butanol, using either cell suspensions or immobilized cells.

Production process

One of the few remaining systems for acetone-butanol production still in use is in South Africa. Stock cultures of *Clostridium acetobutylicum* are stored as spores in sand for up to 30 years. Both the production of the spores and the preparation of the inoculum are critical. Although production fermenters must be inoculated only in the low inoculum ratio of 1:3000, build-up of the inoculum over several stages has the following advantages: greater resistance to contamination, shorter fermentation times, greater yields, greater fermentable sugar concentrations.

After sterilization, the fermenters (12×90 m^3 fermenters) are gassed with CO_2, and before and after inoculation the fermenter contents are stirred with CO_2. The base for the fermentation is molasses plus corn steep liquor with a beginning pH of 5.8–6.0 at 34°C.

The 36-hour fermentation has three phases (Figure 7.5):

1. During the first 18 hours, the pH value decreases to 5.2 due to the formation of acetic acid and butyric acid.
2. During the next 18 hours, the pH value increases through the metabolism of these acids to acetone and butanol. When excess $NADH_2$ is present, the cells take up butyric acid and reduce it to butanol.
3. In the next stage, growth and solvent production stop. The pH value (5.8) remains constant. The products can now be recovered.

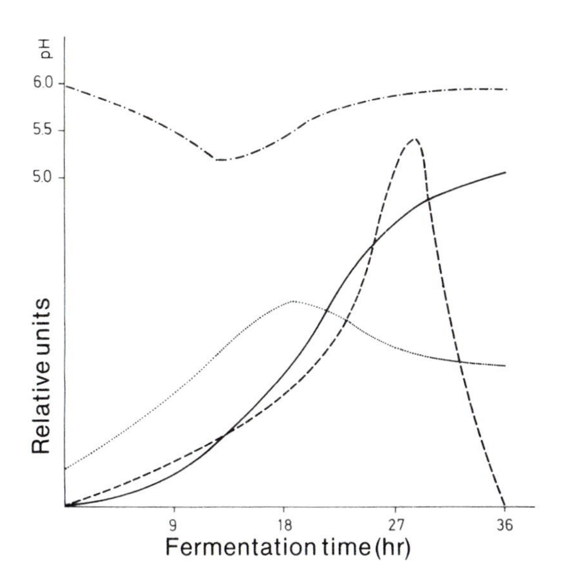

Figure 7.5 The kinetics of the acetone-butanol fermentation (Spivey, 1978), —— solvent, ···· titratable acids, – – – gas production, — · — pH value

Table 7.3 Yields from a 90m³ fermenter

Product	% of converted sugar	Amount produced
Butanol	30 (ratio 6:3:1)	1053 kg
Acetone		526 kg
Ethanol		175 kg
CO_2	50	2900 kg
H_2	2	117 kg

(Spivey, 1978)

Contamination due to bacteriophages and lactobacilli is a common problem and absolute sterility is therefore a necessity. Table 7.3 shows typical yields of a 90 m^3 fermenter containing 5.85 tons of fermentable sugar.

CO_2 produced during the fermentation is recovered and converted to liquid CO_2 or dry ice. Acetone, butanol, and ethanol are recovered through continuous distillation and fractionation. The residue left after the distillation may be dried and utilized as animal feed.

7.4 GLYCEROL

Glycerol has wide uses in commerce and is also a starting material for explosives manufacture. During both world wars, glycerol was produced by microbial processes, but today these processes are no longer of any commercial value. However, the glycerol fermentation is of theoretical interest in demonstrating how modification of fermentation conditions can lead to modification of product type.

Glycerol is formed by yeast along with ethanol during the alcoholic fermentation (Figure 7.6). Normally, the amount of glycerol formed is tiny, but by modifying the fermentation balance the amount of glycerol produced can be greatly increased. This is done in the following way: An intermediate in the ethanol fermentation is acetaldehyde, and if sodium bisulfite is added, an acetaldehyde-sulfite complex is formed. Normally, the $NADH_2$ formed during the first part of glycolysis is reoxidized by acetaldehyde to form ethanol, but if the acetaldehyde is removed in this sulfite complex, the $NADH_2$ is available

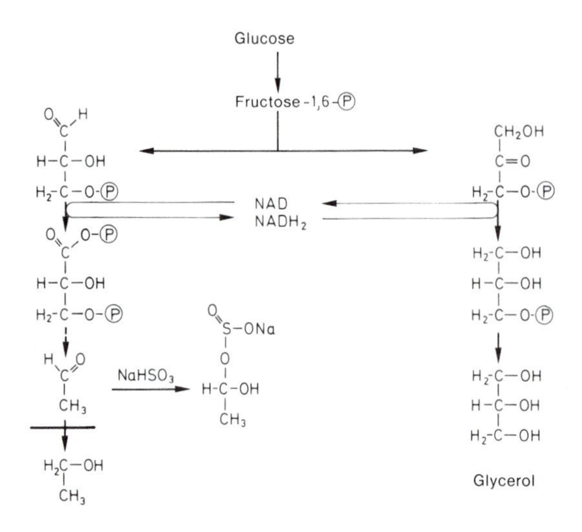

Figure 7.6 Glycerol biosynthesis

for the reduction of dihydroxyacetone phosphate (also formed during glycolysis) via the enzyme glyceraldehyde phosphate dehydrogenase to glycerol phosphate (which is dephosphorylated to glycerol). Through the dismutation of two acetaldehyde molecules into acetic acid and ethanol, some glycerol enrichment is also attained by adding alkali (e.g. Na_2HPO_4).

The balance of the glycerol process is as follows:

Since additional byproducts are formed, yields never exceed 20–30% of the sugar used, in fermenters up to 1000 m^3, during a fermentation running for 2–3 days. Osmotolerant yeasts are best for glycerol production and the strains *Saccharomyces rouxii, Torulopsis magnoliae* and *Pichia farinosa* are commonly used.

An interesting process for glycerol production has been developed in recent years in Israel, using the halophilic alga *Dunaliella salina*. This alga, which is commonly present in hypersaline lakes and salty lagoons, synthesizes glycerol as an intracellular solute which balances the osmotic pressure brought about by the high salt concentration of its environment. The higher the external salt concentration, the higher is the intracellular glycerol concentration. However, if the salt concentration is suddenly reduced, the excess glycerol is excreted by the alga.

In the glycerol production process, the alga is first cultivated photosynthetically in a medium containing a high salt concentration. After growth is completed, the cells are concentrated and transferred to a low-salt medium. During a short incubation period in the low-salt medium, the glycerol is excreted in a fairly concentrated form and purification can be carried out.

REFERENCES

Bringer, S. and H. Sahm. 1985. Jetzt industriereif: Ethanol-Produktion durch Bakterien. (Now ready for large-scale application: ethanol production by bacteria.) BioEngineering 1/85:30-36.

Drawert, F., W. Klisch, and G. Sommer. 1987. Gärungsverfahren-Ethanol Wein, Bier. (Ethanol fermentation processes: wine, beer.), pp. 321-362. In: Präve et al. Handbuch der Biotechnologie. Oldenbourg Verlag, München.

Frick, C. and K. Schügerl. 1986. Continuous acetone-butanol production with free and immobilized Clostridium acetobutylicum. Appl. Microbiol. Biotechnol. 25:186–193.

Jones, D.T. and D.R. Woods. 1986. Acetone-butanol fermentation revisited. Microbiol. Rev. 50:484–524.

Klyosov, A.A. 1986. Review. Enzymatic conversion of cellulosic materials to sugar and alcohol. The technology and its implications. Appl. Biochem. Biotechnol. 12:249 ff.

Kosaric, N., A. Wieczorek, G.P. Cosentino, R.J. Magee, and J.E. Prenosil. 1983. Ethanol fermentation, pp. 257–385. In: Rehm, H.J. and G. Reed (eds.). Biotechnology 3. Verlag Chemie, Weinhem.

Linden, J.C., A.R. Moreira, and T.G. Lenz. 1985. Acetone and butanol, pp. 915–931. In: Moo-Young, M. (ed.). Comprehensive Biotechnology 3. Pergamon Press, Oxford.

Maiorella, B.L. 1985. Ethanol, pp. 861–914. In: Moo-Young, M. (ed.). Comprehensive Biotechnology 3. Pergamon Press, Oxford.

Maiorella, B.L., H.W. Blanch, and C.R. Wilke. 1984. Biotechnology Report. Economic evaluation of alternative ethanol fermentation processes. Biotech. Bioeng. 26:1003–1025.

Meyrath, J. 1979. Das ALCO-FLOC-System, eine kontinuierliche und zuverlässige Methode zur Äthanol-produktion mit extrem hoher Produktivität (The ALCO-FLOC system, a continuous and reliable method for the manufacture of ethanol at extremely high productivity), pp. 107–116. In: Dellweg, H. (ed.), 4th Symp. Technische Mikrobiologie. Verlag Versuchs- und Lehranstalt für Spiritusfabrikation, Berlin.

Misselhorn, K. 1979. Äthanolherstellung unter energiewirtschaftlichen Aspekt (Ethanol production under energy-efficient conditions). Stand der Technik, pp. 47–55. In: Dellweg, H. (ed.), 4th Symp. Technische Microbiologie. Verlag Versuchs- und Lehranstalt für Spiritusfabrikation, Berlin.

Murtagh, J.E. 1986. Fuel ethanol production—the U.S. experience. Proc. biochem. April, 61–65.

Rogers, P.L., K. J. Lee, M.L. Skotnicki, and D.E. Tribe. 1982. Ethanol production by *Zymomonas mobilis*. Adv. Biochem. Eng. 23:37–84.

Schlote, D. and G. Gottschalk. 1986. Effect of cell cycle on continuous butanol-acetone fermentation with *Clostridium acetobutylicum* under phosphate limitation. Appl. Microbiol. Biotechnol. 24: 1–5.

Spencer, J.F.T. and D.M. Spencer. 1978. Production of polyhydroxy alcohols by osmotolerant yeasts, pp. 393–425. In: Rose, A.H. (ed.), Economic microbiology, vol. 1, Primary products of metabolism. Academic Press, London.

Spivey, M.J. 1978. The acetone/butanol/ethanol fermentation. Proc. Biochem. 13: 2–4,25.

Stewart, G.G. and I. Russell. 1985. Modern brewing technology, pp. 335–381. In: Moo-Young, M. (ed.). Comprehensive Biotechnology 3. Pergamon Press, Oxford.

8

Organic acids

8.1 INTRODUCTION

Organic acids find wide use both as additives in the food industry and also as chemical feedstocks. Fermentation processes play a major role in the production of most organic acids. All acids of the tricarboxylic acid cycle can be produced microbiologically in high yields. Some acids derived indirectly from the Krebs cycle, such as itaconic acid, can likewise be produced. Other organic acids are derived directly from glucose (for example, gluconic acid), or they are formed as end products from pyruvate or ethanol (for example, lactic acid and acetic acid).

Except for the production of citric acid which is manufactured entirely by fermentation, there is frequently great competition between microbiological and chemical processes for production of the various organic acids. For some acids (lactic acid, acetic acid) microbiological and chemical methods for preparation are simultaneously in use. For others (α-ketoglutaric acid, malic acid), fermentation processes have been developed which are not used commercially, either due to lack of sufficient need for the acid or for economic reasons. Table 8.1 lists several acids that have been produced microbially or for which processes have been extensively developed. In this chapter, we will discuss the most important processes of commercial interest.

8.2 CITRIC ACID

Citric acid has been known as a natural plant substance since the end of the nineteenth century, and since 1893 scientists have known that it is produced by filamentous fungi. In 1923, the first practical fermentations for the production of this organic acid were started, using microorganisms growing in surface culture. Fermentation using deep-vat fermenters began in the 1930's. Although in South America and Mexico there are some small factories where citric acid is isolated from unripe citrus fruits (lemons contain 7–9% citric acid), today over 99% of the world's output of citric acid is produced microbially.

Table 8.1 Practical microbiological processes for organic acid production

Organic acid	Organism	Commercially produced	Substrate	Process	Yield %	Use of product	References
Fumaric acid	*Rhizopus* sp., *Candida*	Yes	Glucose n-Alkane	3 days, 33°C 7 days, 30°C	65 84	Resin	Buchta, 1983b
Propionic acid	*Propionibacterium*	Yes	Lactose Glucose Starch	8–12 days 30°C	60	Perfume Fungicide	Buchta, 1983b
Malic acid	*Leuconostoc brevis*, *Candida*	No	Fumaric acid n-Alkane	24 h	99 72	Food	Buchta, 1983b
α-Ketoglutaric acid	*Candida*, *Aerobacter*	No	n-Alkane Glutamic acid	40h	67 46		Rose, 1978
5-Ketogluconic acid	*Acetobacter suboxydans*	Yes	Glucose	5–6 days	85	L-Tartaric acid	Buchta, 1983b
2-Ketogluconic acid	*Serratia marcescens*	Yes	Glucose	16 h	95–100	Isoascorbic acid	Buchta, 1983b

The following amounts of citric acid have been produced microbially:

1929	5,000 tons
1953	50,000 tons
1976	200,000 tons
1979	220,000 tons
1982	350,000 tons

The fermentation processes are carried out either in surface culture or in fermenters up to 220 m^3 in volume. Citric acid is marketed as citric acid-1-hydrate or as anhydrous citric acid. Most citric acid (60% of the total) is used in the food and beverage industry. The flavor of fruit juices, fruit juice extracts, candy, ice cream, and marmalade is enhanced or preserved by the addition of citric acid.

The pharmaceutical industry (10% of total usage) uses iron citrate as a source of iron and citric acid as a preservative for stored blood, tablets, ointments, and in cosmetic preparations. In the chemical industry (25% of total usage), citric acid is used as an antifoam agent, as a softener, and for the treatment of textiles. In the metal industry, pure metals are produced as metal citrates.

Citric acid is being used more and more in the detergent industry because it can replace polyphosphates. The higher cost of citric acid as compared to polyphosphates formerly restricted its use; however, detergents containing polyphosphates have been prohibited in some locations and the polyphosphates have been completely replaced with citric acid.

Strains for citric acid production

Many strains excrete traces of citric acid as a metabolite of primary metabolism. Examples are *Aspergillus niger, A. wentii, A. clavatus, Penicillium luteum, P. citrinum, Mucor piriformis, Paecilomyces divaricatum, Citromyces pfefferianus, Candida guilliermondii, Saccharomycopsis lipolytica, Trichoderma viride, Arthrobacter paraffineus,* and *Corynebacterium* sp. However, only mutants of *Aspergillus niger* are used for commercial production. Compared to *Penicillium* strains, the aspergilli produce more citric acid per unit time. Moreover, the production of undesirable side products, such as oxalic acid, isocitric acid, and gluconic acid, can be more efficiently suppressed in these mutants.

Biosynthesis

Citric acid (2-hydroxypropane-1,2,3,-tricarboxylic acid) is a primary metabolic product and is formed in the tricarboxylic acid cycle. Glucose is the main carbon source used for citric acid production. In many organisms, 80% of the glucose used for citric acid biosynthesis is broken down by reactions of the Embden-Meyerhof-Parnas (EMP) pathway and 20% by reactions of the pentose phosphate cycle. During the growth phase, the relationship between these two pathways is 2:1. In citric-acid producers, the enzymes of the EMP-pathway are present throughout the fermentation process and the activity of the pathway is regulated in a positive manner with phosphofructokinase and in a negative manner by pyruvate kinase. When pyruvate is decarboxylated with the formation of acetyl-CoA, the acetate residue is channeled into the tricarboxylic acid cycle. During the idiophase, all enzymes of the Krebs cycle are expressed except α-ketoglutarate dehydrogenase. The citrate synthase activity (condensing enzyme) is increased by a factor of 10 during the production of citric acid, while the activities of the enzymes which catabolize citric acid, such as aconitase and isocitrate dehydrogenase, are sharply reduced as compared to their activity in the trophophase. One of the three isocitrate dehydrogenase isozymes, a mitochondrial enzyme that is specific for NADP, is inhibited by glycerol that accumulates during the spore germination process. In addition, citric acid production is inhibited by high intracellular concentrations of ammonium ion.

Citrate synthase cannot be solely responsible for maintaining the activity of the tricarboxylic acid cycle, since the cycle would cease if citric acid were removed. Distinct sequences which replenish the tricarboxylic acid cycle intermediates (anaplerotic sequences) must thus exist in the production phase, as outlined in Figure 8.1. The first anaplerotic enzyme present in *Aspergillus* is a pyruvate carboxylase (reaction 4, Figure 8.1), which converts pyruvate and CO_2 into oxalacetate, inorganic phosphate, and ADP (while consuming ATP). The reaction is dependent on Mg^{2+} and K^+ ions; acetyl-CoA is not needed for the reaction, in contrast to its requirement in the metabolic reactions of other microorganisms. Pyruvate carboxylase is the key enzyme for citric acid production.

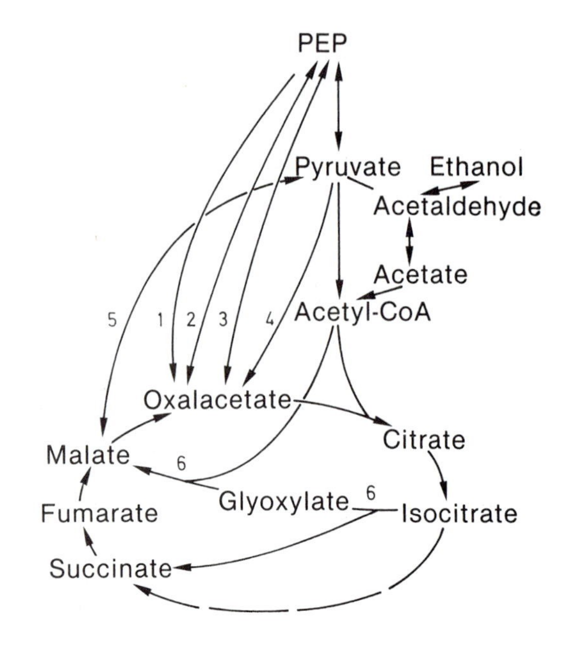

Figure 8.1 Anaplerotic reactions which connect with the tricarboxylic acid cycle

The second anaplerotic sequence involves a phosphoenol pyruvate carboxykinase (reaction 2, Figure 8.1), which converts PEP and CO_2 into oxalacetate and ATP in the presence of ADP. The system requires Mg^{2+} or Mn^{2+} and K^+ or NH_4^+.

If acetate or higher aliphatic compounds such as n-alkanes (C_9–C_{23}) are used as carbon source, a third anaplerotic sequence is present in *Aspergillus niger*. In the absence of glucose, the glyoxylate cycle (reaction 6, Figure 8.1) is operable: isocitrate lyase is induced and malate synthase is present. If glucose is added, the glyoxylate cycle is then repressed, although isocitrate lyase is still partially active.

Yields

In the trophophase, part of the added glucose is used for the production of mycelium and is converted through respiration into CO_2. In the idiophase, the rest of the glucose is converted into the organic acids and during this phase there is a minimal loss through respiration. Table 8.2 gives data on citric acid yields in various processes. The theoretical yield is 123 g citric acid-1-hydrate or 112 g anhydrous citric acid per 100 g sucrose. However, such yields are not obtained in practice, due to the loss during the trophophase.

Nutrient media

The media used for citric acid production have been highly perfected over the many years that the commercial process has been underway.

Carbohydrate source A 15–25% sugar solution is converted during fermentation. A variety of starting materials can be used as carbohydrate sources: Starch from potatoes, starch hydrolysates, glucose syrup from saccharified starch, sucrose of different levels of purity, sugar cane syrup with two-thirds of the sucrose converted into invert sugar, sugar cane molasses, sugar beet molasses.

If starch is used, amylases formed by the producing fungus or added to the fermentation broth hydrolyze the starch to sugars. If hydrolysates or syrup are used, a preliminary treatment with either precipitants or cation exchangers must be carried out to remove cations. For elimination of metals, molasses is generally treated before sterilization with calcium hexacyanoferrate or is purified by cation exchangers. Although the effect of the hexacyanoferrate is primarily to bring about metal precipitation, there are also studies which show a direct influence on the fermentation, since an excess of hexacyanoferrate over that needed to complex metals increases the citric acid yield.

Production strains are optimized based on the carbon sources used. A strain which produces well with one carbohydrate source generally cannot be used with another starting material without a substantial yield reduction.

There are no analytical methods that permit an evaluation of a molasses source for its use in citric acid production; each batch of molasses must be given a preliminary fermentation test. Firms which use molasses as starting material generally optimize molasses treatment and the make-up of the culture media in model fermen-

Table 8.2 Citric acid yields in different fermentation processes

Microorganism	Raw material (Sucrose, %)	Process	Actual yield		Theoretical maximum yield kg%/kg Carbon source	% of Theoretical maximum yield
			kg % Raw material	kg % Carbon source		
A. niger	Beet molasses (50%)	Surface	42	84	61.4	68.4
A. niger	Beet molasses (50%)	Submerged	41	82	61.4	66.8
A. niger	Beet molasses (54%)	Submerged	46	85	66.3	69.4
Yeast	n-Alkane	Submerged	165	82	247.0	66.8
Yeast	Methanol	Submerged	40	45	109.4	36.6
Yeast	Ethanol	Submerged	60	48.4	145.8	41.1

(Siebert and Schulz, 1979)

tations of up to 30 m^3, before the molasses is used in production.

Trace elements Extensive research on the role of trace elements in the citric acid fermentation was begun in the 1940's and has continued as analytical methods have improved. Copper, manganese, magnesium, iron, zinc, and molybdenum are necessary in the ppm range for optimal growth. However, if optimal concentrations are exceeded, there may be a toxic effect.

The role of iron is especially interesting. While optimal growth requires higher iron concentrations (among other things as a cofactor for aconitase), only 0.05–0.5 ppm is needed for maximal production of citric acid. **Optimal iron concentration** is dependent on the starting material used. For instance, with pure sucrose, 2.0 ppm of iron is optimal for growth, whereas when raw materials such as invert sugar or starch hydrolysates are used, growth without acid production is obtained by adding 0.2 ppm iron, due to the fact that the iron content of the raw material solutions is so high. The sensitivity of strains to heavy metals such as zinc, iron, manganese, and copper decreases with a decrease in the temperature. In addition, copper reverses the inhibitory effect of iron (Table 8.3).

Besides the composition of the nutrient solution (sugar content, metal-salt concentration, and phosphate and nitrogen contents) the pH of the medium influences the yield. During the idiophase, the pH must be below 3 in order to suppress oxalic and gluconic acid formation, but during the trophophase, the pH is generally started at 5. In the first 48 hours during the trophophase, the pH falls to below 3.0 as a result of the metabolism of ammonium ions. For better control of yield in submerged cultures, the trophophase and idiophase can be separated by reducing the pH to under 2 once growth has ceased.

Table 8.3 Reversal by copper of the inhibition of citric acid formation by 10 ppm Fe^{3+}

Copper (ppm)	Citric acid % Yield
0	–
5.0	77.8
10.0	77.4
20.0	75.0
30.0	75.6
50.0	74.0

(Noyes, 1969)

Another benefit of the low pH is the decreased risk of contamination, since at these low pH values only *Penicillium* and yeast cause contamination. Especially in surface fermentations, the frequency of contamination is a significant economic factor.

Production processes

Citric acid is produced by both surface and submerged processes. Surface processes can be further subdivided according to the state of the culture medium used: solid or liquid. Two types of submerged processes are used, stirred fermenters or airlift fermenters (Figure 8.2). The surface process employing solid culture medium is in use solely in small plants and a maximum of 500 tons per year are produced, but the other three processes are widely used. Several factors affect the choice of production type: availability of investment capital, energy availability, cost of labor and training, and availability of techniques for measurement and regulation.

Production of inoculum The material used as inoculum for citric acid production is a spore suspension. Spores are produced in glass bottles on solid substrates at 25 °C with incubation times of 10–14 days. Besides the total numbers, the viability of the spore crops is critical. When a submerged fermenter is to be inoculated, the spores are induced to germinate in a preliminary fermentation. A nutrient solution containing 15% sugar (from molasses) is used in this seed fermenter, and to induce the formation of mycelium in the form of pellets, cyanide ions are added. If too little cyanide is added, growth proceeds well but the citric acid yields in the later production-scale fermenter are low. The spores germinate at

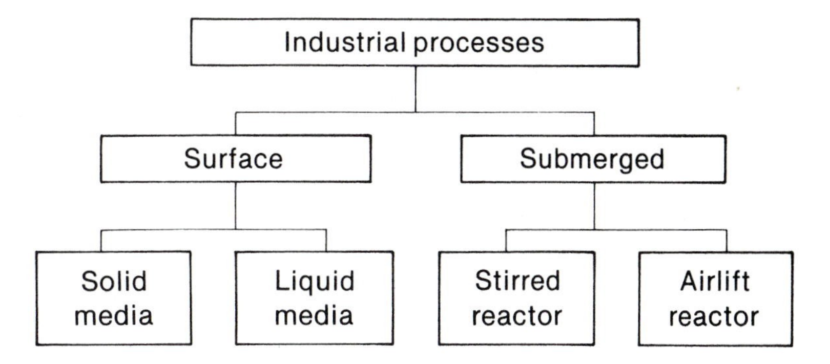

Figure 8.2 Various processes for citric acid manufacture

32°C and form pellets 0.2–0.5 mm in diameter within 24 hours. During this period, the pH falls to 4.3. These pellets are then used as the inoculum for production fermenters. Conversion rate and efficiency in the production fermenter are extremely dependent on the manner in which the spores and pellets are produced.

Surface processes

Surface processes employing solid substrates may use either wheat bran or pulp from sweet potato starch production as a culture medium. In this process, the *A. niger* strains are not as sensitive to trace elements as in the other processes. The pH of the bran is reduced to 4–5 before sterilization; after sterilization the material is inoculated with spores, spread on trays in layers 3–5 cm thick and incubated at 28°C. Growth can be accelerated by adding α-amylases, although the fungus can hydrolyze starch with its own amylase. The solid surface process takes 90 hours, at the end of which the entire solution is extracted with hot water to isolate the citric acid.

Surface processes using liquid nutrient solutions are the oldest production methods and account for 20% of the world's supply of citric acid. These processes are still in use because of the low investment (for the bioreactor), the low energy cost for the cooling system, and the simple technology. The labor costs, however, are substantially greater than for the deep vat processes because of the manpower needed to clean the pipes, trays, and walls of the system.

Table 8.4 shows the composition of a typical nutrient solution; in surface processes, the sucrose is supplied as beet molasses instead of sugar cane molasses. In modern systems, continuous sterilization of the medium is used.

Figure 8.3 shows the layout of a typical fermentation system. The sterilized nutrient solution automatically flows over a distribution system onto the trays. Inoculation is done in the incubation chamber at 30–40°C by blowing dry spores ($2–5 \times 10^7$ spores/m^2) or by spraying spore suspensions.

The temperature is kept constant at 30°C during the fermentation by means of an air current. Ventilation is also important for gas exchange because the rate of citric acid production falls if CO_2 in the atmosphere increases to $> 10\%$. Within 24 hours after inoculation, the germinating spores form a thin cover of mycelium on the

Table 8.4 Composition of nutrient solution for a surface culture

Substrate	g/l Substrate
Sucrose	160–200, ca. 320–400 g molasses
NH_4NO_3	1.6–3.2, Not needed with molasses
CaH_2PO_4	0.3–1.0
$MgSO_4 \cdot 7\ H_2O$	0.2–0.5, Not needed with molasses
$ZnSO_4$	0.01–0.10
Calcium hexacyanoferrate	0.4–2.0

(Schultz and Rauch, 1975)

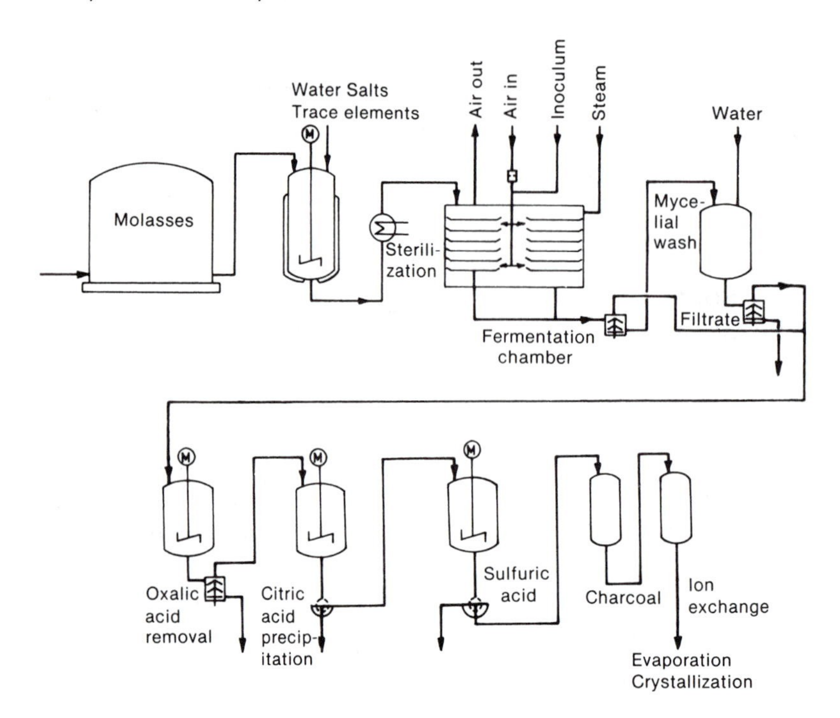

Figure 8.3 Block diagram of a citric acid process using the surface method

surface of the nutrient solution. As a result of the uptake of ammonium ions, the pH in the culture liquids falls to 1.5–2.0. Figure 8.4 shows the course of the fermentation during the first 60 hours. After 30 hours, the **idiophase** begins. If too much iron is present, oxalic acid is produced and a yellowish pigment is formed which later hinders the recovery process. The fully developed mycelium floats as a thick convoluted white layer on the nutrient solution. Through evaporation, the temperature can be maintained constant, but the culture then loses 30–40% of its original volume. The fermentation is stopped after 8–14 days.

Yields from the surface process using liquid nutrient solution amount to 1.2 to 1.5 kg citric acid monohydrate/m^2 of fermentation surface per hour. The heat production is 12,500 kJ/kg of citric acid produced.

During recovery, the mycelium and nutrient solution are removed from the chambers. Due to its volume, the mycelium must be carefully washed in sections. In some cases, mechanical presses are also used to obtain more citric acid from the cells.

Submerged processes

Eighty percent of the world's supply of citric acid is produced by submerged processes. The submerged process, although taking a longer time (8 days), has several advantages: lower investment for construction by a factor of 2.5, 25% lower total investment, and lower labor costs. The disadvantages are the greater energy costs and the more sophisticated control technology which requires more highly trained personnel. Three factors are especially important for production in submerged processes: a) quality of the material used to construct the fermenter, b) mycelium structure, c) oxygen supply.

Fermenters for citric acid production must be either protected from acids or constructed of stainless steel. At pH values between 1–2 the heavy metals leached from normal steel fermenter walls can inhibit the formation of citric acid.

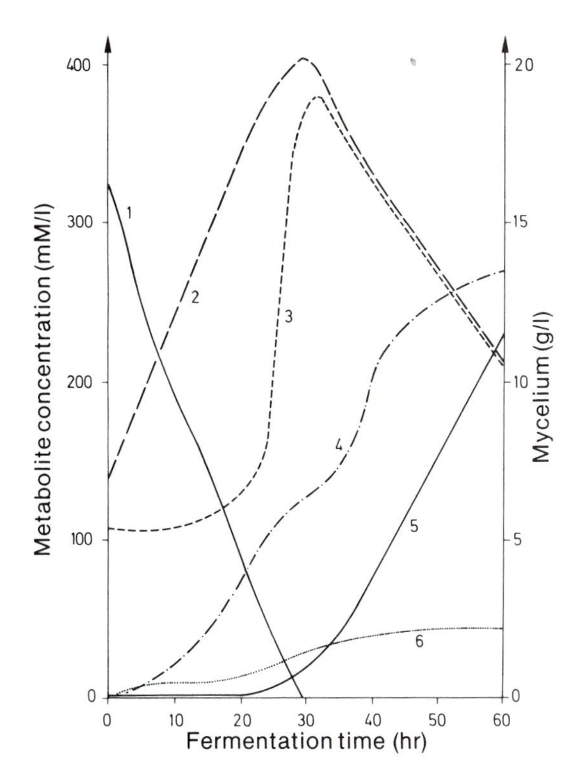

Figure 8.4 Kinetics of the citric acid fermentation (Siebert and Schulz, 1979)
1, Sucrose; 2, Glucose; 3, Fructose; 4, Mycelium; 5, Citric acid; 6, Gluconic acid

With small fermenters (up to 1000 l), even stainless steel chambers must have plastic liners because of the large surface/volume ratio. In large fermenters, the lining is not necessary if stainless steel is used.

The structure of the mycelium that forms in the submerged culture during the trophophase is vital to a successful production process. If the mycelium is loose and filamentous, with limited branches and no chlamydospores, little citric acid is produced in the idiophase. Mycelium for optimal production rates consists of very small solid pellets. The ratio of iron to copper in the medium determines the mycelium structure. Since *Aspergillus* is genetically unstable, only a minimum number of preliminary culture steps should be performed. In some cases, production fermenters are inoculated directly with spores.

Although *Aspergillus niger* requires relatively little oxygen, it is sensitive to oxygen deficiency. Throughout the entire fermentation period, there must be a minimum oxygen concentration of 20–25% of the saturation value. Short interruptions in the oxygen supply cause production to cease irreversibly. The aeration rates in deep vat fermenters should be 0.2–1 volume/volume/minute (vvm) during the acid production phase. Due to the low viscosity, stirring is not necessary. Thus, although some plants use stirred fermenters, air-lift bioreactors can also be used.

Most fermenters for citric acid production are constructed in the range of 120–220 m^3. Submerged systems are somewhat less sensitive than surface systems to variability in the medium composition and in molasses composition (which is notoriously variable), so that they can be operated much more reproducibly than surface systems.

Foaming is a problem in submerged fermentations. A foam chamber 1/3 the size of the fermenter volume is needed in both air-lift and stirred bioreactors. Antifoam agents, such as lard oils, can be added at frequent intervals, and mechanical anti-foam devices can also be used.

Continuous fermentation for citric acid production has not yet been accomplished. The complexity of pellet formation and the control requirements of the idiophase could only be satisfied by a multistage system with fermenters connected in a series; for economic reasons, however, such a system would not be sensible, even if the microbiological and sterility problems could be solved.

Citric acid production with alkane-utilizing microorganisms Several processes for the production of citric acid from C_9–C_{23} hydrocarbons have been described, using both yeast and bacteria. The best producer is *Candida lipolytica,* which can be employed in batch, semicontinuous, or continuous fermentations. Compared to carbohydrate fermentations, the specific yields are measured as higher with hydrocarbons (145% g citric acid/g paraffin), but there are difficulties with the low solubility of the alkanes and the increased amount of isocitric acid produced, which must later be separated from the citric acid (Table 8.5). Studies by Ermakova et al. (1986) and Finogenova et al. (1986) have focussed on the regulation of anaplerotic pathways and the proportion of citric acid versus isocitric acid in *Candida lipolytica* when different substrates are used. In some cases, the proportion of isocitric acid can be up to 50% of the citric acid content. Figure 8.5 shows citric acid production from hydrocarbons by both batch and semicontinuous fermentations.

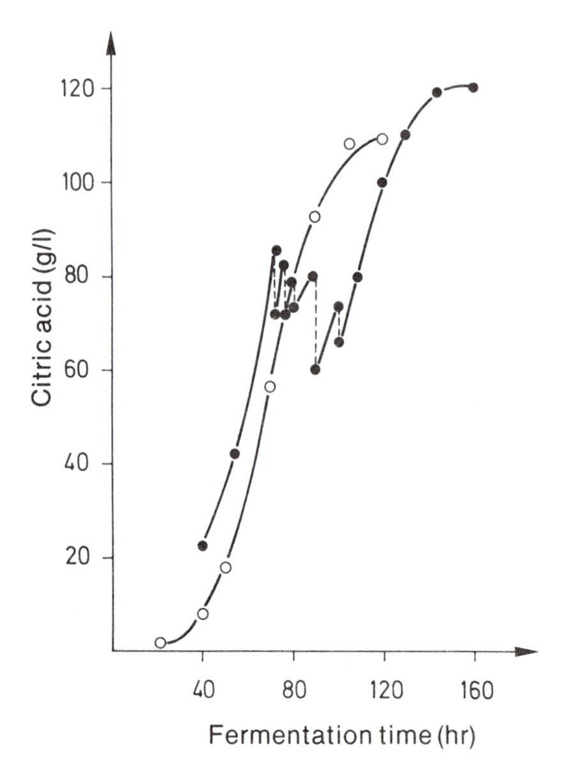

Figure 8.5 Citric acid production with *Candida lipolytica* ATCC 8661 in batch (○ — ○) and semicontinuous (● — ●) fermentation (Gledhill et al., 1973)

Product Recovery

If oxalic acid should be formed as a side product due to suboptimal fermentation control, it is precipitated as calcium oxalate at a low pH, leaving behind the citric acid in solution as the monocalcium citrate. Rotating filters or centrifuges are then used to separate the mycelium and the precipitated calcium oxalate from the liquid. At pH 7.2 ± 0.2 and 70–90°C, citric acid is precipitated, which can in turn be separated by means of rotating filters, and dried. For some uses the citric acid is further purified by adding sulfuric acid to dissolve the citric acid, with a calcium sulfate precipitate forming. The subsequent recovery steps include treatment with activated carbon, treatment with cation and anion exchangers, and crystallization as citric acid or citric acid monohydrate. Above 40°C citric acid crystallizes as the anhydrous acid, and below 36.5°C as the monohydrate.

The required purity of the product depends on its intended use. Citric acid for addition to foods must of course be purer than that needed for industrial purposes.

Table 8.5 Citric acid and isocitric acid production in relation to substrate

Substrate	Citric acid g/l	Isocitric acid g/l	Isocitric acid %
Glycerol	125.5	11.4	8.3
Glucose	89.4	7.0	7.3
n-Paraffin	92.0	49.0	34.8

(Treton et al., 1978)

8.3 GLUCONIC ACID, GLUCONOLACTONE, AND GLUCOSE OXIDASE

Gluconic acid is used in the manufacture of metal, leather, and food; sodium gluconate is used as a sequestering agent in many detergents; δ-gluconolactone functions as a baking powder additive; and calcium gluconate is used in medicine. The production of gluconic acid from glucose by means of glucose oxidase is a simple reaction which can be carried out by many microorganisms (Figure 8.6). The gluconolactone formed in the first step hydrolyzes either spontaneously or enzymatically to gluconic acid. Dur-

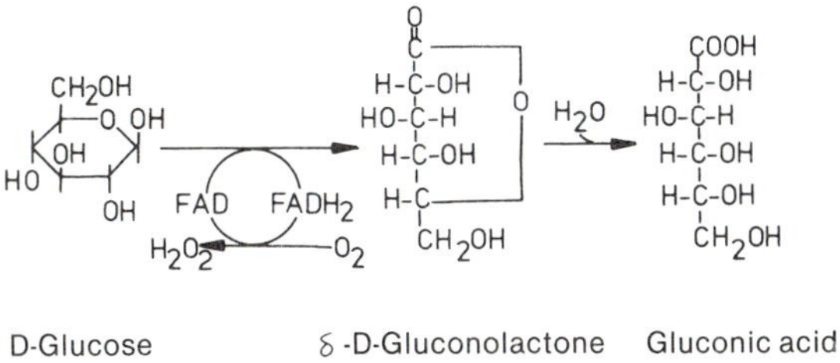

Figure 8.6 Production of gluconic acid from glucose

ing the transfer of hydrogen from $FADH_2$ to oxygen, hydrogen peroxide (H_2O_2) is produced and is immediately split into water by the enzyme catalase. Because the activity of the enzyme leads to the formation of hydrogen peroxide, which has antimicrobial activity, glucose oxidase has also been identified as an antibiotic in fermentation broths and has thus been known under the names notatin and penicillin B.

Gluconic acid has had a long history in industrial microbiology. In 1911 Alsberg described the production of gluconic acid with *Pseudomonas*. The first process with a fungus, a surface process using *Penicillium luteum-purpurogenum*, was begun in 1928. In this process the yields amounted to 80–87% of theoretical.

Today only submerged processes are used, with either the fungus *Aspergillus niger* or the bacterium *Acetobacter suboxydans*. In these processes, gluconic acid, sodium and calcium gluconate, and glucose oxidase are manufactured. Other organisms which have been optimized to produce gluconic acid but which have not been used commercially include the fungi *Penicillium*, *Scopulariopsis*, *Gonatobotrys*, *Endomycopsis*, and *Pullularia* and the bacteria *Vibrio* and *Pseudomonas*.

A process for immobilizing cells or glucose oxidase has been described which leads to yields of 93% when pure oxygen is used. There is considerable competition between the microbiological process and chemical methods, which also produce high yields.

Table 8.6 Relationship between gluconic acid production and pressure

Pressure (bar)	Yield by weight
1	42.5
3	80.4
4	82.4
5	81.3
6	86.1

(May et al., 1934)

Production

The fermentation, carried out at pH 4.5–6.5, requires a growth medium in which both phosphorus and nitrogen are limiting. If calcium gluconate is to be produced, no more than 13–15% glucose can be added as substrate because of the low solubility of calcium gluconate (4 g/l at 30 °C); higher substrate concentrations would allow higher calcium gluconate levels, which would spontaneously crystallize, making purification difficult. In the production of sodium gluconate (solubility 396 g/l), a glucose concentration as high as 28–30% can be used.

The fermentation runs 20 hours at 28–30 °C with a high aeration rate (1–1.5 vvm). By raising the pressure in the system, the oxygen solubility and therefore the gluconic acid yields can be increased (Table 8.6). Commercial yields today are 90–95%.

8.4 ACETIC ACID

The production of acetic acid from alcoholic liquids has been known as long as the production of wine (about 10,000 years). The Romans and Greeks, who used diluted vinegar as a refreshing drink, produced vinegar by leaving wine open to air. Vinegar was produced only for local consumption until the Middle Ages. The first industrially manufactured vinegars were produced in flat open vats. These were slow processes, in which a film of bacteria floated on the surface of the wine. In the nineteenth century, surface fermentations were developed into more rapid processes. One of these, the **trickling generator pro-**

cess is still used today. Beginning in 1949 submerged processes were developed. Both types of process, the trickling generator and the submerged fermenter, are used worldwide, the older methods still being used because of the better flavor of the product. The submerged process has been extensively improved by the H. Frings Company in Germany. As an example of the scale of the process, in 1980 production of vinegar with 10% acetic acid amounted to 356×10^6 liters in the European Economic Community, 469×10^6 liters in the United States and 1600×10^6 liters worldwide (excluding China and the Soviet Union).

Although acetic acid is produced by many fermentative bacteria, only members of a special group, the acetic acid bacteria, are used in commercial production. The acetic acid bacteria can be divided into two genera, *Gluconobacter* and *Acetobacter*, the first group oxidizing ethanol solely to acetic acid and the second group (the overoxidizers) able to oxidize ethanol first to acetic acid and then further to CO_2 and H_2O. Members of the genus *Acetobacter* are Gram-negative and acid tolerant. Strains used commercially belong to the species *Acetobacter aceti*, *A. pasteurianus* or *A. peroxidans*. Members of the genus *Gluconobacter* are not overoxidizers. The species *Gluconobacter oxydans* (formerly *Acetomonas oxydans*) and several subspecies of this species are used commercially. Mixed cultures appear during production, even when the inoculum is assumed to be pure, particularly in surface processes. Production strains seem to lose their high-yielding character if they are transported to a new location on agar. To avoid this problem, production strains are shipped to new vinegar plants in transportable small fermenters.

Biosynthesis

Acetic acid production is an **incomplete oxidation** rather than a true fermentation, because the reducing power which is produced is transferred to oxygen. The first oxidation step from ethanol leads to acetaldehyde with a NAD- or NADP-specific alcohol dehydrogenase. Then there is a hydration to acetaldehyde hydrate and a second oxidation with acetaldehyde dehydrogenase to acetic acid (Figure 8.7). During the oxidation, 1 mole of acetic acid is produced from 1 mole of ethanol. From 1 liter of 12% (v/v) alcohol, 1 liter of 12.4% (w/v) acetic acid is produced.

For optimal production, sufficient oxygen is required, which is reduced by means of the respiratory chain. Six ATP are produced per mole of acetic acid produced. If sufficient oxygen is not available, at high acetic acid and ethanol concentrations the cells die. At a concentration of 5% acetic acid plus ethanol, 34% of the bacteria die after a 2-minute interruption of aeration, whereas at a total concentration of 12% acetic acid plus ethanol, the same killing occurs after only 10–20 seconds.

Both acetic acid and ethanol must be present for optimal growth of *Acetobacter*. Ethanol supply is critical and with less than 0.2% (v/v) in solution the death rate increases. However, the maximal ethanol content should not exceed 5% in conventional processes. Today high-yielding strains produce 13–14% acetic acid.

Production of vinegar

Starting materials with low alcohol content, such as wine, whey, malt, or cider, do not require any additional components to constitute a complete nutrient solution. However, if potato or grain spirits or technical alcohol is used, nutrients must be added in many cases to obtain optimal growth and acetic acid production (Table 8.7). In submerged fermentation, concentrations of nutrients must be five times higher due to the increased biomass which is taken out of the reactor with the vinegar. The inoculum can be either pure cultures or vinegar from previous batches.

Surface processes Despite the success of the submerged process, the **trickling generator** is still widely used in vinegar production (Figure 8.8), especially for making household vinegar. The wooden bioreactor has a total volume of up to

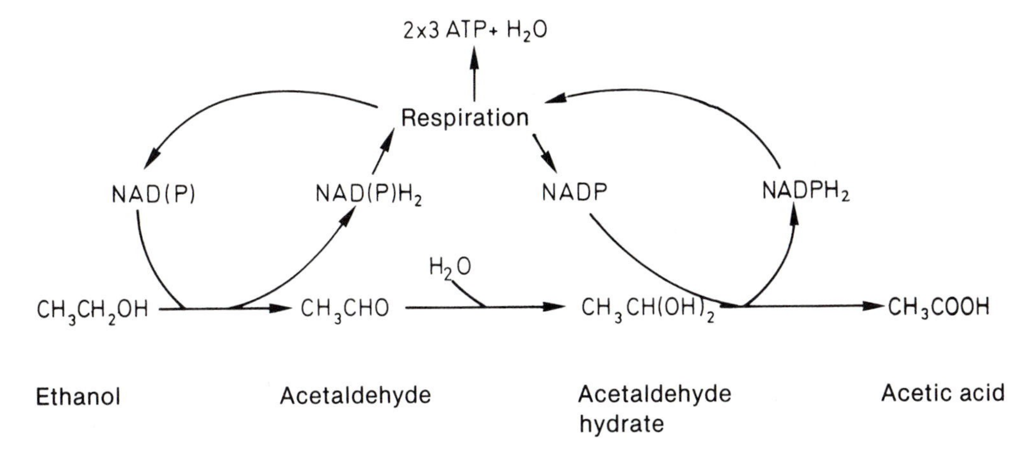

Figure 8.7 Oxidation of ethanol to acetic acid

60 m³ and is filled with beechwood shavings. The starting material is sprayed over the surface and trickles through the shavings containing bacteria into a basin in the bottom, where the partially converted solution is cooled and pumped back up to the top.

Of the alcohol added, 88–90% is converted to acetic acid in the trickling generator process. The rest of the alcohol is used for primary metabolism or it escapes with the waste air. The temperature in the upper part of the system is about 29°C and is about 35°C in the lower part. The time needed to produce 12% acetic acid in this process is about three days.

Submerged processes Fruit wines and special mashes with low total concentrations of alcohol were first used in submerged processes. With such low-yielding processes, aeration was not critical, but in present-day high-yielding processes, which produce 13% acetic acid in quantities of up to 50 m³, aeration must be highly regulated. The fermenters (Figure 8.9) resemble other bioreactors (see Chapter 5). The tanks are constructed of stainless steel and are stirred from the bottom. The aeration apparatus consists of a suction rotor with the incoming air coming down

Table 8.7 Additives for acetic acid fermentation

	Gram/m³ nutrient solution	
	Surface process	Submerged process
Sugar solution from grain hydrolysate	200	1000
Ammonium phosphate	96	480
Magnesium sulfate	24	120
Calcium citrate	24	120
Calcium pantothenate	0.24	1.2

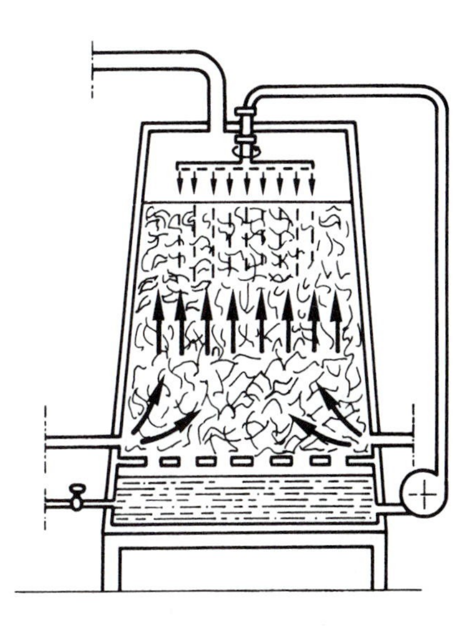

Figure 8.8 Diagram of a trickling generator for vinegar

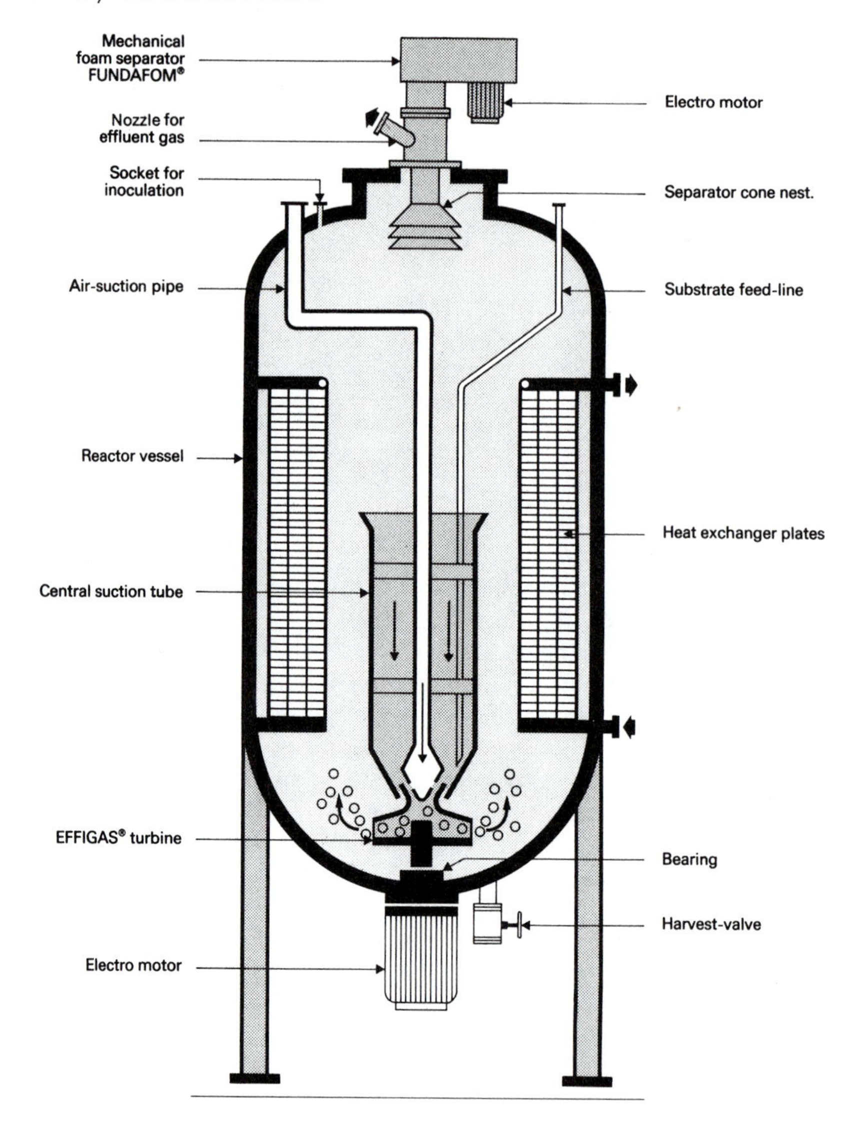

Figure 8.9 Submerged fermenter for acetic acid production (Chemap)

through a pipe from the top of the fermenter. Heat exchangers control the temperature and mechanical foam eliminators must usually be installed. Household vinegar (13% acetic acid) is produced in a semicontinuous, fully automatic process, under continuous stirring and aeration, beginning with a starting material that contains 7–10 g acetic acid/100 ml and 5% ethanol. The ethanol concentration is measured continuously during the process and when the concentration sinks to 0.05–0.3% (about 35 hours), 50–60% of the solution is removed and replaced with a new mash containing 0–2 g acetic acid/100 ml and 10–15% ethanol. Fully continuous processes with yields of 98% at 40°C have also been described.

With the submerged process, the **production rate** per m^3 is 10 times higher than with the surface fermentation and about 5% higher than with the trickling generator process. Other advantages are: lower capital investment per production amount, only 20% of the plant area required for the installation, capability of conversion to other mashes in a short time, and low personnel cost due to fully automatic control.

Recovery The acetic acid obtained in the submerged process is turbid due to the presence of bacteria and the product must be clarified by filtration. Plate filters and filter aids are generally used. Once the filtrate is obtained, $K_4(Fe(CN)_6)$ is used to decolorize the final product when desirable.

8.5 LACTIC ACID

The first microbial production of an organic acid was the production of lactic acid, carried out in 1880. Today chemical methods for the production of lactic acid are very competitive at the same cost as biological processes. Two kinds of lactic acid bacteria are recognized, heterofermentative and homofermentative. The heterofermentative organisms produce a great number of byproducts and are not suitable for commercial purposes. With homofermentative organisms, as little substrate as possible is converted into cell material plus byproducts and as much as possible is metabolized into lactic acid. The biosynthesis of lactic acid from glucose proceeds via glyceraldehyde-3-P, 1,3-di-P-glycerate and pyruvate. The reducing power produced during the oxidation of glyceraldehyde phosphate is transferred with an NAD-dependent lactate dehydrogenase to pyruvate, which is reduced stereospecifically to L(+) or (D−) lactic acid (Figure 8.10).

Theoretically, two moles of lactate are produced from 1 mole of glucose. Over 90% of this theoretical yield is actually attained in practice.

The organisms used are *Lactobacillus delbrueckii* and *L. leichmannii* when glucose is used as substrate, *Lactobacillus bulgaricus* with whey, and *Lactobacillus pentosus* with sulfite waste liquor. These organisms are facultative rather than obligate anaerobes, and therefore the bioreactors need not be run with complete oxygen exclusion.

In addition to 12–13% glucose and $(NH_4)_2HPO_4$ (0.25%), the **fermentation medium** must contain B vitamins. Production is carried out in 25–120 m^3 fermenters at 45–50°C with an excess of $CaCO_3$ added to maintain the pH between 5.5 and 6.5. The fermentation time is about 72 hours. Since free lactic acid is toxic to the organism, procedures for removing the product continuously have been studied. In one approach, electrodialysis has been used. In another procedure, continuous culture was done in a membrane reactor, leading to the production of 80 g/l·h.

Only L+ lactic acid is produced commercially. When the fermentation has completed, the broth is heated to dissolve the calcium lactate. The heated broth is filtered and the calcium precipitated by addition of sulfuric acid. After concentration of the lactic acid, it is further purified. Total microbial production per year amounts to about 20,000 metric tons.

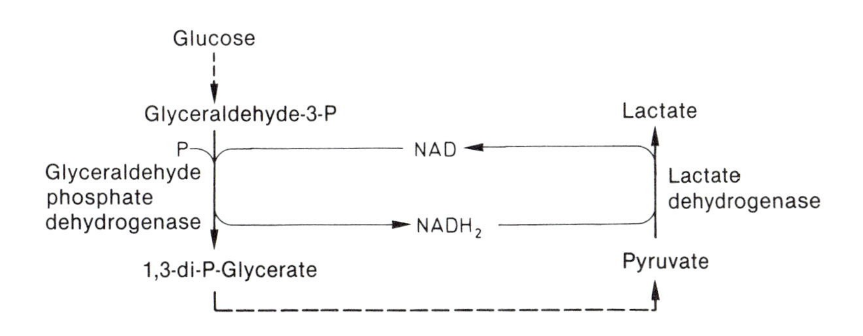

Figure 8.10 Lactic acid production from glucose

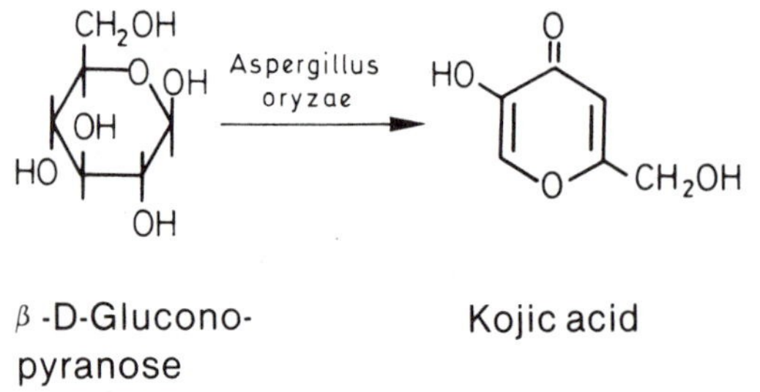

β-D-Glucono-pyranose → (Aspergillus oryzae) → Kojic acid

Figure 8.11 Biosynthesis of kojic acid

8.6 KOJIC ACID

Kojic acid (5-hydroxy-2-hydroxymethyl-4-pyrone) has found minimal commercial use as a feedstock in the plastics industry. For a long time, it has been produced by the direct fermentation of glucose. Yields of 70–90% can be obtained from glucose with *Aspergillus flavus* and *A. oryzae* (Figure 8.11). This acid can also be produced from pyruvate, glycerol, acetate, or ethanol. The production using glucose is very similar to that of the citric acid and itaconic acid processes. By changing the fermentation parameters, such as stirring or aeration rate, some strains can be converted from citric acid or itaconic acid production to kojic acid production.

8.7 ITACONIC ACID

In 1931, itaconic acid was first shown to be a metabolic product of *Aspergillus itaconicus*. In the same decade, it was discovered that some strains of *A. terreus* also excrete itaconic acid. Mutants of both strains are used today for commercial production.

At present there are only a few commercial systems which produce itaconic acid: 2 in Western Europe, one each in America, USSR and Japan. The main use of itaconic acid is in the plastic industry. Itaconic acid forms copolymers with its esters and other monomers; these are used in the paper industry for wall paper and other paper products and in the production of adhesives. An itaconic acid acrylonitrile copolymer is more readily dyed than some other polymers.

Itaconic acid is produced by way of the tricarboxylic cycle from cis-aconitic acid via decarboxylation. The enzyme cis-aconitic acid decarboxylase has been purified and characterized (Figure 8.12). Another biosynthetic pathway from pyruvate through citramalic acid, citraconic acid, and itatartaric acid also results in itaconic acid.

The undesirable by-products **succinic acid** and **itatartaric acid** are produced from itaconic acid (Figure 8.13). Calcium inhibits the enzyme itaconic acid oxidase and hence calcium additions increase the yield of the desired product. By the

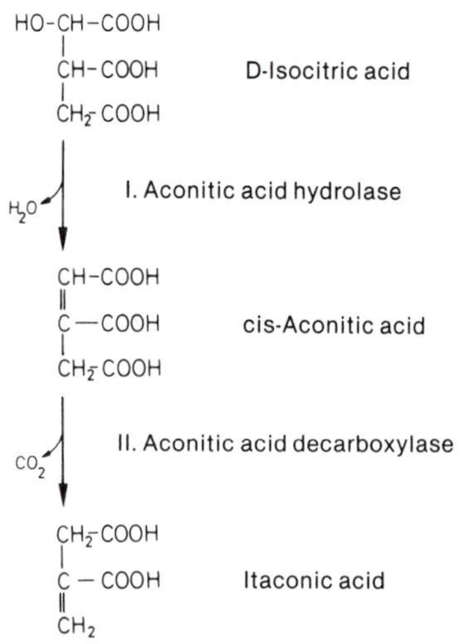

D-Isocitric acid → I. Aconitic acid hydrolase ($-H_2O$) → cis-Aconitic acid → II. Aconitic acid decarboxylase ($-CO_2$) → Itaconic acid

Figure 8.12 Biosynthesis of itaconic acid out of the tricarboxylic acid cycle

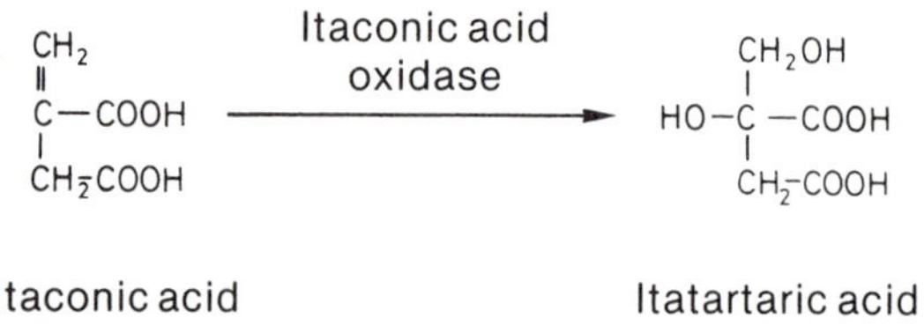

Itaconic acid → (Itaconic acid oxidase) → Itatartaric acid

Figure 8.13 Destruction of itaconic acid by conversion to itatartaric acid

use of immobilized cells, production has been increased to about 0.73 g/l·h.

A. terreus is used only in batch submerged fermentation at present. With a 15% sucrose solution, conversion into itaconic acid can be achieved at 78% of the theoretical yield.

REFERENCES

Adiga, P.R., K. Sivarama Sastry, V. Venkatasubramanyam, and P.S. Sarma. 1961. Interrelationships in trace-element metabolism in *Aspergillus niger*. Biochem. J. 81: 545–550.

Buchta, K. 1983.a. Lactic acid, pp. 409–417. In: Rehm, H.J., and G. Reed (eds.). Biotechnology 3. Verlag Chemie, Weinheim.

Buchta, K. 1983b. Organic acids of minor importance, pp. 467–478. In: Rehm, H.J., and G. Reed. (eds.). Biotechnology 3: Verlag Chemie, Weinheim.

Currie, J.N. 1917. The citric acid fermentation of *Aspergillus niger*. J. Biol. Chem. 31: 15–37.

Ebner, H., and H. Follmann. 1983. Acetic acid, pp. 387–407. In:Rehm, H.J., and G. Reed (eds.). Biotechnology 3. Verlag Chemie, Weinheim.

Ermakova, I.T., N.V. Shishkanova, O.F. Melnikova, and T.V. Finogenova. 1986. Properties of *Candida lipolytica* mutants with the modified glyoxylate cycle and their ability to produce citric and isocitric acid. I. Physiological, biochemical and cytological characteristics of mutants grown on glucose or hexadecane. Appl. Microbiol. Biotechnol. 23: 372–377.

Finogenova, T.V., N.V. Shishkanova, I.T. Ermakova, and I.A. Kataeva. 1986. Properties of *Candida lipolytica* mutants with modified glyoxylate cycle and their ability to produce citric and isocitric acid. II. Synthesis of citric and isocitric acid by *C. lipolytica* mutants and peculiarities of their enzyme systems. Appl. Microbiol. Biotechnol. 23: 378–383.

Fukui, S. and A. Tanaka. 1980. Production of useful compounds from alkane media. Adv. Biochem. Eng. 17: 1–35.

Gledhill, W.E., I.D. Hill. and P.H. Hodson. 1973. Citrate production from hydrocarbons by use of a nonsterile, semicontinuous cell recycle system. Biotechnol. Bioeng. 15: 963–972.

Greenshields, R.V. and E.L. Smith. 1974. The tubular reactor in fermentation. Proc. Biochem. 9: 11–17, 28.

Hardy, G.P.M.A., M.J. Teixeira de Mattos, O.M. Neijssel, and D.W. Tempest. 1987. Effect of the growth conditions on gluconate and 2-ketogluconate production from glucose by three pseudomonads in continuous culture. Poster 9th ISCC "Continuous culture in biotechnology and environment conservation." 19–24. July, Hradec Kralove, CSSR.

Hongo, M., Y. Nomura, and M. Iwahara. 1986. Novel method of lactic acid production by electrodialysis fermentation. Appl. Environ. Microbiol. 52: 314–319.

Ju, N. and S.S. Wang. 1986. Continuous production of itaconic acid by *Aspergillus terreus* immobilized in a porous disk bioreactor. Appl. Microbiol. Biotechnol. 23: 311–314.

Legisa, M. and M. Mattey. 1986. Glycerol as an initiator of citric acid accumulation in *Aspergillus niger*. Enzyme Microb. Technol. 8: 258–259.

Linko, P. 1981. Immobilized live cells. pp. 716–717. In: Moo-Young, M. (ed.). Advances in Biotechnology I. Pergamon Press, Toronto.

Lockwood, L.B. 1979. Production of organic acids by fermentation, pp. 355–387. In: Peppler, H.J. and D. Perlman (eds.), Microbial technology, vol. 1. Academic Press, New York.

May, O.E., H.T. Herrick, A.J. Moyer, and P.A. Wells. 1934. Gluconic acid production by submerged mould growths under increased air pressure. Ind. Eng. Chem. 26: 575–578.

Mehaia,, M.A., and M. Cheryan. 1986. Lactic acid from acid whey permeate in a membrane recycle bioreactor. Enzyme Microb. Technol. 8: 289–292.

Milsom, P.E. and J.L. Meers. 1985a. Citric acid, pp. 665–681. In: Moo-Young, M. (ed.). Comprehensive Biotechnology 3. Pergamon Press, Oxford.

Milsom, P.E. and J.L. Meers. 1985b. Gluconic and itaconic acids, pp. 681–700. In: Moo-Young, M. (ed.). Comprehensive Biotechnology 3. Pergamon Press, Oxford.

Noyes, R. 1969. Citric acid production processes. Noyes Dev. Corp., Park Ridge, USA.

Röhr, M., C.P. Kubicek, and J. Kominek. 1983. Citric acid, pp. 419–453. In: Rehm, H.J. and G. Reed (eds.). Biotechnology 3. Verlag Chemie, Weinheim.

Rose, A.H. 1978. Production and industrial importance of primary products of microbial metabolism, pp. 1–30. In: Rose, A.H. (ed.). Primary products of metabolism. Academic Press, New York.

Schulz, G. and H. Rauch. 1975. Citronensäure (Citric acid), pp. 626–636. In: Ullmanns Encyklopädie der technischchen Chemie., Vol. 9.

Siebert, D. and H. Hustede. 1982. Citronensäure-Fermentation. Biotechnologische Probleme und Möglichkeiten der Rechnersteuerung. (Biotechnological problems and possibilities for computer control of the citric acid fermentation.) Chem. Ing. Technik 54: 659–669.

Siebert, D. and G. Schulz. 1979. Citric acid production by fermentation. Intern. Microbiol. Food Ind. Congr., Paris.

Sodeck, G., J. Modl, J. Kominek, and W. Salzbrunn. 1981. Production of citric acid according to the submerged fermentation process. Proc. Biochem. 16: 9–11.

Treton, B., M.-T. Le Dall, and H. Heslot. 1978. Excretion of citric and isocitric acids by the yeast *Saccharomycopsis lipolytica*. Europ. J. Appl. Microbiol. Biotechnol. 6: 67–77.

9

Amino acids

9.1 INTRODUCTION

The taste-enhancing properties of **glutamic acid** were discovered in 1908 in Japan, and commercial production of sodium glutamate from acid hydrolysates of wheat and soy protein began soon after. In 1957, L-glutamic acid was discovered as a product in the spent medium of *Corynebacterium glutamicum*, and this organism subsequently became the major source of sodium glutamate. This development was an enormous boost to the fermentation industry in Japan, and fermentative processes for the production of amino acids as well as nucleotides (see Chapter 10) have been almost exclusively Japanese developments.

Thirteen of the 17 major manufacturers are situated in Southeast Asia (Japan, Korea, Taiwan), the foremost Japanese firms being Ajinomoto Co., Kyowa Hakko Kogyo Co. and Tanabe Seiyaku Co. The most important western manufacturers of sodium glutamate are Stauffer Chemical Co. (USA) and Orsan S. A. (France). The most well-known manufacturer of L-lysine is Eurolysine Co. (France) and of L-aspartic acid the two French firms Orsan S. A. and Rhône-Poulenc S. A.

In 1985 the world production of biochemically produced amino acids was 460,000 tons, of which 370,000 tons per year were sodium glutamate and 70,000 were L-lysine. It is anticipated that increased demand will result in yearly increases of around 10%.

Further increases in the production of amino acids are expected because of the following developments:

- The isolation of improved production cultures through the addition of recombinant DNA and protoplast fusion techniques to the currently used mutation methods to isolate derepressed strains.
- Combining chemical synthesis with microbial and enzymatic processes.
- Adaptation of existing processes to the use of less expensive starting materials.

- Optimization of scale-up conditions.

9.2 COMMERCIAL USES OF AMINO ACIDS

Amino acids have extensive industrial applications. About 66% of the amino acids produced are used in the food industry, 31% as feed additives, in 4% in medicine and cosmetics and as starting materials in the chemical industry. In the **food industry** amino acids are used alone or in combination to enhance flavors. The flavor-enhancing effect of sodium glutamate has been mentioned; sodium aspartate and D,L-alanine are added to fruit juices to round off the taste, and glycine is added to foods containing sweeteners. L-Cysteine improves the quality of bread during the baking process and acts as an antioxidant in fruit juices. L-Tryptophan combined with L-histidine also acts as an antioxidant and is used to keep powdered milk from getting rancid. Aspartame® (L-aspartyl-L-phenylalanine methyl ester), which is made from L-phenylalanine and L-aspartic acid, is used as a low-calorie sweetener in soft drinks.

Plant proteins are often deficient in essential amino acids such as L-lysine, L-methionine, L-threonine, or L-tryptophan. The lysine contents of some foods and feeds are given in Table 9.1. Lysine is added to bread in Japan, and in some countries soy products are enhanced by the addition of methionine. L-Lysine and DL-methionine are used to upgrade animal feeds, and the addition to foods of L-threonine and L-tryptophan is being developed. In view of the scarcity of nutritious food in the Third World, there will be an ever-increasing need to supplement plant proteins with those essential amino acids which occur at suboptimal levels. Aspartame® (L-α-aspartyl-L-phenylalanine methyl ester) is a major product in the soft-drink industry, serving as an artificial sweetener in so-called "diet" soda. The sweetness level of aspartame in aqueous solution is 200-times higher than a 3–4% sucrose solution. As the starting material for the industrial production of aspartame, L-phenylalanine and L-aspartic acid are used, both of which are products of industrial microbiology. The significance of aspartame can be seen from the fact that during the period from 1981 to 1985 L-phenylalanine production rose from 50 to over 3000 metric tons, and similar rates of increase occurred for aspartic acid. By means of genetic engineering it has been possible to obtain an organism capable of producing an alternating copolymer of phenylalanine and aspartic acid, which can be converted into asp-phe dipeptide by enzymatic splitting. The latter is esterified to produce aspartame.

Table 9.1 L-lysine content of several foods and feeds

Food/feed	Lysine content (%)	Food/feed	Lysine content (%)
Corn	0.21	Soy meal	2.90
Oats	0.50	Dry yeast	3.40
Barley	0.40	Skim milk powder	2.50
Wheat	0.60	Meat-meal	2.60

Many amino acids are used in **medicine**, particularly as ingredients of infusion solutions in post-operative treatment.

In the **chemical industry**, amino acids are used as starting materials for the manufacture of polymers, such as polyalanine fibers and lysine isocyanate resins. Poly-γ-methylglutamate is used as a surface layer in the manufacture of synthetic leather, and the N-acyl derivatives of some amino acids are used in the manufacture of cosmetics and as surface-active substances. Urocanic acid, used as a suntanning agent, is produced by the biotransformation of histidine. Another amino acid, glycine, is used as a starting material for the production of the herbicide glyphosate, and threonine serves a similar purpose for azthreonam (see Section 13.2).

9.3 METHODS OF PRODUCTION

Amino acids that are currently in commerical production are summarized in Table 9.2. Fermentation processes have been developed for all amino acids except glycine, L-cysteine and L-cystine, but not all are in commercial use. In the

Table 9.2 World production of amino acids; production processes and applications

Amino acid	Annual production (metric tons)	Production methods	Application
L-Alanine	130	1, 3c	Flavor enhancer
D,L-Alanine	700	2	Flavor enhancer (beverage production)
L-Arginine	1000	1, 3a	Infusions Therapy (liver diseases) Cosmetics
L-Aspartic acid	4000	1, 3c	Therapy Flavor enhancer Aspartame® production
L-Asparagine	50	1,2	Therapy
L-Cysteine	700	1	Bread production Antioxidant Therapy (bronchitis)
L-Glutamic acid	370,000	3a	Flavor enhancer
L-Glutamine	500	3a	Therapy (ulcer)
Glycine	5000–6000	2	Component of sweeteners
L-Histidine	200	3a,1	Therapy (ulcer)
L-Isoleucine	150	3a	Infusions
L-Leucine	150	1,3a	Infusions
L-Lysine	70,000	3a, 3c	Feed additive Infusion solution
DL-Methionine	70,000	2	Feed additive
L-Methionine	150	3c	Therapy
L-Ornithine	50	3a, 3c	Liver therapy
L-Phenylalanine	3000	3a, 3c	Infusions Therapy Aspartame® production
L-Proline	100	3a	Infusions
L-Serine	50	3a, 3b	Cosmetics
L-Threonine	160	3a	Feed additive
L-Tryptophan	200	3a, 3c	Infusions
L-Tyrosine	100	1, 3c	Infusions Precursor for L-DOPA synthesis
L-Valine	150	3a, 3c	Infusions

Production methods: 1. Extraction of protein hydrolysates; 2. Chemical synthesis; 3a. Direct fermentation; 3b. Microbial transformation of precursors; 3c. Use of enzymes or immobilized cells
(Soda et al., 1983; Kinoshita, 1987)

commercial manufacture of amino acids there are three rival processes:

1. The **extraction of amino acids from protein hydrolysates**. This method is used to obtain L-cysteine, L-cystine, L-leucine, L-asparagine and L-tyrosine.
2. **Chemical synthesis**. The production of glycine, D,L-alanine, D,L-methionine and D,L-tryptophan always involves chemosynthesis. Chemical synthesis is cheaper than microbial production, but the chemical product is the optically inactive mixture of the D- and L-isomers.
3. **Microbiological production**, discussed below.

There are three approaches to microbiological production. The first, outlined in Table 9.3, is by **direct fermentation of amino acids** using different carbon sources, such as glucose, fructose, molasses, starch hydrolysate, n-alkanes, ethanol, glycerol, acetate, propionate, etc. Fermentations using methanol as a starting material have been studied extensively because of the low cost of this substrate, but have not yet been used commercially due to the low yields.

The second approach is by **converting inexpensive intermediate products** via biosynthesis. For example glycine, which is inexpensive, can be converted to L-serine (see Table 9.4).

The third approach is by the use of **enzymes or immobilized cells**, sometimes in continuous processes involving enzyme-membrane reactors (see Section 5.3). The most common reactions of this type are:

- Use of a stereospecific amino acylase from *Aspergillus oryzae* to selectively split a chemically synthesized a D,L-amino acid that had been chemically acetylated. In this way, the L form of the amino acid is selectively liberated (see Figure 9.1). A German company, Degussa AG, has been using an enzyme-membrane reaction technique of this type for many years to produce L-alanine, L-methionine, L-phenylalanine, L-tryptophan, and L-valine (monthly production of 5–20 metric tons).

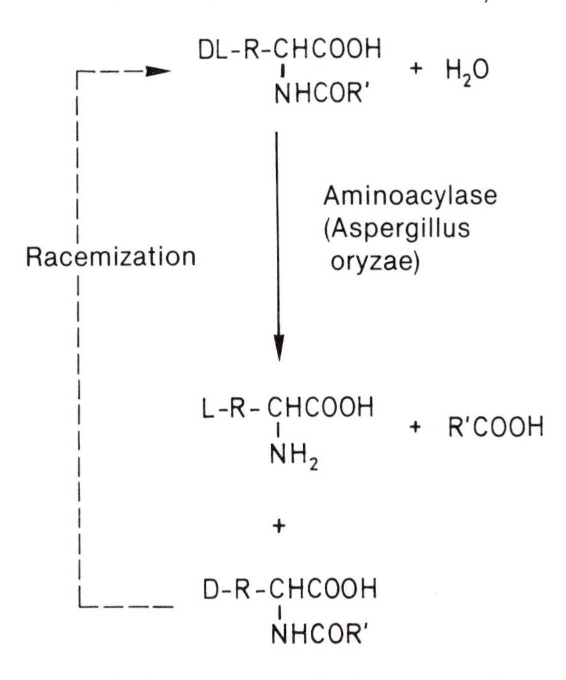

Figure 9.1 Use of an enzyme for the separation of a racemic mixture in the manufacture of L-amino acids (R= side chain of an amino acid; R'= e.g. $-CH_3$)

- L-α-amino acids can be produced by reductive amination of α-ketoacids using amino acid dehydrogenases. The most commonly used dehydrogenase is an enzyme from *Bacillus* strains such as *B. megaterium*, which requires NADH as coenzyme. L-alanine, L-leucine, and L-phenylalanine are produced in this way either with immobilized cells or in enzyme-membrane reactors in combination with a NADH regeneration system.
- D,L-5-monosubstituted hydantoins, which are easily produced chemically, can be converted by means of stereospecific enzymatic cleavage to optically active D- or L-amino acids. The hydantoin cleavage by either D- or L-hydantoinase leads to the production of D- or L-N-carbamoyl-amino acids. The carbamoyl residue is chemically or enzymatically removed with the corresponding car-

Table 9.3 Production of amino acids by fermentation

Amino acid	Strain used	Genetic characteristics	Yield (g/l)	Carbon source
D,L-Alanine	*Microbacterium ammoniaphilum*	ArgHx^r	60	Glucose
L-Alanine	*Pseudomonas* No. 483	Wild type	17.5	Glucose
L-Arginine	*Serratia marcescens* AT 428 (*aru argR2 argA2*)	Transduction Canavanine^r	50	Glucose
L-Aspartic acid	*Escherichia coli* K1-1023*		56	Fumaric acid
L-Glutamic acid	*Corynebacterium glutamicum*	Wild type	100	Glucose
	Brevibacterium flavum		98	Acetate
	Arthrobacter paraffineus		82	n-Alkanes
L-Glutamine	*Corynebacterium glutamicum*	Wild type of glutamic acid producer	58	Glucose, with high biotin and NH_4Cl content
L-Histidine	*Serratia marcescens* L 120 (pSH 368)	rDNA	40	Sucrose
L-Isoleucine	*B. flavum*	Eth^r	30	Acetate
L-Leucine	*Brevibacterium lactofermentum*	Ile$^-$ Met$^-$ TA^r	28	Glucose
L-Lysine	*B. lactofermentum* AJ 11204	AEC^{r}Ala$^-$ CCL^{r}ML^{r}FP^s	70	Glucose
	B. flavum	Hom^{leaky} Thre$^-$	75	Acetate
L-Methionine	*C. glutamicum* KY 9276	Thr$^-$ Eth^r MetHx^r	2	Glucose
L-Ornithine	*C. glutamicum*	Arg$^-$	26	Glucose
L-Phenylalanine	*B. lactofermentum*	5MT^{r}PFP^{r}Decr Tyr^{r}Met$^-$	25	Glucose
L-Proline	*C. acetoacidophilum*	Uncharacterized mutant	108	Glucose + Glutamic acid
L-Serine	*B. lactofermentum*	SG^r	4.5	Glucose
L-Threonine	*E. coli* VL 344 (pYN7)	rDNA	55	Sucrose
L-Tryptophan	*E.coli* JP 4114	rDNA	23.5	Glucose
L-Tyrosine	*C. glutamicum* Pr-20	Phe$^-$ PFP^r PAP^r PAT^r TyrHx^r	18	Glucose
L-Valine	*Brevibacterium lactofermentum* No. 487	TA^r	31	Glucose

*Technique used for the production of L-aspartic acid between 1960 and 1973. Resistance: AEC S-(β-Aminoethyl)-L-cysteine; ArgHx Arginine hydroxamate; CCL, α-Chlorcaprolactam; Dec, Decoyinine; Eth Ethionine; FP, β-Fluoropyruvate; MetHx Methionine hydroxamate; ML, γ-Methyl-L-lysine; 5 MT 5-Methyltryptophan; PAP p-Amino- phenylalanine; PAT p-Aminotyrosine; PFP p-Fluorophenylalanine; SG Sulfaguanidine; TA 2-Thiazolalanine; TyrHx Tyrosine hydroxamate. Auxotrophs: Ala, alanine; arg, arginine; hom, homoserine; ile, isoleucine; met, methionine; phe, phenylalanine; thr, threonine; tyr, tyrosine. *aru:* block in arginine degradation; *argR:* biosynthetic enzyme is derepressed; *argA:* N-acetylglutamate synthesis is insensitive to feedback inhibition by arginine.

Table 9.4 Enzymatic production of amino acids

Amino acid	Enzyme	Enzyme-producing microorganism	Reaction	Yield (g/l)	Efficiency (%)
L-Alanine	Aspartate decarboxylase (RC)	*Pseudomonas dacunhae*	L-Aspartate→L-Alanine+CO_2	600	100
L-Aspartic acid	Aspartase (IC)	*Escherichia coli*	Fumarate +NH_4^+→ L-Aspartic acid	560	99
L-Cysteine	Cysteine desulfhydrase (RC)	*Bacillus sphaericus*	β-Chloro-L-alanine+Na_2S→ L-Cysteine+NaCl+NaOH	70	80–85
L-DOPA (L-3,4-Di-hydroxyphenyl-alanine)	Tyrosinase (RC)	*Erwinia herbicola*	Pyrocatechol + Pyruvate+NH_4^+→ L-DOPA + H_2O	59	95
L-Leucine	L-Leucine dehydrogenase (EMR)	*Bacillus sphaericus*	α-Ketoisocapronic acid+NH_4^++ NADH→L-Leucine+H_2O+NAD^+	42 $g \cdot l^{-1} d^{-1}$	max. 99.7
L-Lysine	D-Amino-caprolactam racemase, L-Amino-caprolactam hydrolase (RC)	*Achromobacter obae*, *Cryptococcus laurentii*	DL-Aminocaprolactam+H_2O→ L-Lysine	100	100
L-Phenyl-alanine	L-Phenylalanine dehydrogenase (EMR)	*Brevibacterium* sp.	Phenylpyruvate+NH_4^+ →L-Phenylalanine+H_2O	37 $g \cdot l^{-1} d^{-1}$	
	Acetamidocinnamate amidohydrolase (RC)	*Bacillus sphaericus*	Acetamido cinnamic acid→ L-Phenylalanine	8	94
	Phenylalanine ammonia lyase (RC)	*Rhodotorula glutinis*	trans-Cinnamic acid+NH_4^+ →L-Phenylalanine	18	70
	Aspartate phenyl-alanine transaminase (EMR)	*E. coli*	L-Aspartate+Phenylpyruvate →L-Phenylalanine+Oxalacetate	30	98
	D-Hydroxyisocaproate dehydrogenase	*Lactobacillus casei*	D,L-Phenyllactate →Phenylpyruvate		
	L-Hydroxyisocaproate dehydrogenase	*Lactobacillus confusus*			
	L-Phenylalanine dehydrogenase (EMR)	*Rhodococcus* sp.	Phenylpyruvate+NH_4^+ →L-Phenylalanine+H_2O	28 $g \cdot l^{-1} d^{-1}$	
L-Serine	Serine hydroxymethyl transferase (CCE)	*Klebsiella aerogenes*	Glycine+HCHO →L-Serine	450	88
L-Tryptophan	Serine hydroxymethyl transferase	*Klebsiella aerogenes*	Glycine+HCHO →L-Serine	200	95 based on indole
	Tryptophanase (EMR)	*E. coli*	L-Serine+Indole →L-Tryptophan+H_2O		
L-Tyrosine	Serine hydroxymethyl transferase	*Klebsiella aerogenes*	Glycine+HCHO →L-Serine		
	β-Tyrosinase (RC)	*Erwinia herbicola*	L-Serine+Phenol →L-Tyrosine+H_2O	26	61 based on glycine

Aida et al. (1986); Groeger and Sahm, 1987; Schmidt et al., (1987)
EMR, enzyme/membrane reactor; IC, immobilized cells; RC, resting cells; CCE, crude cell extract.

bamolyase. L-hydantoinases have been found in only a few bacteria, for example *Flavobacterium ammoniagenes*, whereas D-hydantoinases are found in numerous bacteria, actinomycetes, yeasts, and some strains of *Aspergillus*. In addition to the use of this approach for the production of L amino acids, it can also be used for the manufacture of D-N-carbamoyl amino acids and D amino acids, which find a role as side-chain precursors for the production of semi-synthetic penicillins and cephalosporins.

In some manufacturing processes, several enzymatic reactions are coupled. For example, L-alanine is produced by sequential conversion of fumarate to L-aspartic acid with immobilized cells of *E. coli* and *Pseudomonas dacunhae*. This process, which has been in commercial operation by Tanabe Seiyaku in Japan since 1982, involves the connection of two reactors in series.

Enzymes or immobilized cells are used in the production of L-alanine, L-aspartic acid, L-dihydroxy phenylalanine (DOPA), L-lysine, L-methionine, L-phenylalanine, L-tryptophan, L-tyrosine, and L-valine. The most important approaches used are summarized in Table 9.4. The current interest in the production of L-phenylalanine as a precursor for aspartame production has led to the development of numerous competing biotransformation reactions, which have not as yet replaced the direct fermentation process.

9.4 STRAINS FOR AMINO ACID PRODUCTION

Although microorganisms which excrete glutamic acid are easily obtained from nature, it has been more difficult to obtain natural isolates which excrete other amino acids for large-scale production. Extensive screening has led to the isolation of only a few strains which excrete D,L-alanine and L-valine. Figure 9.2 shows the overall biosynthetic pathways for the other amino acids. Natural isolates rarely excrete these amino acids, since cell metabolism regulates production. In strain development, several methods of eliminating the regulation control are available.

- The use of **auxotrophic mutants** which can no longer form the regulatory effector or corepressor (usually the end product), and excrete the intermediate of the biosynthetic pathway which is just in front of the block.
- The use of **regulatory mutants**. For instance, if an excess of the end product of an unbranched biosynthetic pathway is to be produced, mutants with a feedback-insensitive key enzyme are selected either from antimetabolite-resistant strains or from a population of revertants from auxotrophs.
- Use of **genetic recombination** to combine auxotrophy or regulatory mutants from different strains into a single hybrid strain. In *Serratia marcescens* transduction with phage PS20 has been used for creating suitable hybrids. Protoplast fusion has been used in *Corynebacterium glutamicum, Brevibacterium lactofermentum,* and *B. flavum*.
- Use of **recombinant DNA technology** to amplify the gene dosage of the rate-limiting enzyme. This procedure leads to yield increases if the amount of enzyme synthesis rises in parallel with the gene copy number. This approach has been used with *E. coli*, albeit without significant advantages. In this case, the cloning vector pBR322 was used to increase the gene dosage of genes for the production of the following amino acids: L-glutamic acid, L-histidine, L-lysine, L-phenylalanine, L-proline, L-threonine, and L-valine. The yields were in most cases significantly lower than the yields of commercial production strains of coryneform bacteria. However, cloning systems developed for these commercial strains have permitted more success for several Japanese companies. Kyowa Hakko has used recombinant DNA technology to create hybrid strains of *Corynebacterium glutamicum* for the production of L-cysteine, L-histidine, L-isoleucine, L-phenylalanine, L-serine, and

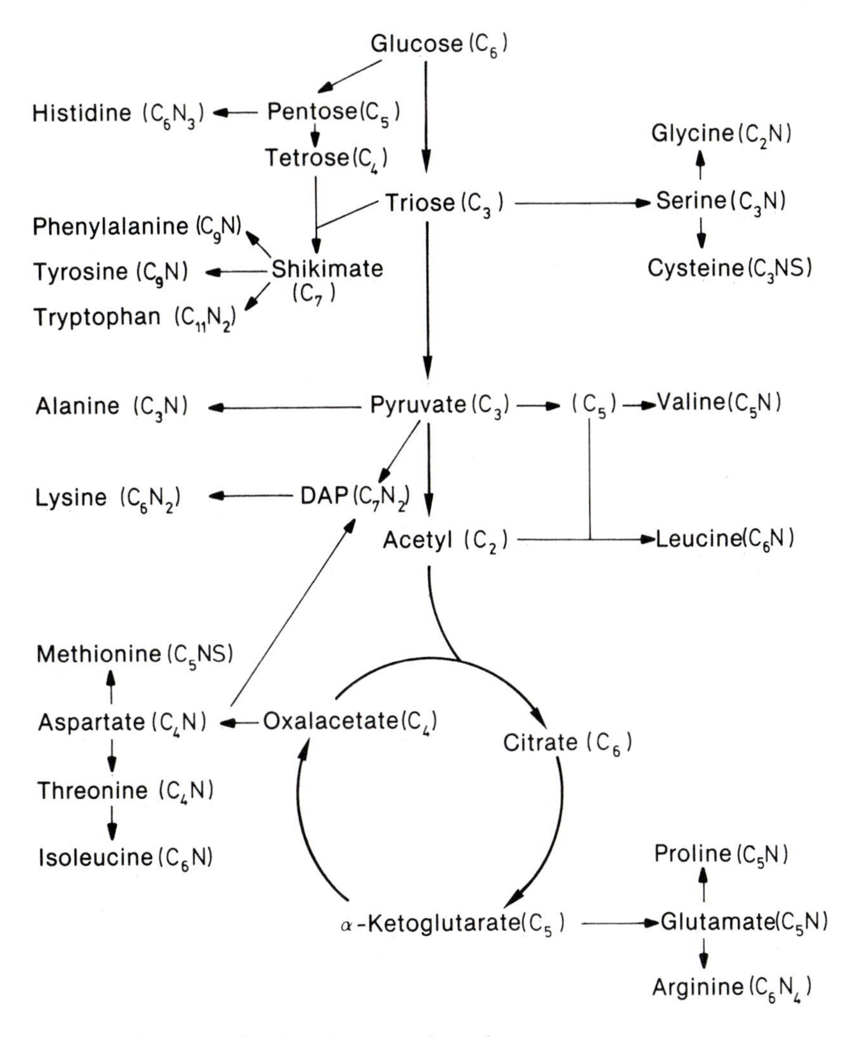

Figure 9.2 Biosynthesis of amino acids using glucose as the carbon source

L-threonine; Ajinomoto has modified *Brevibacterium lactofermentum* for the production of L-histidine, L-phenylalanine, L-proline, L-threonine, and L-tyrosine; Tanabe has altered *Serratia marcescens* for the production of L-histidine, L-proline, and L-threonine.

Mutants producing excess amounts of amino acids have also been isolated by combining several of the above methods.

Table 9.3 lists the strains available for the microbial production of amino acids, their genetic characteristics, and the yields obtained. Processes using various carbon sources have been developed for different production strains. In this list, only the processes with the highest published yields have been given and the actual industrial yields are assumed to be significantly higher.

9.5 PROCESS CONTROL

Amino acid production runs for 2–4 days in batch processes in vessels containing up to 450 m^3.

Methods using continuous processes have been developed, but have not yet been implemented in commercial plants.

Infections due to bacteriophages may cause considerable losses in production. Phage-resistant mutants and chemical agents which inhibit phage reproduction may be used to avoid phage infection. Intensive research on the maintenance of pure cultures (sterilization of equipment, media, and air) is underway; the sterilization of nutrient solutions usually takes place continuously.

Because of the high rate of sugar breakdown and the high respiratory activity during the fermentation, a high oxygen requirement is exhibited. The optimal aeration rates for individual processes depend on the strain, the substrate, and the biosynthetic pathway. Excess heat must be simultaneously dissipated, since the processes are carried out at temperatures between 28–38°C. The pH is kept constant between 6.8–8.0, depending on the process; gaseous NH_3 is frequently used for pH control and is simultaneously metabolized as a nitrogen source. During the fermentation of L-glutamic acid, such critical parameters as aeration rate, temperature, pH, and antifoam dosage are all automatically regulated and in a few cases are under computer control. Biosensors have been developed which permit continuous measurement of the amino acid concentration in the fermenter. For instance, a glutamic acid sensor has been prepared using the glumatine synthase enzyme obtained from a thermophilic organism, *Bacillus stearothermophilus*.

9.6 PRODUCT RECOVERY

Centrifugation is used at the end of the fermentation process to separate the cell material. The amino acids are obtained after acidification through precipitation at the isoelectric point, ion exchange chromatography, electrodialysis, or extraction with organic solvents.

9.7 PRODUCTION OF INDIVIDUAL AMINO ACIDS

The demand for amino acids is increasing. While **L-lysine** and **D,L-methionine** can be manufactured economically for use as feed additives, the demand for **L-threonine** and **L-tryptophan** cannot yet be filled due to the low yields of existing processes.

In order of their importance, the processes for the production of L-glutamic acid, L-lysine and L-tryptophan are discussed below. Further details for these and other amino acids are available in the references at the end of this chapter.

9.8 L-GLUTAMIC ACID

Production strains

L-glutamic acid is manufactured predominantly by microbial means, although it is also manufactured chemically. Japanese researchers began developing a direct fermentation process because the D,L-glutamic acid which is formed by chemical synthesis is the racemic mixture. In screening about 2,000 microorganisms on different media, L-glutamic acid production was found to occur in a wide variety of bacteria, streptomycetes, yeasts, and fungi. The isolation of *Corynebacterium glutamicum* (synonym: *Micrococcus glutamicus*) was accomplished in 1957. It was immediately used industrially by Kyowa Hakko because of its high excretion of glutamic acid. Other industrially important strains with L-glutamic acid excretion of at least 30 g/l belong to the genera *Corynebacterium, Brevibacterium, Microbacterium,* or *Arthrobacter*. Morphologically and physiologically, these glutamic acid-producing strains resemble *C. glutamicum*: They are usually Gram-positive, nonsporulating, nonmotile bacteria. Moreover, all glutamic acid producers require biotin, lack or show little activity of α-ketoglutarate dehydrogenase, and show increased activity of glutamate dehydrogenase. In addition, some *Brevibacterium* and *Corynebacterium* mutants have lower isocitrate lyase activity. Further strain development has led to the isolation of mutants which overproduce glutamic acid even in the presence of high concentrations of biotin. For instance, a lysozyme-sensitive mutant of *C. glutamicum* is able to convert 40% of the added carbon source to L-glutamic acid even in the presence of 100 $\mu g/l$ biotin.

Successful hybrids have also been constructed by cloning *Brevibacterium* or *Corynebacterium* DNA into *Brevibacterium* or *Corynebacterium* recipients.

Biosynthesis of glutamic acid

The glucose carbon source is broken down into C_3 and C_2 fragments by glutamic-acid-producing microorganisms through the Embden-Meyerhof-Parnas (EMP) pathway and the pentose-phosphate-cycle, and the fragments are channeled into the tricarboxylic acid (TCA) cycle. The EMP pathway is more common under conditions of glutamic acid production. The key precursor of glutamic acid is α-ketoglutarate, which is formed in the TCA cycle via citrate and isocitrate and then converted into L-glutamic acid through reductive amination with free NH_4^+ ions (Figure 9.3). This last step is catalyzed by the NADP-dependent glutamate dehydrogenase. The $NADPH_2$ required at this stage of the reaction is furnished through the preceding oxidative decarboxylation of isocitrate to α-ketoglutarate by the enzyme isocitrate dehydrogenase. The $NADPH_2$ is then regenerated by the reductive amination of α-ketoglutarate:

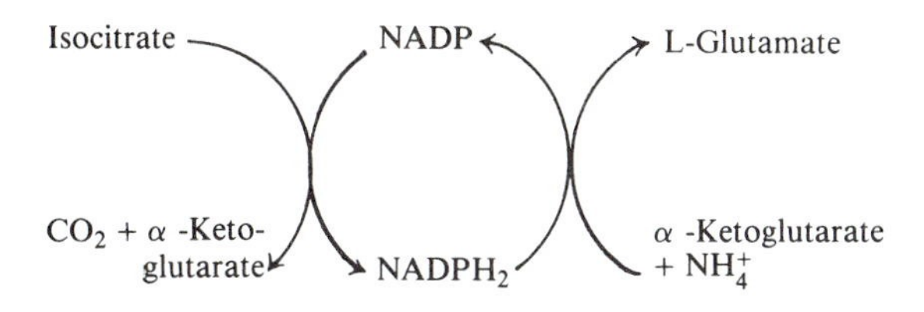

The strain used commercially for glutamic acid production has a block in α-ketoglutarate dehydrogenase. In the absence of NH_4^+ ions, α-ketoglutarate accumulates because of the interruption of the TCA cycle. Thus efficient anaplerotic sequences are necessary to provide TCA cycle intermediates which are required for other cell reactions. Studies on the pathway of biosynthesis have been carried out using labeled compounds. During glutamic acid formation in the presence of $^{14}CO_2$, the α-carboxyl group of glutamate is labeled radioactively. Oxalacetate carboxylase and the NADP-dependent malate enzyme are involved with the CO_2 fixation process. The malate enzyme catalyzes the carboxylation of pyruvate to malate (Figure 9.3). These anaplerotic sequences complete the TCA cycle with C_4 dicarboxylic acids. Malate is then transformed via oxalacetate into citrate and isocitrate, which either serve as preliminary stages to the glutamic acid formation or are channeled into the glyoxylate cycle. Particularly with acetate as a carbon source, the energy gain and the formation of intermediates are carried out chiefly via the glyoxylate cycle for *C. glutamicum* (Figure 9.4). Thus there is competition between a) the isocitrate lyase reaction, which forms succinate and glyoxylate (necessary for optimal growth) and b) the isocitrate dehydrogenase reaction which leads to the key precursor α-ketoglutarate. The stoichiometry for glutamic acid formation from glucose or acetate as a carbon source would be as follows:

$$C_6H_{12}O_6 + NH_3 + 1.5\ O_2 \rightarrow C_5H_9O_4N + CO_2 + 3\ H_2O,$$

$$3\ C_2H_4O_2 + NH_3 + 1.5\ O_2 \rightarrow C_5H_9O_4N + CO_2 + 3\ H_2O$$

One mole of glutamic acid is produced from 1 mole of glucose or from 3 moles of acetate. Experiments with resting cells have shown that the actual conversion rate is between 50–70 mole%. Some part of the yield reduction is due to the reversibility of the malate enzyme reaction and the decarboxylation of oxalacetate to CO_2.

Effect of permeability on glutamic acid production

Production and excretion of excess glutamic acid is dependent upon cell permeability. Increased permeability in glutamic-acid-producing bacteria can be attained in several different ways:

- Through biotin deficiency
- Through oleic acid deficiency in oleic acid auxotrophs

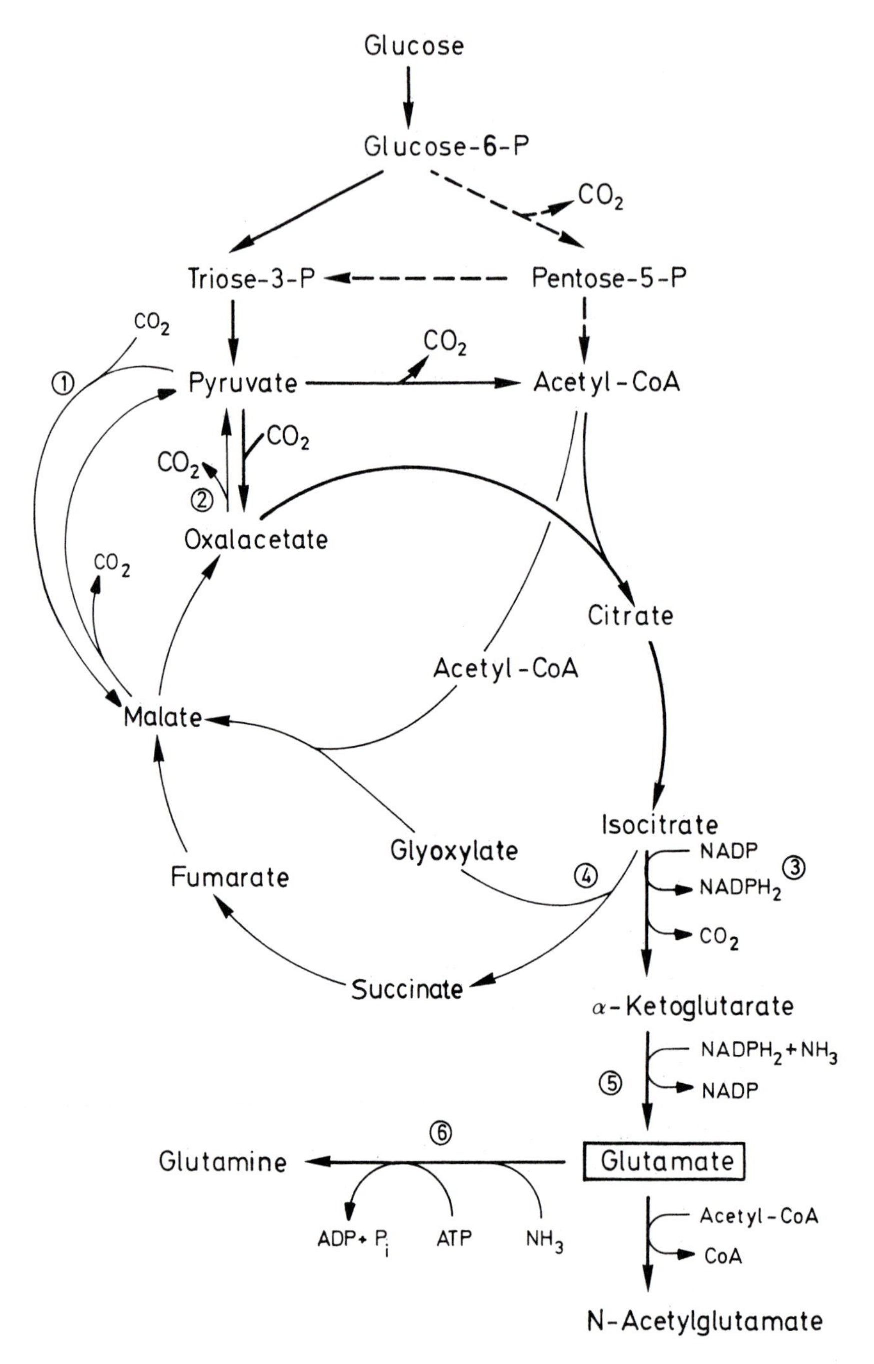

Figure 9.3 Biosynthesis of L-glutamic acid using glucose as the carbon source. Glyoxylic acid cycle, thin lines; pentose phosphate cycle, broken lines. 1, Malic enzyme; 2, Oxalacetate carboxylase; 3, Isocitrate dehydrogenase; 4, Isocitrate lyase; 5, Glutamate acid dehydrogenase; 6, Glutamine synthetase
(Modified according to Kinoshita and Nakayama, 1978)

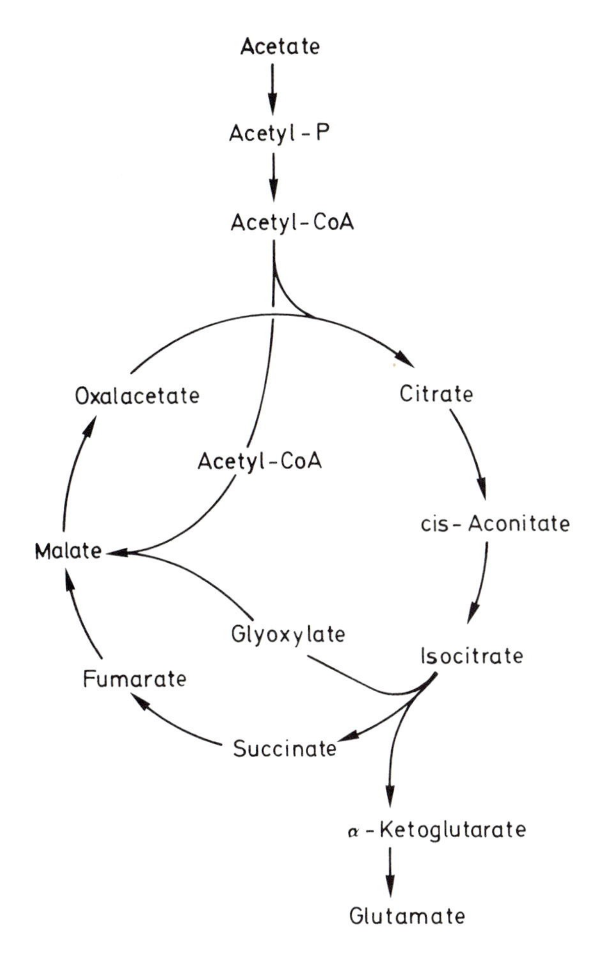

Figure 9.4 Biosynthesis of L-glutamic acid using acetate as the carbon source

- Through the addition of saturated fatty acids (C_{16}–C_{18}) or fatty acid derivatives
- Through the addition of penicillin
- Through glycerol deficiency in glycerol auxotrophs

All glutamic-acid-producing strains have a growth requirement for **biotin**, an essential coenzyme in fatty acid synthesis. In the presence of biotin concentrations >5 μg/l, increased oleic acid synthesis results in a high phospholipid content of the cell membrane. Cells with high phospholipid content are incapable of excreting glutamic acid; up to 25–35 μg L-glutamic acid/mg dry weight is accumulated intracellularly, so that biosynthesis is stopped because of feedback inhibition. On the other hand, growth in a biotin-deficient medium causes membrane damage through reduction of phospholipid synthesis, leading to a changed ratio of saturated to unsaturated fatty acids. Under these conditions, intracellular glutamic acid can be excreted. It has been postulated that a biotin-containing acetyl-CoA-carboxylase, which catalyzes the first stage of fatty acid synthesis from acetyl-CoA to malonyl-CoA, is involved in fatty acid biosynthesis. Saturated fatty acids (C_{16}–C_{18}) repress this acetyl-CoA-carboxylase. Hence it is understandable that oleic acid auxotrophs or strains to which saturated fatty acids have been added synthesize cell membranes with lower phospholipid content; such strains also excrete L-glutamic acid in the presence of biotin.

The addition of penicillin in the logarithmic growth phase promotes significant excretion of L-glutamic acid, even in the presence of biotin. Penicillin is added to fermentations in media containing large amounts of biotin 8–12 hours after inoculation of the fermenter. A penicillin concentration is selected (between 5–300 units/ml) so that the bacterial growth rate is reduced to a level corresponding to the rate in low-biotin media.

The use of penicillin or saturated fatty acids makes possible the commercial use of inexpensive culture medium components, such as sugar cane or sugar beet molasses, which otherwise cannot be utilized due to their high biotin content. The discovery of the role of cell permeability in the production of glutamic acid has thus made possible some rational approaches to the industrial production of this important amino acid.

Conditions of manufacture

Under optimal culture conditions, glutamic-acid-producing bacteria convert about 50–60% of the added carbon source to L-glutamic acid. If less favorable fermentation conditions are used so that low glutamic acid production is obtained, an increase of cell mass and an excretion of lactate, succinate, α-ketoglutarate, glutamine and N-acetylglutamine have been observed. Factors which affect glutamic acid fermentation are described below.

Carbon sources A wide variety of carbohydrates can be used as carbon sources in the fermentation process. Among the monosaccharides, glucose and sucrose are frequently used, and fructose, maltose, ribose, and xylose find some role. Of the unrefined carbohydrate sources, sugar cane and sugar beet molasses are most important, but starch hydrolysates are also frequently employed. Since molasses has a high biotin content (0.4–1.2 mg/kg for cane molasses; 0.02–0.08 mg/kg for beet molasses), penicillin or fatty acid derivatives (e.g., Tween 60) must be added to the fermentation when these inexpensive carbon sources are used. In Europe, only beet molasses is considered an inexpensive carbon source (proportion of manufacturing cost: 26%) but in other areas of the world cane molasses is chiefly used. In Japan, where acetate is inexpensive and readily available in large quantities, extensive studies have been carried out to use this carbon source but for industrial production, cane sugar molasses or starch hydrolysate are still the main carbon sources used.

Processes using methanol, ethanol, acetaldehyde, or n-alkanes have also been developed, but the cost-effectiveness of these processes depends largely on the price of petroleum. Figures 9.5 and 9.6 show the progress of typical fermentations with glucose and acetate as carbon sources.

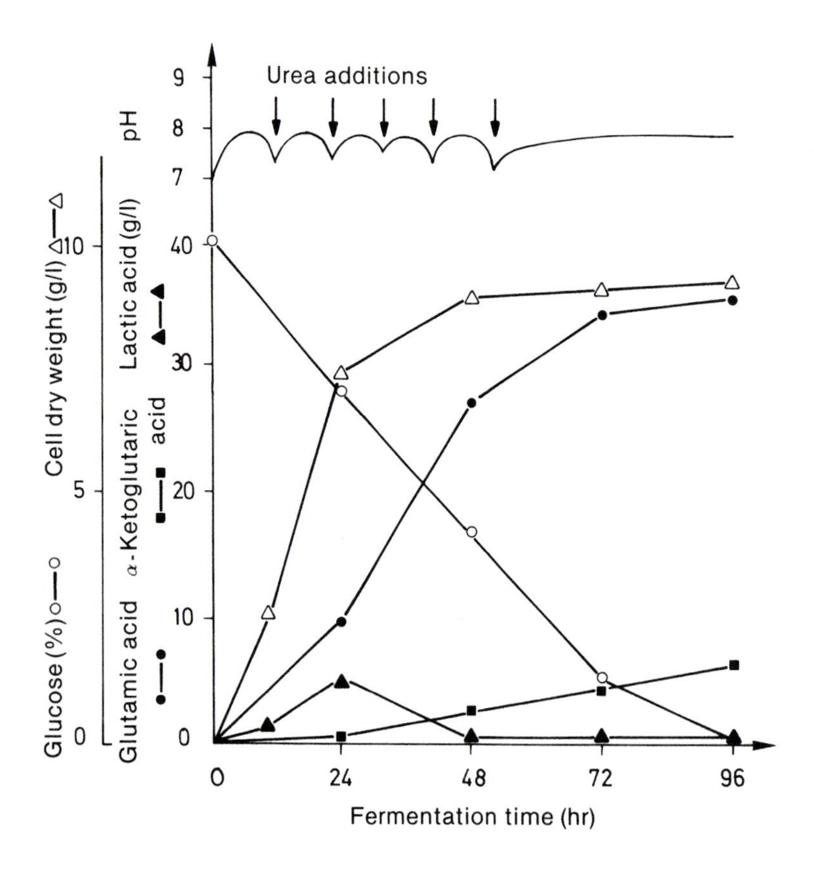

Figure 9.5 Production of L-glutamic acid with *Corynebacterium glutamicum* No. 541 using glucose as the carbon source

Nitrogen sources In addition to ammonium salts, ammonia (gaseous or in aqueous solution) can be used as a nitrogen source. In the industrial manufacture of glutamic acid, ammonia feeding permits pH control and obviates the problem of ammonia toxicity. Most glutamic acid-producing bacteria possess urease activity, so that urea is also frequently used as a nitrogen source. In the acidic pH range with excess ammonia, glutamine is produced instead of glutamic acid.

Growth factors The optimal biotin concentration is dependent on the carbon source used. In media with 10% glucose, it is 5μg/l, in media with lower glucose concentrations it is considerably lower, and for acetate it is between 0.2–1.0 μg/l. Some strains require L-cysteine as an additional growth factor; for media based on an n-alkane, supplementation with thiamine may be necessary.

O_2 Supply Optimal glutamic acid yields are obtained at a K_d value of 3.5×10^{-6} mole·O_2/atm·min·ml. The oxygen concentration should be neither too low nor too high. Under oxygen deficiency, excretion of lactate and succinate occurs, whereas excess oxygen in the presence of an ammonium ion deficiency causes growth inhibition and production of α-ketoglutarate; in both cases glutamic acid yields are low.

Production processes

A typical fermentation from glucose with *Brev-*

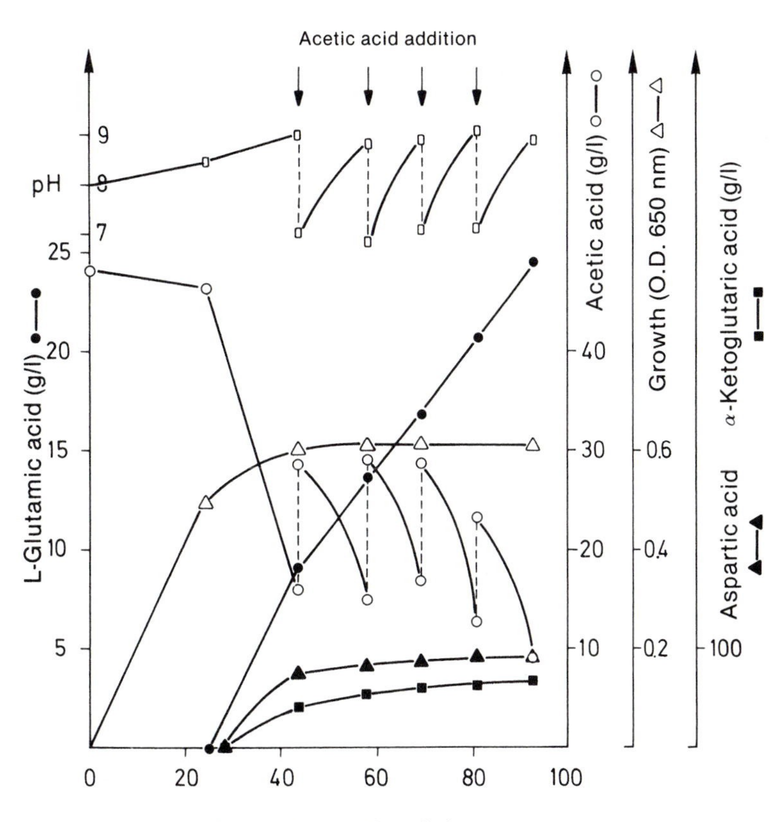

Figure 9.6 Production of L-glutamic acid with *Brevibacterium flavum* No. 2247 using acetate as the carbon source (Tsunoda et al., 1961)

ibacterium divaricatum (NRRL B-231) runs as follows (Miescher, 1975):

- Seed culture: glucose 40 g; K_2HPO_4 1.0 g; $MgSO_4 \cdot 7\ H_2O$ 0.5 g; yeast extract 1.0 g; urea 8 g; tap water 1 l; 16 h incubation at 35°C.
- Main culture: glucose 121 g; ammonium acetate 5 g; molasses from starch saccharification 6 g; KH_2PO_4 1.2 g; K_2SO_4 1.2 g; $MgSO_4$ (anhydrous) 6 g; $FeSO_4 \cdot 7\ H_2O$ 6 ppm; $MnSO_4 \cdot H_2O$ 6 ppm; antifoam agent Hodag K-67 0.1 ml; tap water 1 l. Inoculum volume: 6%.

At the beginning of the fermentation, 0.65 ml/l of oleic acid is added. The pH is set at 8.5 with ammonia and is automatically maintained at 7.8 during the course of the fermentation. After beginning growth of the culture (about 14 hours), the temperature is increased from 32–33°C to 38°C. After metabolism of the glucose down to a level of 0.5–2%, glucose feeding is done until the fermentation is completed; 160 g/l are fed on the average. Aeration is controlled so that the CO_2 content in the exhaust gas does not exceed 4.5 vol%. The glutamic acid content is analyzed hourly. As a rule, the fermentation is stopped after 30–35 hours with a glutamic acid yield of about 100 g/l. If molasses from starch saccharification is substituted for glucose, the glutamic acid yield is 94 g/l after 36 hours. The glutamic acid yields with different carbon sources are listed in Table 9.5.

9.9 L-LYSINE

Lysine is an amino acid essential for animal and human nutrition. It occurs in plant proteins only in low concentrations; addition of lysine can therefore increase the quality of plant foods. The market for lysine is increasing. Lysine is produced today only by microbial processes and a variety of approaches for its production have been developed.

Lysine production via diaminopimelic acid

Lysine-histidine double auxotrophic mutants of *Escherichia coli* (ATCC 13002) produce diaminopimelic acid (DAP) on a molasses medium with a yield of 19–24 g/l. The entire fermentation solution, including the cell material, is subsequently incubated with *Aerobacter aerogenes* (ATCC 12409) at 35°C. After 20 hours, the DAP has been quantitatively decarboxylated to L-lysine. One complication with this procedure is that the DAP formed during the fermentation is a mixture of meso- and LL-forms; since only meso-DAP can be decarboxylated into L-lysine,

Table 9.5 Processes for glutamic acid production with different carbon sources

Carbon source	Organism	Yield (g/l)
Sugar beet molasses	*C. glutamicum*	>100
Glucose + Ammonium acetate	*Brevibacterium divaricatum*	100
Acetate	*B. flavum*	98
Ethanol	*Brevibacterium* sp. 136	59
n-Alkanes	*Arthrobacter paraffineus*	62
	Corynebacterium hydrocarboclastus	84
	C. alkanolyticum (Glycerol$^-$)	72
Benzoic acid	*Brevibacterium* sp.	80
Methanol	*Methylomonas methylovora* M12-4	7

the LL-DAP must be transformed into the meso form by racemization before the decarboxylation step.

Conversion of DL-α-amino caprolactam

The enzymatic transformation of DL-amino caprolactam into L-lysine takes place according to the following scheme:

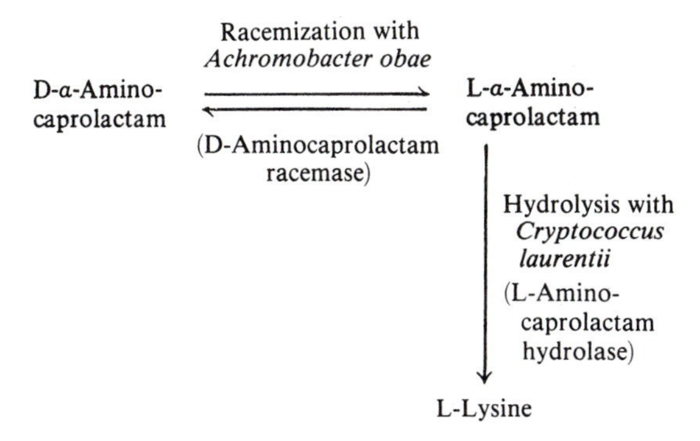

A 10% DL-amino caprolactam solution (pH 8.0) is added to 0.1% (w/v) acetone-dried cells of *Cryptococcus laurentii* and of *Achromobacter obae*. A conversion efficiency of 99.8% is obtained at 40°C after 24 hours. The enzymatic process has been commercialized in Japan by the Toray Company, which has a capacity of over 4000 tons per year. In the Toray process, the enzyme D-α-aminocaprolactam racemase from *Achromobacter obae* has been expressed in a lysine auxotrophic mutant of *E. coli*.

Direct fermentation

Direct fermentation processes are now used world-wide for the production of L-lysine.

Production strains Efficient L-lysine producers are found among glutamic-acid-producing mutants of *Corynebacterium* and *Brevibacterium* which are homoserine auxotrophs or among methionine-threonine double auxotrophs. High-lysine-producing strains are also found among organisms resistant to the lysine antimetabolite S-(β-aminoethyl)-L-cysteine (AEC). The most important lysine-excreting strains are listed later in this chapter (Table 9.7). The development of high-yielding strains by mutation to auxotrophy and to antimetabolite resistance has been carried out with *B. lactofermentum*.

Protoplast fusion between high-yielding strains and wild strains of *B. lactofermentum, Corynebacterium*, and *Brevibacterium* mutants has led to strains with improved growth properties or higher efficiencies.

Cloning studies with *E. coli*, using plasmid pBR322 as a vector, have shown that only in transformed strains that contain the *dapA* gene does a significant increase in lysine production occur (6.5 g/l). The enzyme coded by *dapA*, dihydrodipicolinate synthase (DDPS) is therefore indicated as a rate-limiting step for lysine biosynthesis. Studies have also been carried out on the transformation of *C. glutamicum* with plasmid pAC2 as vector for the DDPS-encoding gene.

Yeasts, such as *Candida periculosa, Saccharomyces cerevisiae* or *Saccharomycopsis lipolytica* have been studied extensively for lysine production. Lysine accumulates intracellularly in these organisms at concentrations up to 20% of the dry weight. However, these yeasts cannot be used successfully in industrial processes because lysine is not excreted into the medium.

Biosynthesis and regulation Lysine is synthesized in microorganisms either via the diaminopimelic acid pathway or the aminoadipic acid pathway. However, in any single organism, only one of the two alternatives is used: Bacteria, actinomycetes, cyanobacteria (blue-green algae), some phycomycetes, and protozoa use the DAP pathway. Some phycomycetes, all ascomycetes, all basidiomycetes, and eucaryotic algae use the aminoadipic acid pathway.

Biosynthesis via the DAP pathway in bacteria is shown in Figure 9.7. Although two organisms may use the same pathway, the manner in which this pathway is regulated may differ, as shown by the comparison of *Escherichia coli* and the lysine-producer *Corynebacterium glutamicum* in

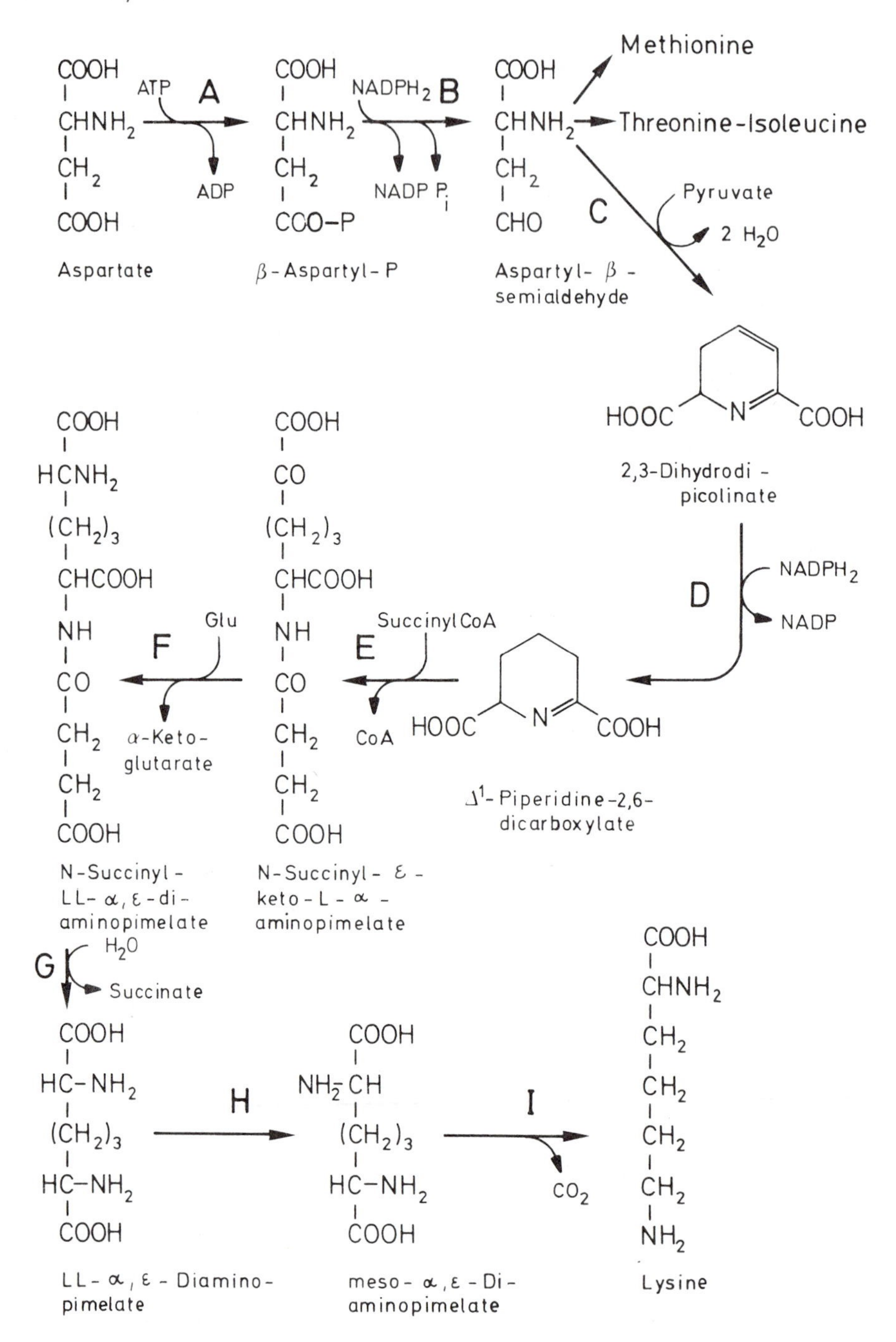

Figure 9.7 Biosynthesis of L-lysine via the DAP pathway. A, Aspartokinase; B, Aspartic semialdehyde dehydrogenase; C, Dihydrodipicolinate reductase; D, Dihydrodipicolinate synthase; E, Succinylketoaminopimelate synthase; F, Succinyldiaminopimelate aminotransferase; G, Succinyldiaminopimelate desuccinylase; H, Diaminopimelate epimerase; I, meso-Diaminopimelate decarboxylase

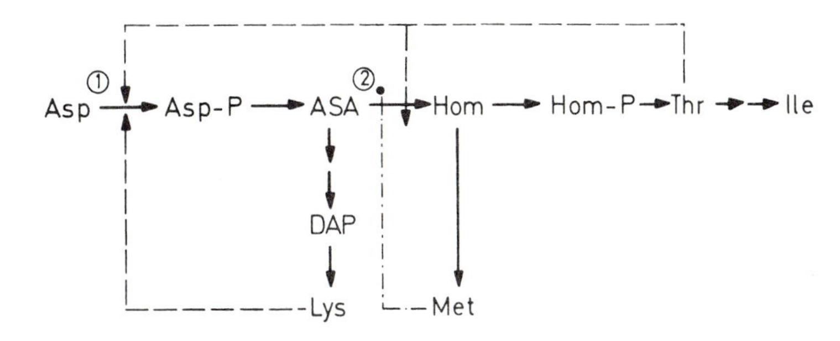

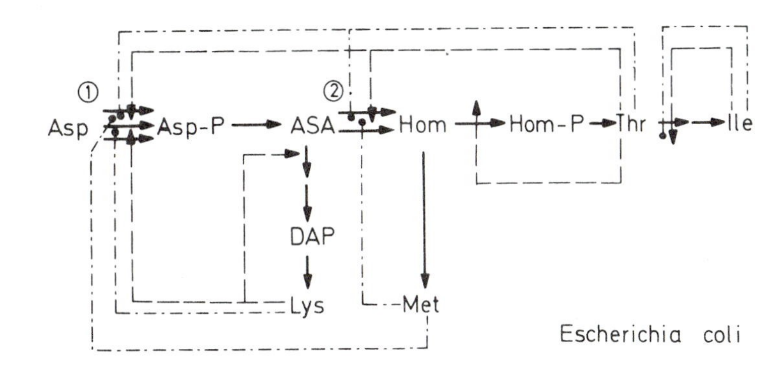

Figure 9.8 Regulation of L-lysine and L-threonine biosynthesis in *Escherichia coli* and *C. glutamicum*. 1, Aspartokinase; 2, Homoserine dehydrogenase. Dashed line, feedback inhibition; dashed and dotted line, repression. (From Kinoshita and Nakayama, 1978)

Figure 9.8. In *Escherichia coli* three distinct regulatory processes are involved:

- Two isoenzymes of homoserine dehydrogenase exist which are repressed by L-methionine or L-threonine.
- There are three isoenzymes of aspartokinase, one showing repression by L-methionine, the second showing multivalent repression by L-threonine and L-isoleucine in addition to feedback inhibition by L-threonine, and the third showing feedback inhibition and repression by L-lysine.
- Dihydrodipicolinate synthase, the first specific enzyme of lysine biosynthesis, shows feedback inhibition due to L-lysine.

For all three of these enzymatic reactions, regulatory mechanisms must be eliminated to obtain the overproduction of L-lysine which is necessary for its commercial preparation.

In contrast to *Escherichia coli*, the regulatory mechanism for lysine-producing strains, such as *Corynebacterium glutamicum* or *Brevibacterium flavum*, is much simpler (Figure 9.8). There is only one aspartokinase and one homoserine dehydrogenase. Aspartokinase is regulated via multivalent feedback inhibition from L-threonine and L-lysine, as the experimental data in Table 9.6 show.

In both organisms, feedback inhibition due to L-threonine and repression due to L-methionine regulate the homoserine dehydrogenase. Biosynthesis from aspartate semialdehyde to L-lysine has already been accomplished in wild strains without regulatory control. Thus good lysine-producers are classified among three mutant types:

- The flux of aspartate semialdehyde to threonine is reduced in homoserine auxotrophs which have a block in the homoserine de-

Table 9.6 Effect of various amino acids on aspartokinase activity

Added amino acid (10 μmol/ml each, L-form)	Relative aspartokinase activity (%) *C. glutamicum* (Homoserine⁻)	*B. flavum*
Control	100	100
Threonine	123	99
Lysine	94	80
Homoserine	94	102
Methionine	109	102
Threonine + Lysine	52	44
Homoserine + Lysine	95	74
Threonine + Methionine	129	114

(Nakayama, 1972)

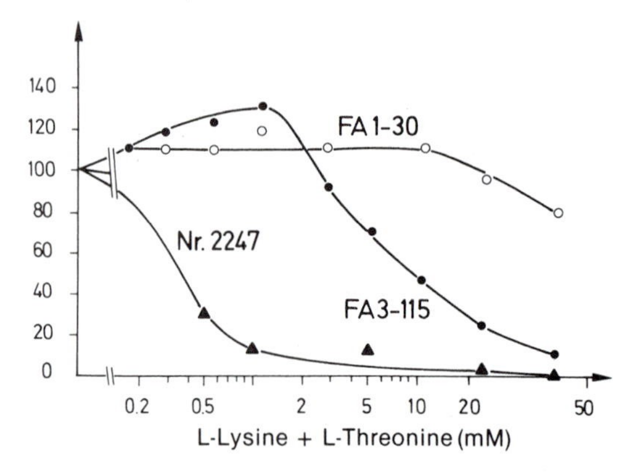

Figure 9.9 Effect of lysine and threonine on aspartokinase activity. No. 2247: wild type *Brevibacterium flavum*. FA 1-30 and FA 3-115: AEC^r mutants (From Hirose et al., 1978)

hydrogenase. Because of the low threonine content in the cell, the multivalent feedback inhibition of aspartokinase is prevented despite increased lysine formation.

- The same effect is obtained with a mutant in which homoserine dehydrogenase is supersensitive to feedback inhibition through threonine.
- In comparison to the wild type, the aspartokinase of AEC-resistant mutants is insensitive to multivalent feedback inhibition, as Figure 9.9 shows. Excess aspartate semialdehyde flows into lysine biosynthesis, because the conversion to L-threonine is prevented through feedback inhibition. In this simply regulated system, strains with increased lysine formation can be developed in just a few mutation steps.

Lysine excretion is accomplished through active transport. In some organisms, a loss of lysine through decarboxylation to cadaverine occurs, but this destruction does not occur in *Corynebacterium* or *Brevibacterium* strains.

Conditions for commercial production Sugar cane molasses is primarily used as a carbon source in industrial production; acetate, ethanol or alkanes can also be used. Gaseous ammonia or ammonium salts are used as nitrogen sources; urea is also used if the producing microorganism has urease activity. Growth factors L-homoserine or L-threonine and L-methionine must be added, but in suboptimal concentration to avoid undesirable regulatory effects. Soy protein hydrolysates or other inexpensive protein sources are frequently used. The biotin content in the medium must be over 30 μg/l for optimal lysine production. The biotin content in sugar cane molasses is usually high enough, but if sugar beet molasses or starch hydrolysates are used, biotin must be added. Table 9.7 shows lysine production from different carbon sources.

A typical fermentation is shown in Figure 9.10. The lysine yield is 44 g/l and the conversion rate in relation to the sugar used approaches a maximum of 30–40%.

A typical fermentation based on sugar cane molasses with *C. glutamicum* (Strain No. 901; (Hom⁻) takes place as follows:

- First seed culture: glucose 20 g; peptone 10 g; meat extract 5 g; NaCl 2.5 g; tap water 1 liter.
- Second seed culture: sugar cane molasses 50 g; $(NH_4)_2SO_4$ 20 g; corn steep liquor 50 g; $CaCO_3$ 10 g; tap water 1 l.
- Main culture: Sugar cane molasses 200 g

Table 9.7 Lysine production on media with different carbon sources

Carbon source	Organism	Genetic characteristics	Yield (g/l)
Glucose	*C. glutamicum*	Hom^- Leu^-	34
Glucose	*C. glutamicum*	Hom^- Leu^- AEC^r	39
Glucose	*B. lactofermentum*AJ11204	$AEC^rAla^-CCL^rML^rFP^s$	70
Acetate	*B. flavum*	Hom^{leaky} Thr^-	75
Ethanol	*B. lactofermentum*	AEC^r	66
n-Alkanes (C_{14}–C_{18})	*Nocardia alkanoglutinosa*	AEC^r	29

(ACE^r: Resistance to S-(β-aminoethyl)-L-cysteine; CCL, α-chlorcaprolactam; ML, γ-methyl-L-lysine; FP^s, β-fluoropyruvate sensitive. Auxotrophs: Ala, alanine; Hom, Homoserine; Leu, Leucine; Thr, Threonine)

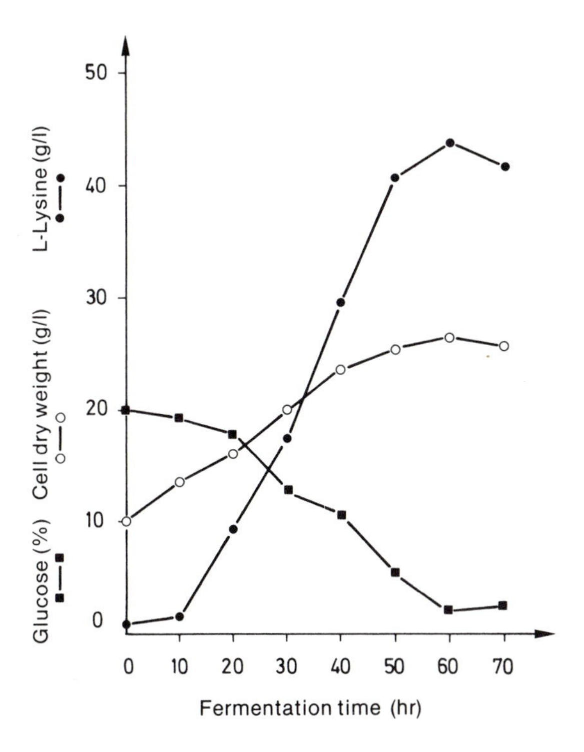

Figure 9.10 Production of L-lysine with *Corynebacterium glutamicum* No. 901 using glucose as the carbon source (From Nakayama, 1972)

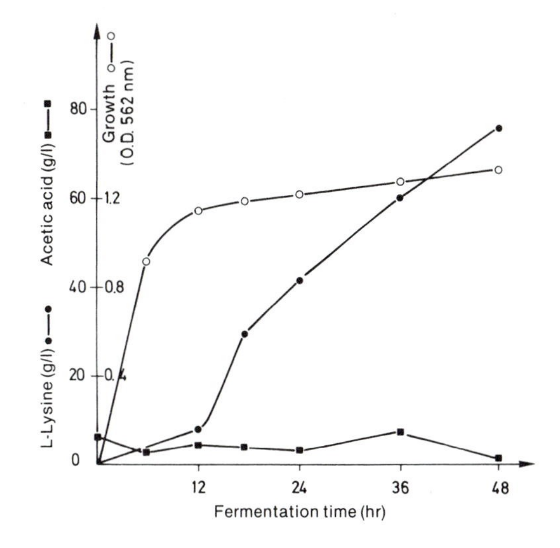

Figure 9.11 Production of L-lysine with *Brevibacterium flavum* using acetate as the carbon source (From Tanaka et al., 1971)

(based on sugar content); soy protein hydrolysate 18 g; tap water 1 l. The pH is kept neutral with aqueous ammonia. Duration of fermentation, 60 hours. Impeller speed 150 rpm; aeration 0.6 vvm; temperature 28°C.

Figure 9.11 shows a fermentation by *Brevibacterium flavum* (Hom^{leaky}, Thr^-) with acetate as the carbon source. The composition of the production medium is as follows: acetic acid 7 g; KH_2PO_4 0.4 g; $MgSO_4 \cdot 7\,H_2O$ 0.4 g; $FeSO_4 \cdot 7\,H_2O$ 0.01 g; soy protein hydrolysate 35 g; glucose 30 g; biotin 50 μg; thiamine HCl 40 μg; tap water 1 l. The pH is kept at 7.4 through acetic acid feeding (60% acetic acid and ammonium acetate in molar ratio of 100:25; 3% glucose). After 48 hours at 33°C the lysine content is 75 g/l; a conversion rate of 29% is obtained based on acetic acid and glucose used.

9.10 L-TRYPTOPHAN

Tryptophan has traditionally been produced through chemical synthesis or through fermentative or enzymatic conversion of chemically synthesized intermediates. The procedures have involved primarily the use of immobilized cells or immobilized enzymes. To optimize the process, recombinant DNA technology has been used. The disadvantage of the enzymatic process is the high cost of the starting materials, indole, serine, pyruvate, or anthranilic acid. Because of this, extensive work has been carried out to develop a direct fermentation of L-tryptophan. Because of the complex regulatory mechanism of tryptophan biosynthesis, these studies have not led to titers sufficiently high so that L-tryptophan could be used in the food industry, although direct fermentation can be used to produce this amino acid for medical uses.

Table 9.8 Production of tryptophan by transformation of precursors

Organism	Precursor	Amount used (g/l)	Yield (g/l)
Hansenula anomala	Anthranilic acid	4.2	5.7
Bacillus subtilis (Anthranilate⁻)	Anthranilic acid	5.0	5.5
Candida utilis	Anthranilic acid	4.2	6.4
Claviceps purpurea	Indole	1.3	1.5
Escherichia coli	Indole +	3.0	
	DL-Serine	6.0	5.2
Bacillus subtilis ($5FT^rAG^r$)	Anthranilic acid	15.75 (feeding)	15.6

(5FT, 5-fluorotryptophane; AG, 8-azaguanine)
From Kinoshita and Nakayama (1978), with additions.

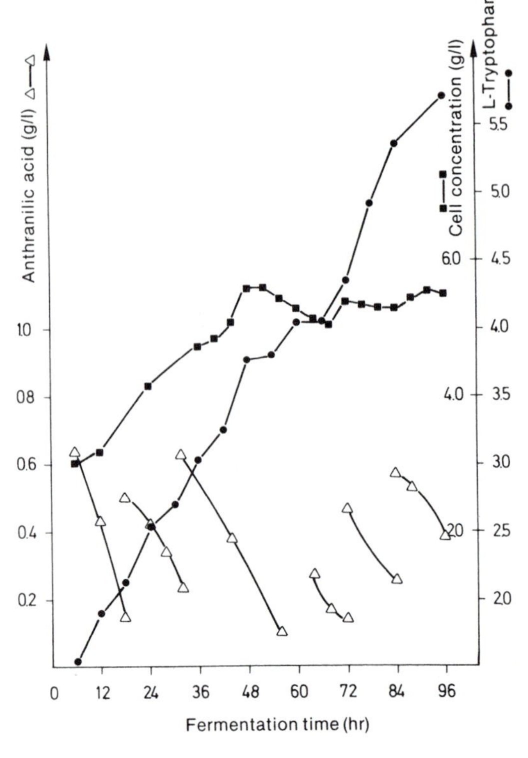

Figure 9.12 Production of L-tryptophan from anthranilic acid by *Hansenula anomala*. Cultured on the following medium (%): glucose 5; NH_4NO_3 0.3; KH_2PO_4 0.05; $Na_2HPO_4 \cdot 12\ H_2O$ 0.2; $NaH_2PO_4 \cdot 2\ H_2O$ 0.1; $MgSO_4 \cdot 7\ H_2O$ 0.05; NaCl 0.01; $FeSO_4 \cdot 7\ H_2O$ 0.001; yeast extract 0.1; anthranilic acid 0.2; $CaCO_3$ 2.0 (From Terui, 1972)

Table 9.9 Enzymatic processes for the production of L-tryptophan

Enzyme	Organism	Reaction
D-Tryptophan hydantoin racemase L-Tryptophan hydantoin hydrolase N-Carbamoyl-L-tryptophan hydrolase	*Flavobacterium aminogenes*	DL-Tryptophan hydantoin + $H_2O \rightarrow$ L-Tryptophan + CO_2 + NH_3
Tryptophanase	*Proteus rettgeri* *Achromobacter liquidum* *E. coli*	Indole + Pyruvate + $NH_3 \rightarrow$ L-Tryptophan + H_2O Indole + L-Serine $\rightarrow$ L-Tryptophan + H_2O
Tryptophan synthetase	*E. coli*	Indole + L-Serine $\rightarrow$ L-Tryptophan + H_2O

Fermentative conversion of intermediates stages

The most important fermentation processes are listed in Table 9.8. Figure 9.12 shows a typical fermentation sequence with *Hansenula anomala*. Starch hydrolysate or molasses is used as a carbon source and the feeding of sugar/NH_4NO_3-solution and anthranilate begins after the culture is grown.

Enzymatic process for tryptophan

Table 9.9 lists enzymatic processes for L-tryptophan production. The methods include either the stereoselective hydrolysis of hydantoin compounds derived by chemical synthesis or the bioconversion of intermediates.

In the **hydantoinase technique,** an enzyme

Table 9.10 Mutants for the production of tryptophan by direct fermentation with glucose

Organism	Genetic characteristics	Yield (g/l)
Bacillus subtilis	$5FT^r$, Arg^- or Leu^-	6
*Enterobacter cloacae*TA 599	$5MT^rTHA^rStr^rTyr^-$	10
Corynebacterium glutamicum Px-115-97	Phe^- Tyr^- $5MT^r$ $TrpHx^r$ $6FT^r$ $4MT^r$ PFP^r PAP^r $TyrHx^r$ $PheHx^r$	12
*Brevibacterium flavum*S-225	$5FT^rAZ^rSG^rPFP^rTyr^-$	19

Resistance: 5FT 5-fluorotryptophan; PFP p-fluorophenylalanine; 5MT 5-methyltryptophan; TrpHx Tryptophan hydroxamate; 6FT 6-fluorotryptophan; 4MT 4-methyltryptophan; PAP p-aminophenylalanine; TyrHx Tyrosine hydroxamate; PheHx Phenylalanine hydroxamate; AZ azide; SG sulfaguanidine; THA thienylalanine; Str streptomycin.
Auxotrophy: Arg arginine; Leu leucine; Phe phenylalanine; tyr tyrosine.

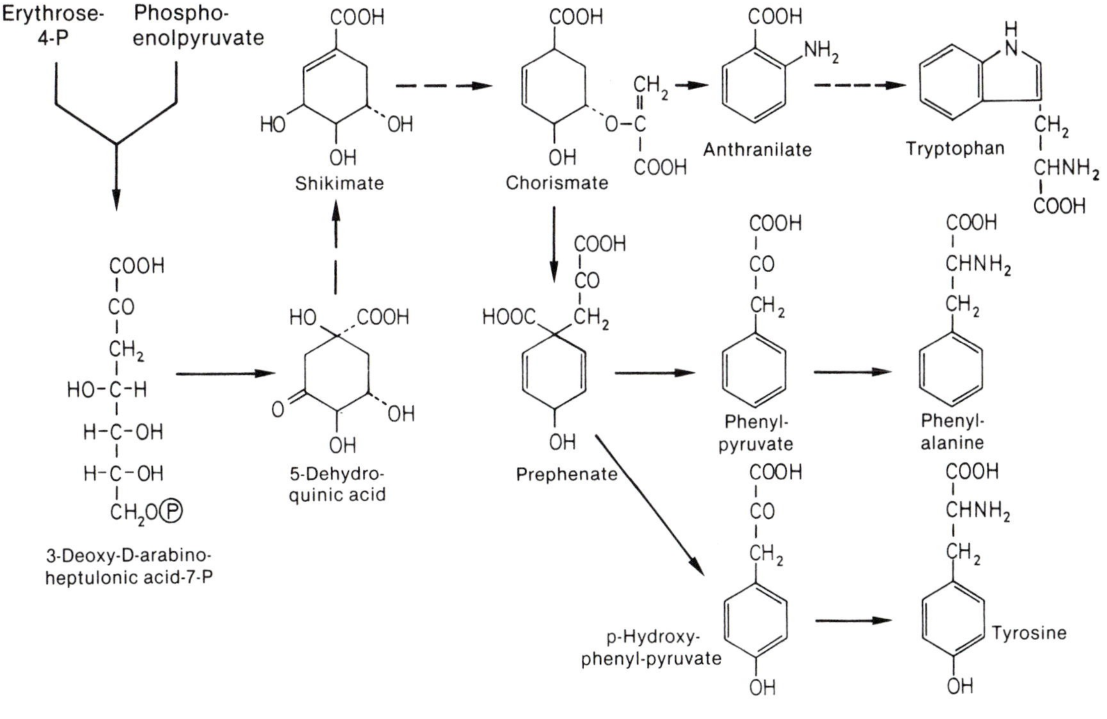

Figure 9.13 Pathways for the biosynthesis of tryptophan, phenylalanine and tyrosine

from *Flavobacterium aminogenes* which hydrolyzes D,L-tryptophan hydantoin is used. This strain has a reduced tryptophan-degrading ability and the enzyme is induced by the addition of D,L-tryptophan hydantoin to the medium. A 100% conversion of the hydantoin to L-tryptophan is obtained (starting material, 5% D,L-tryptophan hydantoin; 40°C; 100 hours), the tryptophan formed being removed from the reaction mixture by complex formation with inosine.

The most widely used commercial process is the **tryptophanase technique,** which involves the conversion of indole using *Proteus rettgeri.* The reaction mixture contains, per liter culture solution, 60 g indole (dissolved in 100 ml methanol), 80 g sodium pyruvate, 80 g ammonium acetate, 0.01 g pyridoxal phosphate, and 1 g sodium sulfate. After incubation at 34°C for 48 hours, 75 g of L-tryptophan is produced. By precipitation of the tryptophan with inosine as it is formed, the yield can be even higher (83.3 g/l), resulting in a molar conversion efficiency of 96%. In a related process, an enzyme reactor containing tryptophanase from *E. coli* has been used to convert indole and L-serine, resulting in a tryptophan concentration of 200 g/l and an efficiency of 95%.

Another process involves the use of the enzyme **tryptophan synthetase** which transforms indole and L-serine to L-tryptophan. In a mixed culture of *E. coli* and *Pseudomonas putida,* using D,L-serine as the starting material, the racemization of the D-form of tryptophan leads to yields of 23 g/l (81% efficiency).

Direct fermentation processes

The possibilities for a direct synthesis of L-tryptophan starting with inexpensive carbon sources are outlined in Table 9.10.

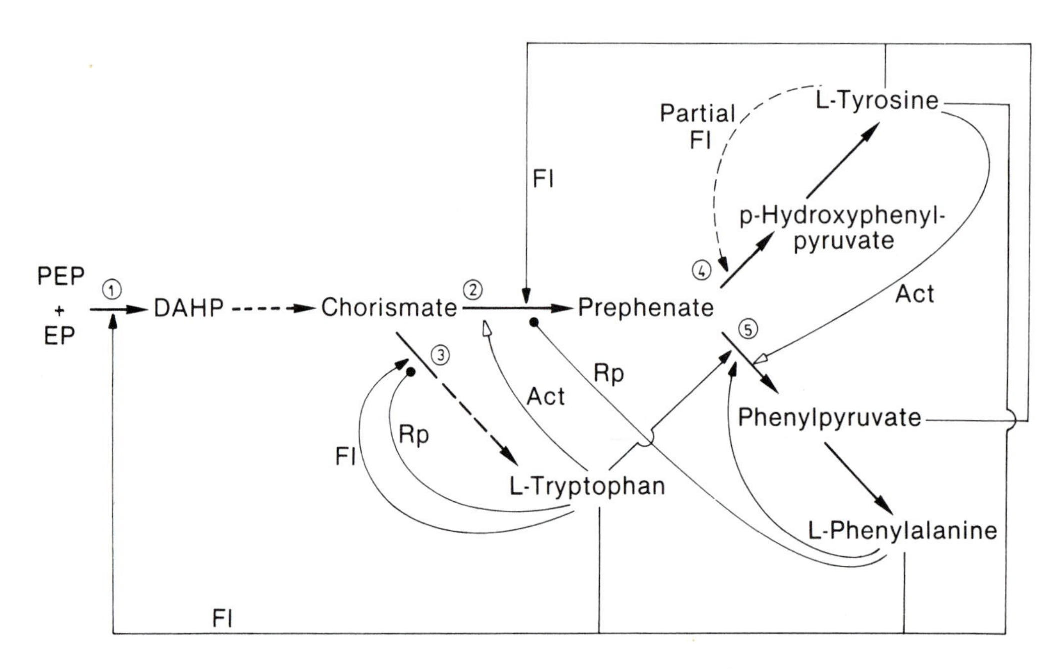

Figure 9.14 Regulation of aromatic amino acid biosynthesis in *Corynebacterium glutamicum.* The main sites of regulation are numbered, and the lines and symbols indicate the type of regulation. 1, DAHP synthetase; 2, Chorismate mutase; 3, Anthranilate synthase; 4, Prephenate dehydrogenase; 5, Prephenate dehydratase.
FI, Feedback inhibition; Rp, Repression; Act, Activation.
EP, Erythrose-4-P; PEP, Phosphoenolpyruvate; DAHP, 3-Deoxy-D-arabinoheptulonic acid-7-P.
(From Nakayama et al., 1976)

The tryptophan biosynthetic pathway illustrated in Figure 9.13 has been found in many microorganisms, but there are considerable differences in the regulation of the individual organisms. In *Corynebacterium glutamicum*, for instance, tryptophan biosynthesis is regulated by the activity of DAHP synthetase and anthranilate synthase (Figure 9.14). DAHP synthetase, which catalyzes the condensation of erythrose-4-phosphate and phosphoenol pyruvate, shows concerted feedback inhibition by phenylalanine and tyrosine; tryptophan increases the inhibition so that the activity of DAHP synthetase is inhibited up to 90% in the presence of all three amino acids. Anthranilate synthase, the first enzyme of tryptophan biosynthesis after the branch point, shows feedback inhibition and repression due to L-tryptophan. The DAHP synthetase of *C. glutamicum* Px-115-97 is completely sensitive to regulation by phenylalanine and tyrosine. The maximum yield is determined not only by the effect of regulatory mechanisms but also by the specific activities of the individual enzymes of the biosynthetic chain. More yield increases are to be expected from the application of recombinant DNA technology to the development of derepressed strains of tryptophan producers, following the methods that have already been perfected with *E. coli.* The genes for the biosynthetic pathway from anthranilic acid to tryptophan are present in a single cluster. A tryptophanase-negative strain with a deletion of the *trp* operon has been transformed with a plasmid containing the *trp* operon in which the gene for anthranilate synthase was resistant to feedback inhibition by tryptophan. The *E. coli* strain Tna (pSC101 *trp* I15–14), whose plasmid contains five copies per genome, forms 4–5 g/l tryptophan when using glucose as carbon source. By the addition of a multicopy plasmid with the *trp* operon, it has also been possible to obtain yields of 23.5 g/l with *E. coli* strain JP 4114.

REFERENCES

Aida, K., I. Chibata, K. Nakayama, K. Takinami, and H. Yamada (eds.). 1986. Progress in industrial microbiology. Vol. 24. Biotechnology of amino acid production. Kodansha Ltd., Tokyo.

Groeger, U., and H. Sahm.1987. Microbial production of L-leucine from α-ketoisocaproate by *Corynebacterium glutamicum*. Appl. Microbio. Biotechnol. 25: 352–356.

Hamilton, B.K., H.Y. Hsiao, W.E. Swann, D.M. Anderson, and J.J. Delente. 1985. Manufacture of L-amino acids with bioreactors. Trends Biotechnol. 3: 64–68.

Hirose, Y., K. Sano, and H. Shibai. 1978. Amino acids. Ann. Rep. Ferm. Proc. 2: 155–189.

Hirose, Y., H. Enei and H. Shibai. 1985. L-Glutamic acid fermentation. pp. 593–600. In: Moo-Young, M. (ed.). Comprehensive biotechnology. Pergamon Press, Oxford.

Kamiryo, T., S. Parthasarathy, and S. Numa. 1976. Evidence that acyl coenzyme A synthetase activity is required for repression of yeast acetyl coenzyme A carboxylase by exogenous fatty acids. Proc. Natl. Acad. Sci. USA 73: 386–390.

Kinoshita, S. 1987. 30 years of amino acid fermentations. Vortrag. 4th European Congress on biotechnology, Amsterdam.

Kinoshita, S. and K. Nakayama. 1978. Amino acids. pp. 209–261. In: Rose, A.H. (ed.). Economic microbiology, Vol. 2. Academic Press, London.

Martin, J.F., R. Santamaria, H. Sandoval, G. del Real, L.M. Mateos, J.A. Gil, and A. Aguilas. 1987. Cloning systems in amino acid-producing *Corynebacteria.* Bio/Technol. 5: 137–146.

Nakanishi, T., T. Azuma, T. Hirao, K. Hattori, and M. Sakurai. 1986. Process for producing L-lysine by fermentation. U.S. patent 4,623,623.

Nakayama, K. 1972. Lysine and diaminopimelic acid. pp. 369–397. In: Yamada, K., S. Kinoshita, T. Tsunoda and K. Aida (eds.). The microbial production of amino acids. Kodansha, Tokyo.

Nakayama, K. 1985. Lysine, pp. 607–620. Tryptophan, pp.621–631. In: Moo-Young, M. (ed.). Comprehensive biotechnology, Vol. 3. Pergamon Press, Oxford.

Schmidt, E., D. Vasic-Racki, and C. Wandrey. 1987. Enzymatic production of L-phenylalanine from the racemic mixture of D, L-Phenyllactate. Appl. Microbiol. Biotechnol. 26: 42–48.

Soda, K., H. Tanaka, and N. Esaki. 1983. Amino acids, pp. 479–530. In: Rehm, H.J. and G. Reed (eds.). Biotechnology, Vol. 3. VCH Publishers, Deerfield Beach, FL.

Syldatk, C, D. Cotoras, A. Möller, and F. Wagner. 1986. Microbial, enantioselective hydrolysis of D, L-l-5-monosubstituted hydantoins for the production of D- and L-aminoacids. BTF-Biotech-Forum 3: 9–18.

Tanaka, K., T. Suzuki, and S. Okumura. 1971. Production of sugars and amino acids from hydrocarbons and petrochemicals by microorganisms, pp. 165–170. 8th World Petroleum Congress, Proceedings No. 5. Applied Science Publ., London.

Teh, J., D. Leigh, G. Allen, H. Burrill, P. Cowan, and H. Camakaris. 1985. Direct production of tryptophan by *Escherichia coli* from simple sugars. Biotech 85 Asia 3: 399–402.

Terui, G. 1972. Tryptophan, pp. 515–531. In Yamada, K., S. Kinoshita, T. Tsunoda, and K. Aida (eds.), The microbial production of amino acids. Kodansha, Tokyo.

Tsuchida, T., K. Miwa, S. Nakamori, and H. Momose. 1984. Method for producing l-glutamic acid by fermentation. U.S. Patent 4,427,773.

Tsunoda, T., I. Shiio, and K. Mitsugi. 1961. Bacterial formation of L-glutamic acid from acetic acid in the growing culture medium. (1) Cultural conditions. J. Gen. Appl. Microbiol. 7: 18–29.

10

Nucleosides, nucleotides, and related compounds

10.1 INTRODUCTION

Microbially produced nucleosides and nucleotides have found wide use as flavor-enhancers for food. The use of synthetic flavor-enhancers developed first in the Orient and has spread throughout the world. Since the beginning of the seventeenth century, the Japanese have used different substances (fungi, dried fish, seaweed, etc.) to enhance the flavor of foods. L-Glutamic acid (see Chapter 9) and the histidine salt of inosinic acid were identified in 1908 and 1913 as flavor-enhancing components of these substances. Further studies showed that a number of other substances, all purine ribonucleoside-5′-monophosphates, also improve flavor. In descending order of effectiveness, these are: guanylic acid (5′-GMP), inosinic acid (5′-IMP) and xanthylic acid (5′-XMP). Substances which have no effect are 5′-AMP, and the 2′ and 3′ isomers, as well as the 5′-deoxyribonucleotides, nucleosides and pyrimidine nucleotides.

The process of producing **5′-IMP** and **5′-GMP** through enzymatic hydrolysis of yeast RNA was developed in Japan in 1959 and was commercialized in 1961. The development of direct fermentation processes for nucleotide production was accomplished at about the same time, also in Japan. The discovery of synergism between the flavor-enhancing properties of sodium glutamate and the sodium salts of 5′-IMP and 5′-GMP was another research advance. $IMP{\cdot}Na_2$, and $GMP{\cdot}Na_2$, as well as mixtures of sodium and calcium salts of both nucleotides, are approved food additives and are frequently used in combination with sodium glutamate in prepared food products. To enhance the flavor of soups and sauces, for example, as little as 0.005–0.01% 5′-ribonucleotide additives is sufficient; moreover, addition of these additives can help eliminate undesirable properties, such as the metallic taste of cans. The amount of seasoning mixture consumed per meal is 0.1–0.2 g; the mixture contains 8–12% 5′-IMP and 1.5–2.0% 5′-GMP added in combination with sodium glutamate (based on the Na glutamate content). Both hydrolytic RNA-splitting and direct fermentation processes are about equally used in Japanese commercial production, for an annual production of 3000 tons.

In addition to their use in the food industry, nucleotides, nucleosides and related compounds are being studied for **therapeutic purposes**. The antibiotic and cytostatic effects of purine analogs (8-azaguanine, 6-mercaptopurine) are being tested, particularly for cancer chemotherapy. 9-β-D-arabinofuranosyladenine is effective against herpes simplex and herpes zoster in humans.

The substances which are microbially produced in Japan and their applications are given in Table 10.1. The following Japanese companies are producers of 5′-ribonucleotides and nucleosides: Ajinomoto Co., Asahi Chemical Ind., Kyowa Hakko Kogyo Co., Takeda Chemical Ind. Ltd. and Yamasa Shoyu Co. Ltd.

Table 10.1 Nucleotides, nucleosides, and related compounds manufactured in Japan

Substance	Amount manufactured (tons/year)	Application
5′-IMP	2000	Flavor enhancer
5′-GMP	1000	Flavor enhancer
Inosine	25	Heart ailments
ATP	6	Muscular dystrophy
CDP-Choline	2	Stroke
FAD	Small amounts	Liver or kidney diseases
NAD	Small amounts	Liver or kidney diseases
Adenine	11	Leukopenia
Adenosine	5	Coronary deficiencies, angina pectoris, arteriosclerosis, high blood pressure
5′-AMP	?	Circulation problems, rheumatism
cAMP	?	Diabetes, asthma, cancer therapy, biochemical reagent
Orotic acid	20	Liver diseases
6-Azauridine	?	Cancer chemotherapy
Ara A (9-β-D-arabino-furanosyladenine)	?	Antiviral

(Yamada, 1977, with additions)

10.2 STRUCTURE AND BIOSYNTHESIS

The structures of the economically relevant compounds are shown in Figure 10.1.

As seen in Figure 10.2, the biosynthesis of purine-5′-ribonucleotides proceeds from 5′-phosphoribosylpyrophosphate (PRPP), which is formed from ribose-5-phosphate and ATP. The first complete purine derivative, inosinic acid (5′-IMP), is produced after several reaction steps and 5′-IMP is then the precursor of 5′-AMP, 5′-XMP and 5′-GMP. AMP biosynthesis involves the action of adenyl succinate lyase (step M, Figure 10.2), a bifunctional enzyme which also catalyzes reaction H to AICARP.

In GMP biosynthesis, IMP is oxidized (reaction N) to XMP and aminated (reaction O) to GMP.

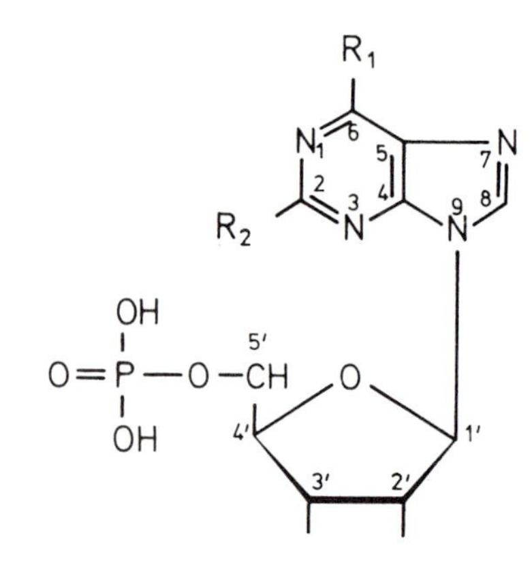

Nucleotide	R_1	R_2
5′-AMP	$-NH_2$	-H
5′-GMP	-OH	$-NH_2$
5′-IMP	-OH	-H
5′-XMP	-OH	-OH

Figure 10.1 Structure of the purine nucleotides

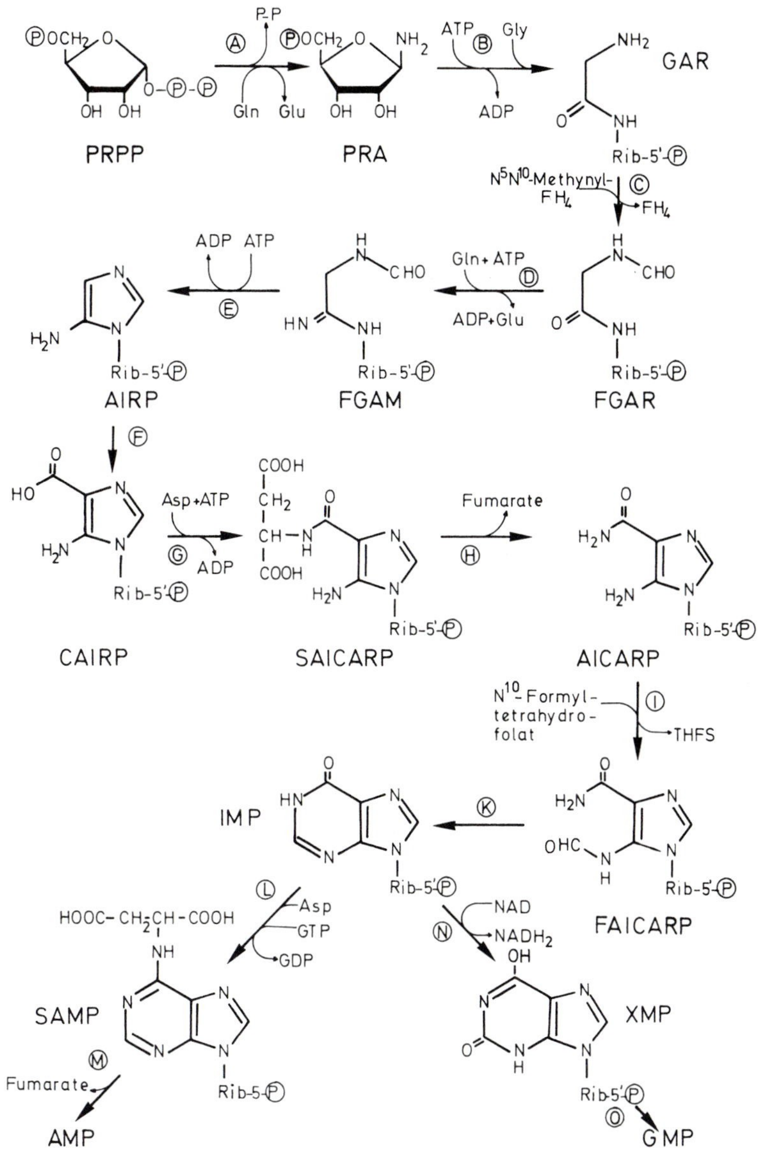

Figure 10.2 Biosynthesis of AMP and GMP.

PRPP	= 5-Phospho-α-D-ribosylpyrophosphate;
PRA	= 5-Phospho-β-D-ribosylamine;
GAR	= 5′ Phosphoribosylglycinamide;
FGAR	= 5′-Phosphoribosyl N′-Formylglycinamide;
FGAM	= 5′-Phosphoribosyl-N′ formylglycinamidine;
AIRP	= 1-(5′-Phosphoribosyl)-5 aminoimidazole;
CAIRP	= 1-(5′-Phosphoribosyl)-5 aminoimidazole-4-carboxylate;
SAICARP	= 1-(5′-Phosphoribosyl)-4-(N-Succinocarboxamide)-5-aminoimidazole;
AICARP	= 5-Amino-1-(5′-phosphoribosyl)-imidazole-4-carboxamide;
FAICARP	= 5-Formamido-1-(5′-phosphoribosyl)-imidazole-4 carboxamide;
SAMP	= Adenylsuccinate

If an excess of AMP or GMP is formed, GMP can be transformed through GMP reductase into IMP, and AMP through AMP deaminase into IMP, or through several steps into AICARP.

10.3 REGULATION

The overproduction of purine nucleotides in wild type bacteria is prevented by feedback regulation. For commercial production, the regulatory mechanisms must be partially eliminated. One approach is to isolate auxotrophic mutants and add the required end product in growth-limiting concentration, so that feedback inhibition and repression are prevented. Another alternative is to isolate mutants with resistance to purine analogs.

The regulation of GMP and AMP synthesis in *Bacillus subtilis* (Figure 10.3) has several peculiarities. IMP, the precursor of AMP and GMP, is transformed mainly into GMP. IMP dehydrogenase, the first enzyme of this biosynthetic pathway, has a 10- to 30-fold greater specific activity than adenyl succinate synthetase. Feedback inhibition and repression of IMP dehydrogenase is caused by GMP and XMP; the last enzyme of the chain, GMP synthetase, is only slightly affected by GMP. Under these conditions the production of AMP is increased. AMP synthesis is chiefly regulated by feedback inhibition of the first common enzyme, PRPP amidotransferase. There is **asymmetric control** by AMP, but the regulatory effect of GMP is considerably lower: 50% of the inhibition of PRPP amidotransferase is accomplished by 0.2 mM AMP or 2 mM GMP. Both enzymes of AMP biosynthesis, adenyl succinate synthetase and adenyl succinate lyase, are regulated by AMP.

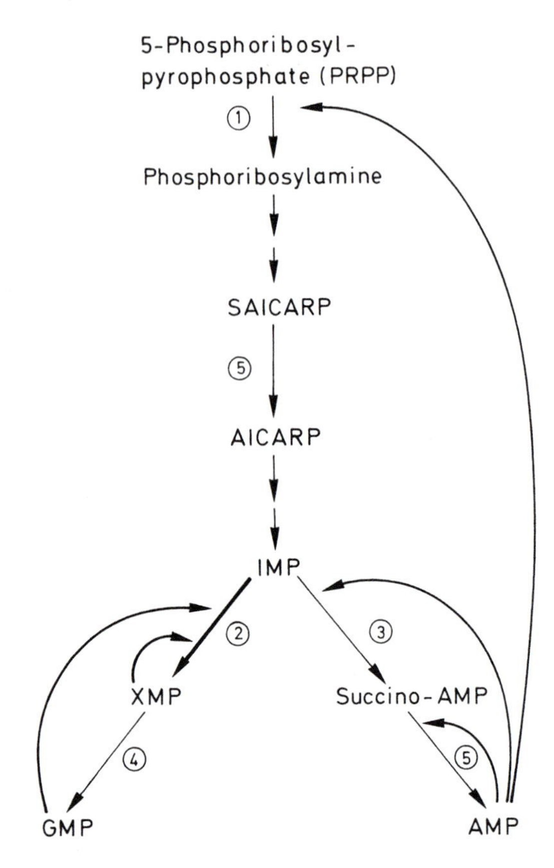

Figure 10.3 Regulation of purine nucleotide biosynthesis in *Bacillus subtilis*. 1, PRPP amidotransferase; 2, IMP dehydrogenase; 3, Adenylsuccinate synthetase; 4, GMP synthetase; 5, Adenylsuccinate lyase (From Shiio, 1979)

10.4 PRODUCTION

Different methods are used to produce nucleotides:

1. **Enzymatic** or **chemical breakdown** of nucleic acids through
   a. hydrolysis of yeast RNA by microbial enzymes
   b. breakdown of cellular RNA by endogenous cell enzymes, leading to excretion of 5′-mononucleotides
   c. chemical hydrolysis of yeast RNA into nucleosides with subsequent chemical phosphorylation.
2. **Direct fermentation** using mutants with a block in nucleotide biosynthesis, in which endproduct regulation is eliminated (as discussed above).
   a. manufacture of nucleosides by fermentation and their chemical phosphorylation
   b. direct manufacture of 5′-nucleotides

through fermentation

c. microbial conversion of bases of nucleosides to 5′-nucleotides.

Methods 1a, 1c, 2a and 2b are used for industrial production of 5′-IMP and 5′-GMP.

Production of 5′-IMP and 5′-GMP by enzymatic hydrolysis of RNA

The breakdown of RNA was the first method used to produce 5′-nucleotides commercially. As late as 1975, 50% of the 5′-nucleotides in Japan were still obtained using this method. The multistage process consists of the following steps: 1) growth of yeast cells containing a high RNA content; 2) extraction of RNA from cell material; 3) production of enzymes used in RNA hydrolysis; 4) enzymatic hydrolysis of RNA; and 5) isolation and purification of 5′-IMP and 5′-GMP.

Obtaining RNA The nucleic acid content of different microorganisms varies considerably:

Microorganisms	DNA content (%)	RNA content (%)
Bacteria	0.37–4.5	5 –25
Yeast	0.03–0.52	2.5–15
Fungi	0.15–3.3	0.7–28

The RNA of the cell consists of 5% messenger RNA, 10–15% transfer RNA and 75–80% ribosomal RNA. Because of their relatively low DNA content, yeasts are the best sources of RNA. Moreover, there are well-developed fermentation and recovery techniques for yeast. The yeast *Candida utilis* is most commonly used, followed by *Saccharomyces cerevisiae*. The RNA content of yeast cells is very dependent on culture conditions. A high proportion of RNA is found in the early logarithmic phase, particularly in media with a low carbon/nitrogen ratio. The RNA content of cells growing on molasses or glucose can be further increased by adding zinc ions (0.25 ppm) and phosphate (0.15%). The yeast fermentation is generally done in continuous culture in an airlift fermenter (aeration rate, $K_d > 20 \times 10^{-6}$ g·mole·O_2/atm·min·ml), using either sulfite waste liquor or molasses. In Japan, a large production plant may produce 10,000–20,000 tons of yeast cells per year.

The cell mass produced in this process is 35 g/l, with an RNA content of 10% (sulfite waste liquor) or 15% (molasses). The cell material is separated and dried by means of yeast separators, and the RNA is then extracted with hot alkaline saline (5–20% NaCl, 8 h, 100°C). After separating the cell residue, the RNA is precipitated by adding HCl or ethanol and is dried. The crude preparation, which has an RNA content of 70–90% with a molecular weight distribution of 10,000–150,000, can be used directly for enzymatic hydrolysis.

Another process for the production of nucleotides makes use of acidophilic, methanol-utilizing bacteria such as *Methylobacter acidophilus*.

Enzymatic hydrolysis As Figure 10.4 shows, several distinct enzymatic reactions can be used for the hydrolysis of RNA. For optimal production of 5′-IMP and 5′-GMP, organisms must be used which have higher activities for the enzymes which catalyze steps a–d (Figure 10.4) than for steps e–g, in order to avoid the production of undesirable substances. Two organisms, *Penicillium citrinum* and *Streptomyces aureus*, have been found to contain enzymes which meet these requirements, and enzymatic RNA hydrolysis on an industrial scale is carried out exclusively with these two organisms.

One process makes use of hydrolysis with nuclease P_1 from *Penicillium citrinum*. A pigment-free mutant from *P. citrinum* is cultivated on wheat bran for 5 days at 30°C. The aqueous extract contains the thermostabile nuclease P_1, which quantitatively splits the 3′–5′-phosphodiester linkage of polynucleotides and the 3′-phosphomonoester linkage of mono- and oligonucleotides. The undesired phosphomonoesterase activity, which produces 3′-nucleotides, can be almost completely inactivated by heat treatment. To eliminate effects of the residual phosphatase activity (optimal temperature

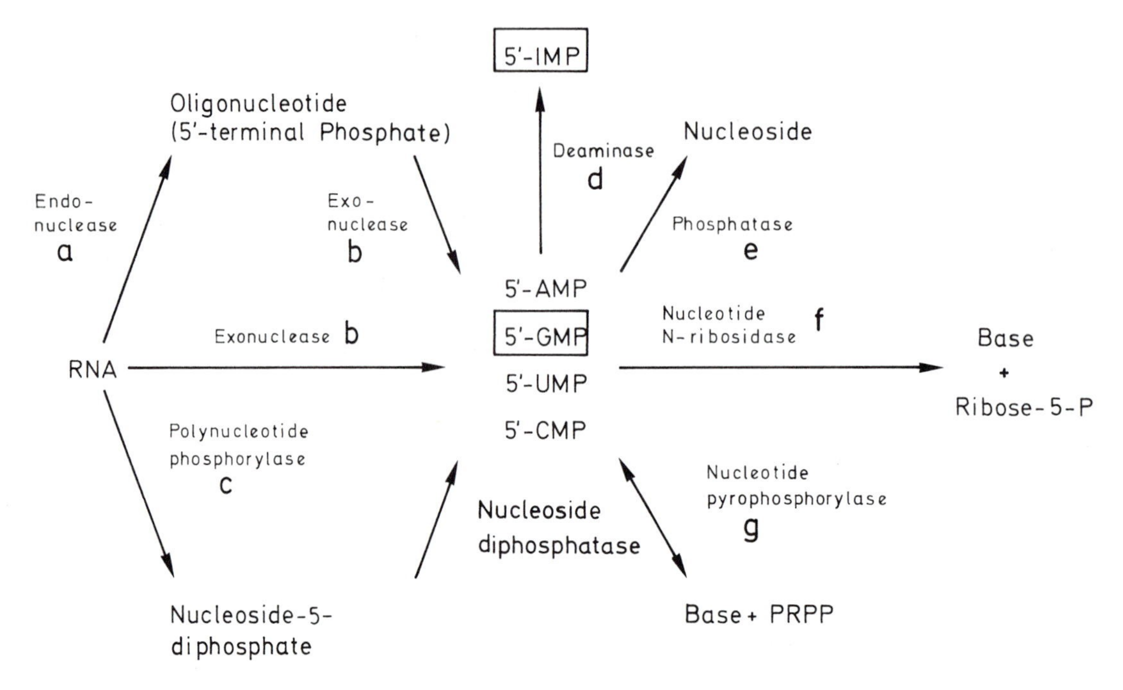

Figure 10.4 Enzymes which are involved in the production of GMP and IMP via RNA hydrolysis (see text for explanation)

45°C), hydrolysis of a solution of 2% RNA is performed in 4 hours at pH 5.0 at the optimal temperature of nuclease P_1 (65°C). In this reaction, 5′-GMP, 5′-AMP, 5′-CMP, and 5′-UMP are produced. The 5′-AMP is deaminated by adenyl deaminase from *Aspergillus oryzae* to 5′-IMP.

A process with immobilized nuclease P_1 has also been described.

Separation of the 5′-nucleotides formed through this enzymatic hydrolysis is performed by means of anion exchange chromatography.

Another process involves hydrolysis with nucleases from *Streptomyces aureus. S. aureus* forms a mixture of endo- and exonucleases; any 5′-AMP is thus deaminated by AMP deaminase to 5′-IMP. By isolating appropriate mutants, researchers have attempted to reduce the amounts of a 5′-nucleotidase and an unspecific alkaline phosphatase which occur in the enzyme mixture produced by *S. aureus*. Table 10.2 shows the enzyme activities of different mutants compared with the wild type. Mutant A-5 is the best mutant because of its clearly reduced phosphatase activity. The production of the enzymes takes place during a 30-hour incubation in a submerged culture ($K_d = 2.72 \times 10^{-5}$ g·Mol·O_2/atm·min·ml) at 28°C on a medium containing starch hydrolysate, soy meal, and corn steep liquor. Typical progress of a 6 m^3 fermentation of *Streptomyces aureus* is given in Figure 10.5. Because of the high 5′-nucleotidase activity of the *S. aureus* enzymes, RNA hydrolysis is carried out at a different temperature and pH. A 0.5–1.0% RNA solution is hydrolyzed for 10 hours at 42–65°C in a pH range between 7.0–8.0.

The purification of 5′-GMP and 5′-IMP is done by means of a preliminary adsorption with activated carbon, then ion exchange chromatography, followed by a fractional precipitation with methanol.

Table 10.2 Enzyme production by *Streptomyces aureus* mutants relative to the wild type

Strain	Endonuclease	Exonuclease	AMP deaminase	5′-Nucleotidase	Alkaline phosphatase
Wild type	100	100	100	100	100
K-1	310	180	110	160	20
A-5	400	210	150	180	10
S-8	100	80	100	60	10

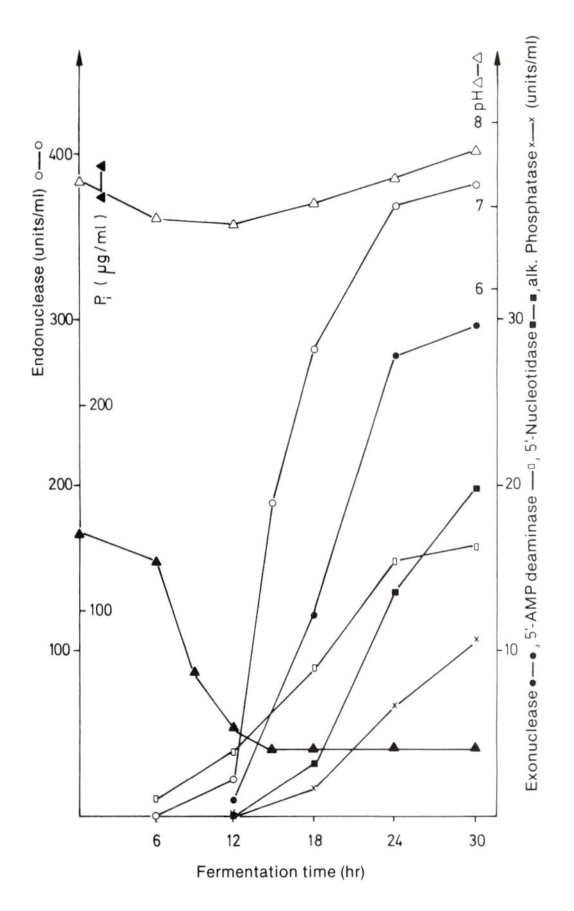

Figure 10.5 Fermentation process for nuclease production with *Streptomyces aureus*

Production of 5′-IMP and 5′-GMP through chemical hydrolysis of RNA

In the chemical hydrolysis of RNA under alkaline conditions, 5′-nucleotides are not produced directly, but a mixture of 2′- and 3′-nucleotides is produced via an intermediate 2′,3′-cyclic phosphodiester. However, nucleosides can be produced quantitatively from RNA by heating to 130°C for 3–4 hours in calcium hydroxide. The chemical phosphorylation of these nucleosides to 5′-IMP and 5′-GMP then can be done, after the deamination of adenosine to inosine.

Production of 5′-IMP by direct fermentation

5′-IMP can be produced fermentatively by various methods:

- Through microbial production of inosine and subsequent chemical phosphorylation to 5′-IMP
- Through direct fermentation to 5′-IMP
- Through fermentative production of adenosine or 5′-AMP and subsequent chemical or enzymatic conversion into 5′-IMP
- Through microbial conversion of chemically synthesized hypoxanthine into 5′-IMP.

For economic reasons only, the first two processes are used commercially.

Fermentative production of inosine The cell membrane is not permeable to nucleotides but is permeable to the dephosphorylated derivatives, the nucleosides. When a cell synthesizes 5′-IMP, it cannot be excreted through the cytoplasmic membrane; however, it can be dephosphorylated by the cell and then can be excreted, so that it accumulates in the medium as inosine. Because of the regulatory effect of AMP and ADP on PRPP amidotransferase in *Bacillus subtilis*, the adenine nucleotide content of the cells must be kept at a low level to produce IMP. This is accomplished through a block in the AMP biosynthesis (absence of SAMP-synthetase). Reduced nucleosidase activity, absence of IMP-dehydrogenase and/or an increase in 5′-nucleotidase activity also favor the accumulation of inosine. Inosine-producing microorganisms were first found among adenine auxotrophs of different strains of the genera *Bacillus, Brevibacterium, Corynebacterium, Streptomyces,* and *Saccharomyces. Corynebacterium petrophilum* (Ade^-) accumulates 1.6 g/l inosine on a medium with n-alkanes (C_{12}–C_{16}). Various strains of *Bacillus subtilis, Brevibacterium ammoniagenes,* and *Microbacterium* have been developed for commercial production, as shown in Table 10.3.

In a typical fermentation process with *Bacillus,* starch hydrolysate is most commonly used as the carbon source, but glucose can also be used. Dry yeast or crude RNA serves as the source of adenine. Ammonia is added both to regulate the pH and as a nitrogen source. The optimal pH range is 6.0–6.2, the optimal temperature 30–34°C. Maximal inosine accumulation is obtained with an oxygen transfer coefficient of $K_d = 5.8\text{–}7.0 \times 10^{-6}$ g·Mol·O_2/atm·min·ml. At the same time, the CO_2 content of the medium must be low. The progress of an inosine fermentation with *Brevibacterium ammoniagenes* is illustrated in Figure 10.6.

The final stage is the phosphorylation of inosine. At pH 11, inosine is precipitated from a culture filtrate and crystallized. The chemical phosphorylation of inosine is carried out using trialkyl phosphate with PCl_3. The amount of undesired 2′(3′)5′-phosphodiester can be reduced to < 10%; under these conditions the 5′-monoester content is 90%.

Table 10.3 Mutants used for the inosine fermentation

Mutants	Genetic characteristics	Inosine production (g/l)
Bacillus subtilis strain		
No. 2	Ade^-	0.21
B-4	Ade^-His^-	4.46
C-30	$Ade^-His^-Tyr^-$	10.5
RDA-16	Ade^- Trp^- Red^- Dea^- $8AG^r$	18.0
Brevibacterium ammoniagenes		
KY 13714	Ade^- $6MG^r$ Gua^-	13.6
KY 13761	Ade^- $6MG^r$ $6MTP^r$ Gua^-	30.0
Microbacterium sp.		
No. 250	Bio^- $6MP^r$ $8AG^r$ MSO^r $6TG^r$	35.0

Auxotrophy: Ade Adenine; Bio Biotin; Gua Guanine; His Histidine; Tyr Tyrosine; Trp Tryptophan; Red^- GMP reductase is absent; Dea^- AMP deaminase is absent.
Resistance: 8AG 8-Azaguanine; 6MG 6-Mercaptoguanine; 6MTP 6-Methylthiopurine; MSO Methionine sulfoxide; 6TG 6-Thioguanine; 6MP 6-Mercaptopurine
(Furuya, in Ogata et al., 1976)

Direct fermentation of 5′-IMP Mutants for the direct fermentation of 5′-IMP should have the following properties:

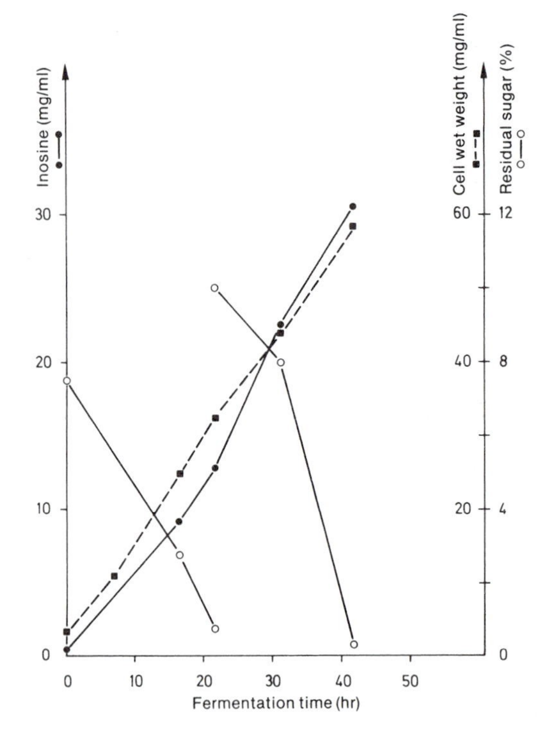

Figure 10.6 Inosine fermentation with *Brevibacterium ammoniagenes* KY 13761 (From Kotani et al., 1978)

- Absence of SAMP synthetase to eliminate AMP regulation at the level of PRPP amidotransferase,
- Low activity of the enzymes which break down 5′-IMP,
- Increased permeability of the cytoplasmic membrane to bring about 5′-IMP excretion.

Mutants with reduced nucleotidase were first isolated among inosine-producing *B. subtilis* strains. Mutants able to grow on adenine but not on AMP showed a lower incidence of nucleotide breakdown. A further inhibition of nucleotidase activity was obtained in mutant A-1-25, which is listed in Table 10.4 along with the other 5′-IMP producers. Industrial production of 5′-IMP is carried out with *Brevibacterium ammoniagenes*. The isolation of permeability mutants whose IMP-accumulation is not affected by Mn^{2+} was an important advance. In the mutants KY 7208 and KY 13102, which are sensitive to Mn^{2+}, the optimal Mn^{2+} content of the medium is 0.01–0.02 mg/l; at higher levels, IMP production decreases. Figure 10.7 illustrates the relationship of the IMP accumulation of both mutants to the Mn^{2+} concentration.

In the fermentation by strain KY 13102 (Figure 10.8), the production of hypoxanthine continues during the first 2–3 days. Then it decreases

Table 10.4 Microorganisms producing 5′-IMP

Mutants	Genetic characteristics	5′-IMP Yield (g/l)
B. subtilis A-1-25	Ade^- Nuc^-	0.6
Corynebacterium glutamicum	Ade^- $6MP^r$	2.0
Brevibacterium ammoniagenes		
KY 7208	Ade^-	5.0
KY 13102	Ade^-	12.8
KY 13105	Ade^- Mn^{2+} insensitive	19
KY 13369	Ade^- Mn^{2+} insensitive Gua^-	20–27

Ade^- Adenine-requiring; Nuc^- Nucleotidase-negative; $6MP^r$ 6-Mercaptopurine-resistant.

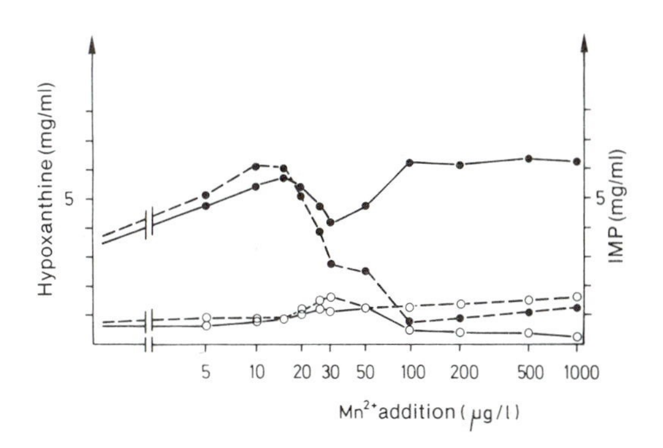

Figure 10.7 Effect of Mn^{2+} on IMP and hypoxanthine production. Mutant KY 13102 (Mn^{2+} sensitive): ●- - -● IMP; ○- - -○ Hypoxanthine. Mutant KY 13105 (Mn^{2+} resistant): ●—● IMP; ○—○ Hypoxanthine

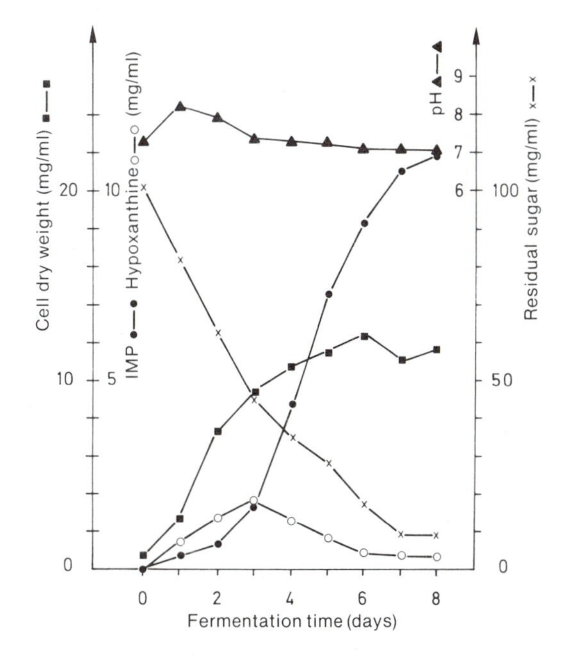

Figure 10.8 5′-IMP fermentation with *B. ammoniagenes* KY 13102 (From Furuya et al., 1968)

as IMP accumulation begins. With Mn^{2+} as a limiting factor, abnormal cells with reduced permeability are formed. They excrete hypoxanthine, which is extracellularly phosphorylated to 5′-IMP, as shown in Figure 10.9. Moreover, 5′-IMP is synthesized *de novo* in the subsequent fer-

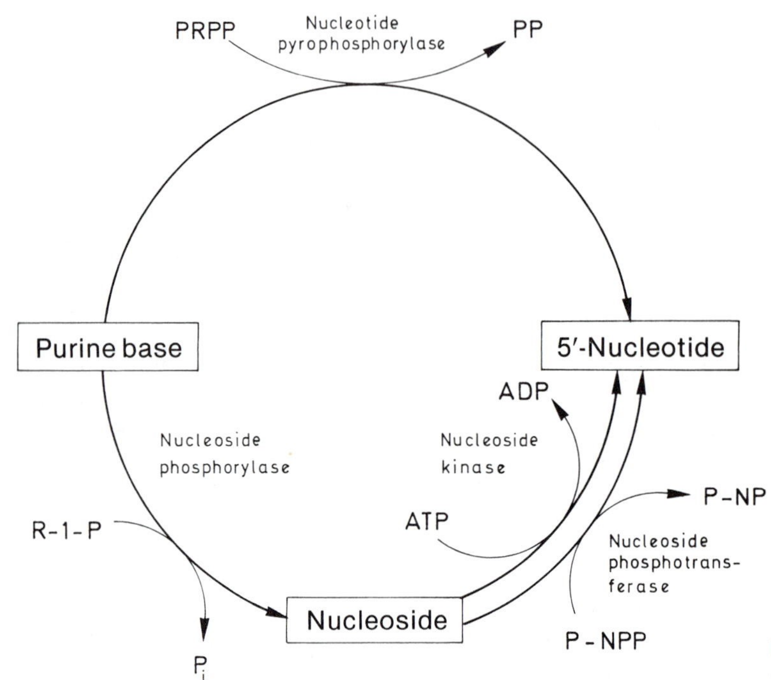

Figure 10.9 Extracellular conversion of purine bases to purine nucleotides

mentation phase and is excreted directly into the culture medium.

For fermentations with *B. ammoniagenes*, it is important that the phosphate and $MgSO_4$ concentrations be kept at 2 and 1%, respectively. The typical composition of the culture medium for Mn^{2+}-insensitive *B. ammoniagenes* is as follows (measured in g/l): KH_2PO_4 10.0; K_2HPO_4 10.0; $MgSO_4 \cdot 7\ H_2O$ 10.0; $CaCl_2 \cdot 2\ H_2O$ 0.1; $FeSO_4 \cdot 7\ H_2O$ 0.01; urea 6.0; cysteine 0.02; thiamine 0.05; nicotinic acid 0.05; Ca pantothenate 0.01; biotin 3×10^{-5}; $MnCl_2 \cdot 4\ H_2O$ 0.001; adenine 0.04; glucose 100.0; pH 8.3.

The Ajinomoto Company has developed a process for IMP production that uses a nitrosoguanidine-induced mutant of *Brevibacterium ammoniagenes*. Inosine (50–2000 mg/l) or hypoxanthine (25–1000 mg/l) is added during the fermentation. This process greatly increases the amount of 5′-IMP production from sugars such as glucose, fructose, sucrose, or starch hydrolysate.

Fermentative production of 5′-GMP

The excretion of sizable amounts of 5′-GMP in wild type strains is rare, due to regulatory effects primarily at the levels of PRPP amidotransferase, IMP dehydrogenase, and GMP synthetase. Moreover, 5′-GMP can be broken down by 5′-nucleotidases or nucleosidases or can be converted to IMP by GMP reductase. Various means of production have therefore been developed.

- Fermentation from AICAR (5-amino-4-imidazole carboxamide riboside, Figure 10.2), which is chemically converted to 5′-GMP.
- Microbial production of guanosine with subsequent chemical phosphorylation to 5′-GMP.
- Fermentative production of xanthine or 5′-XMP and enzymatic conversion into 5′-GMP.
- Direct fermentation of 5′-GMP.

Only the first two processes are used for industrial production of 5′-GMP. In mutants used in production, intracellular accumulation is as

GMP, but the nucleoside guanosine is actually excreted.

Production from AICAR AICAR is commercially produced as the starting material for production of 5'-GMP and as an intermediate for the enzymatic production of 5-amino-4-imidazole carboxamide (AICA), which is used as the starting material for the chemical synthesis of purine derivatives.

Escherichia coli, B. subtilis, B. megaterium, and *Brevibacterium flavum* have been examined as AICAR producers. Purine auxotrophs exhibit especially significant AICAR production, and the purine auxotroph *B. megaterium* No. 366 (ATCC 15117) and its mutants have been used in industrial production. Under optimal fermentation conditions, this strain produces 16 g/l AICAR from 80 g/l glucose. The genetic properties that make this strain suitable for AICAR accumulation are: 1. The strain is a purine auxotroph and has a block in the AICARP formyltransferase reaction, which causes AICARP to be converted into FAICARP (see Figure 10.2). 2. The strain has no enzyme activities causing the hydrolysis of AICA-riboside. 3. The enzymes which catalyze AICAR biosynthesis, especially PRPP amidotransferase, are insensitive to regulation by intracellular purine nucleotides.

Starch hydrolysate is the carbon source used with mutant *B. megaterium* No. 366. Fifty percent of the total AICAR synthesis occurs after the metabolism of glucose is complete. As an intermediate step in glucose breakdown, gluconic acid accumulates and is then subsequently metabolized during AICAR synthesis. For maximal AICAR production, the medium must contain a purine source, such as dry yeast or yeast RNA, in suboptimal concentration. Soy protein hydrolysate and NH_4Cl are used as nitrogen sources and mineral salts ($MgSO_4 \cdot 7\ H_2O$, $FeSO_4 \cdot 7\ H_2O$, $MnSO_4 \cdot H_2O$) are also used. Gaseous ammonia is used both as a nitrogen source and pH regulator.

The AICAR yield during the fermentation is directly affected by the sporulation process; sporulating cultures have a strongly reduced AICAR production. Sporulation can be suppressed by inhibitors such as butyric acid, or by reducing the O_2 supply. If good aeration is provided during the period about 8–12 hours after the start of the fermentation, sporulation occurs and AICAR production is reduced. However, if O_2 supply is suboptimal, especially during this period, sporulation is suppressed and AICAR production is increased.

The final step is the chemical conversion of AICAR to 5'-GMP. The AICARP formed is excreted into the fermentation medium as the dephosphorylated compound, AICAR. The isolation of AICAR can be accomplished with a 90% yield through several purification stages. For the transformation into 5'-GMP, AICAR is first converted to guanosine in several chemical steps and then phosphorylated.

Production of guanosine by direct fermentation Another cost-effective process for the production of 5'-GMP is the production of guanosine by fermentation, followed by chemical phosphorylation. Guanosine-excreting strains should have the following properties: 1. SAMP synthetase-negative, 2. GMP reductase-negative, 3. reduced nucleosidase activity, and 4. enzymes of GMP biosynthesis unregulated, particularly PRPP amidotransferase, IMP dehydrogenase, and GMP synthetase.

Microorganisms which possess the ability to excrete guanosine include *Bacillus subtilis, B. pumilus, B. licheniformis, Corynebacterium petrophilum, C. guanofaciens,* and *Streptomyces griseus.* Mutants of *B. subtilis,* also an inosine producer, are chiefly used for commercial production (Table 10.5).

IMP dehydrogenase is regulated by 5'-GMP. *B. subtilis* AJ 1993 is a mutant which is resistant to the purine analog 8-azaguanine. It shows IMP dehydrogenase activity three times greater than the initial strain, as well as reduced GMP reductase activity. Mutation to resistance to methionine sulfoxide (a glutamine analog) in mutant MG-1 causes another increase in the IMP dehydrogenase activity. In mutants which are re-

Table 10.5 Guanosine-excreting *B. subtilis* mutants

Mutants	Genetic characteristics	Guanosine production (g/l)
B. subtilis AJ 1993	$Ade^- Red^- 8AG^r$	4.3
No. 30-12	$Ade^- Trp^- Red^- 8AX^r$	5
MG-1	$Ade^- His^- Red^- MSO^r Psi^r Dec^r$	16

For the mutant designations, see Table 10.3; Resistance: 8AX 8-azaxanthine; Psi Psicofuranine; Dec Decoyinine (Psicofuranine and Decoyinine are inhibitors of GMP synthetase)

sistant to nucleoside antibiotics such as psicofuranine and decoyinine (which inhibit production of GMP), the GMP synthetase activity is clearly increased, while regulation by GMP is completely eliminated. The high guanosine excretion of mutant MG-1 is thus due to several factors: adenine auxotrophy, the absence of GMP reductase, the elimination of the feedback regulation by GMP, and the increased activity of IMP dehydrogenase and GMP synthetase.

The production medium for *B. subtilis* AJ 1993 has the following composition (g/l): glucose 70; adenine 0.3; soy protein hydrolysate 0.48; NH_4Cl 15; KH_2PO_4 1; $MgSO_4 \cdot 7\ H_2O$ 0.4; $FeSO_4 \cdot 7\ H_2O$ 0.01; $MnSO_4 \cdot H_2O$ 0.01; $CaCO_3$ 25. The fermentation temperature is 30°C. Under these conditions, 4.3 g/l guanosine and 3.1 g/l inosine are produced.

Production of adenosine and adenine nucleotides

The production of adenosine and adenine nucleotides has gained significance due to the use of these substances as starting materials for the synthesis of medically important adenine derivatives.

For adenosine production, the *Bacillus subtilis* mutant P 53-18 ($His^- Thr^- Xan^- 8AX^r$) was isolated. This mutant has a block in adenosine deaminase, GMP reductase, and IMP dehydrogenase. An adenosine accumulation of 16 g/l was obtained under optimal fermentation conditions. Progress of a typical fermentation is given in Figure 10.10.

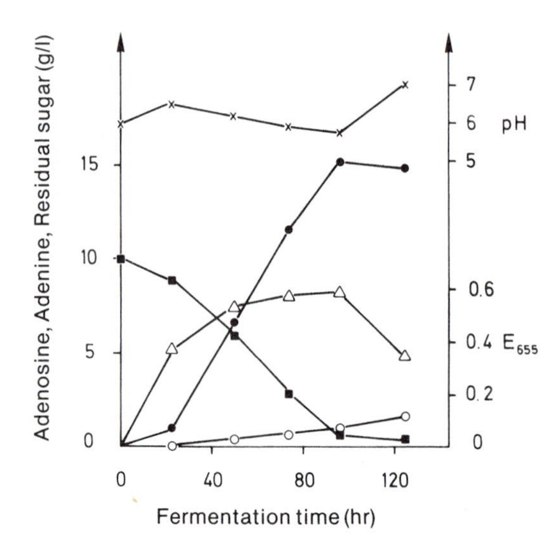

Figure 10.10 Adenosine fermentation with *Bacillus* sp. strain P 53-18. ● — ● Adenosine; ○ — ○ Adenine; ■ — ■ Residual sugar; △ — △ Growth (Absorbance at 655 nm); crosses, pH (From Haneda et al., 1971)

Another organism, a xanthine auxotroph of *Brevibacterium ammoniagenes*, carries out a direct synthesis of 5′-AMP to a level of 2.1 g/l.

Adenosine triphosphate, ATP, can be obtained through phosphorylation of adenine, adenosine, and ATP by various microorganisms in the presence of glucose or methanol plus phosphate. Protoplasts of the yeast *Candida boidinii* produce ATP at a yield of 100 g/l (77.4% efficiency) from a mixture of adenine and methanol.

Cyclic adenosine-3′,5′-monophosphate, cAMP, can be produced either chemically or microbially. cAMP production has been shown in a variety of bacteria, including *E. coli*, *Corynebacterium murisepticum*, *Microbacterium* sp., and *Brevibacterium liquefaciens*. A mutant of *Microbacterium* sp. resistant to 8-azaguanine, azaserine, methionine sulfoxide, and 6-mercaptoguanine has been shown to synthesize cAMP *de novo*, producing as much as 16.6 g/l in a medium containing glucose and inorganic salts. If 3 g/l of adenine or hypoxanthine is added, the production is increased to 20.5 or 21.3 g/l respectively.

Studies on cloning the gene for adenyl cyclase in *E. coli* and *Salmonella typhimurium* have been undertaken. Although the activity of the enzyme in *S. typhimurium* increased by a factor of 20, there was only a 60% increase in cAMP formation.

Fermentative production of other substances related to nucleotides

Other substances related to nucleotides are produced by fermentation in Japan for **medical purposes**, although the market for such products is limited (see also Table 10.1). Production strains and yields of these processes are listed in Table 10.6.

An enzyme/membrane reactor has been developed for the production of **coenzymes NADP/NADPH**, substances that find considerable value in analytical biochemistry.

A group of pharmacologically interesting compounds are the **purine arabinosides.** Substances which possess antibiotic, antiviral, or antitumor activity include guanosine derivatives and adenosine analogs such as arabinosyladenine. A major site of action of these compounds is the enzyme S-adenosylhomocysteine hydrolase, a key enzyme in biochemical systems for biological transmethylation. S-adenosylmethione and S-adenosylhomocysteine, intermediates in the transmethylation pathway, have been studied clinically as sedatives and for the treatment of mental illness. Both microbiological and enzymatic processes for the production of these substances are under study.

Over 100 **antibiotics** have been described which are derived from nucleosides. Several representative examples are discussed in Section 13.7.

Table 10.6 Production by fermentation of substances related to nucleic acids

Product	Organism	Yield (g/l)
FAD	*Sarcina lutea*	1
NAD	*Brevibacterium ammoniagenes*	1.9
Coenzyme A	*Brevibacterium ammoniagenes* IFO 12071	2
Orotic acid	*Arthrobacter paraffineus*	20
CDP-choline	*Saccharomyces carlsbergensis*	17

(Ogata et al., 1976; Hirose et al., 1979)

REFERENCES

Dale,B.E. and D.A. White. 1979. Degradation of ribonucleic acid by immobilized ribonuclease. Biotechnol. Bioeng. 21:1639–1648.

Demain, A.L. 1978. Production of nucleotides by microorganisms, pp. 187–208. In: Rose, A.H. (ed.), Primary products of metabolism, vol. 2. Academic Press, London.

Enei, H., H. Shibai, and Y. Hirose. 1985. 5'Guanosine nonophosphate, pp. 653–658. In: Moo-Young, M. (ed.). Comprehensive Biotechnology, Vol. 3. Pergamon Press, Oxford.

Furuya, A., S. Abe, and S. Kinoshita. 1968. Production of nucleic acid-related substances by fermentative processes. XIX. Accumulation of 5'-inosinic acid by a mutant of *Brevibacterium ammoniagenes.* Appl. Microbiol. 16:981–987.

Haneda, K., A. Hirano, R. Kodaira, and S. Ohuchi. 1971. Accumulation of nucleic acid-related substances by microorganisms. II. Production of adenosine by mutants derived from *Bacillus* sp. Agr. Biol. Chem. 35:1906–1912.

Hirose, Y., H. Enei, and H. Shibai. 1979. Nucleosides and nucleotides. Ann. Rep. Ferm. Proc. 3:253–274.

Kotani, Y., K. Yamaguchi, F. Kato, and A. Furuya. 1978. Inosine accumulation by mutants of *Brevibacterium ammoniagenes,* strain improvement and culture conditions. Agr. Biol. Chem. 42:399–405.

Kuninaka, A. 1986. Nucleic acids, nucleotides, and related compounds, pp. 71–114. In: Rehm, H.J. and G. Reed (eds.). Biotechnology, Vol. 4. VCH-Verlagsgesellschaft, Weinheim.

Nakao, Y. 1979. Microbial production of nucleosides and nucleotides, pp. 311–354. In: Peppler, H.J. and D. Perlman (eds.). Microbial technology. Vol. I. Academic Press, New York.

Ogata, K., S. Kinoshita, T. Tsunoda and A. Aida (eds.). 1976. Microbial production of nucleic acid related substances. Kodansha Ltd., Tokyo.

Perlman, D. 1977. Fermentation industries . . . quo vadis? Chem. Techol. 7:434–443.

Schütz, H.J., M.R. Kula, and C. Wandrey. 1986. Ein enzymatischer Weg zur kontinuierlichen Produktion von phosphorylierten Nicotin-Adenin-Dinukleotiden. (An enzymatic process for the continuous production of phosphorylated nicotine adenine dinucleotides.) BTF-Biotech Forum 3:98–102.

Shibai, H., H. Enei, and Y. Hirose. 1978. Purine nucleosides fermentations. Proc. Biochem. Nov. 1978, pp. 6–8, 32.

Shiio, I. 1979. Microbial production of nucleotides. Congrès international microbiologie et industrie alimentaire. Paris, 10, Oct.

Shimizu, S. and H. Yamada. 1984. Microbial and enzymatic processes for the production of pharmacologically important nucleosides. Trends in Biotechnol. 2:137–141.

Yamada, K. 1977. Bioengineering report. Recent advances in industrial fermentation in Japan. Biotechnol. Bioeng. 19: 1563–1621.

11

Enzymes

11.1 INTRODUCTION

The first enzyme produced industrially was the fungal amylase takadiastase, employed as a pharmaceutical agent (for digestive disorders) in the United States as early as 1894. Otto Roehm's patented "laundry process for any and all clothing via tryptic enzyme additives" was announced in 1915. By 1969, 80% of all laundry detergents contained enzymes, chiefly proteases. Along with these, additional enzymes such as lipases, amylases, pectinases, and oxidoreductases were used experimentally in the detergent industry.

Due to the occurrence of allergies among production workers and consumers, the use of proteases in detergents was drastically reduced in 1971 and the world sales fell from $150 million to one-third this amount. Only when special processing techniques, such as microencapsulation, were developed could dustless protease preparations be produced that were risk-free to workers and consumers. Today 95% of laundry detergents in Europe contain proteases, although such detergents are much less widely used in the United States.

The development of **amylases** and **amyloglucosidases** for the production of glucose from starch has led to a new industrial application of enzymes. Also, the use of glucose isomerase for the production of fructose has become widespread since 1967.

Microbial rennin is next in order of significant enzymes. It has been used instead of calf's rennin in cheese production since 1965.

Gross world sales for enzymes in 1987 was $450 million. The following enzymes are currently produced commercially:

- Enzymes used in industry, such as amylases, proteases, catalases, isomerases, and penicillin acylases;
- Enzymes used for analytical purposes, such as glucose oxidase, galactose oxidase, alcohol dehydrogenase, hexokinase, muramidase, and cholesterol oxidase;
- Enzymes used in medicine, such as asparaginase, proteases, lipases, and streptokinase.

These three areas of application require varying levels of quality and quantity (Table 11.1). On a tonnage basis, the most important are the industrial enzymes (about 1200 tons of pure protein/year) (Table 11.2). The United States and Western Europe produce about 40–45% of the world market, and nine firms account for 90% of the total market. Industrial enzymes are produced in 120 m^3 fermenters, while enzymes for analytical or medical purposes can frequently be produced in pilot-plant-sized fermenters.

The low concentration of enzymes which are normally produced by wild strains is a considerable hindrance for enzyme production. Although in 1985 about 2500 enzymes were known, only 250 were marketable, mostly in amounts around 10 g, some even in mg amounts. Figure 11.1 shows the overall picture of commercial enzyme production.

The prospects for enzyme application have improved due to developments in the following areas:

- Microbial genetics. High yields can be obtained by genetic manipulation. For instance, the yeast *Hansenula polymorpha* has been genetically modified so that 35% of its total protein consists of the enzyme alcohol oxidase.
- Optimization of fermentation conditions via induction of enzyme production, use of low cost nutrients, optimal utilization of components in nutrient solution, and introduction of fed-batch fermentations.
- Release of enzymes from cells by means of new cell-breaking methods.

Table 11.2 Production of industrial enzymes

Enzyme	Enzyme preparation tons/year	Sales, percent of total
Bacillus proteases	6200	45
Glucoamylases	4500	13
Bacillus amylases	4200	5
Glucose isomerases	1900	6
Rennin (microbial)	1500	10
Amylases (fungal)	800	4
Pectinases	100	3
Proteases (fungal)	100	2
Others	–	12

(Aunstrup et al., 1979)

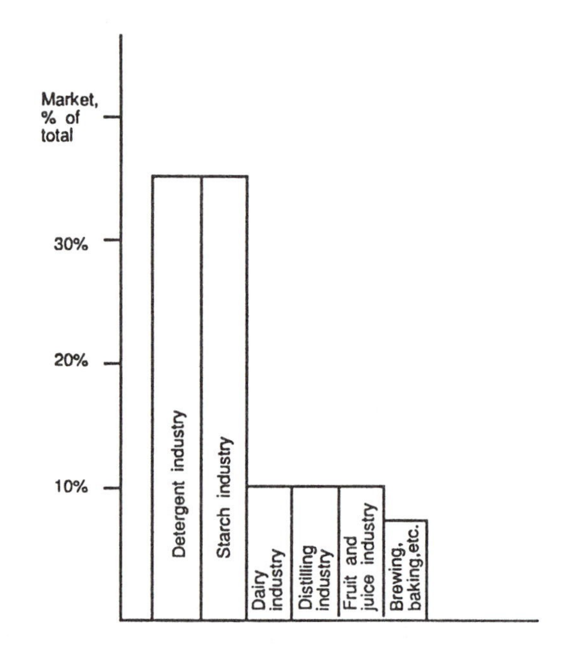

Figure 11.1 Percentage of industrial enzymes used in various markets

Table 11.1 Differences in enzyme qualities

	Industrial enzyme	Analytical enzyme	Clinical enzyme
Scale of application	Tons	Milligrams–grams	Milligrams–grams
Degree of purification	Crude	Pure crystalline	Pure crystalline
Secondary activities	Present	Usually none. If yes, then defined	Only isoenzymes
Origin	Microbial, usually extracellular	Microbial, animal, plant, usually intracellular	Microbial, animal, plant, usually intracellular
Multiple applications	Possible in part	Possible in part	Not yet possible
Production costs	Low	Middle–high	High

- Modern purification processes such as countercurrent distribution, ion-exchange chromatography, molecular-sieve chromatography, affinity chromatography, and precipitation.
- Development of processes for the immobilization of enzymes and for their recycling. The amount of enzyme needed per amount of substrate converted is thus considerably decreased. The proportion of enzyme cost in some processes becomes only a few percent (Figure 11.2).
- Continuous enzyme production in special reactors. The investment cost for a new system can be minimized in a continuous operation, since smaller-sized equipment is used.

In this chapter, we will discuss those reactions which involve more or less purified enzymes (either intracellular or extracellular). Additional enzymatic processes involving living or dead cells will be discussed in Chapter 15 (Transformation).

11.2 AMYLASES

Starch, a glucose polymer, is one of the most widely available plant polysaccharides. Amylases are enzymes which hydrolyze starch. One of the main uses of amylases is in the production of sweeteners for the food industry. The hydrolysis of starch with amylase results first in the production of short-chain polymers called dextrins, then the disaccharide maltose, and finally glucose. Maltose syrup (>80% maltose), which is produced primarily in Japan, is of low viscosity, is weakly hygroscopic, not crystallizable, only slightly sweet, but has good heat stability and does not undergo browning reactions. Glucose is not nearly as sweet as its isomer fructose, so the next step is the conversion of glucose to fructose using the enzyme glucose isomerase (discussed in Section 11.3). Commercial sweeteners based on fructose have some economic and manufacturing advantages over the more widely used

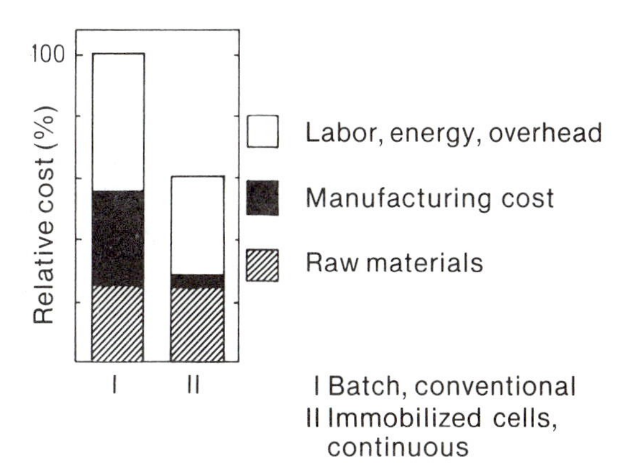

Figure 11.2 Reduction in the cost of L-aspartic acid production through use of cell immobilization techniques (From Chibata and Tosa, 1977)

sweetener sucrose. The most important enzymes in the starch-saccharification process are α-amylases, β-amylases, glucoamylases, glucose isomerases, pullulanases, and isoamylases. Figure 11.3 shows the manner of action of these enzymes on starch.

α-Amylases

α-Amylases (1,4-α-glucan-glucanohydrolases) are extracellular enzymes which hydrolyze α-1,4-glycosidic bonds. These enzymes are **endoenzymes**, splitting the substrate in the interior of the molecule. Their action is not inhibited by α-1,6-glycosidic bonds although such bonds are not split.

α-Amylases are formed by many bacteria and fungi. They are classified according to their starch-liquefying and/or saccharogenic effect, pH optimum, temperature range, and stability. Saccharogenic amylases produce free sugars, whereas starch-liquefying amylases break down the starch polymer but do not produce free sugars. Many organisms produce several α-amylases.

Bacteria which produce α-amylases are: *Bacillus subtilis, B. cereus, B. amyloliquefaciens, B.*

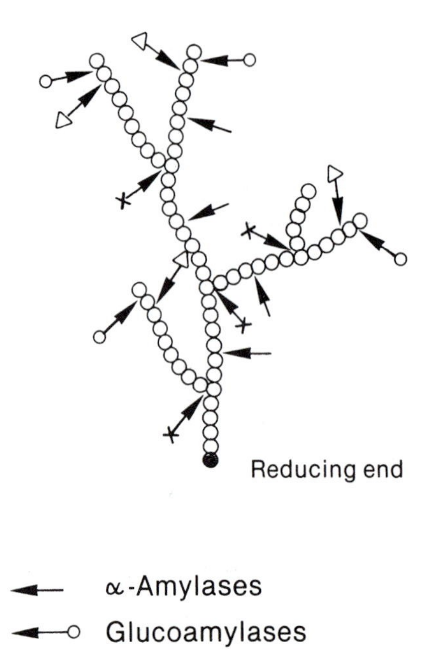

Figure 11.3 Mechanism of starch hydrolysis by amylases

Table 11.3 Molecular weight of some α-amylases from different microorganisms

Organism	Molecular weight $\times 10^3$
Aspergillus oryzae	51– 52
A. niger	58– 61
Bacillus acidocaldarius	68
B. amyloliquefaciens	49
B. subtilis	24–100
Thermomonospora curvata	62

(Fogarty, 1983)

coagulans, B. polymyxa, B. stearothermophilus, B. caldolyticus, B. acidocaldarius, B. subtilis var. *amylosaccharaticus, B. licheniformis, Lactobacillus, Micrococcus, Pseudomonas, Arthrobacter, Escherichia, Proteus, Thermomonospora,* and *Serratia.*

Three very similar strains producing saccharogenic α-amylases are *B. subtilis* Marburg, *B. subtilis* var. *amylosaccharaticus,* and *B. natto.* The strain *B. amyloliquefaciens* differs from these in that it produces a liquefying α-amylase. The substrate concentration determines the extent to which the enzymes act in a liquefying or saccharogenic manner.

Some α-amylase-producing fungi are from the genera *Aspergillus, Penicillium, Cephalosporium, Mucor, Candida, Neurospora,* and *Rhizopus.*

The molecular weights of various α-amylases do not differ considerably (Table 11.3). They all contain a large proportion of tyrosine and tryptophan in the enzyme protein and most require calcium as a stabilizer.

The most important α-amylases are produced by *Bacillus amyloliquefaciens, Bacillus licheniformis,* and *Aspergillus oryzae. Bacillus* amylases are used much more extensively than those of *Aspergillus.* The most important areas of application for these two enzymes are shown in Table 11.4.

Production of bacterial α-amylases Bacterial amylase production involves the function of the normal cell machinery for protein synthesis. This is shown by experiments involving the addition of antibiotics. If the specific antibiotics are used to inhibit protein synthesis in *B. subtilis* during the production of α-amylases, both growth and production of amylase cease (Figure 11.4). Compared to the mRNA of intracellular enzymes (stability about 2 minutes for *Escherichia coli*), the mRNA for the production of extracellular hydrolases has an extremely long lifetime. When actinomycin D is added to the amylase-producing culture to inhibit RNA synthesis, both RNA synthesis and growth cease after 30 minutes but the production of amylase continues (Figure 11.5).

For industrial production, α-amylases are produced either in batch or in fed-batch fermentation. The enzyme formation rate is very low during exponential growth in many production strains, but just before the growth rate decreases and spore formation begins, amylase production increases (Figure 11.6).

The production of α-amylases is regulated by several genes, which have been only partially

Table 11.4 Important applications of α-amylases

Industry	Source		Application
	Bacillus	*Aspergillus*	
Starch industry	+		Liquefaction of starch for production of glucose, fructose, maltose
Milling		+	Modification of α-amylase-deficient flour
Alcohol	+	+	Liquefaction of starch before the addition of malt for saccharification
Baked goods		+	Increase in the proportion of fermentable carbohydrates
Brewing	+		Barley preparation, liquefaction of additives
		+	Improved fermentability of grains, modification of beer characteristics
Paper	+		Liquefaction of starch without sugar production for sizing of paper
Textiles	+		Continuous desizing at high temperatures
Feed industry	+		Improvement of utilization of enzymatically treated barley in poultry and calf raising
Sugar	+		Improvement of filterability of cane sugar juice via breakdown of starch in juice
Laundry and detergent	+		Increase in cleansing power for laundry soiled with starch, additive in dishwasher detergents

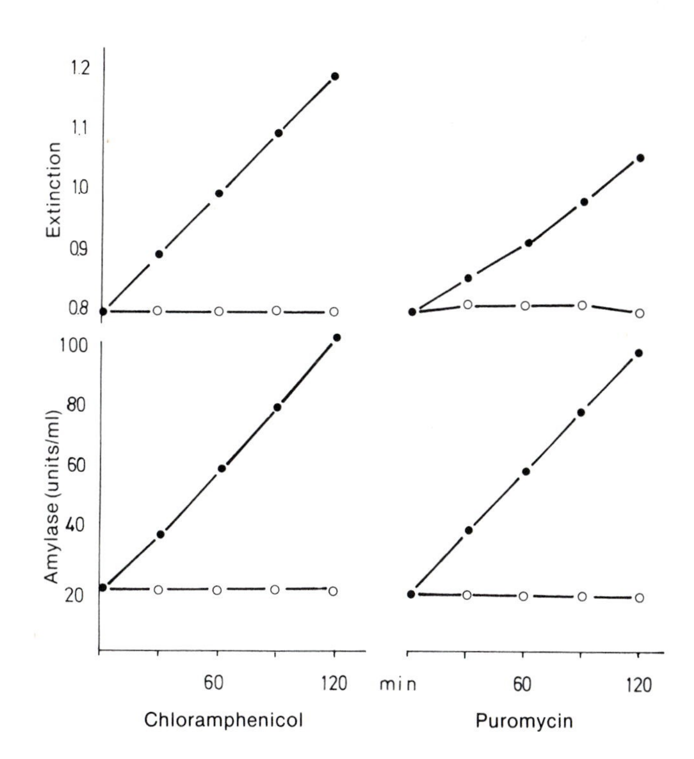

Figure 11.4 Effect of chloramphenicol and puromycin on α-amylase production in *B. subtilis*. ○ — ○ antibiotic added (10 μg/ml chloramphenicol, 100 μg/ml puromycin); • — • control. (From Terui, 1973)

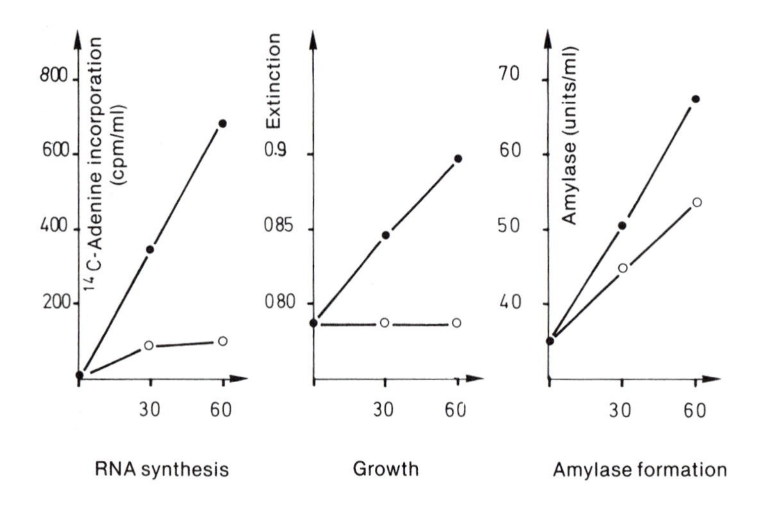

Figure 11.5 Effect of actinomycin D on growth, RNA synthesis, and α-amylase production. ○ — ○ actinomycin D (0.6 μg/ml); • — • control. (From Terui, 1973)

characterized. Whereas single-step mutations increase yields by a factor of 2–7, mutants have been selected after 5 steps which produce yields 250 times greater than the wild strain.

Since starch is a macromolecule, it cannot be taken up by the cells and hence cannot act as an inducer of α-amylase synthesis. It is assumed that a small amount of enzyme is produced constitutively and excreted into the medium where its action on starch leads to the production of low-molecular-weight inducers. **Catabolite repression** is significant in the production of most extracellular enzymes. Glucose promotes the best growth compared to other substrates, but is the least favorable in terms of amylase production. In continuous fermentation with carbohydrate as a limiting factor, growth is inversely proportional to the α-amylase production. When nitrogen is used as a limiting factor with excess glucose, only traces of amylase are produced.

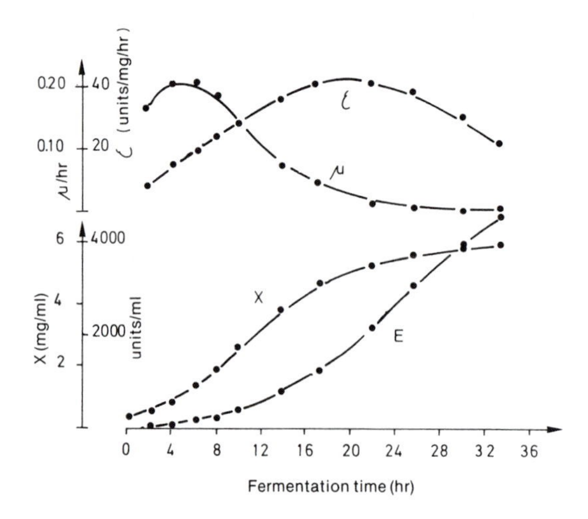

Figure 11.6 Amylase production in *Bacillus amylosolvens*. μ specific growth rate, ξ specific enzyme production rate, X cell concentration, E enzyme concentration (From Terui, 1973)

It seems unlikely that amylase is excreted by pure diffusion, since the molecule is quite large and is hydrophilic. There is good evidence that the biosynthesis as well as the steps leading to excretion of extracellular proteins such as amylases and proteases takes place at the outer membrane of the cell. As it is formed, the protein contains an additional peptide, the *signal peptide*, which participates in the transfer of the nascent protein from the ribosome through the membrane. Only after the protein has passed through the membrane does it fold into its tertiary configuration and the signal peptide is removed.

A **medium** for the production of α-amylases in a 100 m^3 fermenter with *B. subtilis* consists of: 5% starch, 0.56% NH_4NO_3, 0.28% sodium citrate, 0.13% KH_2PO_4, 0.05% $MgSO_4 \cdot 7\ H_2O$,

0.01% $CaCl_2 \cdot 2\ H_2O$, 0.5% peptone, 0.2% yeast extract; pH 6.8.

The effect of temperature on enzyme production and growth rate is given in Figure 11.7. At 45°C, the maximal specific enzyme formation rate is reached with the test strain after 18 hours; however, the maximal amount of enzyme produced (up to 3000 units/ml) is obtained at considerably lower temperatures (27 to 30°C). Thermophilic strains are used in newer processes. *Thermomonospora*, isolated from compost, has a temperature optimum for growth and amylase production at 53°C. Figure 11.8 shows the narrow pH optimum for amylase production by this strain.

α-Amylase production from fungi The production of fungal amylases is constitutive, but as with other enzymes, it is repressed by regulators. For amylase production using *Aspergillus oryzae*, the following nutrient solution can be used: 8% starch, 1.2% $NaNO_3$, 0.1% K_2HPO_4, 0.1% $MgSO_4$, 0.05% KCl, 0.003% $FeSO_4$, 0.08% $Mg(NO_3)_2$, 0.05% $Mg(H_2PO_4)_2$, 2.0% malt extract. The medium composition must be optimized, because as seen in Table 11.5, a marked shift in enzymatic activities is observed when different carbohydrates are used.

The optimal temperature lies in a narrow range between 28–30°C, and the duration of the fermentation process is 3–4 days.

β-Amylases

β-Amylases (α-1,4-glucan-maltohydrolases) are usually of plant origin, but some microbial producers are also known: *Bacillus polymyxa, B. cereus, B. megaterium, Streptomyces* sp., *Pseudomonas* sp., and *Rhizopus japanicus*. Although yields in wild strains are usually low, mutants have been discovered which produce 200 times more enzyme than the wild type.

Bacterial β-amylases have greater heat resistance (>70°C) than plant β-amylases and the pH optimum is also higher (about pH 7.0). In contrast to α-amylase, calcium is not necessary for stabilization and activation of bacterial β-amylase.

In the future, it is likely that β-amylases will be used for the production of maltose syrup.

Glucoamylases

Glucoamylases (α-1,4-glucan-glucohydrolases) act on starch by splitting glucose units from the nonreducing end. Maltose is broken down only slowly, while 1,6-bonds in the branched polysaccharides are hardly attacked. Thus glucose, maltose and limit dextrins are the end products of glucoamylase action.

Microorganisms used to produce glucoamylases are *Aspergillus niger, A. oryzae, A. awamori, Rhizopus niveus, R. delemar, R. formosaensis,* and *R. javanicus*. Production strains in western Europe and in the United States are mutants of *A. niger* and *R. niveus*. Today, industrial fermentation of glucoamylase is carried out almost exclusively in submerged fermenters up to 150 m^3 in volume. Due to the increasing demand for glucoamylase in fructose-syrup production, smaller systems which involve growth on the surface of liquid or solids are no longer economical. However, in genetic studies, strains are frequently isolated which produce higher yields in surface culture than in submerged culture (Table 11.6). Experiments with batch-fed fermentations have been successfully carried out over periods as long as 320 hours.

Frequently, several glucoamylase isoenzymes are produced by one strain; in addition, the enzymes may be modified during the fermentation process. *A. awamori* var. *kawachi* produces three glucoamylases, only one of which hydrolyzes crude corn starch. Due to the action of proteases and glycosidases, which are also present under various experimental conditions, one of the other two enzymes may be formed instead of the glucoamylase which hydrolyzes crude corn starch. These other enzymes can only act on corn starch which has been swollen by heat or chemical treatment. Therefore the changes in glucoamylase activity must be carefully examined

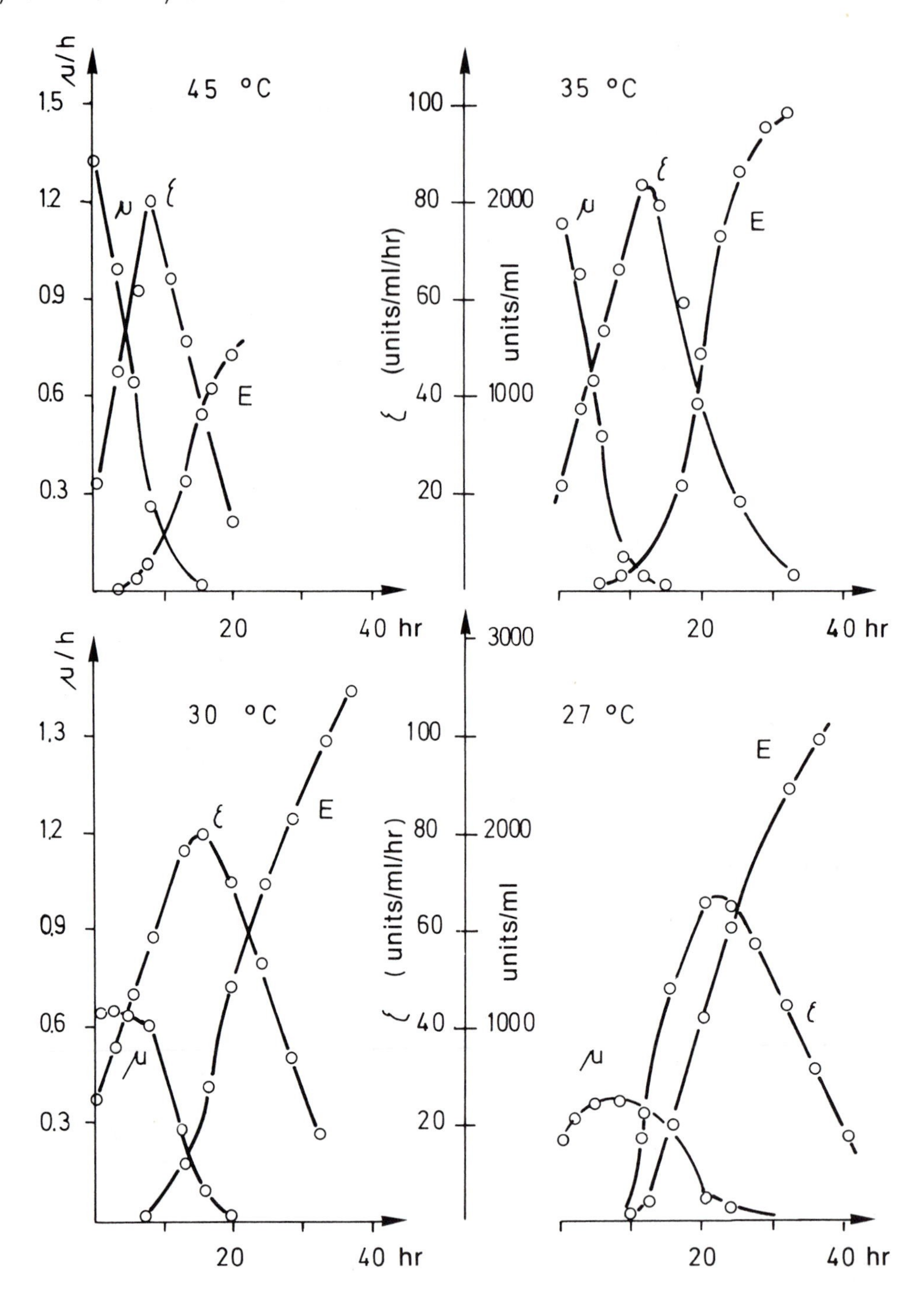

Figure 11.7 Relationship of growth and α-amylase production to temperature. See Figure 11.6 for symbols. (Modified from Terui, 1973)

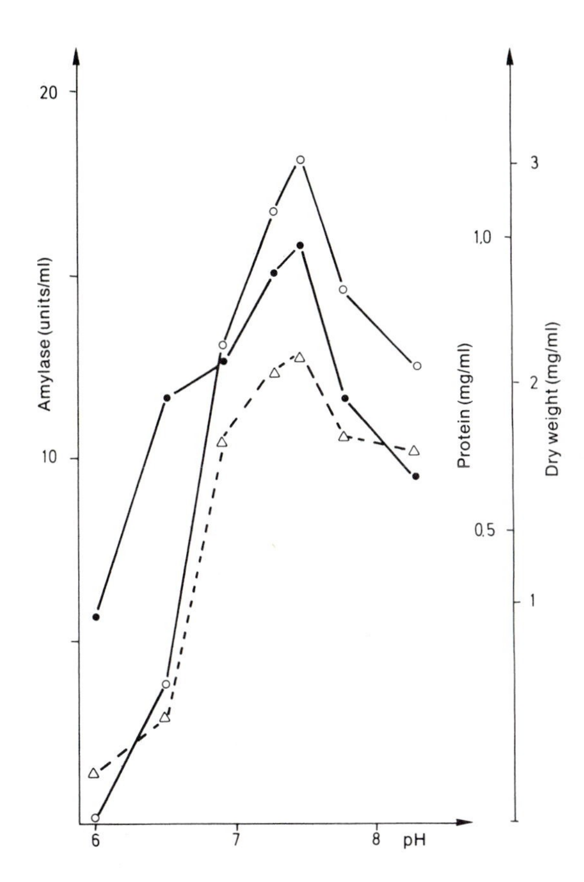

Figure 11.8 Effect of pH on α-amylase production, extracellular protein and growth. ○ — ○ α-amylase; • — • protein; △ — △ cell dry weight (From Glymph and Stutzenberger, 1977)

Table 11.5 Effect of carbon source on amylase and protease production in *Trichoderma viride*

Carbon source	Amylase units/ml	Protease units/ml
Corn starch	235	351
Maltose	179	175
Glucose	52	243
Sucrose	17	350
Lactose	3	175
Control	0	324

(Upton and Fogarty, 1977)

when high-yielding strains are isolated and when fermentation conditions are being optimized.

Table 11.6 Glucoamylase production in some mutants of *Rhizopus formosaensis*

Strain	Submerged culture units/100 ml	Surface culture (liquid) units/100 ml	Surface culture (solid) units/g
R 13-5	295	2745	858
R 13-59	491	3067	1322
R 13-591	1595	3788	2160
R 13-59136	4432	7425	5120

(Lin, 1972)

Either starch or dextrins induce enzyme production; nutrient solutions for enzyme production are generally based on starch. Glucose, glutamic acid, and lactose cause catabolite repression. Fermentation time varies from 3–5 days, depending on the organism, and the fermentation temperature can range from 28–35 °C. In addition to glucoamylase, α-amylase may also be produced in the same fermentation.

The quality of the spores used as inoculum has an effect on the enzyme yield in glucoamylase production with *A. niger*. When spores are produced on a rich agar medium, such as one containing a supplement of malt extract, the glucoamylase formation rate of later cultures decreases with each transfer, although the spore formation rate does not diminish. After 6 transfers only 50% of the original enzyme activity is produced. However, if the inoculum spores are produced on a minimal medium (Czapek Dox), the enzyme formation rate remains constant, but the spore formation rate decreases.

1,6-Glycoside-splitting enzymes

Enzymes which can split the side chains of amylopectin via hydrolysis of α-1,6-bonds are important in commercial processes. These enzymes can be subdivided into two groups: a) those which directly affect amylopectin, such as pullulanases and isoamylases, and b) those which have an indirect effect after enzymatic modification of the substrate. Pullulanases and isoamylases hydrolyze amylopectin. Moreover, pullulan, the α-glucan from the fungus *Pullularia pullulans*, is split by means of pullulanases. This latter reaction cannot be carried out with isoamylases. Table 11.7 shows the effect of pullulanase on starch hydrolysis by β-amylase.

Late in the log phase, *Aerobacter aerogenes* ATCC 15050 excretes pullulanase into the medium. With inducible strains, either maltose or pullulan can be used as a carbohydrate source for enzyme induction, but constitutive mutants are now generally used instead. Both batch fermentations and continuous processes are in use. Several other pullulanase producers besides *Aerobacter aerogenes* have been described: *Bacillus polymyxa, Pseudomonas saccharophila, Streptococcus* sp., and *Streptomyces* sp. Among the isoamylase-producers, *Pseudomonas amyloderamosa* is the most important strain. In addition, there are other strains: *Agrobacterium, Erwinia, Staphylococcus, Serratia, Nocardia, Bacillus, Pediococcus, Lactobacillus,* and *Leuconostoc.*

Starch hydrolysis

In the food industry, starch hydrolysates are used as additives in the manufacture of candies, baked goods, canned goods, and frozen foods. Syrups used can contain maltose or glucose (after saccharification), or fructose (after isomerization, Section 11.3). Starch hydrolysates affect the taste, sweetness, and texture of the food products.

To produce a hydrolysate, starch must first be converted to paste at 65–90°C. During this process, the viscosity of a 20% solution increases significantly to 3000 centipoise. Acid treatment at 120–140°C (pH 1.8–2.0) or treatment with α-amylases cause starch liquefaction in the next step; an alternative method is a combined acid/enzyme liquefaction. The disadvantages of acid treatments are generally low glucose yields, higher purification costs, risk of deterioration, and corrosion of the system used.

Table 11.7 Conversion of 10% potato starch (pretreated with α-amylase) via pullulanase and β-amylase

Type of process	Maltose %	Glucose %	Maltotriose %	Dextrin %
25 units β-Amylase/g starch; 16 hr 45°C	65.5	1.1	3.5	29.9
10 units Pullulanase + 25 units β-amylase/g starch; 16 hr 45°C	90.5	0.4	1.2	7.9

(Hayashibara, 1970)

Enzymatic liquefaction has therefore generally displaced acid treatment. Liquefaction is carried out at 110°C with amylases from *B. subtilis* which are added to the solution before conversion to paste. When α-amylases from *A. niger* or *A. oryzae* are used, the process is carried out at 55°C and the hydrolysis continues to the point of a maltose/maltotriose mixture without the addition of other enzymes.

Starch solution which has been liquefied with α-amylases can be saccharified with glucoamylases at a pH of 4.0–5.0 and a temperature of 55–60°C without an intermediate purification stage. Pullulanases or isoamylases may also be added at this stage to bring about splitting of bonds other than the 1–4 linkages. The specificity of enzymes used and the conditions of conversion determine the composition of the end products.

Figure 11.9 shows a diagram of the glucose syrup production process.

11.3 GLUCOSE ISOMERASES

Glucose isomerase (D-glucose ketoisomerase) causes the isomerization of glucose to fructose. The reaction is reversible and a mixture of glucose and fructose is produced, the ratio of the components depending on the enzyme used and the reaction conditions, such as temperature. The enzyme has become of commercial value because the price of sugar has increased compared to that of starch. In the Western world, the current sucrose consumption rate is 50 kg per person per year. In the United States in 1985, 4×10^6 tons of isosyrup (glucose/fructose mixture) were produced. As an example of European usage, the

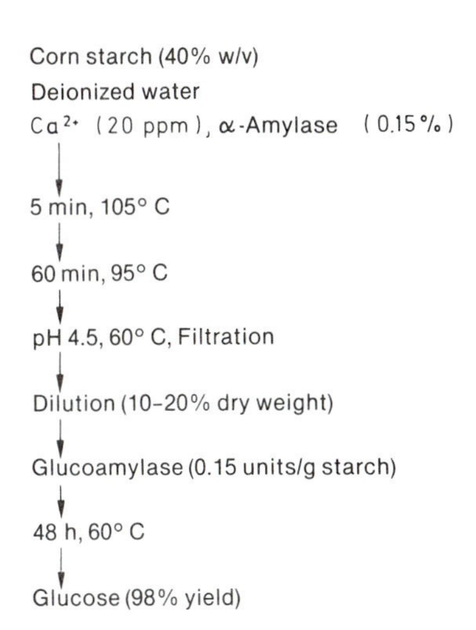

Figure 11.9 Production of glucose solution from starch (Novo Industries, 1977)

size of the German fructose production is estimated at 10,000–20,000 tons per year.

Starch can be converted to glucose by either acid or enzyme hydrolysis (see Section 11.2). The advantage of starch hydrolysis over direct sugar production is that the initial materials used, such as wheat, corn, cassava, and to some extent potatoes, are nonperishable, but sugar beets (the only direct source of sugar in temperate climates) are only available about 100 days per year.

Glucose has 70–75% the sweetening strength of beet sugar (sucrose) but β-D-fructopyranose (fructose), the sweetest monosaccharide, has twice the sweetening strength of sucrose. Thus, processes for the manufacture of fructose are of considerable value. Fructose can be produced biochemically from sucrose or starch by way of glucose production, as shown in Figure 11.10. It can also be produced chemically from glucose at high temperatures under alkaline conditions, but byproducts are formed such as psicose which cannot be metabolized in the body (Figure 11.11). Thus, chemical production of fructose is not acceptable to the food industry.

The glucose isomerases in use today are actually D-xylose-isomerases which have additional activity with D-glucose and D-ribose. The

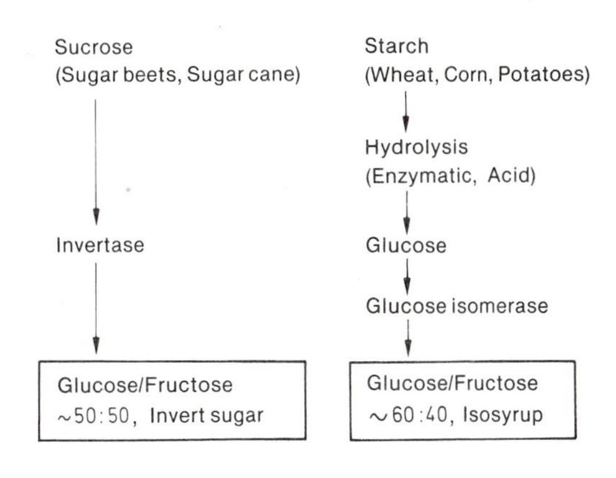

Figure 11.10 Use of biochemical methods for the production of high-fructose syrup

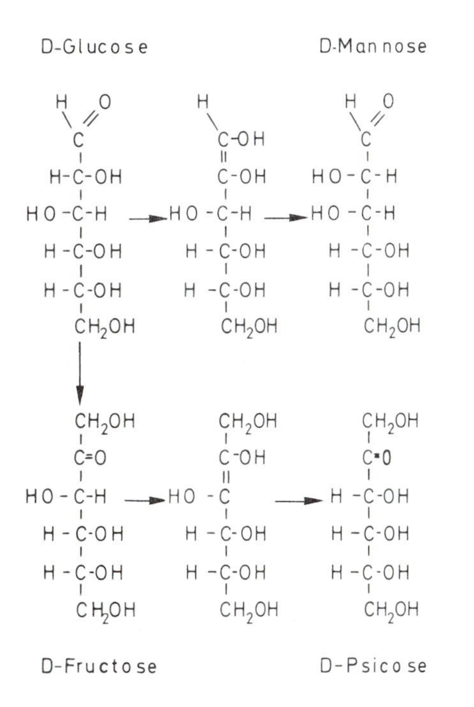

Figure 11.11 Chemical isomerization of glucose to fructose

glucose isomerase used must fulfill the following criteria: low pH optimum to avoid side reactions, high specific activity, high temperature optimum.

A large group of bacteria and some yeasts are known which carry out the direct isomerization

(Table 11.8). There are other bacteria (*Escherichia intermedia, E. freudii,* and *Aerobacter aerogenes*) which have a somewhat different glucose isomerase activity. The enzyme produced by these bacteria is actually a glucose phosphate isomerase. These organisms are not usable industrially, however, since their enzymes require arsenic for maximal activity, and arsenic cannot be used in food production.

The most important glucose isomerase producers are *Bacillus coagulans* (Sweetzyme®, Novo Industri), *Streptomyces rubiginosus* (Optisweet®, Miles-Kali), *Actinoplanes missouriensis* (Ketozyme®, Universal Oil Products; Maxazyme®, Gist Brocades), *Flavobacterium arborescens* (Taka-Sweet®, Miles Laboratories).

Glucose isomerase from *Bacillus*

Novo Industries has developed glucose isomerase from *B. coagulans* for commercial use. This **immobilized enzyme process** is used today in many plants in the United States, Europe, Japan and Korea. Since this glucose isomerase is primarily a xylose isomerase, xylose must be added for induction of the enzyme with all wild strains. Xylose is quite expensive, however, and may be replaced by xylan or wheat bran, which also contain xylose. Constitutive mutants have also been isolated.

Catabolite repression regulates the production of glucose isomerase. Glucose acts as a repressor in both batch process and continuous fermentation. In batch process, *B. coagulans* does not form the enzyme during the log phase. As soon as the glucose content in the nutrient solution approaches zero, growth ceases. In a typical diauxy, additional carbon sources present in the medium are then metabolized, and enzyme production begins. Maximal enzyme activity is obtained after 24 hours' incubation.

Cobalt and magnesium are prerequisites for maximal enzyme production in wild strains, but mutants have been isolated which produce optimal enzyme titers in the absence of cobalt. Either yeast extract or corn steep liquor can be used as the nitrogen source. As in the case of many other fermentations, the choice of a nitrogen source is vital for the yield and the concentration must be optimized for each individual process.

The difficulties that arise when attempting to optimize the nitrogen source for glucose isomerase production are illustrated in Table 11.9 by results for *Actinoplanes missouriensis*. As seen, there is wide variation in yield using various complex nitrogenous substrates. Because it is so difficult to standardize the nitrogen source, the fermentation yield may vary considerably from batch to batch.

Table 11.8 Glucose isomerase producers

Bacillus coagulans
Streptomyces phaeochromogenes
Streptomyces rubiginosus
Streptomyces olivochromogenes
Arthrobacter sp.
Actinoplanes missouriensis
Microbispora rosea
Micromonospora coerula
Microellobospora flavea
Nocardia asteroides
Nocardia dassonvillei
Brevibacterium imperiale
Flavobacterium arborescens

Table 11.9 Effect of organic nitrogen sources on glucose isomerase production in *Actinoplanes missouriensis*

Nitrogen source	Manufacturer	Activity	
		GI/ml	%
Corn steep liquor	Anheuser-Busch	17.8	100
O. M. peptone	Amber Lab.	13.4	75
Casein hydrolysate (Amber EHC)	Amber Lab.	13.0	73
Bacto-soytone	Difco	11.8	66
Yeast extract (BYF-100)	Amber Lab.	10.3	58
Distiller's solubles	National Distillers	7.9	44
Yeast extract (BYF-300)	Amber Lab.	5.8	33
Bacto-peptone	Difco Lab.	5.2	29
Yeast extract (BYF 50 X)	Amber Lab.	4.3	24
Malt extract	Difco Lab.	2.5	14
Atlantic Menhaden peptone (fish)	Haynie Products	1.1	6

(Activity with corn steep liquor = 100%)

In a continuous fermentation process, the growth-limiting substrate must be glucose. For optimal enzyme yields, oxygen limitation is also necessary since microaerophillic conditions inside the cells stabilize the system.

Glucose isomerase from *Streptomyces*

Glucose isomerases from *Streptomyces* strains are produced only in batch processes today. There are mutants which require neither xylose nor cobalt. Production is begun with a spore inoculum of the working strain, which is cultivated for 24–48 hours at 30°C in shaken culture. The culture medium used contains soy meal, yeast extract, glucose, starch, phosphate, and deionized water. A number of such shake flasks are used to inoculate (5% inoculum) the production fermenter (which is 80–150 m³ in volume). *Streptomyces olivaceus* NRRL 3588, produces glucose isomerase at pH 8.5 and between 60–70°C. A typical fermentation with *Streptomyces albus* is shown in Figure 11.12.

Figure 11.13 shows the pH and temperature optima of the glucose isomerase produced in this process.

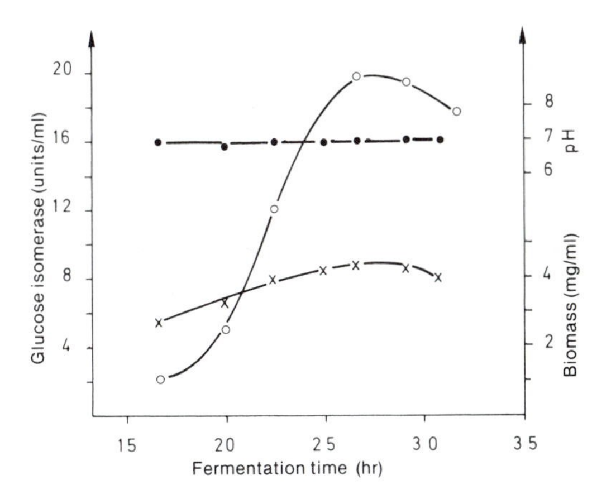

Figure 11.12 Industrial production of glucose isomerase with *Streptomyces albus*. 22 m³ fermenter, temperature 30°C, impeller speed 150 rpm, aeration rate 0.33 vvm. ● — ● pH; × — × cell growth; ○ — ○ glucose isomerase (From Takasaki et al., 1969)

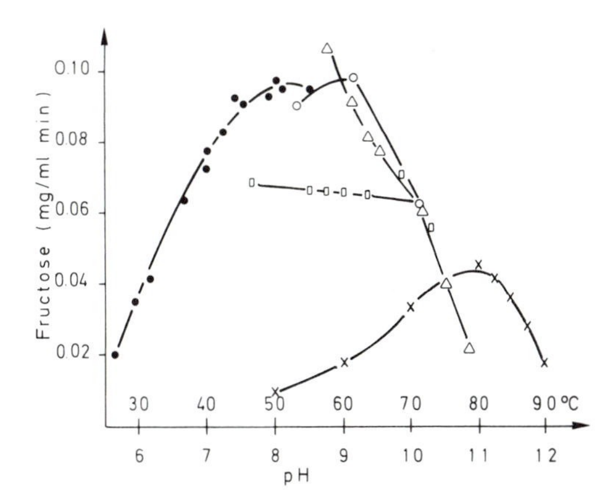

Figure 11.13 Effect of temperature and pH on glucose isomerase activity. The enzyme was derived from *Streptomyces albus*. ● — ● Na_2HPO_4-KH_2PO_4 buffer; ○ — ○ Na_2HPO_4-NaOH buffer; □ — □ NaOH-glycine buffer; △ — △ Na_2CO_3-$NaHCO_3$ buffer; × — × temperature experiment with phosphate buffer 0.05 M (From Takasaki et al., 1969)

Immobilization of glucose isomerase

The commercial process for production of fructose from glucose became feasible only when procedures for immobilization of the enzyme were developed (see Section 11.11), since this permitted the same batch of enzyme to be used many times. Since glucose isomerase is formed intracellularly in most strains, many commercial processes are carried out with immobilized cells or by addition of partly broken cells.

The Novo process uses glutaraldehyde to cross-link cell suspensions of *B. coagulans*. The cross-linked material is processed into 0.3–1.0 mm diameter pellets and used either in a batch reaction or continuously with several bioreactors connected in series. Figure 11.14 shows a system which produces 100 tons of fructose per day with 2 parallel rows, each with three 2.2 m³ columns.

The starting material is glucose syrup with a dry weight of 40–45% containing 93–96% glucose. The solution must first be purified by filtration, carbon treatment, and ion exchange to prevent loss of enzyme activity during isomerization. Magnesium is added as an activator and

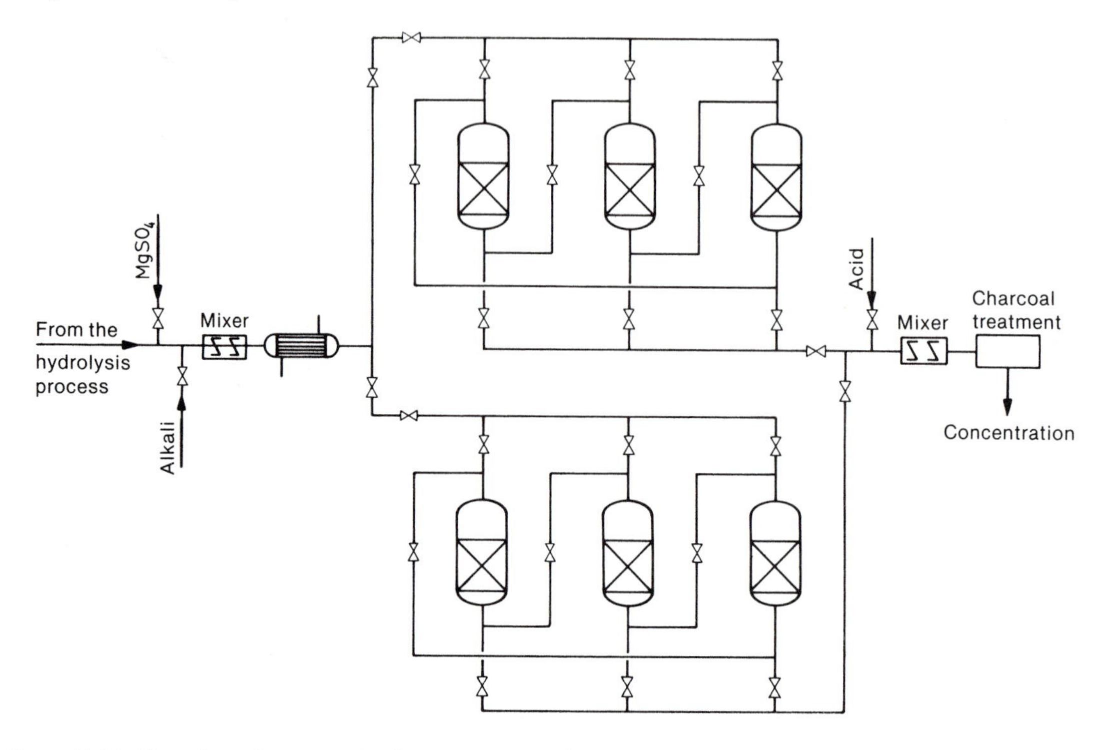

Figure 11.14 Flow chart of a process for fructose syrup production (100 tons/day)

the reaction is run at pH 8.5 and a temperature of 60–65°C. The time required for isomerization is 0.8–4 hours.

When compared to the batch process, at the same conversion temperature, a considerably better product is obtained in a continuous system in 20% of the reaction time and with only 4% of the reactor volume (Table 11.10).

Table 11.10 Comparison of batch and continuous methods for production of fructose

Parameters	Batch process	Continuous conversion
Bioreactor volume	750 m^3	30 m^3
Enzyme consumption	17–20 tons	10 tons
$MgSO_4 \cdot 7\ H_2O$-consumption	4.3 tons	2.2 tons
$CoSO_4 \cdot 7\ H_2O$-consumption	2.2 tons	0
Color formation OD_{420nm}	0.05–0.10	0–0.02
Psicose formation	0.1%	0.1%
Purification	Carbon treatment, cation and anion exchangers	Carbon treatment

(Zittan et al., 1975)

In the recovery process, the converted solution is first acidified and then passed over activated carbon and ion exchange resins to remove salts and colored materials. Fructose syrup is then concentrated to about a 70% solution (dry weight basis), stored, and used in this form in the food industry. Since di- and trisaccharides are not converted by glucose isomerase, a typical product has the following composition:

Fructose	42	%	Dry weight
Glucose	53	%	Dry weight
Oligosaccharide	5	%	Dry weight
Psicose	0.1	%	Dry weight
Ash	0.05–0.1%		Dry weight
Dye value	0.003		
pH	4–5		

11.4 L-ASPARAGINASES

L-Asparaginases (L-asparagine amidohydrolases) are used as **antitumor agents** in the treatment of some leukemias and lymphomas. The principle behind the use of this enzyme as an antitumor agent is as follows: The metabolism of these tumor cells requires L-asparagine, and since asparagine is split into aspartic acid and ammonium ions by asparaginase, these malignant cells can be destroyed.

The enzyme is produced by several bacteria, but for commercial production (80 m^3 fermenters) only mutants of *Escherichia coli* ATCC 9637, *Serratia marcescens* ATCC 60, and *Erwinia carotovora* are used. An appropriate medium is 3% corn steep liquor, 0.6% sodium acetate, and 0.2% ammonium sulfate at pH 7.0, used in reactors up to 80 m^3 in volume. Oxygen supply is critical during the 16- to 40-hour fermentation. Table 11.11 shows that the prerequisite for a high enzyme titer is a low pO_2.

At the end of the fermentation process, the cells are broken open and the enzyme is crystallized.

11.5 PROTEASES

After amylases, the second most important industrial enzymes currently are the proteases. About 500 tons of these enzymes (based on the pure protein) are produced per year. Proteases are used primarily in the detergent industry and in the dairy industry (rennin, see Section 11.6). Other areas in which proteases are used include the pharmaceutical industry, the leather industry, the manufacture of protein hydrolysates, the food industry, the film industry, and waste processing.

Proteases are produced commercially both from bacteria and fungi. An important distinguishing feature is the pH optimum of the protease. The proteases on the market include **alkaline, neutral**, and **acid proteases**.

Akaline proteases

Many bacteria and fungi excrete alkaline proteases. The **most important producers** are *Bacillus* strains such as *B. licheniformis, B. amyloliquefaciens, B. firmus, B. megaterium*, and *B. pumilis; Streptomyces* strains such as *S. fradiae, S. griseus*, and *S. rectus*; and the fungi *Aspergillus niger, A. sojae, A. oryzae*, and *A. flavus*. Enzymes used in detergents are mainly proteases from *Bacillus* strains (Bacillopeptidases). The best known proteases are Subtilisin Carlsberg from *B. licheniformis*, and Subtilisin BPN and Subtilisin Novo from *B. amyloliquefaciens*. These enzymes contain serine at the active site of the molecule and are not inhibited by EDTA (ethylene diamine tetraacetic acid), but are inhibited by DFP (diisopropyl fluorophosphate).

The proteases of this type have many features of value for use as detergent enzymes:

- stability at high temperature
- stability in the alkaline range (pH 9–11)
- stability in association with chelating agents and perborates.

Table 11.11 L-asparaginase formation with *Erwinia aroideae* at 28°C in a small fermenter

Aeration rate vvm	Baffles	Impeller speed rpm	Cell dry weight g/10 l	L-Asparaginase Units/g
0.5	–	300	14	960
1.0	+	300	16	580
1.5	+	300	18	340
2.0	+	300	18	230

(Peterson and Ciegler, 1969)

However, their stability in the presence of surface-active agents is low, thus limiting their shelf life.

Screening Because the enzymes must be stable under alkaline conditions, screening for better producers is done using strongly basic media. Single colonies have been tested on protein-agar plates at pH 10.0. Figure 11.15 shows the results of screening with a large number of strains. As seen, strains of *B. licheniformis* and *B. subtilis* showed optimal growth in the pH range of 6–7, but some new strains had maximal growth rates at pH 8–9 and grew somewhat at pH 11.

Since the enzyme yields of wild-type strains are insufficient for industrial utilization, extensive genetic studies have been carried out to increase the yield. The genes for the formation of several proteases have been cloned. Protein engineering has been used to develop modified *Bacillus* subtilopeptidases with altered amino acid sequences and corresponding changes in enzymatic properties such as substrate specificity, pH optimum, and stability to bleaching agents.

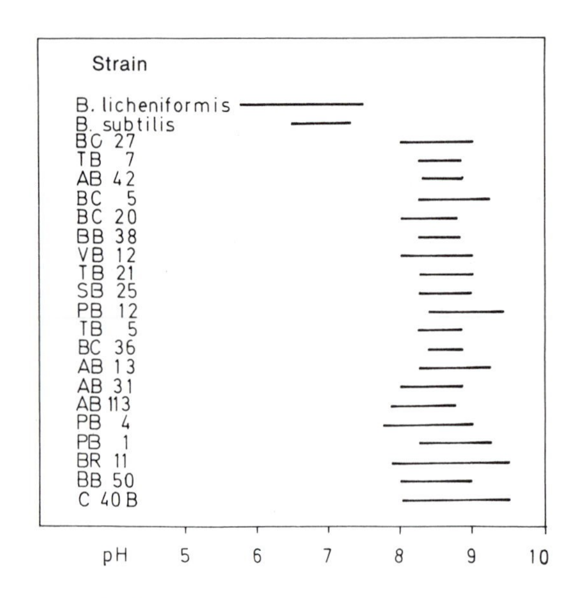

Figure 11.15 Screening for alkaline proteases. Within the indicated pH range, at least 90% of the maximal growth rate is attained. (From Aunstrup et al., 1972)

Fermentation process Sterility is mandatory for protease production, as with other enzyme fermentations. Cultures are stored in the lyophilized state or under liquid nitrogen. Initial growth is carried out in shaken flasks and small fermenters at 30–37°C. For production, 40–100 m^3 fermenters are used. Production of extracellular proteases is chiefly regulated by the medium composition.

The fed-batch process is generally used in order to keep down the concentration of ammonium ions and amino acids, since these nitrogenous materials repress protease production. Although continuous processes have been described, they are not used commercially.

High oxygen partial pressure is generally necessary for optimal protease titers. Aeration rates are 1 vvm and the time span of the fermentation is 48–72 hours, depending on the organism.

Proteases must be converted into particulate form before they are added to detergents, since if dry enzyme powder is inhaled by production workers or users, allergic reactions may result. Enzyme concentrates are marketed in a microencapsulated form. To make a suitable encapsulated product, a wet paste of enzyme is melted at 50–70°C with a hydrophobic substance such as polyethylene glycol and then converted into tiny particles. These solidified spherical particles are not hazardous when added directly to the detergent. A further development is the immobilization of enzymes in fibrous polymers and in granules.

Neutral proteases

Neutral proteases are excreted by both bacteria and fungi. Producing organisms include: *Bacillus subtilis, B. cereus, B. megaterium, Pseudomonas aeruginosa, Streptomyces griseus, Aspergillus oryzae, A. sojae,* and *Pericularia oryzae*.

Neutral proteases are relatively unstable and calcium, sodium, and chloride must be added for maximal stability. The pH range of activity is fairly narrow and these enzymes are not very stable to increased temperatures (Figure 11.16).

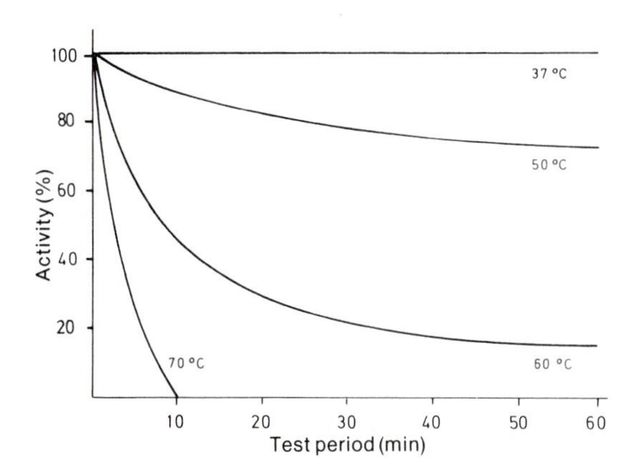

Figure 11.16 Temperature stability of a neutral protease (Neutrase®, Novo, 1978)

The neutral proteases are also quickly inactivated by alkaline proteases. Because of these limitations, they have restricted industrial application, but do find some uses in the leather industry and in the food industry for the manufacture of crackers, bread, and rolls.

Acid proteases

In this category are rennin-like proteases from fungi which are chiefly used in cheese production (see Section 11.6). In addition, other acid proteases which are similar to mammalian pepsin are also on the market. These enzymes have a pH optimum of 2–4. Acid proteases are used in medicine, in the digestion of soy proteiln for soy sauce production, and to break down wheat gluten in the baking industry.

11.6 RENNIN

In earlier centuries, curds for cheese production were produced by the action on milk casein of microorganisms which were naturally present in milk. These organisms produced organic acids, primarily lactic acid, which lowered the pH to the isoelectric point of casein, leading to a natural precipitation of casein in the form of curds which could easily be removed from the whey. Another possibility for curd production, salting out, is rarely used. A third possible method is heat precipitation but this method is used for only a few processes. A fourth method, the addition of the milk-coagulating enzyme rennin, has been known for a long time and is used in most cases today (alone or in combination with pH precipitation).

In modern processes, **starter cultures** are added which convert lactose to lactic acid and lower the pH. The most commonly used strains include *Streptococcus lactis, S. cremoris, S. thermophilus, Lactobacillus helveticus,*and *L. bulgaricus.* These organisms are cultured industrially and the culture selected depends upon the desired curd structure and flavor, a minimal gas production during fermentation, a low sensitivity to bacteriophages, and the extent of acid production.

Rennet has been used for generations as a milk-coagulating enzyme. It is an extract from the fourth stomach of 3- to 4-week-old calves which have been raised on milk. The purified enzyme is called *rennin, chymase,* or *chymosin.* The main binding site of rennin on κ-casein is the phenylalanine residue at location 105 and the methionine at location 106. The kinetics of milk coagulation with microbial rennins has been extensively studied.

Due to the increasing production of cheese (in 1974, six million tons were manufactured worldwide) and a decline in the number of slaughtered calves, intensive research has been underway since 1960 to develop rennin products of microbial origin.

Enzyme systems must meet the following requirements:

- Good coagulation of casein without hydrolysis
- Good odor and structure of the cheese
- No unpleasant odor of its own
- Nontoxic
- Low proteolysis in order to prevent the development of bitterness in the ripening process

- Low lipase activity, to avoid the development of rancidity in the cheese.

A variety of fungi and bacteria have been isolated as producers and have been examined for the above characteristics.

The following are the most important genera of bacteria: *Alcaligenes, Bacillus, Corynebacterium, Lactobacillus, Pseudomonas, Serratia, Streptococcus,* and *Streptomyces*. The *Bacillus* strains (*B. subtilis, B. polymyxa,* and *B. mesentericus*) have been marketed but have not proved commercially successful.

There are several genera of **fungi** which produce milk-coagulating enzymes: *Aspergillus, Candida, Coriolus, Endothia, Enthomophthora, Irpex, Mucor, Penicillium, Rhizopus, Sclerotium,* and *Torulopsis*.

Only three strains of fungi are used worldwide in production. They are subdivided into 2 groups:

Type I	*Mucor pusillus* var. Lindt	Solid surface culture
Type II	*Endothia parasitica*	Submerged culture
	Mucor miehei	Submerged culture

It has not yet been possible to cultivate *Mucor pusillus* successfully by the submerged culture process; in a stirred bioreactor, only about half as much rennin activity is formed as in the surface process. Moreover, there are substantial side products in submerged fermentation, such as proteases and lipases, which prohibit the direct use of enzyme preparations without extensive purification. The enzyme, which is produced at 30°C in 70 hours, is sold commercially as a liquid preparation. The rennin produced by *Endothia parasitica* was the first rennin marketed commercially, placed on the market in 1967. The enzyme is an acid protease, stable at pH 4.0–5.5, with a molecular weight of 34,000–37,500. Within the designated pH range at 50°C, only 30% of the activity is lost within 30 minutes. A nutrient solution for enzyme production contains 3% soy meal, 1% glucose, 1% skim milk, 0.3% $NaNO_3$, 0.05% K_2HPO_4, 0.025% $MgSO_4 \cdot 7\ H_2O$; pH 6.8. The fermentation takes 48 hours at 28°C. At the end of the fermentation, after removal of the mycelium, the extracellular enzyme is concentrated and precipitated in an evaporation process.

Several *Mucor miehei* strains produce usable rennin. The enzyme is stable around pH 4.5 and the molecular weight is 38,000–41,800. A typical medium for production is 4% potato starch, 3% soy meal, 10% ground barley, 0.5% $CaCO_3$. The fermentation takes 5–6 days at 40°C.

After the curd has formed, most of the enzyme remains in the whey and is therefore lost. Because of this, studies have been carried out to immobilize the enzyme, but so far these studies have been unsuccessful. Microbial rennins are actually too temperature stable, remaining active in the curd after precipitation and subsequently causing harmful proteolysis. Research is thus being done to clone the gene for calf rennin into microorganisms. Complementary DNA (cDNA) from calf prorennin has been successfully expressed in *Escherichia coli*, making possible the first commercial production by a microorganism of the calf rennin enzyme.

11.7 PECTINASES

Pectinase preparations contain at least six enzymes, splitting pectins at different sites of the molecule. The basic structure of pectin is α-1,4-linked galacturonic acid, with up to 95% of its carboxyl groups esterified with methanol. Pectinases are classified according to their point of attack on the pectin molecule (Figure 11.17). The methyl ester is split by pectinesterase and the glycosidic bonds of the pectin chain are split by hydrolysis with endo-polygalacturonases or exo-polygalacturonases. Another method of splitting is transelimination by means of bacterial endo-pectate lyases, or fungal exo- and endo-pectin lyase (Figure 11.18).

A number of commercial firms produce fungal pectinases using *Aspergillus niger* or *A. wentii* and one fermentation plant uses *Rhizopus*. Both surface and submerged processes are used. Pec-

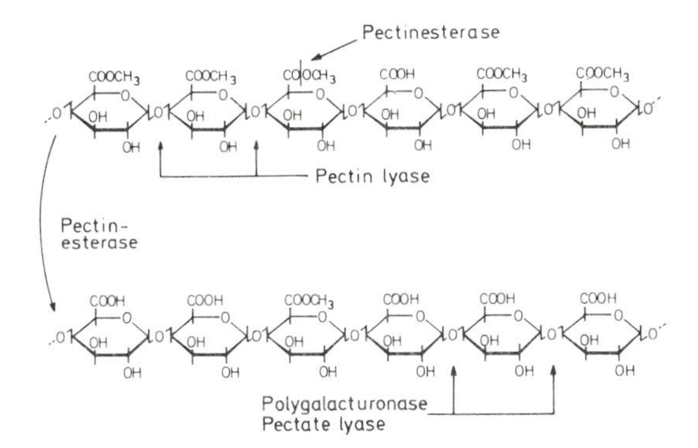

Figure 11.17 Pectin hydrolysis with various individual pectinases

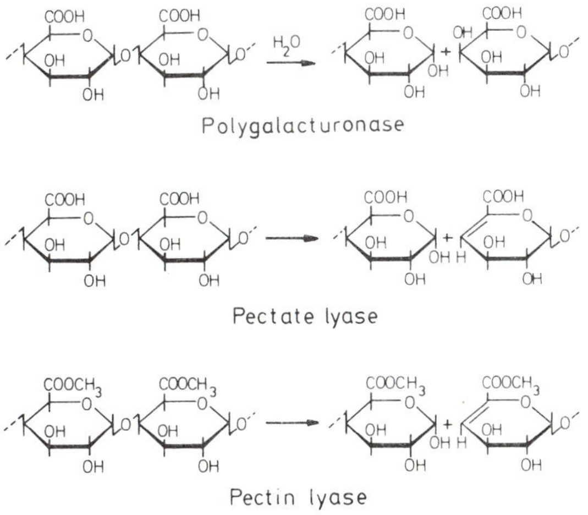

Figure 11.18 Splitting of glycosidic bonds in pectin through hydrolysis with polygalacturonase and through transelimination with pectate lyase and pectin lyase

tinases have been discovered in other fungi and also in bacteria, protozoa, insects and higher plants.

The fermentation with *Aspergillus niger* runs for 60–80 hours in fed-batch cultures at pH 3–4 and 37°C, using 2% sucrose and 2% pectin. The purification of the enzyme is simple: the biomass is removed by filtration or centrifugation, stabilizing agents are added, the enzyme is precipitated with organic solvents, and the crude protein dried.

Pectinases are used primarily to clarify fruit juices and grape must, for the maceration of vegetables and fruits, and for the extraction of olive oil. Pectin-like substances are present at concentrations ranging from 0.2–1% in grapes to 30% in sugar beet pulp. By treatment with pectinase (1–2 hours at 54°C or 6–8 hours at 18°C) the yield of fruit juice during pressing is considerably increased.

11.8 LIPASES

Lipases (glycerol ester hydrolases) split fats (glycerol esters) into di- or monoglycerides and fatty acids (Figure 11.19). They are usually extracellular enzymes. Fungi producing lipases include *Aspergillus, Mucor, Rhizopus, Penicillium, Geotrichum,* and the yeasts *Torulopsis* and *Candida.* Bacteria producing lipases include *Pseudomonas, Achromobacter,* and *Staphylococcus.* Of these organisms, only *Aspergillus, Mucor, Rhizopus,* and *Candida* are used commercially.

In most cases, enzyme production must be induced by adding oils and fats. However, there are some cases in which fats have no effect on lipase production, and with *Penicillium roqueforti,* they actually repress enzyme production. Glycerol, a product of lipase action, represses lipase formation. This is also true for glucose, so that lipase production in *Geotrichum candidum* is first induced at the end of the log phase (around 15 hours). Lipases are generally bound to the cells and hence inhibit an overproduction, but addition of a cation such as magnesium ion liberates the lipase and leads to a higher enzyme titer in the production process. Isoenzymes with

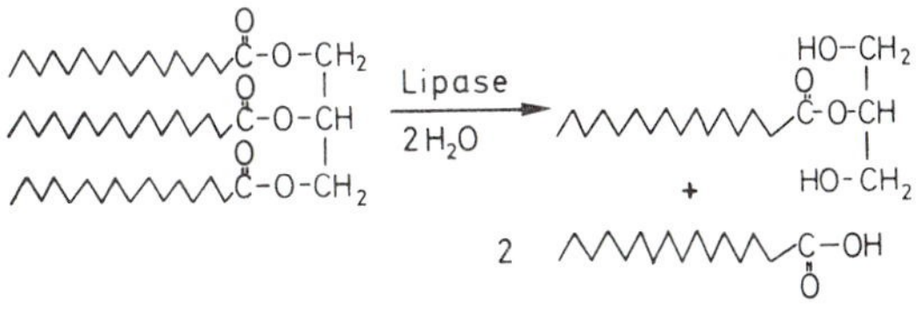

Figure 11.19 Splitting of fats into monoglycerides and fatty acids with lipases

Table 11.12 Temperature and pH optima of several microbial lipases

Organism	Optimal pH	Optimal temperature
Penicillium chrysogenum	6.2–6.8	37
Pseudomonas fragi	7.0–7.2	32
Rhizopus delemar	5.6	35
Aspergillus niger	5.6	35
Penicillium roqueforti	8.0	37
Staphylococcus aureus	8.5	45
Geotrichum candidum	8.2	37
Achromobacter lipolyticum	7.0	37

(Shahani, 1975)

varying pH optima, temperature optima, or substrate specificities are frequently produced.

Temperature and pH optima for lipases are given in Table 11.12.

The commercial use of lipases has been limited. They are primarily marketed for therapeutic purposes as digestive enzymes to supplement pancreatic lipases. The enzymes also find some use in the dairy industry. Since free fatty acids determine the odor and taste of cheese, and the cheese ripening process is affected by lipases, microbial affects during the aging process can be due to lipase action. For instance, in the production of roquefort cheese (blue cheese) spores of *Penicillium roqueforti* are added and serve as a sort of lipase preparation. In the soap industry, the lipase from *Candida cylindraceae* is used to hydrolyze oils. A potentially important future use of lipases is for the synthesis of esters from acids and alcohols in nonaqueous media. Another potential use is improvement of fat quality by the exchange of one fatty acid by another.

11.9 PENICILLIN ACYLASES

As we will describe in Chapter 13, many penicillin antibiotics used in medicine are produced **semisynthetically** by chemical modification of the basic penicillin ring structure. The starting material for chemical modification is the non-acylated thiazolidine-β-lactam ring system of penicillins, 6-aminopenicillanic acid (6-APA, Figure 11.20). 6-APA is excreted in low concentrations by *Penicillium chrysogenum* when the organism is grown in a medium without addition of phenylacetic acid as a precursor. This process for production of the basic ring system cannot currently be commercially utilized due to its high cost. 6-APA can also be produced by the splitting-off of the acyl side chain of microbially produced penicillin G with the enzyme penicillin acylase (also called penicillin amidase or benzylpenicillin amidohydrolase). It is essential that with the enzyme used the β-lactam ring not open (Figure 11.20). Penicillin acylases thus have commercial use in the production of semisynthetic penicillins.

Classification of penicillin acylases

Penicillin acylases are produced by yeasts, fungi, and bacteria. They can be subdivided into 2 types. Type I acylases, also called the fungal type, split phenoxymethyl penicillin (penicillin V) but attack benzyl penicillin (penicillin G) much less efficiently. Known producers of these acylases are *Penicillium* sp., *P. chrysogenum, Aspergillus ochraceus, Trichophyton mentagrophytes, Epidermophyton floccosum, Cephalosporium* sp., and *Fusarium semitectum*. Aside from fungi, the bacterium *Streptomyces lavendulae* also produces this

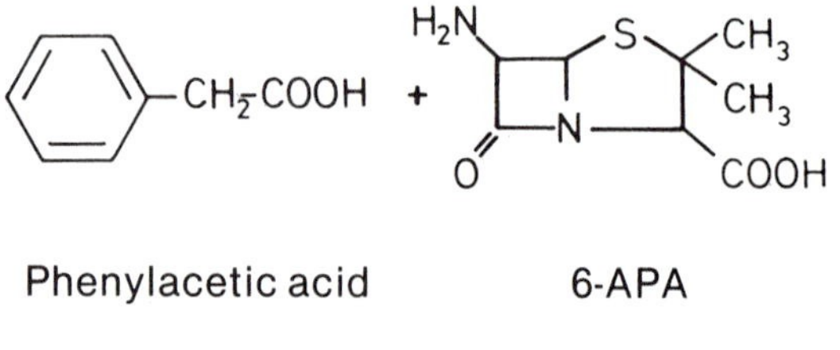

Penicillin acylase

Phenylacetic acid + 6-APA

Figure 11.20 Splitting of penicillin G into 6-aminopenicillanic acid and phenylacetic acid

type of acylase. In addition to penicillin V, penicillin K (heptyl penicillin), dihydro penicillin F (pentyl penicillin), and a number of synthetic penicillins are split. These acylases, which are usually extracellular, have a pH optimum of 10 and a temperature optimum of 50°C.

Type II acylases, also designated as the bacterial type, are produced by *Aerobacter, Alcaligenes, Bordetella, Cellulomonas, Corynebacterium, Erwinia, Escherichia, Flavobacterium, Micrococcus, Nocardia, Proteus, Pseudomonas, Salmonella, Sarcina,* and *Xanthomonas*. The temperature optimum is 40°C and the pH optimum is pH 8, which is lower than that of the Type I acylases. Phenylacetic acid acts as a competitive inhibitor and 6-APA as a noncompetitive inhibitor.

The extracellular and intracellular acylases of various microorganisms differ in substrate specificities. The substrate specificity is determined by the nature of the acyl moiety and not by the 6-APA.

Penicillin acylase from *Escherichia coli*

Penicillin acylases from *Escherichia coli* are used almost exclusively for 6-APA production. The production strains are mutants of *Escherichia coli* ATCC 11105 and *Escherichia coli* ATCC 9637. Intracellular acylase production is induced in wild-type strains by addition of phenylacetic acid. Glucose represses enzyme production and must therefore be present in the culture medium only in low concentrations. The production rate is also strongly affected by the O_2 partial pressure. Although *E. coli* can grow under anaerobic conditions, enzyme yields are low; however, enzyme production is also suppressed by high aeration rates.

A suitable nutrient medium consists of 2% corn steep water at pH 7.0. The inoculum consists of 0.25% of an 18-hour preculture. To induce enzyme formation, after 8 hours of fermentation at 24°C, 0.1% of a sterile ammonium phenylacetate solution is added at hourly intervals for the next 13 hours.

The cells are concentrated 20-fold on a separator and the enzyme is then released by use of a pressure disintegrator (500 kg/cm^2 pressure). The crude enzyme preparation so obtained is subsequently purified by conventional biochemical procedures.

The enzyme yield in commercial production has been substantially improved by optimizing the culture medium and by use of classical genetic techniques. Recombinant DNA techniques promise to increase the yields even more. For instance, by cloning the penicillin acylase gene on a multicopy plasmid, strains have been obtained having a 28-fold increase in enzyme formation rate in noninduced fermentation and a further 6-fold increase in induced fermentation (Table 11.13).

There are 3 methods used commercially for producing 6-APA from penicillin G. A strictly chemical method involves converting penicillin G in a series of steps to N-iminomethoxypenicillin ester, which is hydrolyzed with aqueous ammonia to 6-APA. Two microbial methods are used for the enzymatic conversion, one using a bacterial slurry and the other using carrier-bound penicillin acylase.

Enzymatic splitting with whole bacteria The original process uses cell-bound penicillin acylase. *Escherichia coli* slurry is added as a crude enzyme solution directly to a penicillin G solution in a batch process which has been run at pH 8.0 and 37°C. At the end of the reaction, the cells are removed by filtration and discarded. The filtrate is adjusted to pH 2.0 and the 6-APA separated

Table 11.13 Penicillin G-acylase formation in hybrid strains

Escherichia coli Strain	Specific activity	
	Not induced	Induced
ATCC 11105	0.02	0.12
5 K pHM6	0.56	0.70
5 K pHM7	–	0.44
5 K pHM8	–	0.36
5 K pHM11	–	0.28

(Mayer et al., 1979)

from phenylacetic acid and nonconverted penicillin by extraction with methylisobutyl ketone. 6-APA is then precipitated at its isoelectric point (pH 4.3) and the crystals are washed and dried.

Carrier-bound penicillin acylase Several hundred tons of 6-APA are produced each year using this widely applied process. Its advantages are: conservation of the penicillin acylase by recycling, greater purity of the 6-APA, and fewer losses due to side reactions. There are processes for the immobilization of whole cells, in which acrylamide monomers are polymerized with cells. However, methods in which the purified penicillin acylase is immobilized are more widely used.

In one method, penicillin acylase from *Escherichia coli* is rendered insoluble by formation of a covalent bond with a polymer. Penicillin (100,000 units/ml) can be quantitatively split at 38°C and pH 7.8 in 6 hours when substrate and enzyme are used in a proportion of 5×10^4 units of substrate (1670 International Units = 1 mg penicillin G) per unit of penicillin acylase. One unit of enzyme is defined as the activity which hydrolyzes 1 μMole penicillin G in one minute to 6-APA and phenylacetic acid at 37°C. With this method, 6-APA can be isolated with a recovery of 87% of the theoretical yield and at 97% purity.

Cephalosporin acylases

Cephalosporins are antibiotics with a β-lactam ring similar to but different from that found in the penicillins. Semisynthetic cephalosporins are also produced in a manner analogous to the penicillins. In the past, the side chain L-α-aminoadipic acid was chemically split to produce the ring structure needed for the synthetic process, since microbiological methods resulted only in low yields. Today, however, there is a process which allows the microbiological splitting of cephalosporin C into the 7-acyl side chain and 7-aminocephalosporanic acid (7-ACA). A solution of 35% cephalosporin C can be split in 7 hours using an enzyme from *Pseudomonas* sp. (BN-188), which is closely related to *Pseudomonas putida*.

11.10 LACTASES

Lactase, also called β-galactosidase, splits lactose into glucose and galactose. The enzyme is intracellular in bacteria and yeast, but it is excreted by many fungi. The genetics of some bacterial β-galactosidase systems have been intensively studied, but the enzymes produced commercially are obtained from fungi or yeasts and are less well known. Lactases of commercial interest include those from the fungi *Aspergillus oryzae* and *A. niger*, and those from the yeasts *Kluyveromyces lactis*, *K. fragilis*, and *Torula cremoris* and *Bacillus* sp. (in immobilized form). The enzymes differ mainly in the following:

	Fungi	Yeasts
pH optimum	2.5–4.5	6.0–7.0
Temperature optimum	55°C	35°C

The yeasts are cultured in a submerged process, but the aspergilli must be cultivated on the surface of wheat bran at 30°C. Lactose is generally added to induce enzyme formation.

Lactases are used as digestive enzymes in cases of lactose-intolerance (usually due to lactase deficiency) and as a feed additive to increase the nutritive value of some animal feeds. In the dairy industry, lactases are used to break down lactose in milk. For example, in Europe lactose-low milk is produced by pasteurization and then treating with yeast lactase (4 hours at 35°C, which results in 70–80% hydrolysis of the lactose). The resulting product is then sterilized by the ultra-high temperature process and marketed. Lactases are also used commercially to process whey.

11.11 STABILIZATION OF ENZYMES AND CELLS

Purified enzyme preparations generally cannot be stored for long periods without losing their effectiveness. Native enzymes are subject to inactivation by chemical, physical, and biological factors, and the inactivation can occur either in storage or during use. There is thus a need to stabilize enzymes because of the high cost of enzyme production. Ideally, enzymes should be stable for months with as much activity as possible under varied conditions. Moreover, it should be possible to stabilize enzymes in commercial processes in such a way that they can be used over and over again and so that conversion of substrate to product can be carried out continuously.

If a cofactor such as NAD or an energy-rich substrate such as ATP is needed in the process, it is essential that regeneration be possible. Under such conditions, it is more cost-effective to use immobilized cells rather than immobilized enzymes. The technique for immobilizing cells is actually very analogous to that for immobilizing enzymes.

There are four basic methods of achieving enzyme stability:

1. **Stabilization of soluble enzymes**. Stabilization of soluble enzymes is accomplished by additives or by chemical modification. This improves the stability of enzymes against physical and chemical agents without decreasing their solubility. Such stabilized enzymes can be successfully stored but since they are still soluble they cannot be recycled after use.
2. **Stabilization by cross-linkage (immobilization) of enzyme molecules or cells.** Cross-linked enzyme molecules are linked to each other in such a way that their activity is not affected. They are no longer soluble and can be repeatedly or continuously used.
3. **Bonding to carriers**. In these procedures, the enzyme molecules are not bound to each other, but to a carrier. Immobilization is also possible within the original microbial cell. The fixation is done in such a way that it does not affect the enzyme activity, and the preparation can be used repeatedly or continuously as a carrier-bound enzyme.
4. **Incorporation within semipermeable membranes**. Encapsulated enzymes are separated from the surrounding substrate and product by a semipermeable membrane. Cells with enzymatic activity can also be encapsulated. The activity of the enzyme is not affected as a result of the encapsulation process and the preparation can be used repeatedly or continuously.

Stabilization of soluble enzymes

In some cases, the enzyme cannot be immobilized but must be used in soluble form. Examples include the enzymes used in liquid detergents, as diagnostic reagents, and as food additives. In such cases, some procedure for stabilization of the enzyme is necessary in order to prolong the shelf life of soluble enzymes. Following are some methods of stabilization:

Substrate stabilization The active site of an enzyme is responsible for its specific activity and this site can be stabilized by adding the substrate. For instance, α-amylase is stabilized by adding starch and glucose isomerase is stabilized against heat damage by addition of glucose. Figure 11.21

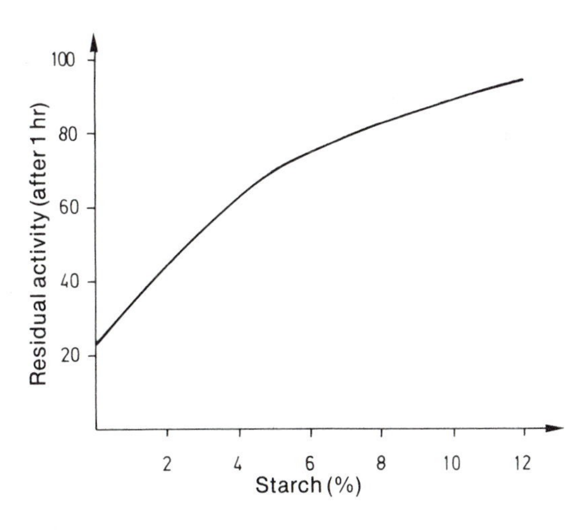

Figure 11.21 Stabilization of α-amylase by its substrate starch (Novo, 1970)

shows the stabilization of a bacterial amylase at pH 7.0 and a temperature of 80°C using starch.

On the other hand, there are enzymes such as pyruvate dehydrogenase whose activity is decreased when substrate is added.

Solvent stabilization Enzymes can also be stabilized by adding solvents; many of these can cause denaturation at high concentrations but afford considerable stabilization at low concentrations. Figure 11.22 shows several examples of the stabilizing effect of solvents on benzylalcohol dehydrogenase.

Stabilization by means of salts Cations such as Ca, Cu, Fe, Mn, Mo, and Zn have an effect on the stability and activity of metalloenzymes. For example, calcium helps to stabilize the tertiary structure of α-amylases of *Bacillus caldolyticus* and proteases.

Stabilization by means of polymer additives Natural or synthetic polymers such as gelatin, albumin, fatty alcohol ethylene oxide adducts, or polyethylene glycols can increase the heat stability of enzymes.

Stabilization by chemical means In addition to enzyme-bonding to soluble carriers, stabilization can be accomplished by chemical modification without a loss of solubility. One method is the formation of enzymes with polyamino side chains, such as the production of polytyrosyl or polyglycyl enzymes. Another method is acylation with acetyl, formyl, propionyl, or succinyl groups. Asparaginase in blood can be considerably protected by acylation. In other enzymes, such as α-amylases, the temperature sensitivity and sensitivity to proteases can be reduced.

By means of bifunctional or multifunctional additives, enzymes can also be polymerized in such a way that they remain soluble.

Stabilization by means of immobilization

The bonding of an enzyme to another enzyme or to a carrier must take place without changing the three-dimensional structure at the active site of the molecule. Neither the substrate specificity nor the specificity of the reaction must be lost as a result of the immobilization process.

Functional groups on the enzyme molecule that are suitable for use in the immobilization process are free α-, β-, or γ-carboxyl groups, α- or β-amino groups, and phenyl, hydroxyl, sulfhydryl, or imidazole groups of the appropriate amino acids. The groups used must not be critical for the activity of the enzyme.

The three major approaches to enzyme immobilization are illustrated in Figure 11.23 and are summarized below.

Cross-linked enzymes Several examples of this method for immobilizing enzymes can be given. Glutaraldehyde is commonly used as a polymerizing agent, and Figure 11.24 shows how this

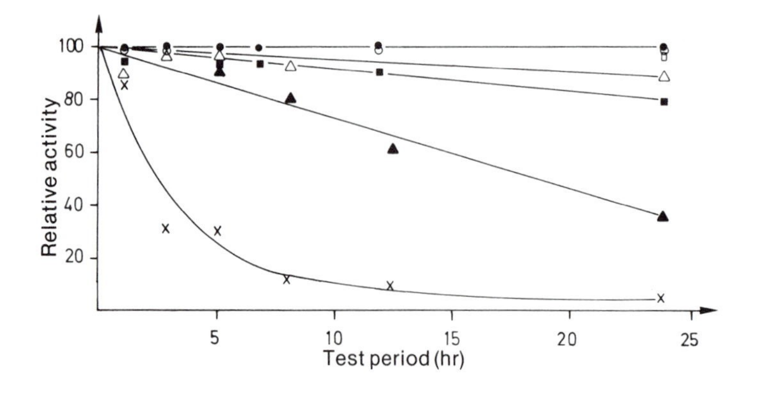

Figure 11.22 Stability of benzyl alcohol dehydrogenase in organic solvents. × — × without solvent; △ — △ 2% acetone; □ — □ 5% acetone; ○ — ○ 10% acetone; ▲ — ▲ 2% ethanol; ■ — ■ 5% ethanol; ● — ● 10% ethanol (From Katagiri et al., 1967)

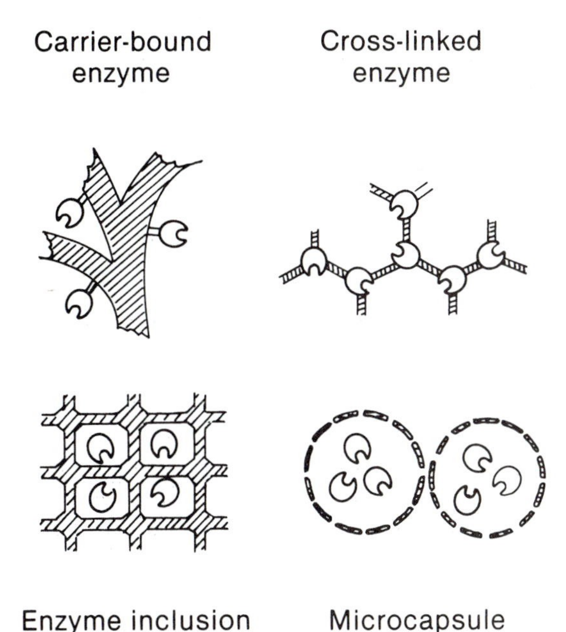

Figure 11.23 Procedures for the immobilization of enzymes (From Chibata, 1978)

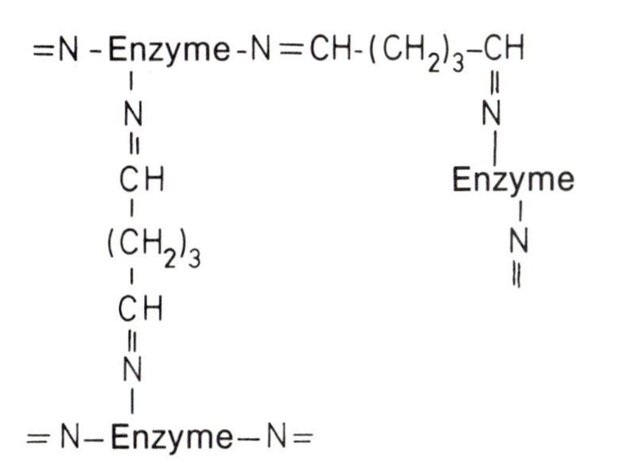

Figure 11.24 Cross linkage of enzymes with glutaraldehyde

compound reacts with amino groups of the enzymes (for example, amylases and alcohol dehydrogenase) in the formation of Schiff bases. Immobilization of cells is also frequently accomplished with glutaraldehyde. Another cross-linking agent is hexamethylene diisocyanate, which reacts with enzymes by peptide bond formation (Figure 11.25).

Carrier-bound enzymes The bonding of an enzyme to a carrier through adsorption, ionic bonding, or covalent bonding is the oldest method of enzyme immobilization, and it is the most frequently used commercial method. The structure of the carrier, the type of functional bonding sites, the ratio of hydrophilic to hydrophobic groups, the particle size, and the surface/volume ratio must be suitable for the individual enzyme and for its eventual commercial use.

Bonding by means of adsorption causes the least damage or conformation alteration of the protein. Fixation is easily accomplished and the method is applicable to a wide variety of enzymes. In addition, the carrier can be regenerated in most cases. However, the bonding strength is low, and this is a disadvantage of adsorption procedures; any temperature changes in the bioreactor cause elution of enzyme from the carrier and hence activity loss. Both organic and inorganic carriers can be used for the adsorption process and several types are listed in Table 11.14.

Ion exchangers can be used as enzyme carriers, the enzyme being held to the carrier by ionic bonds. This method overlaps in part with the adsorption method. Ionic bonding is an easily accomplished method with low activity loss, al-

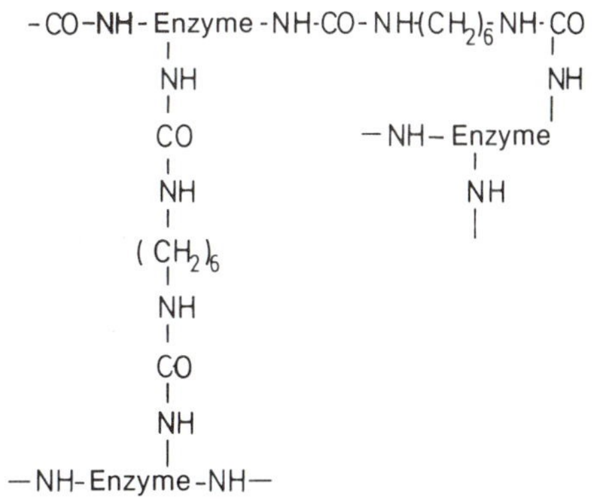

$$
\begin{array}{l}
-CO-NH-Enzyme-NH-CO-NH(CH_2)_6NH\cdot CO \\
\qquad NH \qquad\qquad\qquad\qquad\qquad NH \\
\qquad CO \qquad\qquad -NH-Enzyme \\
\qquad NH \qquad\qquad\qquad\quad NH \\
\qquad (CH_2)_6 \\
\qquad NH \\
\qquad CO \\
\qquad NH \\
-NH-Enzyme-NH-
\end{array}
$$

Figure 11.25 Cross linkage of enzymes with hexamethylene diisocyanate

Table 11.14 Adsorption carriers for immobilization of enzymes

Type of carrier	Carrier	Enzyme used
Inorganic	Aluminium oxide	Glucoamylase
	Bentonite	Invertase
	Glass	Lipase
	Ca phosphate gel	Aspartase
Organic	Activated carbon	Glucose oxidase
		α-Amylase
		β-Amylase
		Glucoamylase
		Invertase
	Starch	α-Amylase
	Tannin aminohexyl cellulose	Glucose isomerase
		Aminoacylase

though changes in pH or ionic strength can result in enzyme loss from the carriers. A variety of commercially available ion exchangers can be used as carriers.

Immobilization by covalent bonding has the advantage of strong bonding forces. In this procedure, the side chain of one or more amino acids of the protein is modified so that it forms a covalent link with the carrier. The preparation of a covalently bonded enzyme is expensive and difficult to carry out because amino acids of the active center may also become involved in the reaction. Covalent bonding is the most widely used commercial process. The kinds of covalent bonds used include peptide bonds and diazo bonds, as well as alkylation and isourea bond formation with BrCN-activated carbohydrates.

For enzyme bonding to occur, the functional groups of the carrier must be activated. For the production of peptide bonds with primary amino groups of the enzyme protein, the carboxyl groups of the carriers can be activated through the production of azide or anhydride derivatives. An example of covalent linking is that shown with carboxymethyl cellulose (CM cellulose) in Figure 11.26. Carboxymethyl cellulose is first converted to the methylester and subsequently treated with hydrazine. The resulting hydrazide then reacts with sodium nitrite and the corresponding azide derivative is produced which reacts with the amino groups of the enzyme.

$$\text{Cellulose-O-}CH_2\text{-}COOH \xrightarrow[HCl]{CH_3OH} \text{Cellulose-O-}CH_2\text{-}COO\text{-}CH_3 \xrightarrow{NH_2\text{-}NH_2}$$

$$\text{Cellulose-O-}CH_2\text{-}CO\text{-}NH\text{-}NH_2 \xrightarrow[HCl]{NaNO_2} \text{Cellulose-O-}CH_2\text{-}CO\text{-}N_3$$

$$\xrightarrow{\text{Enzyme}} \text{Cellulose-O-}CH_2\text{-}CO\text{-}NH\text{-Enzyme}$$

Figure 11.26 Immobilization of enzymes with carboxymethylcellulose

```
           CO-NH2                   NH2-CO
           |                            |
-CH2-CH-CH2-CH-CH2-CH-CH2-CH-CH2-CH-
     |            |      |
     CO           CO-NH2 CO
     |                   |
     NH                  NH
     |                   |
     CH2    Enzyme       CH2
     |                   |
     NH                  NH     NH2
     |                   |      |
     CO        NH2-CO    CO     CO
     |            |      |      |
-CH2-CH-CH2-CH-CH2-CH-CH2-CH-CH2-CH-
            |
            CO-NH-
```

Figure 11.27 Enzyme incorporation into a polyacrylamide gel

Encapsulation of enzymes

When enzymes are **physically enclosed** in gels, microcapsules, or fibrous polymers, there must be pores which are so small that the enzyme molecules cannot be washed out, yet which are still large enough to permit the unimpeded diffusion of low-molecular-weight substrates and products across the barrier. Unfortunately, there may be considerable yield loss in the production of an encapsulated enzyme, but the finished products are very stable.

Gel enclosure Cross-linked, water-insoluble polymers such as polyacrylamide, polyvinyl alcohol or starch can be used to surround the enzyme. The method most frequently used involves incorporation into a polyacrylamide gel (Figure

11.27). Acrylamide and the enzyme are polymerized within minutes and at room temperature into a gel by means of a cross-linking compound (e.g. NN'-methylene bisacrylamide), a starter (sodium persulfate), and an accelerator (β-dimethylamino propionitrile). The gel can then be cut up and used in a batch or column process.

Microcapsules The advantage of immobilization of enzymes using microcapsule preparations is that free enzyme or free cells can be surrounded by a semipermeable polymer membrane. Production takes place by means of a polymerization reaction at the surface of aqueous enzyme droplets which are suspended in a non-water-soluble organic phase. Microcapsulated enzymes have not yet found commercial application, but they appear promising for future clinical and analytical uses.

Immobilization in fibrous polymers This method of immobilization is similar to the method using microcapsules. An emulsion of aqueous enzyme solution and organic water-immiscible solvent is extruded into a precipitation bath in which the fibrous polymer forms and encloses the enzyme. Cellulose triacetate and other cellulose derivatives serve as suitable polymers.

Industrial application of immobilized enzymes and cells

The first large-scale applications of immobilized cells were optimized empirically and have been in use for many years. The best example is the use of the *trickling filter* for sewage treatment (see Chapter 17). A modern application of this procedure is the immobilization of cultures of anaerobic bacteria on sintered glass particles, which makes it possible to increase the rate of anaerobic sewage treatment. For instance, in the effluent from a cellulose processing plant, 84% of the chemical oxygen demand (COD) was removed in a 12 hour holding time with a loading of 88 kg COD/m^3/day. Another well-known example of the use of immobilized cells is in the trickling generator used for vinegar production.

A number of processes have also been introduced in more recent times. Among these are the use of immobilized cells containing **glucose isomerase** for the production of high-fructose syrup (see Section 11.3). The half-life of the immobilized isomerase in large-scale operation ranges from 70–120 days. World-wide production with this method amounts to 7.5×10^6 tons per year, of which about half is in the United States. **Glucose amylase** for the production of glucose from starch is now frequently used in an immobilized state. The enzyme **β-galactosidase** from yeast is now being immobilized in some countries (for instance, Italy) for the splitting of lactose to glucose plus galactose. Some studies are under way to use this process to produce a sweet syrup from whey. A large-scale process for the immobilization of **invertase** has been developed for the splitting of sucrose to glucose plus fructose. The enzyme **α-galactosidase** splits the trisaccharide raffinose to galactose plus sucrose. Raffinose is present in sugar beet juice at a concentration of around 0.1% and reduces the yield of sucrose during sugar beet production because of its effect on the crystallization process. A large-scale process is under study to immobilize the α-galactosidase from the fungus *Absidia* sp. for use in the sugar beet industry. A number of processes for **amino acid production** have been developed using immobilized enzymes or cells (see Section 9.3).

In addition to the food industry, immobilized systems have also been developed for fine-chemical production. The process using penicillin acylase has already been discussed in Section 11.9. In another example, malic acid is being produced from fumaric acid using the fumarase of *Brevibacterium ammoniagenes*. In this process, an enzyme/membrane reactor has been developed which carries out the following reaction:

$$HOOC{-}CH{=}CH{-}COOH + H_2O \rightarrow HOOC{-}CHOH{-}CH_2COOH$$

In this process, a 1 m^3 column reactor has been

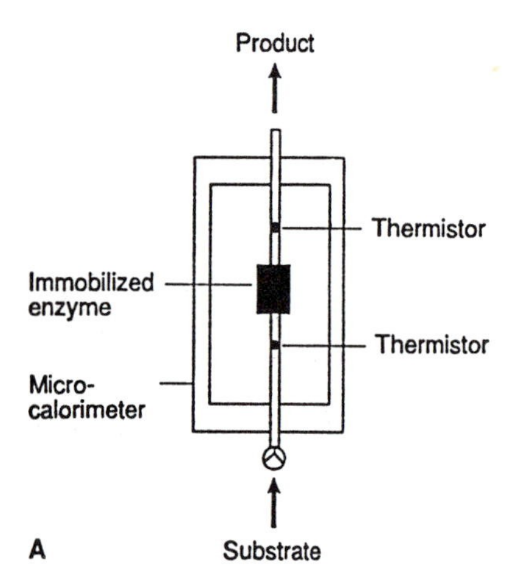

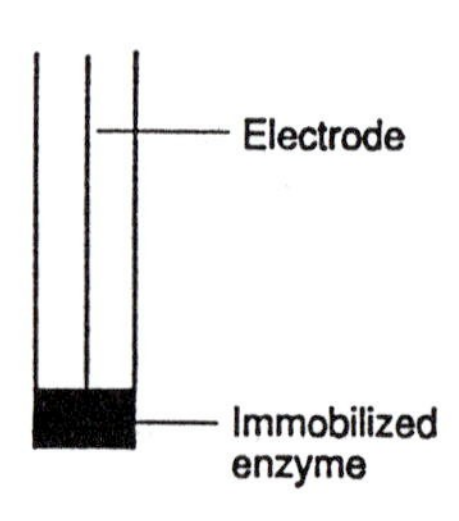

B

Figure 11.28 Biochemical assays using immobilized enzymes or cells. A. Enzyme/thermistor. B. Enzyme electrode

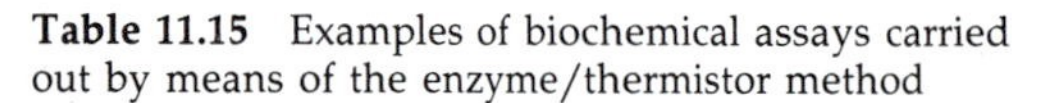

Table 11.15 Examples of biochemical assays carried out by means of the enzyme/thermistor method

Substrate	Enzyme	Concentration (mMol/l)
Ascorbic acid	Ascorbic acid oxidase	0.05–0.6
ATP	Hexokinase	1–8
Cholesterol	Cholesterol oxidase	0.03–0.15
Cephalosporin	Cephalosporinase	0.005–10
Ethanol	Alcohol oxidase	0.01–1
Galactose	Galactose oxidase	0.01–1
Glucose	Hexokinase	0.5–25
Lactose	Lactase/Glucose oxidase/Catalase	0.05–10

Table 11.16 Electrode sensors employing immobilized enzymes or cells

Substrate	Strain/enzyme	Measurement principle
Sucrose	Invertase/Mutarotase Glucose oxidase	O_2 uptake
Monoamine	Monoamine oxidase	O_2 uptake
Cholesterol	Cholesterol oxidase	O_2 uptake
Acetic acid	*Trichosporon brassica*	O_2 uptake
Ethanol	*Trichosporon brassica*	O_2 uptake
Nystatin	Yeast	O_2 uptake (as a result of cell death)
Cephalosporin	*Citrobacter* with cephalosporinase	pH measurement

used, with a flow rate of 0.3 h^{-1} at pH 7.5 and 37°C. The half-life of the enzyme in this process was about 160 days.

In Japan, the large-scale production of **ethanol** has been carried out using growing, immobilized cells of the yeast *Saccharomyces*. The cells are immobilized by photochemical reaction with a cross-linked gel. The bioreactor, which produces more than 1.5 mol/l·h ethanol, has been operated successfully over long periods of time. A process for the production of **glycerol** has been developed using immobilized *Saccharomyces cerevisiae*, but is still at the laboratory stage.

Special bioreactors for use in immobilized processes were discussed in Section 5.3.

Uses of immobilized cells and enzymes for analytical biochemistry

An immobilized system can be used in the development of a precise and sensitive biochemical assay. The basic principle is that the enzyme acts on the substrate and a decrease in substrate concentration, increase in product concentration, or change in cofactor concentration can be followed, using one of many assay methods. The reaction can be followed spectrophotometrically, polarimeterically, photometrically, or by mass spectroscopy. Additionally, as shown in Figure 11.28

A, the reaction can be followed calorimetrically by measuring heat production with a thermistor, or as shown in Figure 11.28 B, by measuring potential difference with an electrode.

Table 11.15 gives several examples of substances that can be assayed calorimetrically and Table 11.16 gives some examples in which electrodes are used.

In some cases, a coupled series of reactions can be employed. As an example, the measurement of sucrose can be linked to the measurement of oxygen uptake via the following reaction chain:

$$\text{Sucrose} + H_2O \rightarrow \alpha\text{-D-Glucose} + \text{D-Fructose}$$
$$\alpha\text{-D-Glucose} \rightarrow \beta\text{-D-Glucose}$$
$$\beta\text{-D-Glucose} + O_2 + H_2O \rightarrow \text{D-Glucose-}\delta\text{-lactone} + H_2O_2$$

This particular enzyme electrode, which involves invertase, mutarotase, and glucose oxidase, respectively, functioned for over 85 separate determinations over a 10 day period.

REFERENCES

Aivasidis, A. and C. Wandrey. 1984. Ein "Glasschwamm" als Bakterienspeicher. Abwasserreinigung ohne Sauerstoff (A glass sponge as support for bacterial growth. Sewage purification without oxygen). Bericht der Kernforschungsanlage Jülich Nr. 1900.

Aunstrup, K. 1979. Production, isolation, and economics of extracellular enzymes, pp. 27–69. In: Wingard, L.B., E. Katchalski-Katzir, and L. Goldstein (eds.), Applied biochemistry and bioengineering, vol. 2. Academic Press, New York.

Aunstrup, K., H. Outtrup, O. Andreson, and C. Dambmann. 1972. Proteases from alkalophilic *Bacillus* species, pp. 299–305. In: Terui, G. (ed.), Fermentation technology today. Yamada-Kami, Osaka.

Aunstrup, K., O. Andresen, E.A. Falch, and T.K. Nielsen. 1979. Production of microbial enzymes, pp. 281–309. In: Peppler, H.J. and D. Perlman (eds.), Microbial technology, vol. 1. Academic Press, New York.

Berg, M., A. Boeck, R.D. Schmid, and H. Verbeek. 1976. Enzyme als Waschmittelkomponente (Enzymes in detergents), pp. 155–178. In: Henkel and Cie (eds.), Waschmittelchemie. Hüthig Verlag, Heidelberg.

Bisping, B. and H.J. Reh, 1986. General production by cells of Saccharomyces cerevisiae immobilized in sintered glass. Appl. Microbiol. biotechnol. 23:174–179.

Boyce, C.O.L. (ed.) 1986. Novo's Handbook of practical Biotechnology. Novo Industries AS, Copenhagen.

Brown, G.B. 1976. Chemically aggregrated enzymes, pp. 263–280. In: Mosbach, K. (ed.), Methods in enzymology, vol. 44, Immobilized enzymes. Academic Press, New York.

Carlson, A., G.C. Hill, and N.F. Olson. 1986. The coagulation of milk with immobilized enzymes: a critical review. Enzyme Microbiol. Technol. 8:642–650.

Carlson, A., G.C. Hill, and N.F. Olson. 1987. Kinetics of milk coagulation I–IV. Biotechnol. Bioengineering 29:582–624.

Cheetham, P.S.J. 1980. Developments in the immobilisation of microbial cells and their applications, pp. 189–238. In: Wiseman, A. (ed.), Topics in enzyme and fermentation biotechnology, vol. 4. Ellis Horwood Ltd., Chichester.

Chibata, I. 1978. Immobilized enzymes. John Wiley and Sons, New York.

Chibata, I., and T. Tosa. 1977. Transformations of organic compounds by immobilized microbial cells, pp. 1–27. In: Perlman, D. (ed.), Advances in appl. microbiol. Academic Press, New York.

Chibata, I., T. Tosa, and T. Sato. 1979. Use of immobilized cell systems to prepare fine chemicals, pp. 433–461. In: Peppler, H.J. and D. Perlman (eds.), Microbial technology, vol. 2. Academic Press, New York.

Flickinger, M.C. 1985. Anticancer agents, pp. 231–273. In: Moo-Young, M. (ed.). Comprehensive Biotechnology III. Pergamon Press, Oxford.

Fogarty, W.M. (ed.). 1983. Microbial enzymes and biotechnology. Applied Science Publishers, London.

Fogarty, W.M. and C.T. Kelly. 1983. Enzymatic developments in the production of maltose and glucose, pp. 149–163. In: Lafferty, R.M. (ed.). Enzyme Technology. Springer Verlag, Berlin.

Glymph, J.L. and F.J. Stutzenberger. 1977. Production, purification, and characterization of α-amylase from *Thermomonospora curvata*. Appl. Environ. Microbiol. 34:391–397.

Green, M.L. 1977. Reviews of the progress of dairy science: Milk coagulants. J. Dairy Res. 44: 159–188.

Hartmeier, W. 1986. Immobilisierte Biokatalysatoren (Immobilized biocatalysts). Springer Verlag, Berlin.

Hayashibara. 1970. Verfahren zur Gewinnung hochreiner Maltose (Technique for producing highly pure maltose). German Patent 1,935,760.

Hüper, F. 1973a. Verfahren zur Herstellung von 6-Aminopenicillansäure durch enzymatische Spaltung von Penicillinen (Technique for the production of 6-amino penicillanic acid by the enzymatic breakdown of penicillin). German Patent 2,157,970.

Hüper, F. 1973b. Trägergebundene Penicillinacylase. (Immobilized penicillin acylase). German Patent 2,157,972.

Kaufmann, W., K. Bauer, and H.A. Offe. 1959. Verfahren zur Herstellung von 6-Acylaminopenicillansäure auf enzymatischen Weg (Enzymatic procedures for the production of 6-amino penicillanic acid). German Patent 1,149,361.

Katagiri, M., S. Takemori, K. Nakazawa, H. Suzuki, and K. Akagi. 1967. Benzylalcohol dehydrogenase, a new alcohol dehydrogenase from *Pseudomonas* sp. Biochim. Biophys. Acta. 139: 173–176.

Lee, S.M., J.T. Ross, M.E. Gustafson, M.H. Wroble, and G.M. Muschik. 1986. Large scale recovery and purification of L-asparaginase from Erwinia carotovora. Appl. Biochem. Biotechnol. 12:229–247.

Lin, C.-F. 1972. Production of glucoamylase by a temperature-sensitive mutant of *Rhizopus formosaensis* R 13-5 and its properties, pp. 327–332. In: Terui, G. (ed.), Fermentation technology today. Yamada-Kami, Osaka.

Linhardt, R.J., P.M. Galliher, and C.L. Cooney. 1986. Review. Polysaccharide lyases. Appl. Biochem. Biotechnol. 12:135–176.

Mahajan, P.B. 1984. Penicillinacylases, an update. Appl. biochem. biotechnol. 9:537–554.

Marconi, W. and F. Morisi. 1979. Industrial applications of fiber-entrapped enzymes, pp. 219–258. In: Wingard, L.B., E. Katchalski-Katzir, and L. Goldstein (eds.), Applied biochemistry and bioengineering, vol. 2. Academic Press, New York.

Marshall, R.O. and E.R. Kovi. 1957. Enzymatic conversion of D-glucose to D-fructose. Science 125: 648–649.

Mayer, H., J. Collins, and F. Wagner. 1979. Cloning of the penicillin G acylase gene of *Escherichia coli* ATCC 11105 on multicopy plasmids, pp. 459–470. In: Timmis, K.N. and A. Pühler (eds.), Plasmids of medical, environmental, and commercial importance. Elsevier/North-Holland Biomedical Press.

Moss, M.O. 1977. Enzymic alterations of penicillins and cephalosporins, pp. 110–131. In: Wiseman, A. (ed.), Topics in enzyme and fermentation technology, vol. 1. Ellis Horwood, Chichester.

Novo Industries. 1977. Production of high-purity glucose syrups. U.S. Patent 4,017,363.

Novo 1978. Novo Enzyme Information JB163cGB.

Oestergaard, J. and S.L. Knudsen. 1976. Use of Sweetzyme in industrial continuous isomerization. Various process alternatives and corresponding product types. Die Stärke 28: 350–356.

Peterson, R.E. and A. Ciegler. 1969. L-Asparaginase production by *Erwinia aroideae*. Appl. Microbiol. 18: 64–67.

Shahani, K.M. 1975. Lipases and esterases, pp. 181–217. In: Reed, G. (ed.), Enzymes in food processing. Academic Press, New York.

Suzuki, S. and I. Karube. 1981. Bioelectrochemical sensors based on immobilized enzymes, whole cells and proteins. Appl. Biochem. Bioengineering 3:145–174.

Takasaki, Y., Y. Kosugi, and A. Kanbayashi. 1969. *Streptomyces* glucose isomerase, pp.561–589. In: Perlman, D. (ed.), Fermentation advances. Academic Press, New York.

Terui, G. 1973. Kinetics of hydrolase production by microorganisms, pp. 377–395. In: Sterbacek, Z. (ed.), Microbial engineering. Butterworths, London.

Thorbek, L. 1976. Influence of propagation methods on enzyme formation by *Aspergillus niger*, p. 249. In: Dellweg, H. (ed.), Abstracts 5th Int. Ferment. Symp. Verlag Versuchs- und Lehranstalt für Spiritusfabrikation, Berlin.

Upton, M.E. and W.M. Fogarty. 1977. Production and purification of thermostable amylase and protease of *Thermomonospora viridis*. Appl. Environm. Microbiol. 33: 59–64.

van Tilburg, R. (ed.). 1983. Engineering aspects of biocatalysts. In: Industrial starch conversion technology. Delftse Universitaire Press, Delft.

Verhoff, F.H., G. Boguslawski, O.J. Lantero, S.T. Schlager, and Y.C. Jao. 1985. Glucose isomerase, pp. 837–859. In: Moo-Young, M. (ed.). Comprehensive Biotechnology. Pergamon Press, Oxford.

Ward, O.P. 1985. Proteolytic enzymes. pp. 789–818. In: Moo-Young, M. (ed.). Comprehensive Biotechnology III, Pergamon Press, Oxford.

Wingard, L.B., W. Katchalski-Katzir, and L. Goldstein (eds.). Applied Biochemical Bioengineering. Volume 1, Immobilized enzyme principles (1976); volume 2, Enzyme technology (1976); volume 3, Analytical applications of immobilized enzymes and cells (1981); volume 4, Immobilized cells (1983). Academic Press, New York.

Woodward, J. (ed.). 1985. Immobilized cells and enzymes, a practical approach. IRL Press, Oxford.

Zittan, L., P.B. Poulsen, and S.H. Hemmingsen. 1975. Zweetzyme–a new immobilized glucose isomerase. Die Stärke 27:236–241.

12

Vitamins

12.1 INTRODUCTION

Microorganisms can be used for the commercial production of certain vitamins, such as thiamine, riboflavin, folic acid, pantothenic acid, pyridoxal, vitamin B_{12}, and biotin. Also, microorganisms (including algae) synthesize β-carotene, which is provitamin A. Ergosterol (provitamin D_2), produced with a yeast fermentation, can occur in certain *Saccharomyces* strains at concentrations as high as 0.1–10% of the cell weight. In addition to direct fermentation, certain vitamins can also be produced by combined chemical/microbiological means, the microorganism carrying out specific reaction steps (so-called *biotransformation*, see Chapter 15). Biotransformation plays an important role in the production of ascorbic acid and tocopherol. In Japan, a wide variety of vitamins are produced by microbial fermentation. However, on a worldwide basis only the production of vitamin B_{12}, riboflavin, and ascorbic acid have any major economic significance. The microbial production of β-carotene, although possible, is not economic under large-scale conditions.

12.2 VITAMIN B_{12}

Occurrence and economic significance

In 1926, G. R. Minot and W. B. Murphy determined that liver extracts cure human pernicious anemia, a discovery for which they were awarded the Nobel Prize in medicine in 1934. Independently, E. L. Ricke and L. Smith isolated crystalline vitamin B_{12} from liver extracts in 1948. Vitamin B_{12} (cyanocobalamin) is a vitamin that is synthesized in nature exclusively by microorganisms; because it is required by animals it is present in every animal tissue in very low concentrations (e.g. 1 ppm in the liver). Although the substance isolated from tissues is cyanocobalamin, in the cell only the coenzyme form (adenosyl- or methylcobalamin) is present, cyanocobalamin not appearing until the product recovery stage. The vitamin B_{12} needs of animals are covered by food intake or by absorption of vitamin B_{12} produced by intestinal microorganisms. However, humans obtain vitamin B_{12} only from food, since the B_{12} synthesized by microorganisms in the large intestinal tract cannot be assimilated.

The concentrations of vitamin B_{12} which are present in animal tissues are too low for use in commercial production. Activated sludge from sewage treatment contains 4–10 mg B_{12}/kg, but isolation from this source is expensive due to the problem of separating the various B_{12} analogs. Chemical synthesis is also impractical, since it requires 70 reaction steps. Vitamin B_{12} was first obtained commercially as a byproduct of streptomycete fermentations for the production of the antibiotics streptomycin, chloramphenicol, or neomycin, with a yield of about 1 mg/l. As the demand for vitamin B_{12} increased, fermentation processes were developed with higher-yielding strains. Commercial production is currently carried out entirely by fermentation. The most important manufacturers of vitamin B_{12} are: Farmitalia S.p.A. (Italy); Glaxo Lab., Ltd. (England); Merck & Co., Inc. (United States); Rhône Poulenc S.A. (France); Roussel UCLAF (France); G. Richter Pharmaceutical Co. and Chinoin (both Hungary). The current annual world production of vitamin B_{12} is estimated at about 12,000 kg. Approximately 3500 kg cyanocobalamin, 2000 kg hydroxocobalamin, 1000 kg coenzyme B_{12} and a small amount of methylcobalamin is supplied to the pharmaceutical industry; the remainder goes to the animal feed industry. For swine and poultry feeds, 10–15 mg vitamin B_{12} is added per ton of feed, since animal protein can be replaced with less-expensive vegetable protein if the vegetable protein is fortified with vitamin B_{12}.

Structure

Corrin, the base structure of vitamin B_{12}, consists of a tetrapyrrole ring which differs from the porphyrin ring system in that the methene bridge between rings A and D is missing. The C-1 of ribose from cobamide (see structural formula, Figure 12.1) is linked to the heterocyclic base 5,6-dimethylbenzimidazole, which in turn occupies another coordination site of the cobalt next to CN in the final cyanocobalamin product. Vitamin B_{12} analogs having other heterocyclic bases (purines or substituted benzimidazoles) are either spontaneously produced by microorganisms or are produced after the addition of these substances to the culture medium. However, they are less biologically effective in vertebrates than the corresponding 5,6-dimethylbenzimidazole compounds. It is noteworthy that vitamin B_{12} derivatives with a purine base are ineffective in humans.

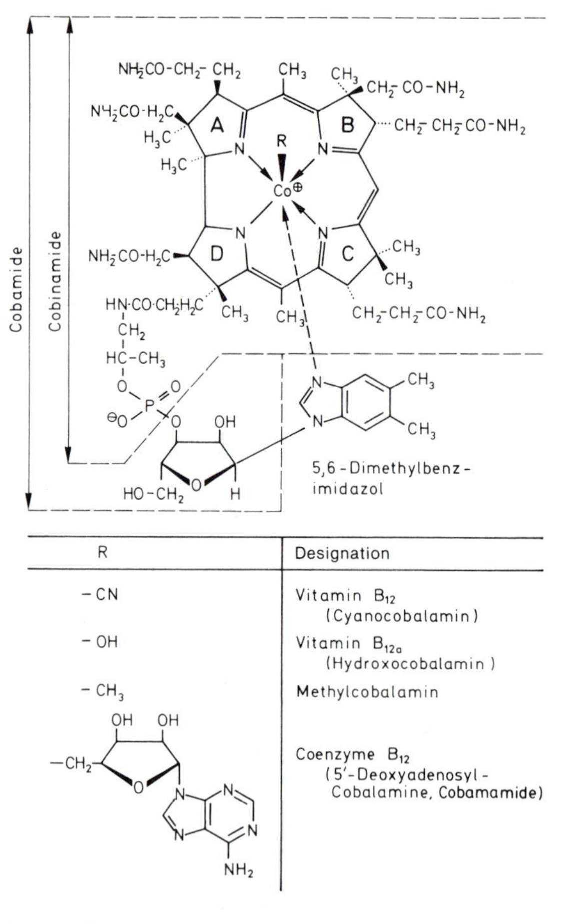

R	Designation
– CN	Vitamin B_{12} (Cyanocobalamin)
– OH	Vitamin B_{12a} (Hydroxocobalamin)
– CH_3	Methylcobalamin
(5'-deoxyadenosyl structure)	Coenzyme B_{12} (5'-Deoxyadenosyl-Cobalamine, Cobamamide)

Figure 12.1 Structure of cobalamins

Biosynthesis

The biosynthesis of vitamin B_{12}, which runs parallel with the biosynthesis of porphyrins and chlorophyll up to the formation of uroporphyrinogen III, is diagramed in Figure 12.2. The conversion of uroporphyrinogen III into the corrin ring system is not yet fully understood.

Vitamin B_{12} production based on media containing carbohydrates

Most of the B_{12} fermentation processes use glucose as a carbon source. Several producing strains are known, listed here with their yield of B_{12}: *Bacillus megaterium* (0.45 mg/l); *Butyribacterium rettgeri* (5 mg/l); *Streptomyces olivaceus* (3.3 mg/l); *Micromonospora* sp. (11.5 mg/l); *Klebsiella pneumoniae* (0.2 mg/l). Higher yields have been obtained from *Propionibacterium freudenreichii* (19 mg/l), *Propionibacterium shermanii* (30–40 mg/l), and (in a process using sugar cane molasses) *Pseudomonas denitrificans* (60 mg/l).

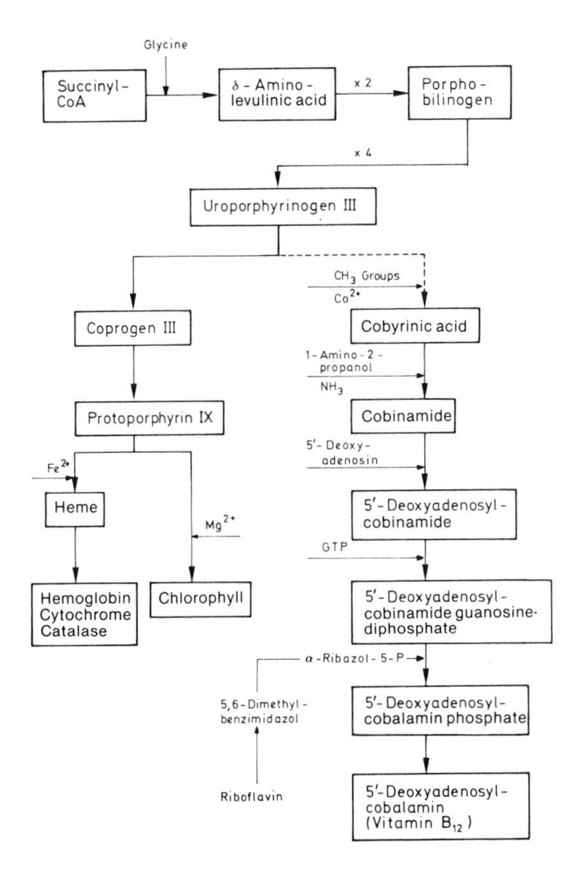

Figure 12.2 Biosynthesis of cobalamins

The use of modern genetic engineering techniques is being applied to vitamin production also. A hybrid strain called *Rhodopseudomonas protamicus*, made by the protoplast fusion technique between *Protaminobacter ruber* and *Rhodopseudomonas spheroides*, produces 135 mg/l vitamin B_{12} using glucose as carbon source without the addition of 5,6-dimethylbenzimidazole.

Processes with Propionibacteria *Propionibacterium freudenreichii* ATCC 6207 and *P. shermanii* ATCC 13673, as well as other variants and mutants, are used in a two-stage process with added cobalt (10–100 mg/l). In a preliminary anaerobic phase (2–4 days), 5'-deoxyadenosylcobinamide is mainly produced; in a second, aerobic phase (3–4 days) the biosynthesis of 5,6-dimethylbenzimidazole takes place, so that 5'-deoxyadenosylcobalamin (coenzyme B_{12}) can be produced. Only traces of other cobamides are synthesized in this process. As an alternative to the two-stage batch process in a fermenter, both stages can also be operated continuously in two tanks operated in cascade fashion.

During the recovery process, the cobalamins, which are almost completely bound to the cell, are brought into solution by heat treatment (10–30 min at 80–120°C, pH 6.5–8.5). They are then converted chemically into the more stable cyanocobalamin. The raw product (80% purity) is used as a feed additive. Additional purification stages yield a medically usable preparation (95–98% purity). The total yield is 75% of the theoretical value.

Processes with Pseudomonas *Pseudomonas denitrificans* has been found to be the most productive species among the different vitamin B_{12}-producing pseudomonads. In this one-stage process, vitamin B_{12} is produced during the entire fermentation. Cobalt and 5,6-dimethylbenzimidazole must be added as supplements. It has also been found that additions of the compound betaine result in increased yield; sugar beet molasses is used as a low-cost betaine source. Although the mode of action is not known, betaine is assumed to cause an activation of biosynthesis or an increase in membrane permeability. The process shown in Figure 12.3 is one of the most cost-effective processes. After 12 years of strain development, the yields from this process have been increased from 0.6 mg/l to 60 mg/l.

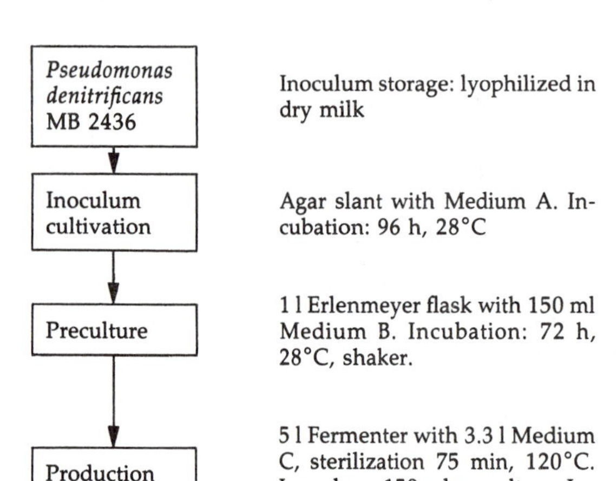

Medium A (g/l): Sugar beet molasses 60; Yeast extract 1; N-Z-Amine 1; $(NH_4)_2HPO_4$ 2; $MgSO_4 \cdot 7\,H_2O$ 1; $MnSO_4 \cdot H_2O$ 0.2; $ZnSO_4 \cdot 7\,H_2O$ 0.02; $Na_2MoO_4 \cdot 2\,H_2O$ 0.005; Agar 25; Tap water; pH 7.4

Medium B (g/l): Medium A, without agar

Medium C (g/l): Sugar beet molasses 100; Yeast extract 2; $(NH_4)_2HPO_4$ 5; $MgSO_4 \cdot 7\,H_2O$ 3; $MnSO_4 \cdot H_2O$ 0.2; $Co(NO_3)_2 \cdot 6\,H_2O$ 0.188; 5,6-Dimethylbenzimidazol 0.025; $ZnSO_4 \cdot 7\,H_2O$ 0.02; $Na_2MoO_4 \cdot 2\,H_2O$ 0.005; Tap water; pH 7.4

Figure 12.3 Laboratory scale production process for vitamin B_{12} using *Pseudomonas denitrificans*. (Merck and Co., Inc., 1971)

Vitamin B_{12} production with other carbon sources

In the last 10 years, studies have been underway with a variety of strains to determine if various alcohol and hydrocarbon substrates could be used for vitamin B_{12} production (Table 12.1). Hydrocarbon and higher-alcohol fermentations resulted in low yields, but those using methanol seem to have considerable promise.

12.3 RIBOFLAVIN

Occurrence and economic significance

Riboflavin (also called lactoflavin or vitamin B_2) was first isolated from whey by Kuhn, György and Wagner-Jauregg in 1933; the structure was confirmed by synthesis by Kuhn and Karrer in 1935. Riboflavin is present in milk as free riboflavin, but is present in other foods (liver, heart, kidney, or eggs) as part of flavoproteins which contain the prosthetic groups FMN (flavin mononucleotide) or FAD (flavin adenine dinucleotide).

A riboflavin deficiency in rats causes stunted growth, dermatitis, and eye damage. Ariboflavinosis is a disease in humans caused by riboflavin deficiency. Appearing as a type of dermatitis, this disease can be counteracted by administering riboflavin at a daily dosage of 1 mg. Riboflavin is available in various preparations for human and veterinary application. In the United States riboflavin, as well as thiamine and nicotinic acid, is frequently added to flour for use in the production of vitamin-enriched bread.

Riboflavin is produced industrially by several processes:

- Chemical synthesis, primarily for pharmaceutical use (20% of world-wide production).
- Biotransformation of glucose to D-ribose by mutants of *Bacillus pumilus* and subsequent chemical conversion of ribose to riboflavin (about 50% of world-wide production).

Table 12.1 Vitamin B_{12} formation with alcohols and carbohydrates as carbon sources

Microorganism	Carbon source		Vitamin B_{12} yield (mg/l)
Bacteria from activated sludge	Methanol	0.3–1.2% (w/v)	35
Methanobacterium soehngenii	Methanol	0.8% (w/v)	8.0
Methanobacillus omelianski	Methanol	0.8% (w/v)	8.8
Protaminobacter ruber	Methanol	1.2% (w/v)	2.5
Methanosarcina barkeri	Methanol, fed-batch culture	0.8% (w/v)	42
Arthrobacter sp.	Isopropanol, fed-batch culture		1.1
Arthrobacter hyalinus	Isopropanol	1% (v/v)	0.56
Nocardia MT 2003	1,2-Propanediol	5% (w/v)	0.78
Klebsiella sp. 101	Ethanol	2% (v/v)	0.02
	1,2-Propanediol	2% (v/v)	0.06
	Methanol	2% (v/v)	0.16
	Methanol, fed-batch culture	0.5% (v/v)	1.3
Pseudomonas sp. ATCC 14718	Ethanol	1% (v/v)	0.05
	1,2-Propanediol	2% (v/v)	0.12
	Methanol	2% (v/v)	0.16
	Methanol, fed-batch culture	0.7–0.8% (v/v)	3.2
Strain XF	Methanol	2% (v/v)	0.23
Pseudomonas ovalis	n-Decane	1% (v/v)	0.08
Pseudomonas aureofaciens	n-Dodecane	1% (v/v)	0.04
Nocardia gardneri	n-Hexadecane	2% (w/v)	4.5
Mixed culture of *Corynebacterium* sp. and *Rhodopseudomonas spheroides*	n-Alkanes	10% (w/v)	2.3

(Florent and Ninet, 1979; Kamikubo et al., 1978)

- Direct fermentation (about 30% of worldwide production).

Total production of riboflavin on a worldwide basis is around 2000 tons per year.

Direct fermentation Riboflavin is synthesized by many microorganisms, including bacteria, yeasts and fungi. High-yielding organisms which could be used in production are listed in Table 12.2. However, only the two ascomycetes listed are important in commercial production. Originally *Eremothecium ashbyii* was used (yield 2 g/l), but after 1946, a process with genetically stable *Ashbya gossypii* was introduced. The riboflavin titer of this process is 10–15 g/l. Despite this yield, there is stiff competition between this microbiological process and the two other processes.

Table 12.2 Microorganisms making riboflavin and the effect of iron on biosynthesis

Microorganism	Riboflavin formation (mg/l)	Optimal iron concentration (mg/l)
Clostridium acetobutylicum	97	1–3
Mycobacterium smegmatis	58	No effect
Mycocandida riboflavina	200	No effect
Candida flareri	567	0.04–0.06
Eremothecium ashbyii	2480	No effect
Ashbya gossypii	6420	No effect

(Perlman, 1979)

Structure

Riboflavin is an alloxazine derivative which consists of a pteridine ring condensed to a benzene ring. The side chain consists of a C_5-polyhydroxy group, a derivative of ribitol. The structure of riboflavin (6,7-dimethyl-9-(D-1′-ribityl)-isoallox-

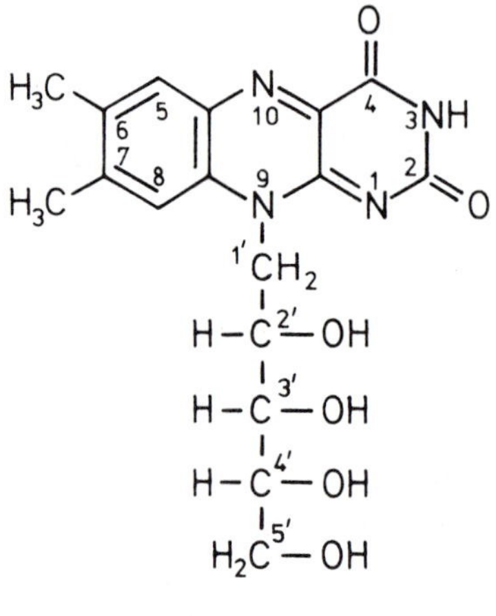

Figure 12.4 Structure of riboflavin

azine) is given in Figure 12.4. The isoalloxazine ring acts as a reversible redox system, as described in textbooks of biochemistry and microbiology.

Biosynthesis

The biosynthetic pathway shown in Figure 12.5 is postulated on the basis of experiments done primarily with yeast and *Ashbya gossypii*. The effect of iron on riboflavin synthesis is not yet understood. The overproduction of riboflavin by *Eremothecium ashbyii* and *A. gossypii* is not affected by iron, but riboflavin production in clostridia and yeasts is inhibited even by very low concentrations of iron. In clostridia, 1 ppm iron causes a 75% inhibition of synthesis. It is assumed that an iron-flavoprotein acts as a repressor of riboflavin synthesis. The overpro-

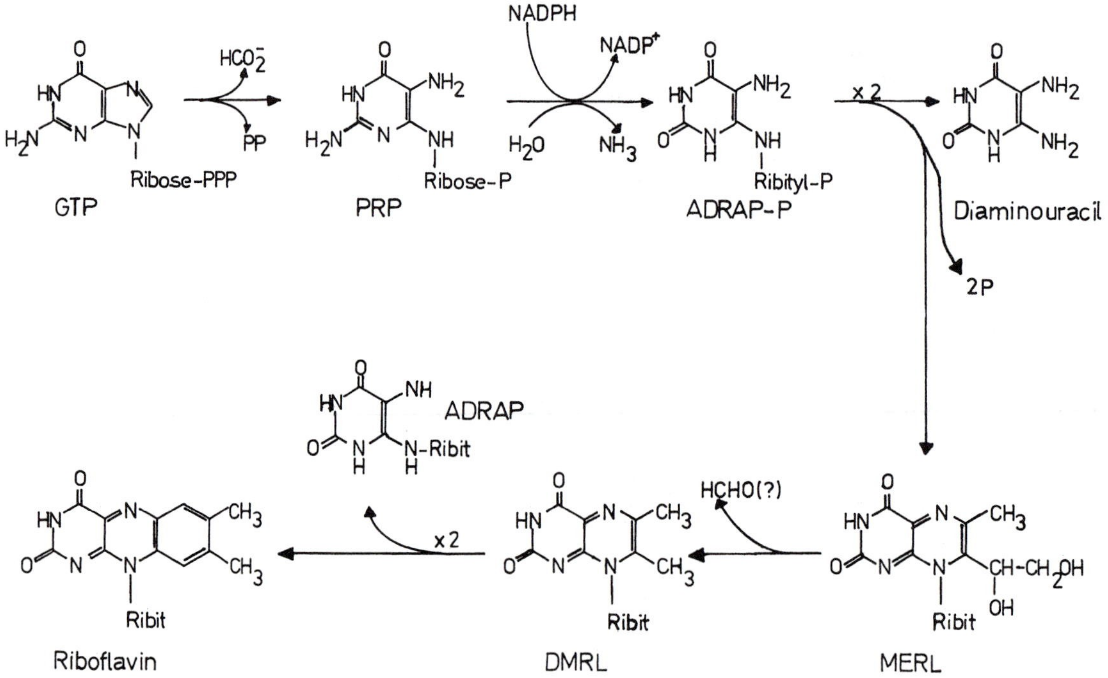

Figure 12.5 Biosynthetic pathway for riboflavin (from Brown and Williamson, 1982). Abbreviations: GTP, Guanosine triphosphate; PRP, 2,5-Diamino-6-keto-4-(5′-phosphoribosylamino)-pyrimidine; ADRAP, 5-Amino-2,5-dioxy-4-(5′-phosphoribitylamino)-pyrimidine; MERL, 6-Methly-7-(1′,2′-dihydroxyethyl)-8-ribityllumazine; DMRL, 6, 7-Dimethyl-8-ribityllumazine; Ribit, Ribitol.

duction by *E. ashbyii* and *A. gossypii* could be explained by the presence of constitutive riboflavin-synthesizing enzymes.

Production process

The fermentative production is currently carried out with *Ashbya gossypii* NRRL Y-1056. The high yields of more than 10–15 g/l were attained by strain development as well as by optimization of the nutrient solution, the cultivation of inoculum, and the fermentation conditions. Originally the fermentation used a medium with glucose and corn steep liquor; sucrose and maltose were other suitable carbon sources. When lipids were also used as energy sources, yields markedly increased. Riboflavin production on what is now the basic medium (corn steep liquor 2.25%, commercial peptone 3.5%, soy bean oil 4.5%) has been further stimulated by the addition of different peptones, glycine, distiller's solubles, or yeast extract. By simultaneous feeding of glucose and inositol the rate of formation of riboflavin can be further increased. A careful sterilization of the culture medium is critical for high yields, as is the use of a small inoculum (0.75–2%) of a 24- to 48-hour-old actively growing culture. The fermentation takes 7 days with an aeration rate of 0.3 vvm at 28°C. For foam control, silicone antifoam is applied at first, and soy bean oil, which is also metabolized, is added later.

Riboflavin is present both in solution and bound to the mycelium in the fermentation broth. The bound vitamin is released from the cells by heat treatment (1 hour, 120°C) and the mycelium is separated and discarded. The riboflavin is then further purified.

Production of riboflavin with an aliphatic hydrocarbon (C_{10}–C_{18}) as carbon source has been reported using *Pichia guilliermondii*, but yield information was not given. Using *Pichia miso*, 51 mg/l riboflavin was obtained on a medium with n-hexadecane, corn steep liquor, and urea. Studies have been carried out on the use of a methanol-utilizing organism, for example, *Hansenula polymorpha*, for the formation of riboflavin in batch or continuous culture.

Crystalline riboflavin preparations of high purity have been produced using *Saccharomyces* fermentation with acetate as sole carbon source.

Table 12.3 Production processes for several carotenoids

Carotenoid	Organism	Composition of medium	Length of fermentation (days)	Yield (mg/l)
β-Carotene	Mixed culture of *Blakeslea trispora* NRRL 2456(+) and NRRL 2457(−)	see Figure 12.9	8	3000
Lycopene	*Streptomyces chrestomyceticus* subsp. *rubescens*	Starch, soy meal, $(NH_4)_2SO_4$	6	500
	Mixed culture of *Blaskeslea trispora* NRRL 2895(+) and NRRL 2896(−)	Cotton seed meal, corn meal, soy bean oil; addition of triethylamine after 48 hours (suppresses carotenoid cyclization)	5	400
Zeaxanthin	*Flavobacterium* sp.	Glucose, corn steep liquor, palmitate ester, methionine, pyridoxine, Fe^{2+} salts		335
Mixture of various carotenoids	*Mycobacterium phlei*	Sugar beet molasses, urea; exposure to light required	7	300
	S. chrestomyceticus	Glucose, yeast extract	5	500
	Yeast	Sugar, ethanol or alkanes		< 5

(Ninet and Renaut, 1979)

12.4 β-CAROTENE

Occurrence and economic significance

Carotenoids are found in many animal and plant tissues, but originate exclusively from plants or microbes. β-Carotene (provitamin A) is converted into vitamin A in the intestinal mucous membrane and is stored in the liver as the palmitate ester. In human beings, a deficiency of vitamin A (daily requirement 1.5–2 mg) causes night blindness and changes in the skin and mucous membranes; it is particularly necessary for the normal biosynthesis of mucopolysaccharides.

The demand for β-carotene is about 100 tons per year, primarily for use as a food coloring agent. The demand for provitamin A is very low. Other carotenoids, such as lycopene or xanthophylls, which do not have a provitamin A activity, are used as food coloring agents (e.g. margarine, cheese, egg products), for improving color in meat or egg yolk in the poultry industry (citrinaxanthine, 3–5 tons per year) or in the salmon industry (astaxanthine). Chemical synthesis is currently the only commercially feasible means of obtaining carotenoids; xanthophylls may also be obtained by extraction of plant material. Carotenoids are produced by many microorganisms (including algae), but fermentative production is not economical in the present market, since the yields of currently available processes (Table 12.3) do not compete with cost-effective chemical synthetic processes. However, depending on the market it appears that a competitive process might be successful, using the halophilic alga *Dunaliella salina* (see Borowitzka et al., 1986).

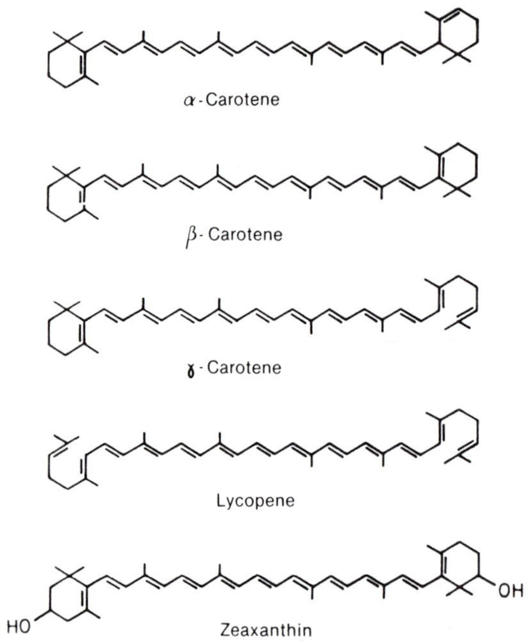

α-Carotene

β-Carotene

γ-Carotene

Lycopene

HO Zeaxanthin OH

Figure 12.6 Structures of several carotenoids that can be produced by fermentation

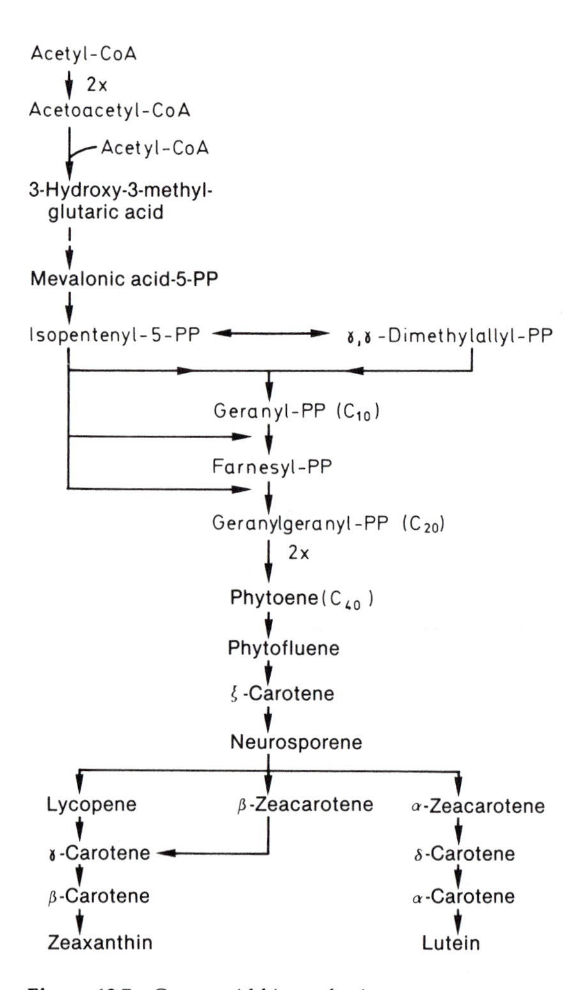

Figure 12.7 Carotenoid biosynthesis

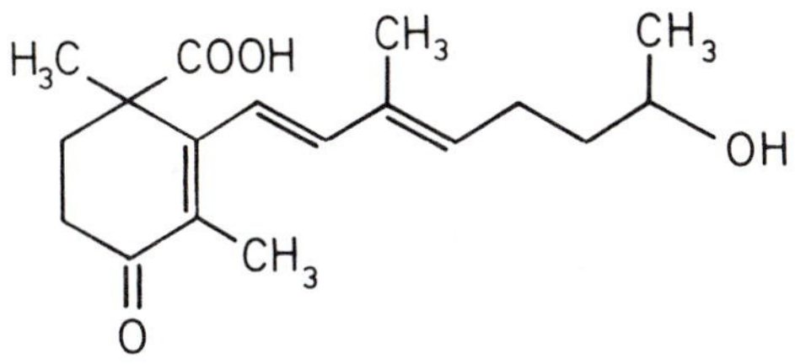

Figure 12.8 Structure of trisporic acid C

Structure

Carotenoids are highly unsaturated isoprene derivatives. Naturally occurring carotenoids are tetraterpenoids consisting of 8 isoprene residues. Over 400 naturally occurring carotenoid compounds are known today. Figure 12.6 shows the impact on process development. When cultures of both sexual forms, (+) and (−) strains, are mixed, a significant increase in carotene production. structures of some carotenoids produced by microbial fermentation. Only compounds with the β-ionone structure (the ring structure found at each end of the β-carotene molecule) are effective as provitamin A. Thus, 2 molecules of vitamin A can be formed from β-carotene; only one molecule of vitamin A can be formed from α- and γ-carotene since each of these molecules has one β-ionone ring.

Biosynthesis

Biosynthesis of β-carotene has been studied mainly in plants and fungi and takes place as shown in Figure 12.7. β-Carotene and related carotenoids are produced primarily by fungi and algae, xanthophylls by bacteria and algae.

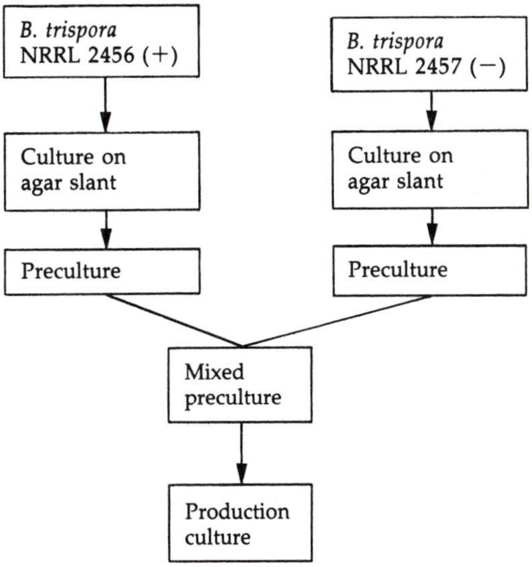

B. trispora NRRL 2456 (+) / *B. trispora* NRRL 2457 (−)	Inoculum storage: spores in sterile soil
Culture on agar slant / Culture on agar slant	Incubation: 168 h, 27°C
Preculture / Preculture	2 l Erlenmeyer flask with 400 ml Medium A. Incubation: 48 h, 26°C, shaker.
Mixed preculture	170 l Fermenter with 120 l Medium A. Inoculum: 400 ml of each preculture. Incubation: 40 h, 26°C; stirring 170 rpm; aeration 1.1 vvm
Production culture	800 l Fermenter with 320 l Medium B. Sterilization 55 min, 122°C. Inoculum: 32 l of mixed preculture. Incubation: 185 h, 26°C; stirring 210 rpm; aeration 1.3 vvm

Medium A (g/l): Corn steep liquor 70; Corn starch 50; KH_2PO_4 0.5; $MnSO_4 \cdot H_2O$ 0.1; Thiamin HCl 0.01; Tap water

Medium B (g/l): Distiller's soluble 70; Corn starch 60; Soy bean meal 30; Cottonseed oil 30; Antioxidant 0.35; $MnSO_4 \cdot H_2O$ 0.2; Thiamin HCl 0.5; Isoniazid 0.6; Kerosene 20 ml; Tap water; pH 6.3. Isoniazid and kerosene are sterilized separately. After 48 h 1 g/l β-ionone and 5 ml/l kerosene are added; glucose feeding (total addition 42 g/l) until the end of the fermentation.

Figure 12.9 Production process for β-carotene using *Blakeslea trispora* (From Ninet and Renaut, 1979)

Production processes for β-carotene

The highest yields have been obtained with the *Blakeslea trispora* process. The observation that carotene production occurs during the process of zygospore formation in this organism has had an impact on process development. When cultures of both sexual forms, (+) and (−) strains, are mixed, a significant increase in carotene production in the (−) strain is achieved. Production is induced by trisporic acids (Figure 12.8), a mixture of closely related substances. Trisporic acids are not precursors of β-carotene but rather they act as (+)-gamones (sexual hormones). They are derived biosynthetically from β-carotene, as shown by labeling experiments. The production of β-carotene (Figure 12.9) is carried out in a submerged fermentation using a mixed culture of (+) and (−) strains. The proportions of the two strains need not be equal and since it is the (−) strain which produces the β-carotene, this strain can be present in great excess. Another activator of β-carotene synthesis is isoniazid, particularly in combination with β-ionone. Alone, β-ionone is toxic to the production organism, but in the presence of plant oils it promotes carotene production. The β-ionones themselves are not incorporated into carotene but affect the synthesis of various of the enzymes involved in the biosynthetic pathway. They can be replaced by a number of other substances, such as terpenes or cyclohexane and cyclohexanone and their trimethyl derivatives. The addition of purified kerosene to the medium doubles the yield by increasing the solubility of the hydrophobic substrate.

Because of the low stability of β-carotene within the cells, the addition of an antioxidant is necessary for the fermentation process.

The carotenoid-rich mycelium can be used directly as a feed additive. To obtain pure β-carotene, the mycelium is removed, dehydrated (for example, with methanol), extracted with methylene chloride (75–92% yield), and the crude product is further purified.

REFERENCES

Beytia, E.D. and J.W. Porter. 1976. Biochemistry of polyisoprenoid biosynthesis. Ann. Rev. Biochem. 45: 113–142.

Borowitzka, L.J., T.P. Moulton, and M.A. Borowitzka. 1986. Salinity and the commercial production of beta-carotene from *Dunaliella salina*. Nova Hedwigia 83:224–229.

Brooke, A.G., L. Dijkhulzen, and W. Harder. 1986. Regulation of flavin biosynthesis in the methylotrophic yeast *Hansenula polymorpha*. Arch. Microbiol. 145:62–70.

Brown, G.M. and J.M. Williamson. 1982. Biosynthesis of riboflavin, folic acid, thiamine, and pantothenic acid. Adv. Enzymol. 53:346–353.

Ciegler, A. 1965. Microbial carotenogenesis. Adv. Appl. Microbiol. 7: 1–34.

Florent, J. 1986. Vitamins, pp. 115–158. In: Rehm, H.J. and G. Reed (eds.). Biotechnology, Vol. 4, VCH Publishers, Deerfield Beach, FL.

Florent, J. and L. Ninet. 1979. Vitamin B_{12}, pp. 497–519. In: Peppler, H.J. and D. Perlman (eds.), Microbial technology, vol. I, 2nd edition. Academic Press, New York.

Kamikubo, T., M. Hayashi, N. Nishio, and S. Nagai. 1978. Utilization of non-sugar sources for vitamin B_{12} production. Appl. Environ. Microbiol. 35: 971–973.

Karlson, P. 1980. Kurzes Lehrbuch der Biochemie für Mediziner und Naturwissenschaftler (Short textbook of biochemistry for medicine and biology), 11th edition. Georg Thieme Verlag, Stuttgart.

Malzahn, R.C., R.F. Phillips, and A.M. Hanson. 1959. U.S. Patent 2,876,169.

Mazumder, T.K., N. Nishio, M. Hayashi, and S. Nagai. 1986. Production of corrinoids including vitamin B_{12} by *Methanosarcina barkeri* growing on methanol. Biotechnol. Letters 8:843–848.

Merck and Co., Inc. 1971. French Patent 2,038,828.

Ninet, L. and J. Renaut. 1979. Carotenoids, pp. 529–544. In: Peppler, H.J. and D. Perlman (eds.), Microbial technology, vol. I, 2nd edition. Academic Press, New York.

Perlman, D. 1977. Fermentation industries, quo vadis? Chem. Technol. 7: 434–443.

Perlman, D. 1978. Vitamins, pp. 303–326. In: Rose, A.H. (ed.), Economic microbiology, vol. 2, Primary products of metabolism. Academic Press, London.

Perlman, D. 1979. Microbial process for riboflavin production, pp. 521–527. In: Peppler, H.J. and D. Perlman (eds.), Microbial technology, vol. I, 2nd edition. Academic Press, New York.

Renz, P. 1984. Untersuchungen zur Biosynthese von Vitamin B_{12} (Studies on the synthesis of vitamin B_{12}). GIT Fachz. Lab. 28:884–892.

Speedie, J.D. and G.W. Hull. 1960. U.S. Patent 2,951,017.

Upjohn Co. 1965. Netherlands Patent 64/11 184.

Yamada, K. 1977. Bioengineering report. Recent advances in industrial fermentation in Japan. Biotechnol. Bioeng. 19: 1563–1621.

13

Antibiotics

13.1 INTRODUCTION

Antibiotics are products of secondary metabolism which inhibit growth processes of other organisms even when used at low concentrations. Growth inhibition of one microorganism by another in mixed culture has been known for a long time. The most famous example is the growth inhibition which was observed by Alexander Fleming in 1929, when staphlococcal growth on a petri plate was inhibited by a contaminating *Penicillium notatum* culture. The *Penicillium* contaminant produced the antibiotic penicillin. During World War II, the demand for chemotherapeutic agents to treat wound infections led to the development of a production process for penicillin and the beginning of the era of antibiotic research. This continues to be the most important area of industrial microbiology today. Intensive screening programs in all industrial countries continue to increase the number of described antibiotics: 513 antibiotics were known in 1961, 4076 in 1972, 7650 in 1985, and currently around 8000. In addition, around 3000 antibiotically active substances have been detected in lichens, algae, higher animals, and plants. Each year, about 300 new antibiotically active materials are detected, of which 30–35% are secondary components from fermentations with known antibiotics.

Of the large number of known antibiotics of microbial origin, only 123 are currently produced by fermentation. In addition, more than 50 antibiotics are produced as semisynthetic compounds, and three antibiotics, chloramphenicol, phosphonomycin, and pyrrolnitrin, are produced completely synthetically.

The significance of antibiotic production for the producing strain is unclear. Antibiotic production could be of ecological significance for the life of the organism in nature, but solid research to support this hypothesis is very limited. As secondary metabolites, antibiotics could serve regulatory roles during differentiation, perhaps acting as temporary inhibitory agents. To date, most new antibiotics have been detected strictly by empirical screening methods, with little attention to their possible roles for the producing strains.

The microbial groups producing antibiotics

Antibiotics are produced by bacteria, actinomycetes and fungi; their distribution within the taxonomic groups is shown in Table 13.1.

In the **fungi**, only the antibiotics produced by the Aspergillaceae and Moniliales are of practical importance. The compounds isolated from basidiomycetes have not had any practical use. Only 10 of the known fungal antibiotics are produced commercially and only the penicillins, cephalosporin C, griseofulvin, and fusidic acid are clinically important. In the **bacteria**, there are many taxonomic groups which produce antibiotics. The greatest variety in structure and number of antibiotics is found in the **actinomycetes**, especially in the genus *Streptomyces*. Another important group of substances are the peptide antibiotics, produced by bacteria of the genus *Bacillus*. A number of new compounds have also recently been isolated in other taxonomic groups (see Chapter 2).

The significance of antibiotic production for producer strains is still unclear. The status of the controversy can be found in the references of Zähner and Barabas cited in the bibliography.

Classification of antibiotics

Antibiotics can be classified according to their antimicrobial spectrum, mechanism of action, producer strain, manner of biosynthesis, or chemical structure. There is some overlap in each classification. Table 13.2 shows a simplified classification according to chemical structure. In this chapter, primarily the economically important compounds are covered.

Table 13.1 Numbers of antibiotics produced by major groups of microorganisms

Taxonomic group	Number of antibiotics
Bacteria, other than actinomycetes	950
Actinomycetes	4600
Fungi	1600

Berdy (1985)

Table 13.2 Classification of antibiotics according to their chemical structure. An example of each is given in parentheses.

1. Carbohydrate-containing antibiotics	
Pure sugars	(Nojirimycin)
Aminoglycosides	(Streptomycin)
Orthosomycins	(Everninomicin)
N-Glycosides	(Streptothricin)
C-Glycosides	(Vancomycin)
Glycolipids	(Moenomycin)
2. Macrocyclic lactones	
Macrolide antibiotics	(Erythromycin)
Polyene antibiotics	(Candicidin)
Ansamycins	(Rifamycin)
Macrotetrolides	(Tetranactin)
3. Quinones and related antibiotics	
Tetracyclines	(Tetracycline)
Anthracyclines	(Adriamycin)
Naphthoquinones	(Actinorhodin)
Benzoquinones	(Mitomycin)
4. Amino acid and peptide antibiotics	
Amino acid derivatives	(Cycloserine)
β-Lactam antibiotics	(Penicillin)
Peptide antibiotics	(Bacitracin)
Chromopeptides	(Actinomycins)
Depsipeptides	(Valinomycin)
Chelate-forming peptides	(Bleomycins)
5. Heterocyclic antibiotics containing nitrogen	
Nucleoside antibiotics	(Polyoxins)
6. Heterocyclic antibiotics containing oxygen	
Polyether antibiotics	(Monensin)
7. Alicyclic derivatives	
Cycloalkane derivatives	(Cycloheximide)
Steroid antibiotics	(Fusidic acid)
8. Aromatic antibiotics	
Benzene derivatives	(Chloramphenicol)
Condensed aromatic antibiotics	(Griseofulvin)
Aromatic ether	(Novobiocin)
9. Aliphatic antibiotics	
Compounds containing phosphorous	(Fosfomycins)

(Berdy, 1985)

Applications of antibiotics

Chemotherapeutic antibiotics can be either broad-spectrum antibiotics, active against many organisms, or narrow-spectrum antibiotics, active against only a restricted range of organisms. Most antibiotics are manufactured as antimicrobial agents for chemotherapy, but some have other applications, as outlined below.

Antitumor antibiotics Such antibiotics are clinically used as cytostatic agents (Table 13.3). Although generally toxic, with careful control of the dose certain of these antibiotics are effective in the treatment of certain kinds of tumors.

Antibiotics for plant pathology Antibiotics may be more useful than synthetic chemicals in the control of plant diseases for the following reasons: they may be applied selectively in low concentrations, they are only slightly toxic to warm-blooded animals and beneficial insects, and they are normally easily broken down by soil microorganisms. Initially, medically useful antibiotics were also used on plants (especially streptomycin to combat plant disease caused by *Pseudomonas* sp. and *Xanthomonas oryzae*). Now antibiotics have been developed which are used exclusively for plant application (Table 13.4).

Table 13.3 The most important antitumor antibiotics

Antibiotic	Category	Organism producing
Aclacinomycin	Anthracycline	*S. galilaeus*
Actinomycin C_1, C_3, D	Chromopeptide	*S. antibioticus*
Adriamycin	Anthracycline	*S. peucetius*
Daunomycin	Anthracycline	*S. peucetius*
Chromomycin A_3	C-Glycoside (oligosaccharide with aromatic chromophore)	*S. griseus*
Mithramycin	C-glycoside (oligosaccharide with aromatic chromophore)	*S. plicatus*, *S. argillaceus*, *S. atroolivaceus*
Mitomycin C	Benzoquinone	*S. caespitosus*
Bleomycin A_2, B_2	Glycopeptide	*S. verticillus*
Neocarzinostatin	Peptide	*S. carzinostaticus*

Antibiotics as food preservatives Government regulations in each country directly control the use of antibiotics as food preservatives. The following compounds are available: pimaricin, a fungicide applied to food surfaces; tylosin (effective against *Bacillus* spores) and nisin (effective against clostridia), both used in the canning industry; chlortetracycline, used to maintain freshness in fish, meat and poultry by incorporation into ice (5 ppm) or by addition to an immersion bath (10 ppm). Of these, pimaricin and nisin are currently used commercially.

Antibiotics used as animal growth promoters and in veterinary medicine Animal feed is more efficiently processed in the animal's digestive system if an antibiotic additive is used in subtherapeutic concentrations (1–10 mg/kg feed). Weight gain may also be accelerated. The cause of this growth promotion may be traced to changes in the microflora of the gastrointestinal tract. This is further confirmed by the fact that the growth of animals raised in germ-free conditions is not stimulated by antibiotics. Originally, large quantities of therapeutically useful antibiotics (such as penicillins, tetracyclines, erythromycins, bacitracin, or streptomycin) were added to feed. Because such extensive use of common antibiotics encouraged the risk of rapid development of antibiotic resistance, government regulations have been established to eliminate the parallel use of antibiotics in human medicine and in animal feed. Some new-generation antibiotics which are used solely for nutritional purposes are listed in Table 13.5.

In veterinary medicine, the use of antibiotics is similarly regulated. Antibiotics which are used solely for veterinary work include hygromycin B, thiostrepton, tylosin, and the coccidiostatic agents monensin, lasalocide, and salinomycin.

Antibiotics as tools in biochemistry and molecular biology The use of antibiotics as selective inhibitors has made a vital contribution to the understanding of certain cell functions, such as

Table 13.4 Antibiotics of use in plant pathology

Antibiotic, chemical type in ( )	Organism producing	Uses
Blasticidin S (Nucleoside)	*S. griseochromogenes*	Rice fungicide against *Piricularia oryzae* (rice burn); relatively toxic
Mildiomycin (Nucleoside)	*Streptoverticillium rimofaciens*	Fungicide for mildew
Polyoxin (Nucleoside)	*S. cacaoi* var. *asoensis*	Multi-purpose fungicide
Prumycin (Nucleoside)	*S. kagawaensis*	Fungicide against *Botrytis* and *Sclerotinia* species
Cycloheximide (Amino acid)	*S. griseus*	Leaf fungicide, highly toxic to warm-blooded animals; aids harvesting
Kasugamycin (Aminoglycoside)	*S. kasugaensis*	Rice fungicide against *Piricularia oryzae*
Validamycin (Aminoglycoside)	*S. hygroscopicus* var. *limoneus*	Fungicide against *Rhizoctonia solani* (leaf dropping and stalk diseases); vegetable crops
Tetranactin (Macrotetrolide)	*S. flaveolus*	Insecticide (mites)

Table 13.5 Antibiotics used in animal feed

Antibiotics	Classification	Organism producing
Enduracidin	Peptide antibiotic	*S. fungicidicus*
Mikamycin	Peptide antibiotic	*S. mitakaensis*
Siomycin	Peptide antibiotic	*S. sioyaensis*
Thiopeptin	Peptide antibiotic	*S. tateyamensis*
Thiostrepton	Peptide antibiotic	*S. azureus*
Virginiamycin	Peptide antibiotic	*S. virginiae*
Macarbomycin	Phospho-glycolipid	*S. phaeochromogenes*
Moenomycin	Phospho-glycolipid	*S. bambergiensis*
Quebemycin	Phospho-glycolipid	*S. viridans*
Tylosin	Macrolide	*S. fradiae*
Mocimycin	Heterocyclic compound	*S. ramocissimus*

DNA replication, transcription, translation, and cell wall synthesis. The uses of antibiotics in these areas is discussed in textbooks of biochemistry and microbiology. In addition, some antibiotics (for example, gentamicin) have been added to animal cell cultures to control contamination.

Economic significance of antibiotics

World-wide antibiotic production is over 100,000 tons per year and estimated gross sales for 1980 were $4.2 billion. The annual gross sales in the United States alone is $1 billion, with cephalosporin in the leading position, followed by ampicillin and the tetracyclines. Feed-additive antibiotics are believed to have a world market of $100 million annually.

Objectives of antibiotic research

Before 1960, about 5% of the newly isolated antibiotics were therapeutically useful. In the following years new antibiotics were discovered at an approximately constant rate, but the percentage of the new antibiotics which actually came on the market decreased from 2.6% in 1961–1965 to 1% in 1966–1971. This is primarily because of severe cost increases in development and clinical testing, so that manufacturers produce only those compounds which clearly show promising therapeutic progress. On the average, it takes around 8–10 years to develop a new antibiotic, at an average cost of $10,000,000 to $20,000,000.

Considering the very large number of known compounds, it may seem questionable whether the search for new antibiotics should continue. The reasons for continued research are:

- In many cases the properties of natural antibiotics are not optimal for therapeutic application. The following improvements are needed: greater activity with unchanged or diminished toxicity, decreased side effects, broader antimicrobial range, greater selectivity against certain pathogens, improved pharmacological properties.
- Suitable antibiotics are not available in many fields of human medicine or in nonmedical areas, as shown in Table 13.6. Numerous tests have been made of new and semisynthetic substances, but no significant breakthroughs have been made.
- Since the beginning of chemotherapy, the number of resistant strains has increased. Multiple- and cross-resistance can occur, i.e., if resistance develops to one antibiotic it may simultaneously develop to others having the same mode of operation or uptake mechanism. Careless use of antibiotics has been responsible for much increase in resistance, but even with careful use in chemotherapy, resistance still develops, albeit at a slower rate. Currently, the only alternative for overcoming the resistance problem is the discovery of new antibiotics.

Improved antibiotics can be obtained by modifying known compounds using either chemical or genetic means (mutasynthesis, protoplast fusion, recombinant DNA technology—see Chapter 3). However, antibiotics with entirely new basic structures can be expected only from screening, especially by the use of new test procedures and by research on new groups of microorganisms (see Chapter 2).

13.2 β-LACTAM ANTIBIOTICS

As shown in Figure 13.1, the β-lactam antibiotics can be divided into five distinct classes. The penicillins and the cephalosporins belong to the most effective of all therapeutic agents for the control of infectious diseases. In addition to the development of numerous semisynthetic β-lactams based on the known β-lactam rings, antibiotics with completely new β-lactam ring systems have been isolated in the past few years using new specific and sensitive screening methods.

Penicillins

Penicillin was described by Fleming in 1929. A research group at Oxford under Florey and Chain isolated it from surface cultures of *Penicillium notatum* in 1940 and the first clinical application of penicillin was in 1941. Penicillins are produced by many fungi, particularly *Penicillium* and *Aspergillus* species. Natural penicillins are effective against numerous gram-positive bacteria. They are labile in acid and may be inactivated by split-

Table 13.6 Possible applications for antibiotics

Application	Many products	Some products	Products needed
Medicine	Gram-positive bacteria	Gram-negative bacteria, including multiple resistance Dermatophytes	Systemic mycoses Viruses Protozoa Parasites Tumors
Nonmedical areas	–	Plant pathology (phytopathogenic fungi) Animal nutrition	Plant pathology Phytopathogenic bacteria and viruses Insects and mites Nematodes Food preservatives

(Zähner, 1978)

Basic structure	Antibiotic	Most important producing strains
Penam 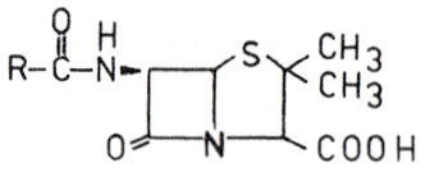	Penicillins	*Penicillium chrysogenum* *Aspergillus nidulans* *Cephalosporium acremonium* *Streptomyces clavuligerus*
Ceph-3-em 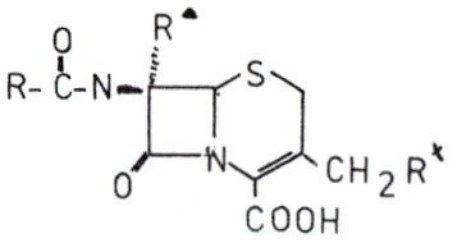	Cephalosporins 7-Methoxycephalosporins	*Cephalosporium acremonium* *Nocardia lactamdurans* *Streptomyces clavuligerus*
Clavam 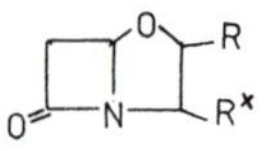	Clavulanic acid	*S. clavuligerus*
Carbapenem 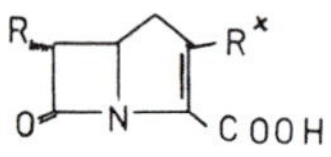	Thienamycins Olivanic acids Epitheinamycins	*S. cattleya* *S. olivaceus* *S. flavogriseus*
Monolactam 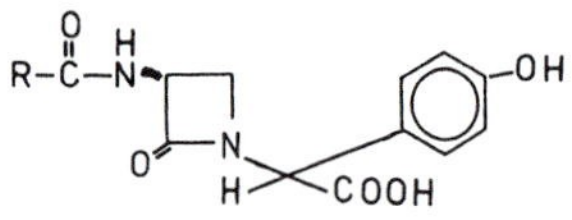	Nocardins	*Nocardia uniformis* subsp. *tsuyamenensis*
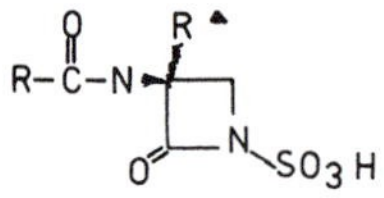	Monobactams	*Gluconobacter* sp. *Chromobacter violaceum* *Agrobacterium radiobacter* *Pseudmonas acidophila* *Pseudomonas mesoacidophila* *Flexibacter* sp. *Acetobacter* sp.

Figure 13.1 The basic structures of the naturally occurring β-lactam antibiotics

ting the β-lactam ring with penicillin-β-lactamases (see Figure 15.11). Because of its low toxicity, large doses of penicillin can be used; only a small percentage of patients develops allergies (0.5–2%).

β-Lactam antibiotics are specific inhibitors of bacterial cell wall (peptidoglycan) synthesis. They combine specifically with the so-called **penicillin-binding protein** (PBP) of the bacterial cell and, by inhibiting the enzyme activity of this protein, bring about cell death. In *E. coli* it has been shown that PBP1a and 1b are transpeptidases involved in the cross-linking of the peptidoglycan, PBP3 plays a role in septum formation, and PBP4 and 5 are carboxypeptidases. A further point of attack of the β-lactam antibiotics seems to involve phospholipid synthesis.

Chemical structure The basic structure of the penicillins is 6-aminopenicillanic acid (6-APA), which consists of a thiazolidine ring with a condensed β-lactam ring. The 6-APA carries a variable acyl moiety in position 6, as illustrated in Figure 13.2. If the penicillin fermentation is carried out without addition of side-chain precursors, the **natural penicillins** are produced. From this mixture, only benzylpenicillin is therapeutically useful; the other compounds must be re-

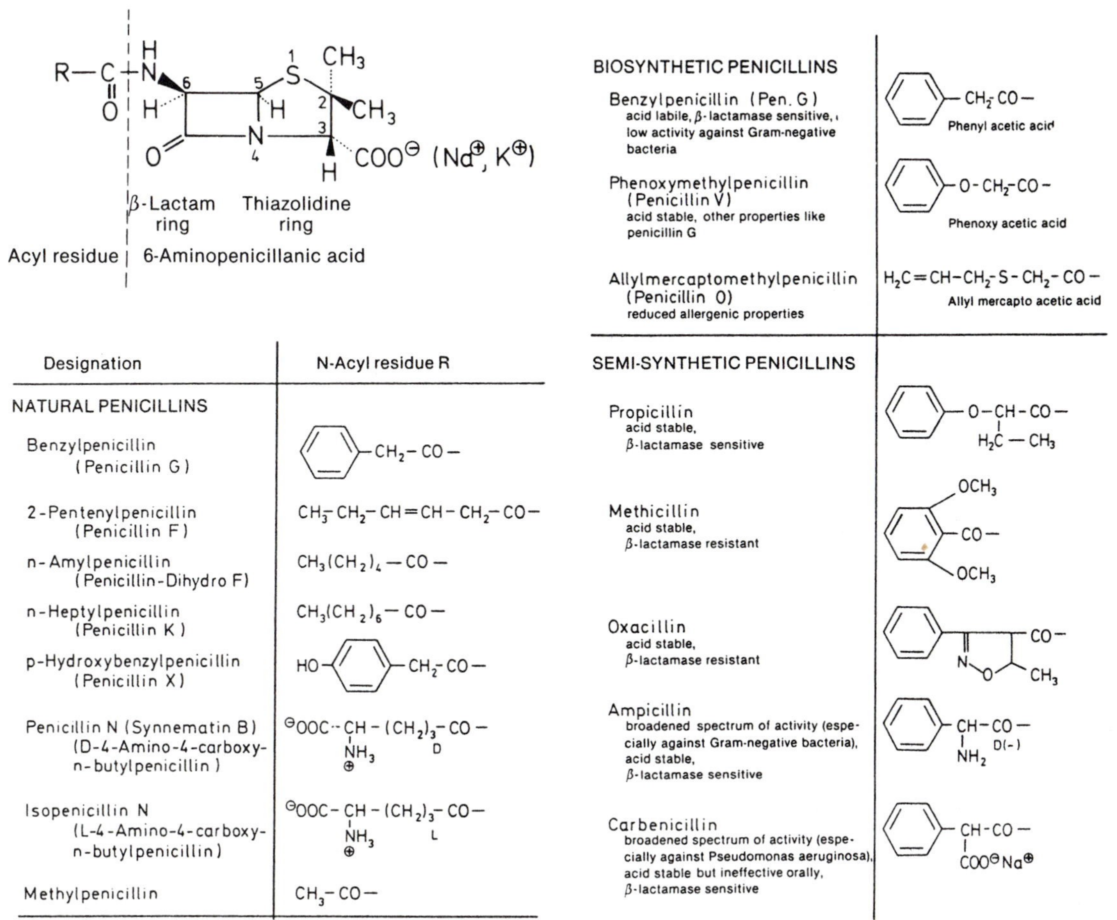

Designation	N-Acyl residue R
NATURAL PENICILLINS	
Benzylpenicillin (Penicillin G)	C_6H_5–CH_2–CO–
2-Pentenylpenicillin (Penicillin F)	CH_3–CH_2–CH=CH–CH_2–CO–
n-Amylpenicillin (Penicillin-Dihydro F)	$CH_3(CH_2)_4$–CO–
n-Heptylpenicillin (Penicillin K)	$CH_3(CH_2)_6$–CO–
p-Hydroxybenzylpenicillin (Penicillin X)	HO–C_6H_4–CH_2–CO–
Penicillin N (Synnematin B) (D-4-Amino-4-carboxy-n-butylpenicillin)	⊖OOC–CH(NH_3⊕)–$(CH_2)_3$–CO– D
Isopenicillin N (L-4-Amino-4-carboxy-n-butylpenicillin)	⊖OOC–CH(NH_3⊕)–$(CH_2)_3$–CO– L
Methylpenicillin	CH_3–CO–
BIOSYNTHETIC PENICILLINS	
Benzylpenicillin (Pen. G) acid labile, β-lactamase sensitive, low activity against Gram-negative bacteria	C_6H_5–CH_2–CO– Phenyl acetic acid
Phenoxymethylpenicillin (Penicillin V) acid stable, other properties like penicillin G	C_6H_5–O–CH_2–CO– Phenoxy acetic acid
Allylmercaptomethylpenicillin (Penicillin O) reduced allergenic properties	H_2C=CH–CH_2–S–CH_2–CO– Allyl mercapto acetic acid
SEMI-SYNTHETIC PENICILLINS	
Propicillin acid stable, β-lactamase sensitive	C_6H_5–O–CH(H_2C–CH_3)–CO–
Methicillin acid stable, β-lactamase resistant	2,6-(OCH_3)$_2$$C_6H_3$–CO–
Oxacillin acid stable, β-lactamase resistant	C_6H_5-isoxazolyl(CH_3)–CO–
Ampicillin broadened spectrum of activity (especially against Gram-negative bacteria), acid stable, β-lactamase sensitive	C_6H_5–CH(NH_2)–CO– D(–)
Carbenicillin broadened spectrum of activity (especially against Pseudomonas aeruginosa), acid stable but ineffective orally, β-lactamase sensitive	C_6H_5–CH(COO⊖Na⊕)–CO–

Figure 13.2 Structure of the known natural penicillins, the most important biosynthetic penicillins, and several semi-synthetic penicillins

moved at the product recovery stage. The fermentation can be better controlled by adding a side-chain precursor, so that only one desired penicillin is produced. Over 100 **biosynthetic penicillins** have been produced in this way. In commercial processes, however, only penicillin G, penicillin V, and very limited amounts of penicillin O have been produced. In order to produce **semisynthetic penicillins**, penicillin G (sometimes penicillin V is used instead) is chemically or enzymatically split to form 6-APA, which is chemically recycled to make yet another penicillin derivative. (See Section 11.9 for a discussion of the enzyme penicillin acylase). Due to their improved characteristics (acid stability, resistance to plasmid or chromosomally coded β-lactamases, expanded antimicrobial effectiveness), semisynthetic penicillins have come to be extensively used in therapy. Because of their broadened action spectrum, these semisynthetic penicillins are sometimes compared with the so-called third generation cephalosporins.

About 38% of the penicillins produced commercially are used in human medicine (see below), 12% in veterinary medicine, and 43% are used as starting materials for the production of semisynthetic penicillins. In the early 1980's, world-wide production of penicillins amounted to 12,000 tons, an exceedingly large amount considering how highly active these agents are.

Biosynthesis and regulation The β-lactam-thiazolidine ring of penicillin is constructed from L-cysteine and L-valine. Biosynthesis occurs in a nonribosomal process by means of a dipeptide composed of L-α-aminoadipic acid (L-α-AAA) and L-cysteine or a breakdown product of cystathionine. Subsequently, L-valine is connected via an epimerization reaction, resulting in the formation of the tripeptide δ-(L-α-aminoadipyl)-cysteinyl-D-valine. The first product of the cyclization of the tripeptide which can be isolated is isopenicillin N, but the biochemical reactions leading to this intermediate are not understood. Benzylpenicillin is produced in the exchange of L-α-AAA with activated phenylacetic acid (Figure 13.3). 6-APA, which is not an intermediary

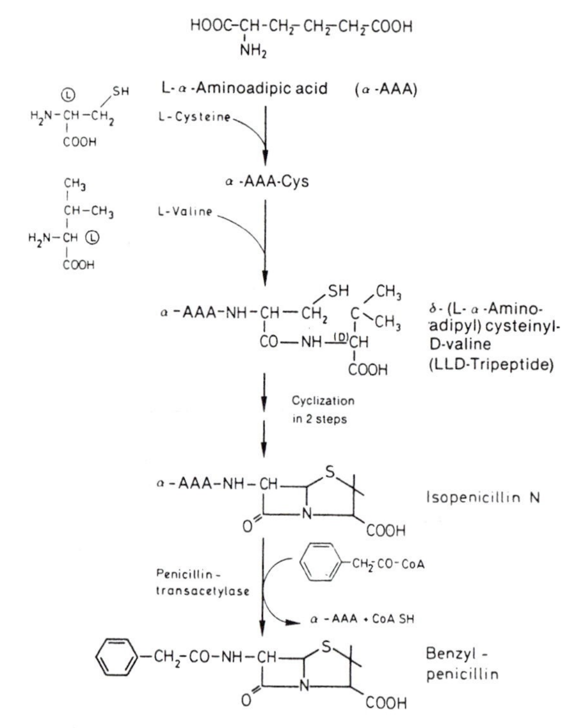

Figure 13.3 Biosynthesis of penicillin in *Penicillium chrysogenum*

product of biosynthesis, is excreted in the absence of a side chain precursor.

Several regulatory mechanisms are known in penicillin biosynthesis. The amino acid lysine is synthesized from a pathway that involves L-α-aminoadipic acid, so that penicillin and lysine share a common branched biosynthetic pathway. Lysine inhibits penicillin synthesis because it is a feedback inhibitor of homocitrate synthase, an enzyme involved in L-α-AAA synthesis. If L-α-AAA is deficient, penicillin cannot be synthesized. However, feedback regulation by lysine does not seem to be a rate-limiting step in penicillin biosynthesis.

Penicillin biosynthesis is affected by phosphate concentration and also shows a distinct catabolite repression by glucose, in addition to a regulation by concentration of ammonium ion, the latter by an unexplained mechanism. Because of the glucose repression, penicillin fermentations were originally done only with the slowly metabolizable sugar lactose.

Strain development The penicillin production of Fleming's isolate was about 2 International units/ml; today's processes yield a penicillin titer of about 85,000 units/ml. This is an increase from 0.0012 g/l to about 50 g/l and well illustrates the value and power of a strain-selection program.

The initial culture improvement came in 1943 with the isolation of *Penicillium chrysogenum* strain NRRL 1951. This organism was better suited for submerged production than the original *P. notatum* strain. Through further mutagenesis, strain Wis Q 176 was isolated, the original strain in the famous Wisconsin culture line. Figure 13.4 shows the genealogy of these strains over a time period of about 10 years. Wis Q 176 was adopted by most penicillin manufacturers and was used as the original strain for the various commercial strain improvement programs, about which little has been published. Newer data have been released only by Panlabs Inc., a corporation which since 1973 has specialized in commercial strain development (Table 13.7).

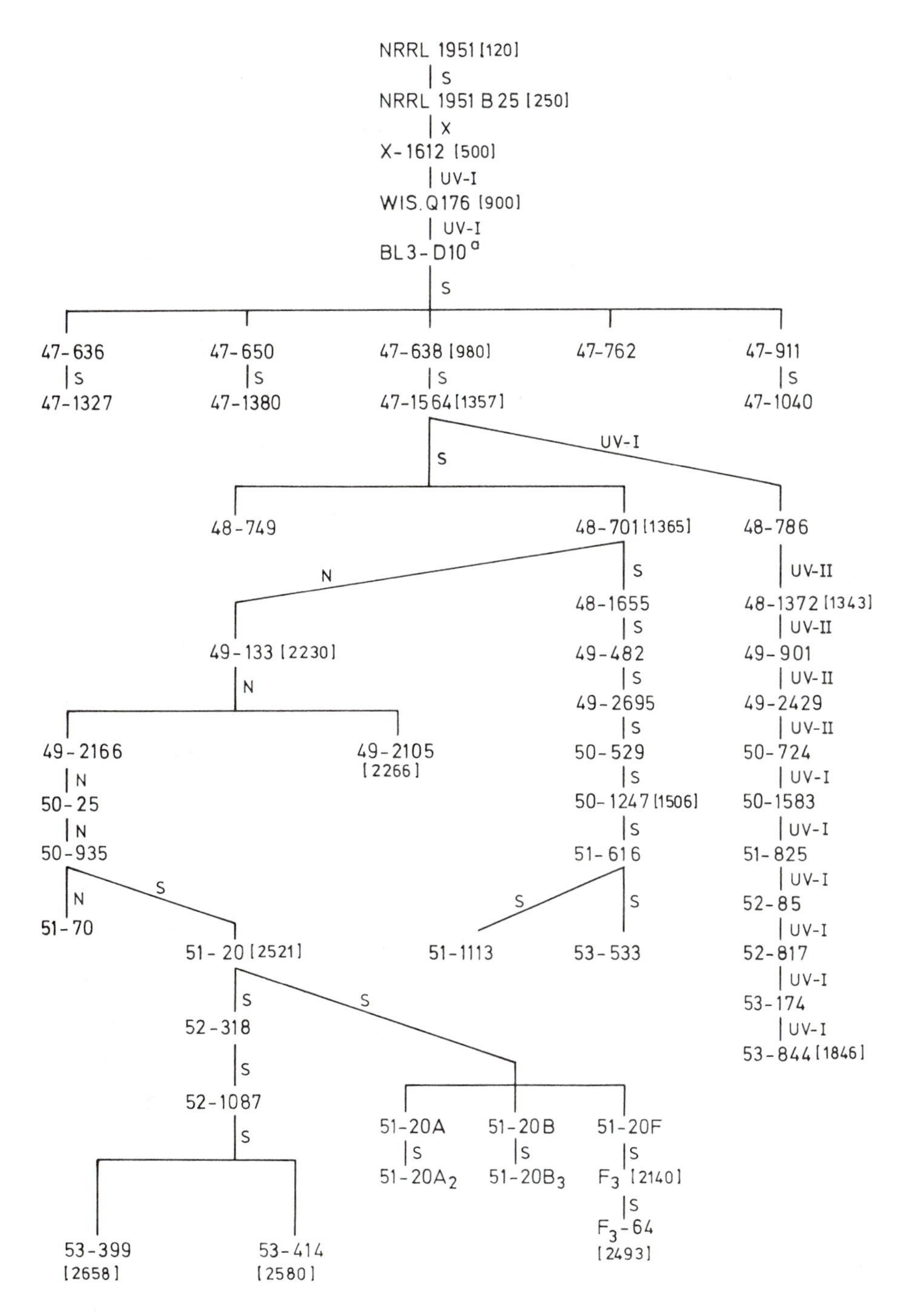

Figure 13.4 Genealogy of the Wisconsin strain of *Penicillium chrysogenum*. S, stages of selection; X, X-ray treatment; UV I, ultraviolet radiation at 275 nm; UV II, ultraviolet radiation at 253 nm; N, treatment with nitrogen mustard. Square brackets show yields in International units/ml. a = pigment-free mutant (From Backus and Stauffer, 1955)

Table 13.7 Increase of penicillin formation in *Penicillium chrysogenum*.

Strain	Penicillin G potassium (mg/ml)	Penicillin yield in glucose equivalents (g Pen.G-K/ g gluc. equiv.) (a)	Productivity P (b)
P- 2	9.0	0.05	0.72
P- 7	16.1	0.09	1.8
P-11	21.6	0.12	2.3
P-13	27.0	0.09	2.6
P-15	29.4	0.12	3.2

(a): (kg fat/oils × 2.5) = kg glucose equivalents. Carbon sources in medium: sugar and plant oils

(b): kg product/1000 l total fermenter capacity × day

Results of strain development and culture medium optimization at Panlabs Inc. (Swartz, 1979)

Yield increases have been the main objective of strain development, but other factors which have an effect on fermentation and efficiency of product recovery have also been optimized. Major advances in processing have resulted through empirical screening of mutants. Until the mid-1960's, the most frequently used mutagens were X-rays, methylbis-(β-chloroethyl)amine (nitrogen mustard), and short-wave ultraviolet radiation. More recently, nitrosoguanidine, alkylating agents, and nitrite have been used as mutagens.

In the early 1970's, strain improvement by using mere mutation had reached its limit. The discovery of a parasexual cycle in *P. chrysogenum* provided a means of utilizing genetic recombination for strain development. Heterozygous diploids were described which had penicillin titers above those of the haploid parent strains. Several such diploids have found use as production strains. However, the highest percentage of strains with increased penicillin production arose from haploid segregants of crosses between mutants of different strain lines.

The protoplast fusion technique has opened up a new approach to the development of high-yielding strains. Yield increases of 8% have been obtained over those obtained by mutation and selection. In addition, some of the resulting strains have had better growth characteristics, such as more stable sporulation and improved growth for inoculum production.

Some of the genes involved in penicillin synthesis in *P. chrysogenum* have been identified by analysis of mutants blocked in penicillin synthesis (*npe*). Twelve *npe* mutants have been mapped into 5 loci (*npe V, W, X, Y,* and *Z*) by means of complementation experiments with heterozygous diploids. Cosynthesis studies showed a defect in the LLD-tripeptide synthesis for *npe X, Y,* and *Z*, a block between the tripeptide and isopenicillin N in *npe W*, and a block between isopenicillin N and penicillin G in *npe V*. The function of *npe Y* was elegantly demonstrated as follows: Protoplasts of the various mutants were fused with liposomes into which the tripeptide had been preloaded, and penicillin synthesis was obtained.

The use of recombinant DNA technology to increase the formation of rate-limiting enzymes via gene amplification or improved transcription has not yet been possible in *P. chrysogenum*, due to the absence of precise biosynthetic data and the absence of good host-vector systems. However, a gene bank for *P. chrysogenum* has been constructed, and a transformation system has been developed.

Production methods Penicillin G and V are produced using submerged processes in 40,000–200,000 liter fermenters. Due to difficulties with the O_2 supply, larger tanks cannot be employed. Penicillin fermentation is an aerobic process with a volumetric oxygen absorption rate of 0.4–0.8 mM/l·min. The required aeration rate is between 0.5–1.0 vvm depending on the strain, on the bioreactor, and on the impeller system; various turbine impellers are used for mixing (120–150 rpm). Some manufacturers use Waldhof fermenters or air-lift fermenters, but this is only possible in mutants which generate low viscosity. Depending on the production strain used, the optimal temperature range is between 25–27°C.

A typical **flow chart** for penicillin production is shown in Figure 13.5.

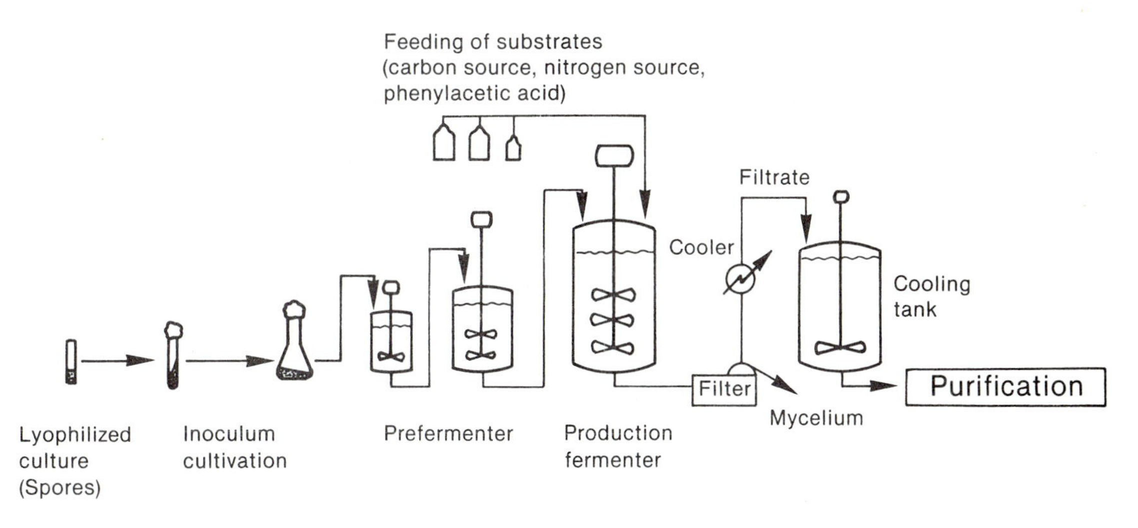

Figure 13.5 Flow chart of the penicillin fermentation (From Swartz, 1979)

The **inoculum** is started using lyophilized spores. Because of the great variability of high-yielding strains, careful strain maintenance is necessary. Spore concentration (optimal 5×10^3/ml) and pellet formation are crucial for the subsequent yield. If an optimal penicillin formation rate is to be achieved, pellets must grow not as compact balls, but in a loose form.

After several stages of growth the production culture is ready. In the typical penicillin fermentation (Figure 13.6), there is a **growth phase** of about 40 hours duration, with a doubling time of 6 hours, during which time the greatest part of the cell mass is formed. The **oxygen supply** in the growing culture is critical, since the increasing viscosity hinders oxygen transfer. This can be resolved by engineering changes in the fermenter or by the use of mutants which develop reduced viscosity. After the growth phase the culture proceeds to the actual **penicillin production phase**. In cases of high penicillin production, growth is sharply reduced ($\mu = 0.01$ h^{-1}). By feeding with various culture medium components, the production phase can be extended to 120–180 hours.

The **medium** of a typical fed-batch culture may vary depending on the strain and usually

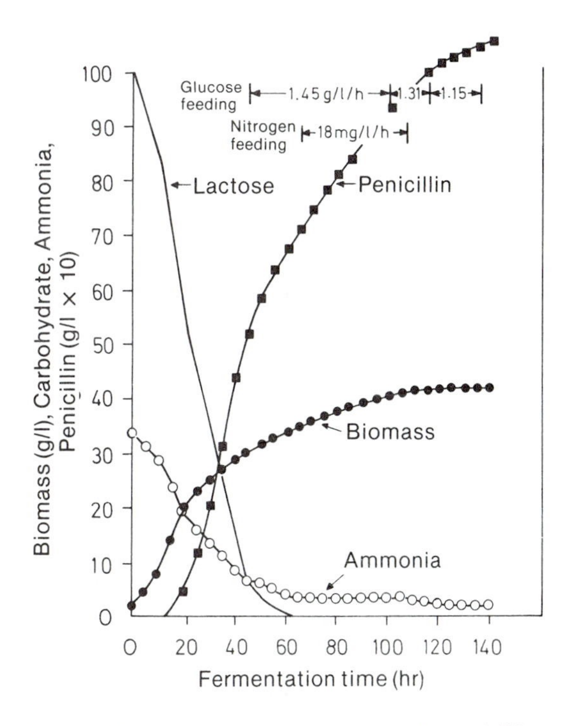

Figure 13.6 Penicillin fermentation with *Penicillium chrysogenum* (From Swartz, 1979)

consists of: corn steep liquor (4–5%, dry weight), which in present processes may be replaced by other nitrogen sources (e.g., a commercial nitrogen source called Pharmamedia); an additional nitrogen source, such as soy meal, yeast extract, or whey; a carbon source (such as lactose); and various buffers. The pH is kept constant at 6.5. Phenylacetic acid or phenoxyacetic acid is fed continuously as a precursor (0.5–0.8% of the total). Processes with glucose or molasses feeding are also successful: feeding rates of 1.0–2.5 $kg \cdot m^{-3} h^{-1}$ with a glucose concentration of 500 $kg \cdot m^{-3}$. About 65% of the metabolized carbon source is used for maintenance energy, 25% for growth, and only 10% for penicillin production. When it is considered that 50% of the production cost derives from the medium ingredients, it is obvious that control of carbon metabolism offers an important point of attack for further optimization of the production process. Yield increases of 25% have been reported by adding glucose and acetic acid. Critical parameters in these fed-batch systems are the rate of sugar utilization and the oxygen supply.

Production with continuous penicillin fermentation has been attempted, but has been difficult due to the instability of the production strains. A "batch fill and draw" system has been suggested as an alternative. In this procedure, 20–40% of the fermentation contents is drawn off and replaced with fresh nutrient solution; this process may be repeated up to 10 times without yield reductions.

Extensive research is being carried out to produce penicillin with immobilized cells. In one laboratory-scale study, the advantage of this approach over the use of a fed-batch system was demonstrated, but this technique has not as yet been introduced commercially.

Penicillin is excreted into the medium and less than 1% remains mycelium-bound. After separation of the mycelium, product recovery is accomplished by means of a two-stage continuous countercurrent extraction of the fermenter broth with amyl or butyl acetate at 0–3°C and pH 2.5–3.0. The yield is around 90%.

Cephalosporins

Cephalosporins are β-lactam antibiotics containing a dihydrothiazine ring with D-α-aminoadipic acid as acyl moiety (Figure 13.7). Cephalosporin C was discovered in culture filtrates of *Cephalosporium acremonium* in 1953. The strain isolated by Brotzu in 1945 was later classified as *Acremonium chrysogenum* and produces several antibiotics: cephalosporin C, penicillin N (a 6-APA derivative with D-α-AAA as side chain), and the steroid cephalosporins P_1–P_5. The cephalosporin antibiotics are also produced by other fungi, such as *Emericellopsis* and *Paecilomyces*. In 1971, in a screening program designed to discover β-lactam antibiotics, the first *cephamycins* (7-methoxy-cephalosporins) were found, which are produced

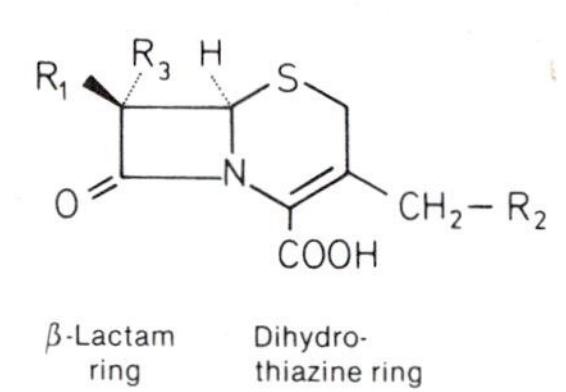

Designation	R_1	R_2	R_3
7-Aminocephalo-sporanic acid (7-ACA)	$-NH_3^+$	$-O-CO-CH_3$	-H
NATURAL CEPHALOSPORINS			
Cephalosporin C	$-CO-(CH_2)_3-(D)CH(NH_3^+)-COO^-$	$-O-CO-CH_3$	-H
Deacetyl-3'-carbamoyl-cephalosporin C		$-O-CO-NH_2$	-H
7-Methoxy-cephalosporin C		$-O-CO-CH_3$	$-OCH_3$
Cephamycin A		$-OCO-C=CH-C_6H_4-O-SO_3H$	$-OCH_3$
Cephamycin B		$-OCO-C(O-CH_3)=CH-C_6H_4-OH$	$-OCH_3$
Cephamycin C		$-O-CO-NH_2$	$-OCH_3$
SEMI-SYNTHETIC CEPHALOSPORINS			
Cephalotin	$C_4H_3S-CH_2CO-HN-$	$-O-CO-CH_3$	-H
Cephalexin	$C_6H_5-CH(NH_2)-CO-NH-$	H	-H

Figure 13.7 Structure of cephalosporin C, cephamycins, and several semi-synthetic cephalosporins

from various *Streptomyces* species, such as *S. lipmanii, S. clavuligerus,* or *Nocardia lactamdurans.*

Cephalosporins are valued not only because of their low toxicity but also because they are broad-spectrum antibiotics, comparable in action to ampicillin. With about 29% of the antibiotic market, the cephalosporins are the single most important group of antibiotics. Therapeutically only semisynthetic derivatives of cephalosporin and cephamycin are used. For oral use, the first generation cephalosporins have been modified so that they have improved stability against β-lactamases. These second generation antibiotics were then further modified so that they were more active against gram-negative bacteria. Further modifications have led to the commercial development of third generation antibiotics with excellent β-lactamase stability and a broadened action spectrum.

Thirteen therapeutically important semisynthetic cephalosporins are commercially produced. These have been synthesized by chemical splitting to form 7-aminocephalosporanic acid (7-ACA) with subsequent chemical acylation, as well as by modifications on the C-3 site. Cephalosporins are as economically significant as penicillins.

Biosynthesis and regulation As in the case of benzylpenicillin, the first stages of cephalosporin biosynthesis (Figure 13.8) proceed from δ-(α-aminoadipyl)-L-cysteinyl-D-valine to isopenicillin N. In the next stage, penicillin N is produced by transformation of the L-α-AAA side chain into the D-form, via the action of a very labile racemase. After ring expansion to deacetoxycephalosporin C by the so-called "expandase" reaction, hydroxylation via a dioxygenase to deacetylcephalosporin C occurs. The acetylation of cephalosporin C by an acetyl-CoA dependent transferase is the end point of the biosynthetic pathway in fungi. In streptomycetes, further transformations occur: cephalosporin C or the carbamoyl derivative of deacetylcephalosporin C is converted in a two-step reaction with molecular oxygen and S-adenosylmethionine to 7-methoxycephalosporin or cephamycin C. In contrast to penicillin, the D-α-AAA moiety cannot be changed with precursor feeding.

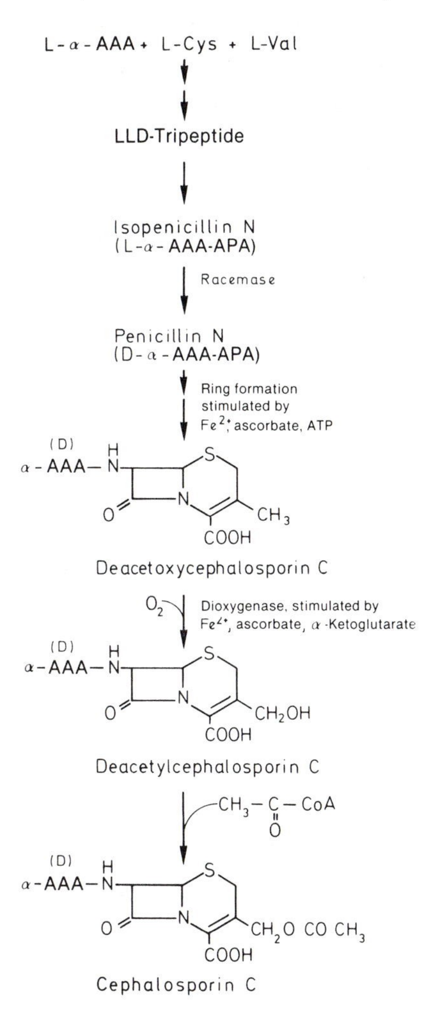

Figure 13.8 Biosynthesis of cephalosporin C by *Cephalosporium acremonium*

Cephalosporin production is affected by phosphate regulation as well as by nitrogen and carbohydrate catabolite regulation. Rapidly metabolizable carbon sources, such as glucose, maltose, or glycerol, reduce the formation of cephalosporin. The repression of the "Expandase" seems to be the most significant effect. The addition of lysine in low concentrations promotes cephalosporin production (Figure 13.9), whereas the inhibition caused by higher amounts of lysine can be overcome by addition of DL-α-AAA. Methionine stimulates cephalosporin synthesis in *C. acremonium*, but not in the streptomycetes.

In the biosynthesis of cephamycins by streptomycetes, it should be noted that α-AAA is not a precursor of L-lysine, because streptomycetes synthesize lysine via the diaminopimelic acid pathway. L-lysine is converted by an L-lysine α-ketoglutarate-6-aminotransferase into α-AAA via L-1-piperidine-6-carboxylate as intermediate.

Strain development Strain development has been conducted with the original Brotzu strain, *C. acremonium* CMI 49 137, using mutation, selection, and parasexual techniques. Mutants have also been isolated with defective sulfur metabolism. Some of these regulatory mutants, which were isolated because they were resistant to sulfur analogs, exhibit an increased cephalosporin production or utilize inorganic sulfate as well as methionine. Parasexual recombination with mycelium of *C. acremonium* is difficult because the compartments of the hyphae are almost always mononuclear. By using protoplasts of different strains in fusion experiments, haploid recombinants have been isolated which showed a 40% increase in cephalosporin C production, as well as increased growth and improved sporulation. Protoplast fusion techniques have been combined with parasexual reproduction to develop an overall map of the biosynthetic genes for cephalosporin C. The gene for isopenicillin N synthetase from *C. acremonium* has been cloned in *E. coli* and an effective transformation system for *C. acremonium* has been developed, making it likely that the key steps in cephalosporin production will become subject to analysis and modification.

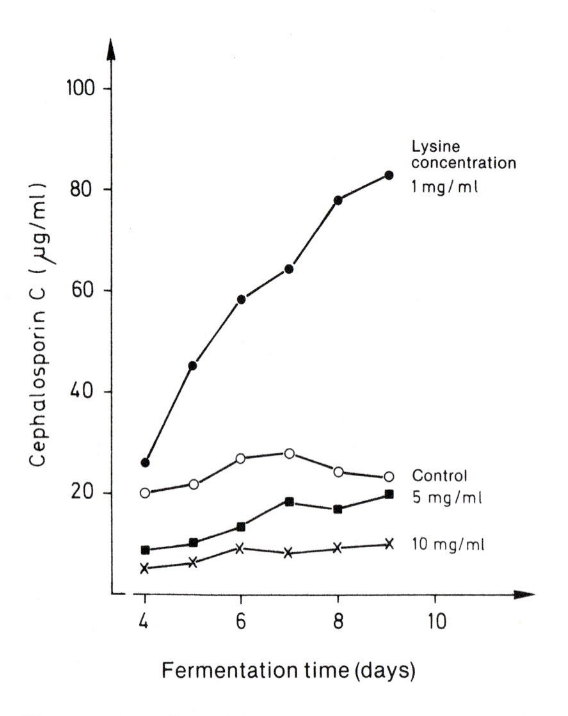

Figure 13.9 Effect of lysine concentration on cephalosporin C formation in *Cephalosporium acremonium* (From Mehta et al., 1979)

The cephalosporin C titer of contemporary high-yielding strains, in which the proportion of penicillin N is reduced, is about 20 g/l.

Production method The fermentation of cephalosporin is similar to that of penicillin. Complex media with corn steep liquor, meat meal, sucrose, glucose and ammonium acetate are used. Fermentations are carried out at as fed-batch processes with semicontinuous addition of nutrients at pH 6.0–7.0 and at temperatures between 24–28°C. In the main growth phase, a high aeration rate is necessary, but during the production phase (48–160 h), O_2 consumption decreases sharply.

The chemical synthesis of cephalosporin by ring-expansion of penicillin has been developed in recent years. This process has become especially attractive because of the low price of the penicillin starting material. An example of how this approach has been commercialized is the use of pencillin V to produce oraspor, an orally active cephalosporin.

New β-lactam ring systems

New β-lactam ring systems have been discovered since the 1970's by the introduction of screening methods such as using as test organisms β-lactam hypersensitive strains or using enzymatic tests to detect β-lactam inhibitors. The basic structures of these new β-lactams are shown in Figure 13.1, and some specific examples are given in Figure 13.10. Because they may be active as either broad-spectrum antibiotics or β-lactamase inhibitors, these new compounds, especially their semisynthetic derivatives, can be expected to find considerable use in chemotherapy.

Clavulanic acid, a β-lactam-oxazolidine ring system, is produced along with cephalosporins by *Streptomyces clavuligerus*. Clavulanic acid is only slightly effective antibacterially but inhibits a wide range of β-lactamases in concentrations under 0.1 μg/ml. The inhibition is irreversible; in combination with β-lactamase-sensitive penicillins and cephalosporins, this compound causes a distinct increase in activity of these antibiotics, even against bacteria which are normally resistant to β-lactam antibiotics. Clavulanic acid has been marketed in combination with amoxycillin.

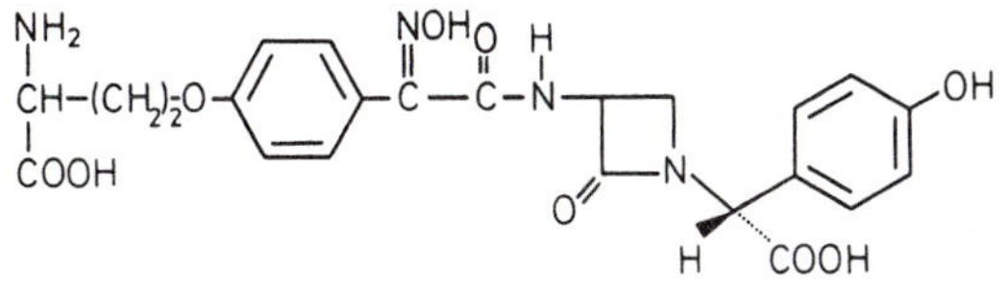

Thienamycins and olivanic acids have a β-lactam-pyrroline ring system in common. Thienamycin, the first in a series of carbapenems with very similar structures, was isolated in 1976 from *Streptomyces cattleya*. Thienamycin acts as a β-lactamase inhibitor and as a broad-spectrum antibiotic against Gram-positive and Gram-negative bacteria, including *Pseudomonas aeruginosa*, but it is extremely unstable.

The semisynthetic N-formimidoylthienamycin, resistant to dehydropeptidases, has good stability. It is the β-lactam with the broadest spectrum and is undergoing clinical trials.

The **olivanic acids** and the closely related epithienamycins show a lower activity against *Pseudomonas aeruginosa* and *Staphylococcus aureus*.

Nocardicins are monocyclic β-lactams. From a mixture of seven components, nocardicin A has been isolated as the most active compound. It is weakly active against gram-negative bacteria, but research is being carried out to improve its activity by formation of chemical derivatives.

Monobactams, which are also monocylcic β-lactams, have been isolated from bacterial cultures. However, they are only weakly antibacterial. Compounds with a 3α-methoxy group are resistant to β-lactamases, whereas the rest of the monobactams show varying degrees of sensitivity to β-lactamases. Azthreonam, a semisynthetic monobactam made from threonine, has low toxicity and excellent activity against gram-negative pathogens; it is in clinical trials.

Nocardicin A

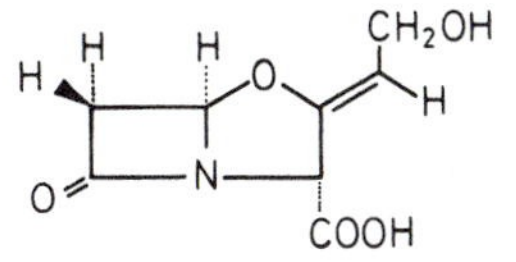

Clavulanic acid

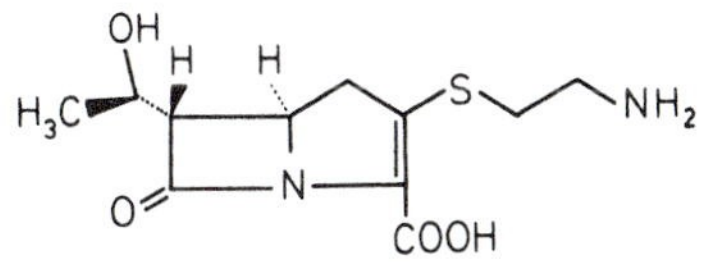

Thienamycin

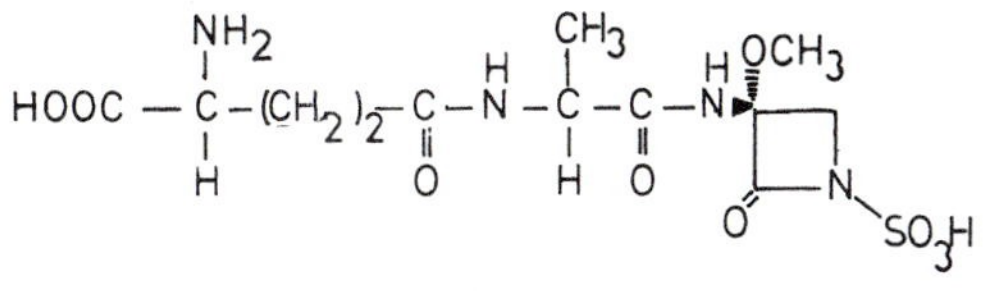

Sulfazecin

Figure 13.10 Newer β-lactam antibiotics

13.3 AMINO ACID AND PEPTIDE ANTIBIOTICS

This group of antibiotics includes amino acid derivatives such as cycloserine and azaserine, most of the β-lactam antibiotics, which have already been covered in Section 13.2, chromopeptide antibiotics, depsipeptide antibiotics, and linear or cyclic peptide antibiotics. Over 400 peptide antibiotics are currently known.

D-cycloserine

Cycloserine, a D-4-amino-3-isoxazolidone (Figure 13.11 A), may be synthetically produced or may be isolated from cultures of *Streptomyces orchidaceus, S. lavendulae, S. garyphalus,* or *S. roseochromogenes.* D-Cycloserine, a D-alanine analog, inhibits cell wall synthesis by inhibiting alanine racemase, the enzyme responsible for production of the D-alanine moiety of the peptidoglycan. Cycloserine is very effective against mycobacteria, particularly *M. tuberculosis,* and is used in the treatment of tuberculosis in combination with isonicotinic acid hydrazide (INH) and rifampicin or streptomycin.

A. Cycloserine

(MeVal = N-Methylvaline, Sar = Sarcosine)

Actinomycin	X_1	X_2
Actinomycin C_1	D-Val	D-Val
Actinomycin C_2	D-Val	D-α-Ileu
Actinomycin i-C_2	D-α-Ileu	D-Val
Actinomycin C_3	D-α-Ileu	D-α-Ileu

B. Actinomycins

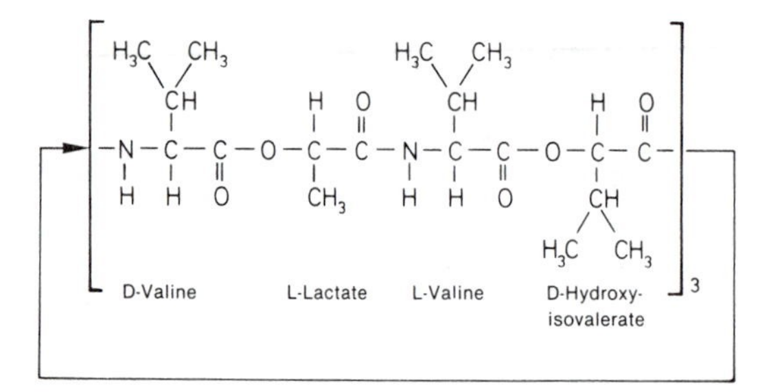

C. Valinomycin

Figure 13.11 Structure of D-cycloserine (A), several actinomycins (B) and valinomycin (C)

Chromopeptide antibiotics

This group includes the actinomycins (Figure 13.11 B), which are produced as a mixture of very similar substances by different streptomycetes, especially *Streptomyces antibioticus* and *S. chrysomallus.* The following actinomycins have been described: A, B, C_1, C_2, i-C_2, C_3, D, F_1, Z_0–Z_5. All actinomycins which have thus far been isolated have the same phenoxazone-chromophore (actinocin), which is linked to two pentapeptide lactones. These two lactones have varied amino acid sequences, and by adding different amino acids to the fermentation broth, 30 distinct biosynthetic actinomycins have been produced.

The production of actinomycin from 2 molecules of 3-hydroxy-4-methylanthranilic acid pentapeptide lactone is catalyzed by phenoxazine synthase (Figure 13.12). This enzyme is produced only after the transition into the idiophase and shows pronounced catabolite regulation by glucose.

The phenoxazone rings of actinomycins intercalate at a GC pair in 5′-T-G-C-A-3′ palindromes of DNA, thus blocking DNA-dependent RNA-polymerases. Actinomycins are very toxic, causing liver and kidney damage, but are beneficial in tumor treatment. The most important actinomycin used in cancer chemotherapy is dactinomycin (actinomycin C_1).

Depsipeptide antibiotics

In depsipeptides, the subunits (amino acids, hydroxy acids) are usually linked with alternating ester and acid amide bonds. **Valinomycin** (Figure 13.11 C), a cyclododecadepsipeptide produced by *Streptomyces fulvissimus,* has no eco-

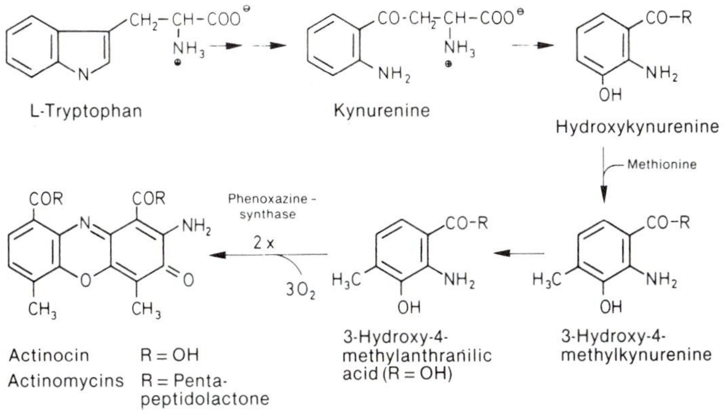

Figure 13.12 Actinomycin biosynthesis in *Streptomyces antibioticus*

nomic significance but finds wide use in biochemical reasearch as an agent for uncoupling oxidative phosphorylation. It is also useful as a potassium carrier because of its selective binding of K^+ ions.

Antibiotics of the **virginiamycin** family also have a depsipeptide structure. They consist of a mixture of structurally different macrocyclic lactone rings (type A and B). Figure 13.13 shows the structure of virginiamycin S (type B).

Antibiotics of the virginiamycin group are chiefly effective against Gram-positive bacteria and are widely used as growth promoters for poultry, swine, and calves. Seventeen compounds are commercially produced; the most important are the mikamycins and virginiamycins, which are produced by *Streptomyces mitakaensis* or *S. virginiae*.

Linear and cyclic peptide antibiotics

The majority of the antibiotics formed by species of the genus *Bacillus* are peptides; other peptide antibiotics are produced by streptomycetes. The molecular weights of peptide antibiotics are between 270 (bacilysin) and 4500 (licheniformin). Some peptide antibiotics are linear (Figure 13.14), but most are cyclic compounds. They may consist of either amino acids alone (often with D-amino acids and rare amino acids) or they may contain other components in addition to the amino acids (Figure 13.15). The peptide antibiotics produced by bacilli are formed at the onset of sporulation and seem to play a regulatory role in the sporulation process. Table 13.8 lists the peptide antibiotics which are produced commercially. Bialophos, a tripeptide formed by *S. hygroscopicus* from two L-alanine and the L-glu-

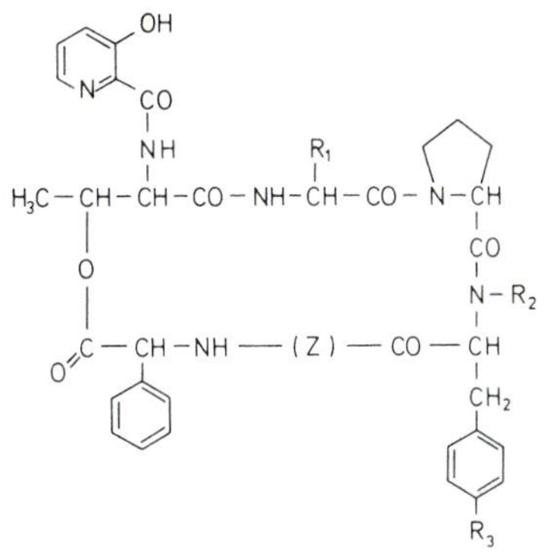

	R_1	R_2	R_3	Z
Virginiamycin S_1	C_2H_5	CH_3	H	4-Ketopipecolinic acid
Virginiamycin S_4	CH_3	CH_3	H	4-Ketopipecolinic acid
Virginiamycin S_2	C_2H_5	H	H	4-Hydroxy-pipecolin acid
Virginiamycin S_3	C_2H_5	CH_3	H	3-Hydroxy-4-keto-pipecolinic acid

Figure 13.13 Structures of the virginiamycin antibiotics

HCO–X–Gly–L-Ala–D-Leu–L-Ala–D-Val–L-Val–D-Val–L-Trp —
1 2 3 4 5 6 7 8 9

—D-Leu–Y—D-Leu–L-Trp—D-Leu–L-Trp–$NHCH_2CH_2OH$
10 11 12 13 14 15

Gramicidin

	X	Y
Valine-Gramicidin A	L-Val	L-Trp
Isoleucine-Gramicidin A	L-Ileu	L-Trp
Valine-Gramicidin B	L-Val	L-Phe
Isoleucine-Gramicidin B	L-Ileu	L-Phe
Valine-Gramicidin C	L-Val	L-Tyr
Isoleucine-Gramicidin C	L-Ileu	L-Tyr

β-Tyr–β-Ser — DAPA — DAHAA– Gly – Spermidine

Edein A

Figure 13.14 Linear peptide antibiotics. Above, gramicidins. Below, edein A (edein B: guanylspermidine). DAPS, Diaminopropionic acid; DAHAS, Diaminohydroxyazelain acid

tamic acid analog phosphinothricin, is being field-tested as an herbicide.

Therapeutic application of peptide antibiotics is limited, due to their toxicity. Gramicidin S, linear gramicidins, tyrocidin, and bacitracin are used topically (for wounds, burns, skin transplants, etc.). Polymyxins are used for infections caused by Gram-negative pathogens (including *Pseudomonas*), and viomycin and capreomycin are used in tuberculosis therapy.

Bacitracin

Bacitracin is the most economically important of the peptide antibiotics. In addition to its use as a topical antibiotic, it is also used as a growth promoter in animal feeds. As a feed supplement, it can be used as the zinc or manganese salt, as the lignin-bacitracin complex, or as bacitracin methylene disalicylate. Total production of bacitracin world-wide was estimated at 500 tons in 1976.

Bacitracin structure Various strains of *B. licheniformis* produce a mixture of bacitracin A, A′, B, C, D, E, F, F_1, F_2, F_3 and G. Bacitracin A is the main component of the mixture (about 70%) and has the greatest biological activity. It is a dodecapeptide consisting of 8 L- and 4 D-amino acids and contains both a cyclic hexapeptide and a thiazoline ring structure (see Figure 13.15).

Mechanism of action Bacitracin inhibits the formation of the bacterial cell wall at the level of peptidoglycan biosynthesis. It prevents the dephosphorylation of C_{55}-isoprenylpyrophosphate to C_{55}-isoprenylphosphate.

Biosynthesis and Regulation Peptide formation can occur by either ribosomal or nonribosomal

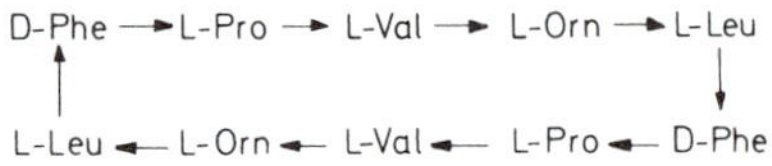

D-Phe → L-Pro → L-Val → L-Orn → L-Leu
L-Leu ← L-Orn ← L-Val ← L-Pro ← D-Phe

Gramicidin S

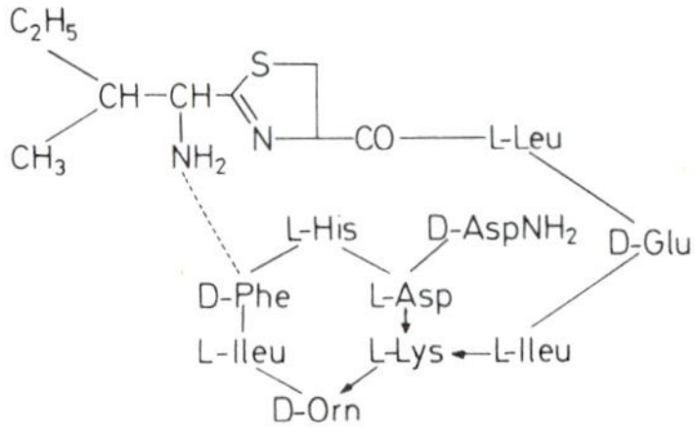

C_2H_5
CH–CH
CH_3
NH_2
S
N
CO — L-Leu
L-His
D-AspNH$_2$
D-Glu
D-Phe
L-Asp
L-Ileu
L-Lys ← L-Ileu
D-Orn

Bacitracin A

NH$_2$
L-DAB → Ⓧ → L-Leu
Ⓡ → L-DAB → L-Thr → L-DAB → L-DAB
NH$_2$ NH$_2$
NH$_2$
L-Thr ← L-DAB ← L-DAB–NH$_2$

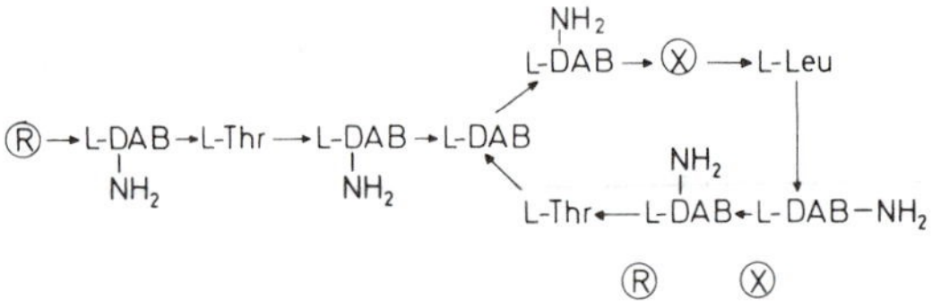

	Ⓡ	Ⓧ
Polymyxin B_1	MOS	D-Phe
Polymyxin B_2	IOS	D-Phe
Polymyxin E_1 (Colistin A)	MOS	D-Leu
Polymyxin E_2 (Colistin B)	IOS	D-Leu

Polymyxins

Figure 13.15 Cyclic peptide antibiotics: gramicidin S, bacitracin A, and the polymyxins.
DAB, Diaminobutyric acid; MOS, 6-Methyloctanoic acid; IOS, Isooctanoic acid

Table 13.8 Commercially produced peptide antibiotics

Antibiotic	Organism producing	Activity
Application: Human Therapy		
Amphomycin	*Streptomyces canus*	G^+
Bacitracin	*Bacillus licheniformis*	$G^+(G^-)$
Capreomycin	*S. capreolus*	My
Gramicidin A	*B. brevis*	G^+
Gramicidin S (J)	*B. brevis*	G^+
Iturin A	*B. subtilis-ituriensiens*	G^+G^-AF
Polymyxin B	*B. polymyxa*	G^-
Tyrothricin (bacitracin-tyrocidin complex)	*B. brevis*	G^+
Tyrocidin	*B. brevis*	G^+
Viomycin	*S. floridae*	My
Application: Feed additive		
Bacitracin	*B. licheniformis*	$G^+(G^-)$
Enduracidin A	*S. fungicidicus*	G^+My
Parvulin	*S. parvulus*	G^+
Siomycin	*S. sioyaensis*	G^+
Thiopeptin	*S. tateyamensis*	G^+
Thiostrepton	*S. azureus*	G^+
Application: Food preservative		
Nisin	*Streptococcus cremoris*	G^+

G^+: Gram-positive; G^-: Gram-negative bacteria; AF: antifungal; My: mycobacteria.
(Perlman, 1979; Kleinkauf and von Döhren, 1985)

mechanisms. The biosynthesis of high molecular-weight antibiotics such as nisin and neocarzinostatin involves ribosomes, mRNA, and tRNA. However, the biosynthesis of most peptide antibiotics does not involve this protein-synthesizing machinery, but takes place instead on a multi-enzyme complex. This nonribosomal peptide synthesis process, which has also been called the "thiotemplate" mechanism, has been described for gramicidin S, tyrocidin, and the linear gramicidins, as well as for bacitracin.

Bacitracin synthetase consists of subunits A, B and C, with molecular weights of 200,000, 210,000 and 380,000, respectively. A simplified version of the process of bacitracin synthesis is given in Figure 13.16. The first step, the activation of amino acids in the presence of ATP and Mg^{2+} to aminoacyl adenylates, takes place at a specific activation site for each amino acid on the enzyme complex. In a transfer reaction, the activated amino acids are then bound to specific thiol groups in thioester linkages. The amino acid sequence of the subsequent peptide is determined by the arrangement of the thiol groups on the enzyme. The racemization to the D-enantiomers probably takes place at this level. The transfer of the growing peptide chain (growth from the N-terminal end toward the C-terminal end) to the next amino acid and to the next enzyme sub-unit is catalyzed by 4′-phosphopantetheine, a cofactor of each sub-unit. The cyclization of the peptide chain takes place in the multi-enzyme complex, but the formation of the thiazoline ring is not yet fully understood.

Production methods Bacitracin was initially obtained from surface cultures; today its production occurs in aerobic submerged culture, like that of all other antibiotic processes. The production process for bacitracin is shown in Figure 13.17.

The yields of this process were increased from about 13 mg/l to about 9 g/l through strain development and medium optimization. A continuous process has also been developed for bacitracin production using immobilized cells.

Chelate-forming peptide antibiotics

Siderochromes and bleomycins are peptide antibiotics which form chelates with metals. **Siderochromes** are iron-containing or iron-binding natural substances containing hydroxamic acid groups, and bring about the formation of ferrihydroxamate complexes (Figure 13.18). Albomycins and ferrimycins have been isolated as antibiotically active siderochromes (sideromycins), but there is no known industrial application for these substances. Sideramines are siderochromes which act as growth stimulants, and ferrioxamine B, a sideramine from *Streptomyces pilosus* ETH 21748, is commercially produced. It is used in human therapy as the desferri compound for the treatment of diseases involving iron storage and also in agriculture in such applications as the elimination of soil iron deficiencies in peach orchards.

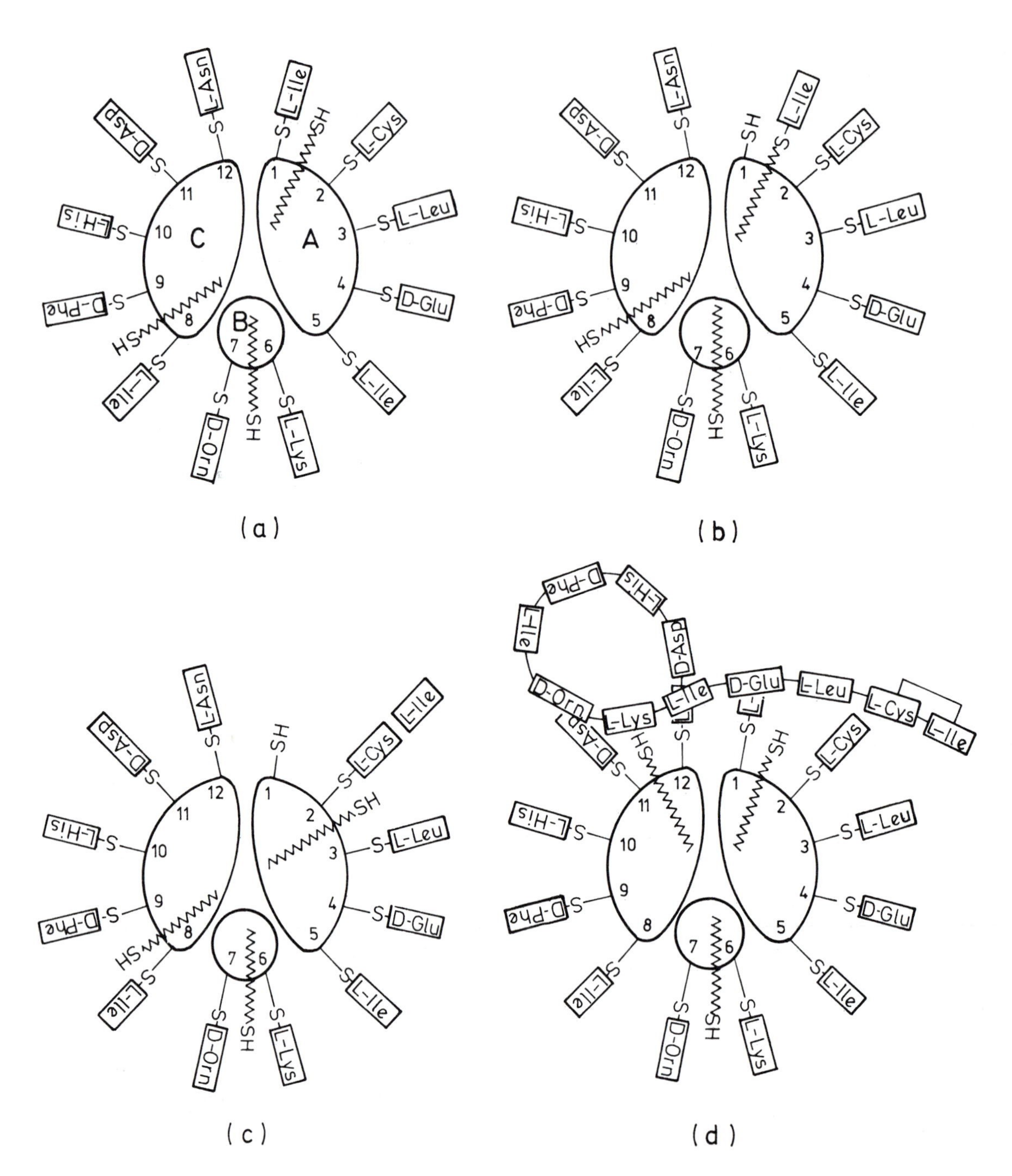

Figure 13.16 Model of nonribosomal peptide synthesis according to the "thiotemplate" model for the formation of bacitracin through the action of bacitracin synthetase in *B. licheniformis* (From Zimmer et al., 1979)

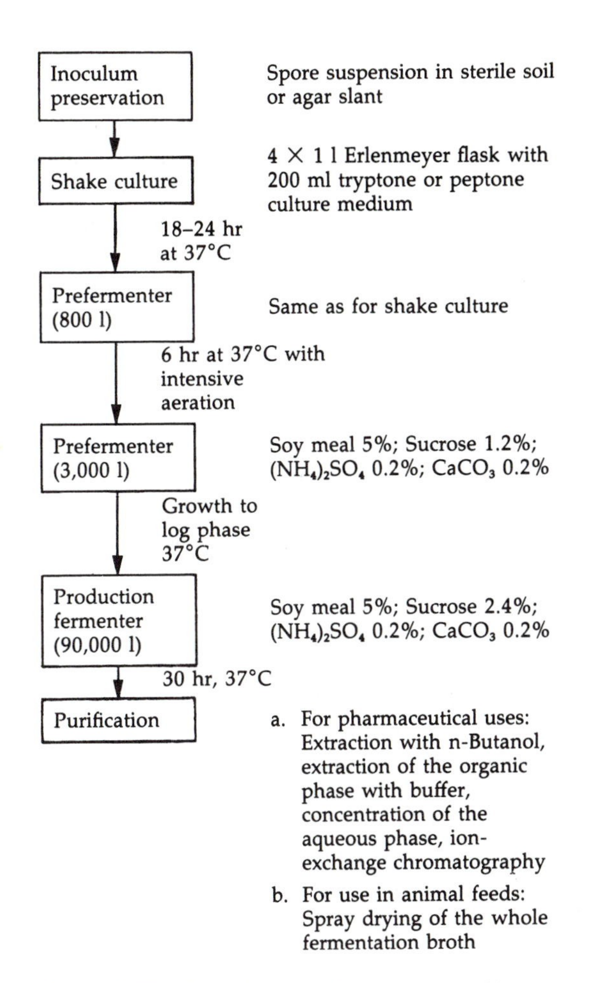

Figure 13.17 Flow scheme for the production of bacitracin by *B. licheniformis* (From Perlman, 1979)

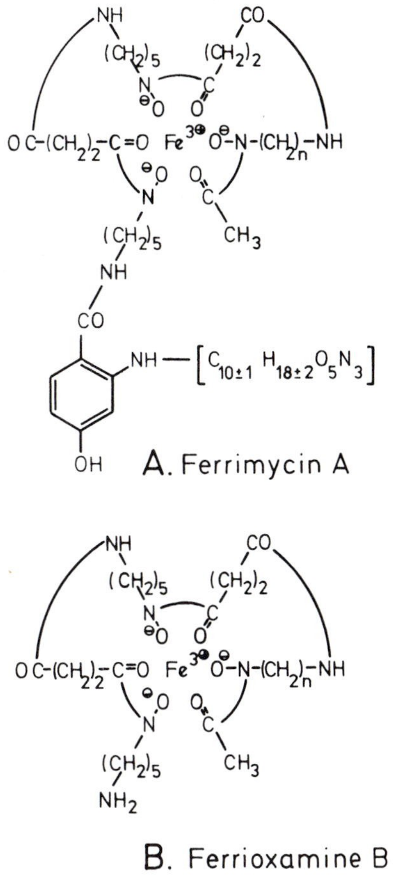

A. Ferrimycin A

B. Ferrioxamine B

Figure 13.18 Structure of siderochromes

Bleomycins are isolated as Cu(II) complexes from the culture broth of *Streptomyces verticillus.* The glycopeptide exhibits some antitumor action and is marketed as a mixture of bleomycin A_2 and B_2.

13.4 CARBOHYDRATE ANTIBIOTICS

Carbohydrate antibiotics include the therapeutically important aminoglycosides, the orthosomycins, and various sugar derivatives.

Glycosides and sugar derivatives

Several commercially important antibiotics from this heterogenous group are listed in Table 13.9.

Nojirimycin (Figure 13.19 A), shows antibacterial activity against *Sarcina lutea* and *Xanthomonas oryzae,* and is of interest because of its high inhibitory activity against α- and β-glucosidases and amylases.

Vancomycin, a broad-spectrum antibiotic, consists of the amino sugar vancosamine, two β-hydroxychlortyrosine moieties, three phenylglycine derivatives, N-methyl leucine, and aspartic acid. It is effective against Gram-positive organisms, particularly staphylococci. However, it has adverse side effects in humans and thus is used only to treat infections caused by pathogens which are resistant to other antibiotics or with patients who are hypersensitive to penicillin. No

Table 13.9 Carbohydrate antibiotics. Sugars, glycosides and other commercially important sugar derivatives

Group	Antibiotic	Organism producing
5-Amino sugar	Nojirimycin	*Streptomyces roseochromogenes* *S. lavendulae* *S. nojiriensis*
Glycopeptide	Vancomycin	*S. orientalis*
Sugar amide	Lincomycin	*S. lincolnensis* var. *lincolnensis*
Phospho-glycolipid	Moenomycin A,C,D,E,F,G,H (= Bambermycin)	*S. bambergiensis* *S. ederensis* *S. ghanaensis* *S. geysiriensis*

A

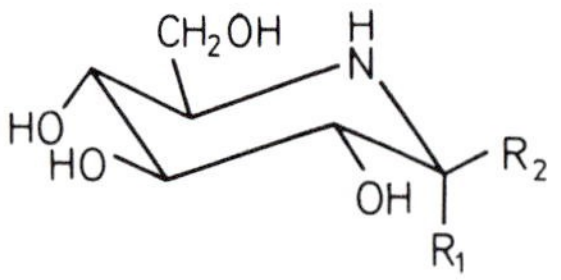

Nojirimycin

(5-Amino-5-deoxy-D-glucopyranose)

R_1 = OH; R_2 = H (60%)
R_1 = H; R_2 = OH (40%)

B

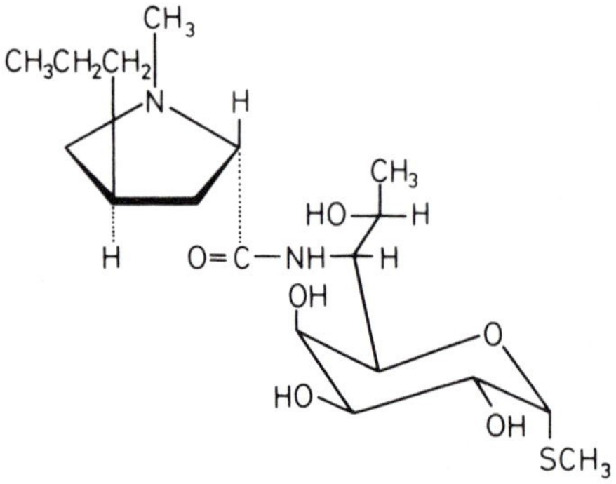

Lincomycin

Figure 13.19 Structure of nojirimycin and lincomycin

development of resistance to vancomycin has yet been observed in therapy.

Lincomycin (Figure 13.19 B) is used in the treatment of staphylococcus, streptococcus, and pneumococcus infections; only minor side effects are observed, but during therapy, resistance is acquired rapidly. Clindamycin (7-chlor-7-deoxy-lincomycin), a semisynthetic derivative with improved pharmacological properties, is principally marketed today. In protein synthesis, the bonding of aminoacyl-tRNA to the A-binding site on the ribosome is inhibited by lincomycin, probably through interaction with the ribosomal protein L6 of the 50S subunit.

Moenomycin is a phosphoglycolipid which shows activity against Gram-positive bacteria as well as Gram-negative bacteria containing a resistance plasmid. Moenomycin A and the related antibiotics macarbomycin and quebemycin are marketed for use as nutritional feed supplements. Phosphoglycolipids inhibit bacterial cell wall synthesis by affecting the transglycosylation reaction in peptidoglycan biosynthesis.

Orthosomycins

The orthosomycins are a class of antibiotics newly discovered in the late 1970's. They are characterized by an oligosaccharide structure and two noncarbohydrate ester groups (Figure 13.20). The following antibiotics have been described: the avilamycins (produced by *Streptomyces viridochromogenes*), the curamycins (*S. curacoi*), the everninomicins (*Micromonospora carbonaceae*), and flambamycin (*S. hygroscopicus*). The avilamycins are being tested for use as animal feed supplements.

Aminoglycoside antibiotics

Aminoglycosides are oligosaccharide antibiotics and consist of an aminocyclohexanol moiety (e.g. deoxystreptamine, streptidine), which is glycosidically linked to other amino sugars. Over 100 aminoglycosides are known, of which those produced industrially as well as some newer com-

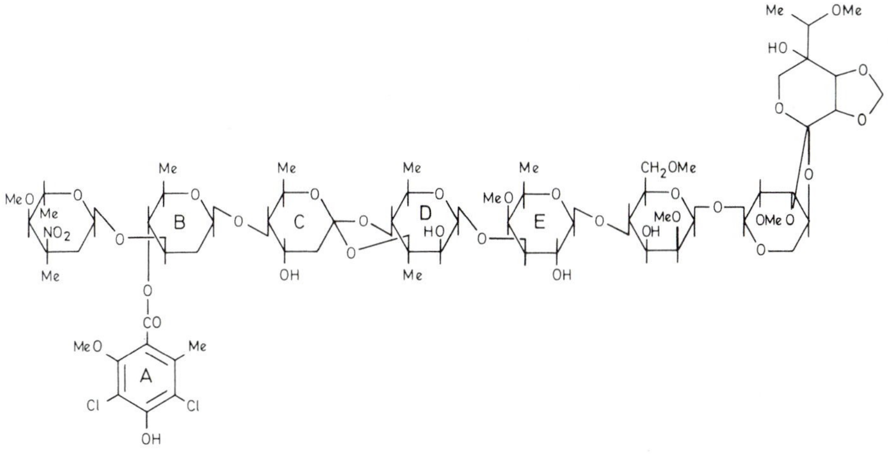

Figure 13.20 Everninomicin B

pounds are listed in Table 13.10; Figure 13.21 shows the most important structures.

Aminoglycosides are primarily used against Gram-negative bacteria in a wide range of applications. Although the semisynthetic cephalosporins are also now widely used for Gram-negative infections, in severe infections the aminoglycosides are often the antibiotics of choice. Although streptomycin and dihydrostreptomycin are broad-spectrum antibiotics, they are primarily used to treat tuberculosis. They are considered reserve antibiotics, mainly due to the fact that resistance can develop rapidly and in a single step by a chromosomal mutation.

In 1976 the world market for aminoglycosides (not including the Soviet Union and China) was estimated at $500 million, with a yearly increase of 10–15%.

All of the aminoglycoside antibiotics cause kidney damage and deafness (nephro- and ototoxicity) as side effects. In addition to resistance development by chromosomal mutation, plasmid-determined resistance due to the formation of inactivating enzymes is known (see Section 15.4). Compounds with reduced toxicity and/or increased effectiveness against resistant organisms are being sought via screening, mutasynthesis, and formation of chemical derivatives. As a result of a screening program, the new group of fortimicins was discovered. This is a group of pseudodisaccharides with a broad spectrum of activity, particularly against aminoglycoside-resistant bacteria. The kanamycin derivatives amikacin and dibekacin (1-N-L(−)-γ-amino-α-hydroxybutyrylkanamycin A and 3′,4′-dideoxykanamycin B), and netilmicin (1-N-ethylsisomicin) are semisynthetic compounds with partial resistance to inactivating enzymes. These compounds have already been introduced for clinical use. The compound 5-*epi*-sisomicin, isolated by mutasynthesis, has been tested clinically because of its action against sisomicin-resistant organisms. However, for cost reasons it is being produced chemically rather than by fermentation.

Besides the antibiotics noted above which are useful for human therapy, kasugamycin and validamycin are used as important antibiotics in rice cultivation.

Mode of action The mode of action of aminoglycosides is on protein synthesis in sensitive bacteria. Streptomycin, the aminoglycoside which has been studied the most, binds to protein S12 of the 30S ribosome and causes misreading of the code and hence inhibition of pro-

Table 13.10 Classification of the most important aminoglycoside antibiotics according to the aminocyclohexanol moiety. Production strain, spectrum of activity, and application are given for commercially produced antibiotics

Group	Antibiotic	Organism producing	Spectrum of activity[a]	Application
I Streptamine derivatives				
Streptidine	Streptomycin	*Streptomyces griseus*	G^+G^-My	Human therapy
	Dihydrostreptomycin	*S. humidus*[b]	G^+G^-My	Human therapy
Actinamine (N,N-dimethylstreptamine)	Spectinomycin[c]	*S. spectabilis* *S. flavopersicus*	G^+	Human therapy (Penicillin-resistant gonococci)
	Bluensomycin	(*S. bluensis*)		
II 2-Deoxystreptamine (DOS) derivatives				
A. 4,5-Disubstituted DOS				
Neomycin group (Pseudotetrasaccharides)	Neomycins B, C	*S. fradiae*	G^+G^-	Human therapy
	Paromomycins I, II	*S. rimosus* forma *paromomycinus*	G^+G^- Protozoa	Human therapy
	Lividomycins A, B	*S. lividus*	G^+G^-	Human therapy
Ribostamycin group (Pseudotrisaccharides)	Ribostamycin	*S. ribosidificus*	G^+G^-	Human therapy
	Butirosins	(*Bacillus circulans*)		
B. 4,6-Disubstituted DOS				
Kanamycin group	Kanamycins A, B, C	*S. kanamyceticus*	G^+G^-My	Human therapy
	Tobramycin (= Nebramycin-factor 6)	*S. tenebrarius*	G^+G^-	Human therapy
	Seldomycins 1, 2, 3, 5	(*S. hofunensis*)		
Gentamicin group	Gentamicins C_1, C_{1a}, C_2	*Micromonospora purpurea* *M. echinospora*	G^+G^-	Human therapy
	Sagamicin = Micronomicin	(*M. sagamiensis*, etc.)	G^+G^-	Human therapy
	Sisomicin	*M. inyoensis*	G^+G^-	Human therapy
	Verdamicin	(*M. grisea*)		
C. Monosubstituted DOS	Hygromycin B	*S. hygroscopicus*	G^+G^-F worms	Feed additive
	Destomycins A, B, C	(*S. rimofaciens*)	G^+G^-My worms	Feed additive
	Apramycin (=Nebramycin-factor 5)	(*S. tenebrarius*)		
III Fortamine derivatives	Fortimicin	*M. olivoasterospora*	G^+G^-	Human therapy
	Sannamycins A, B	(*S. sannanensis* sp. nov.)		
	Sporaricins A, B	(*Saccharopolyspora hirsuta* subsp. *kobensis*)		
	Istamycins A, B	(*S. tenjimariensis* nov. sp.)		
IV Other aminocyclohexanol derivatives	Validamycins	*S. hygroscopicus* var. *limonensis*	G^+G^-F	Plant pathology
	Kasugamycin	*S. kasugaensis*	G^+G^-F	Plant pathology

[a] G^+: antibiotically effective against gram-positive organisms, G^-: against gram-negative organisms; My: against mycobacteria; F: antifungal antibiotics.
[b] Dihydrostreptomycin is almost exclusively produced via chemical reduction of streptomycin.
[c] Spectinomycin has unusual status: it contains one aminocyclitol moiety but no amino sugar.
(Berdy, 1985)

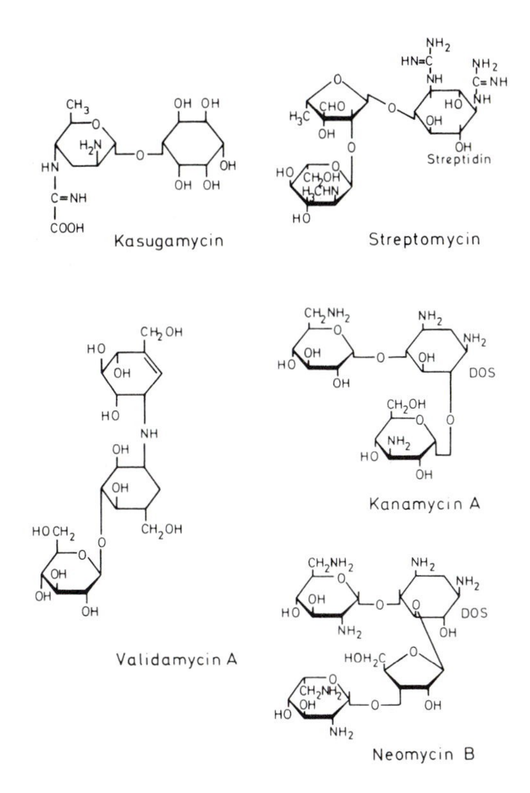

Figure 13.21 Aminoglycoside antibiotics (structure of validamycin according to Suami et al., 1980). For the structure of neomycin/ribostamycin see Figure 15.9; gentamicin complex see Figure 15.10

tein synthesis. S12 is a component of the A site at which the amino-acyl-tRNA and fMet-tRNA bind. 2-Desocystreptamine-containing antibiotics inhibit the translocation process of the peptidyl-tRNA. In addition, aminoglycoside antibiotics cause damage to the bacterial cell membrane, thus bringing about loss from the cell of low-molecular weight constituents.

Biosynthesis of aminoglycosides As shown by labeling experiments, glucose is the precursor of all the subunits of those aminoglycoside antibiotics which have been examined. Although a number of enzymatic reactions have been uncovered, the exact biosynthesis of any aminoglycoside antibiotic is not yet fully understood. Figure 13.22 shows the pathway of **streptomycin biosynthesis**. Over 30 separate enzymatic steps are known. One of the principal components of streptomycin, **streptidine,** is synthesized from glucose-6-phosphate via *myo*-inositol. A number of enzymatic steps, involving oxidation, amination, phosphorylation, carbamidinylation, and dephosphorylation, lead to the formation of streptidine-6-P. The second dephosphorylation step takes place during the final biosynthetic step to streptomycin synthesis. Although the pathway for **streptose** biosynthesis is known (see Figure 13.22), that for **N-methyl-L-glucosamine** is still unclear. The combination of the three subunits involves two steps. Only in the last step is the phosphate moiety removed, leading to the formation of the biologically active streptomycin from the biologically inactive streptomycin-phosphate.

In the DOS-aminoglycosides (see Table 13.10), the biosynthesis of 2-deoxystreptamine occurs from glucose (Figure 13.23).

Regulation of biosynthesis Only few data are available on the induction and regulation of aminoglycoside biosynthesis. Streptomycin-producing strains of *S. griseus* and *S. bikiniensis* show regulation via the so-called factor A, 2 (S)-isocapryloyl-3-(S)-hydroxymethyl-γ-butyrolactone, shown in Figure 13.24. This compound controls streptomycin biosynthesis, streptomycin resistance, and sporulation. It is active at the extremely low concentration of 10^{-9}M. Factor A^- mutants that are unable to produce the antibiotic can be restored to streptomycin production by addition of factor A.

In addition, streptomycin production is influenced by nitrogen and carbohydrate catabolite regulation, as well as phosphate regulation.

Strain development Strain-development programs have been conducted to increase yields for all industrially produced aminoglycoside antibiotics, since yields of wild strains range only from 10–200 μg/ml, which is not sufficient for a cost-effective process. The following yields have been published: kanamycin 2000 μg/ml; tobra-

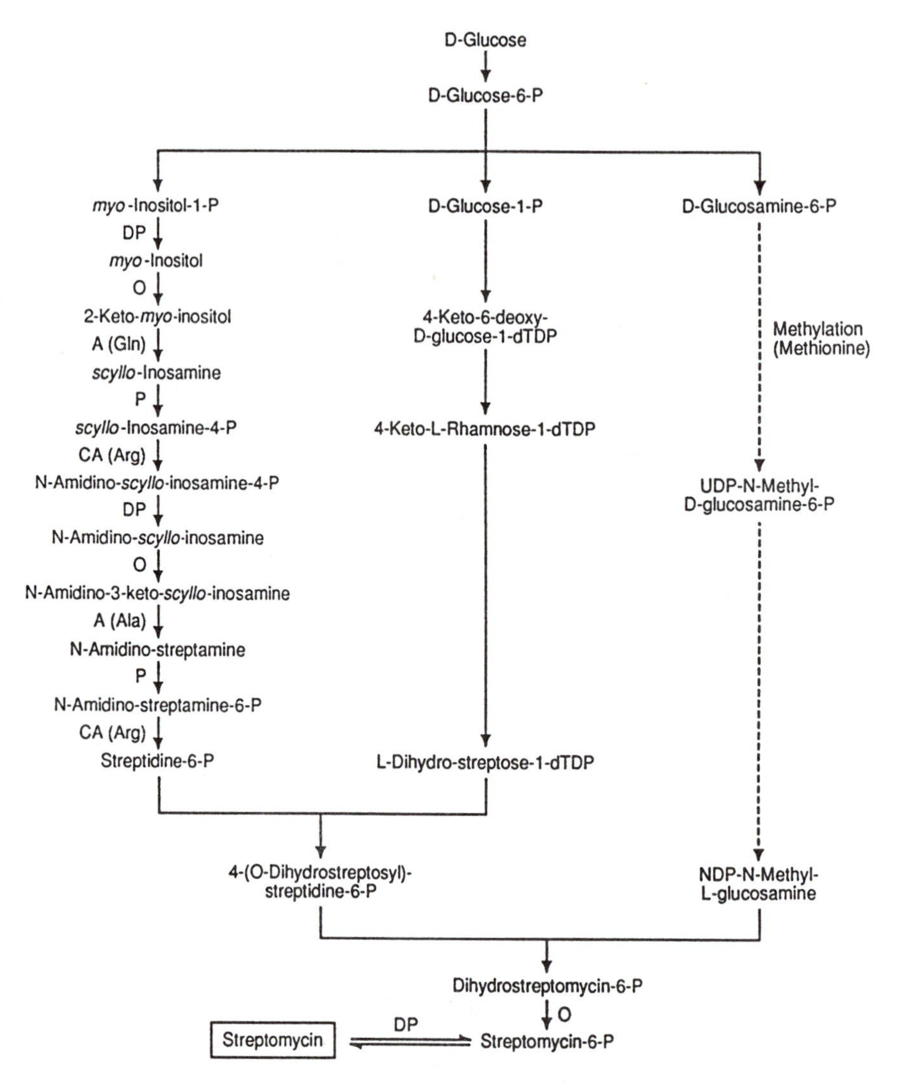

Figure 13.22 Streptomycin biosynthesis (based on Florent, 1985). O, oxidation; A, amination; P, phosphorylation; CA, carbamidinylation; DP, dephosphorylation; Ala, alanine; Arg, arginine; Glu, glutamine.

mycin 1500 μg/ml; neomycin 2000 μg/ml; gentamicin 1400 μg/ml. Streptomycin yields have been increased from 100–200 μg/ml to about 15 mg/ml by strain-improvement techniques. The current yields of production processes are probably far more than the published values in many cases.

Mutation and selection are the principal methods which have been used in strain development; there have also been some cases in which recombination (conjugation and protoplast fusion) has been employed. The strain development of kasugamycin is a typical example. Using the "agar disk" method (see Section 3.3), in one year 650,000 isolates were screened after mutagenesis, resulting in a yield increase from 200 μg/ml to about 7000 μg/ml. Conjugation of auxotrophs produced hybrid strains, whose off-

Figure 13.23 Biosynthesis of 2-deoxystreptamine in *Bacillus circulans* S-11. A, 2-Deoxy-scyllo-inosose; B, 2-Deoxy-scyllo-inosamine (From Fujiwara et al., 1980)

Figure 13.24 Structure of Factor A from *S. griseus*

spring in one case produced up to 26% more kasugamycin (9300 μg/l) than the initial prototroph strain (Figure 13.25). Cell fusion and recombinant DNA methods have been used in several cases to produce higher-yielding strains. Protoplast fusion has been used to increase yield of sisomicin producers and amplification of a resistance gene, 6′-N-acetyltransferase, has increased production of kanamycin and neomycin.

Production methods Aminoglycoside antibiotics are produced in fermenters with volumes up to 150,000 l. Optimal oxygen supply is required (aeration rate 0.5–1.0 vvm) and temperatures are held between 28–30°C with the pH in the neutral range. The length of fermentation is between 4–7 days, depending on the strain.

Glucose in combination with starch or dextrin serves as the carbon source. Originally, the antibiotic butirosin could be produced only with glycerol as carbon source, due to catabolite repression by glucose, but the isolation of mutants not subject to catabolite regulation has made it possible to use glucose. In several processes, continuous feeding of glucose has been successfully used. Soy meal is an ideal nitrogen source for aminoglycoside production because of its slow catabolism. NaCl (1–3 g/l) must be added to the streptomycin fermentation. The production of gentamicin, sisomicin, and fortimicin A is significantly accelerated by addition of Co^{2+}, which is active in the methylation steps during biosynthesis. Addition of Zn^{2+} increases the yield of nebramycin and addition of Mg^{2+} stimulates kanamycin and neomycin production.

Aminoglycoside processes are typical secondary metabolite fermentations with tropho- and idiophases, as shown in Figure 13.26 for the butirosin fermentation.

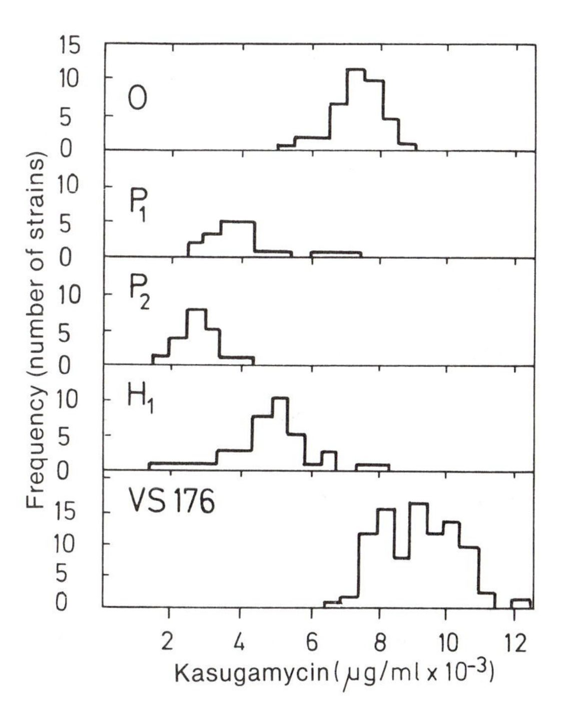

Figure 13.25 Effect of genetic recombination on kasugamycin formation. O, prototrophic original strain (kasugamycin formation, 7300 μg/ml; P_1, auxotrophic *(ilv)* mating partner (4400 μg/ml); P_2, auxotrophic *(ser)* mating partner (2900 μg/ml); H_1, prototrophic hybrid strain *(ilv* × *ser)* (5100 μg/ml); VS176, selection from H_1 (9300 μg/ml) (From Ichikawa et al., 1971)

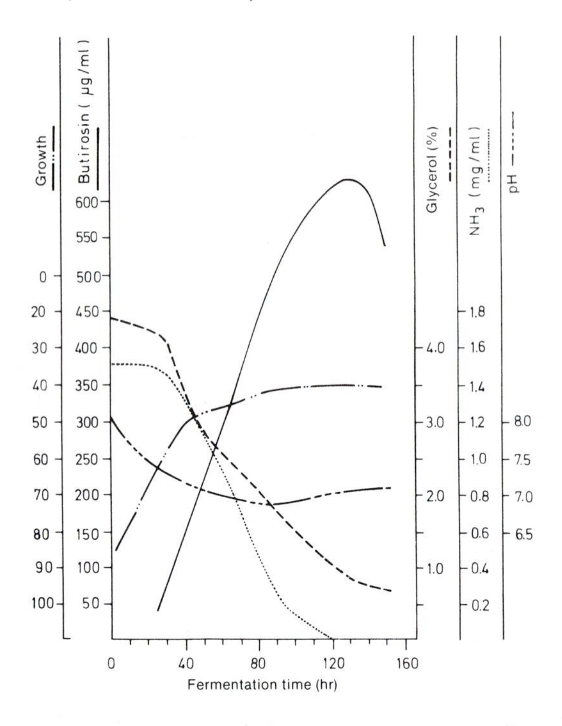

Figure 13.26 Butirosin formation with *Bacillus circulans*. Medium: glycerol 4%; soy meal 1%; Wilson's peptone 1.75%; NH_4Cl 0.4%; $CaCO_3$ 0.5% (Howells et al., 1972)

Most of the aminoglycoside antibiotics are excreted and are present in the culture supernatant, from which they are removed by adsorption to ion-exchange columns. Cell-bound aminoglycosides, such as gentamicin, sisomicin, and fortimicin, must first be released by acidification to pH 2–2.5 with H_2SO_4.

13.5 MACROCYCLIC LACTONE ANTIBIOTICS

The macrolide group of antibiotics consists of compounds with large lactone rings. In addition to those antibiotics conventionally called **macrolides,** this group includes **polyenes, macrotetrolides,** and a heterogeneous group of atypical macrolides.

Macrolide antibiotics

Macrolides are hydrophobic, usually basic compounds. They consist of 12-, 14-, 16-, or 17-membered lactone rings with 1–3 sugars glycosidically linked with the aglycone (lactone ring) and with each other. The sugars are amino sugars and/or 6-deoxyhexoses.The structures of erythromycin and oleandomycin are examples of 14-membered macrolides (Figure 13.27); the 16-membered macrolides leucomycin and tylosin are illustrated in Figure 13.28. Macrolides are produced by 1–3% of all actinomycete isolates; the compounds listed in Table 13.11 are commercially available.

Macrolides are very effective against Gram-positive bacteria, particularly staphylococci, and mycoplasmas and were frequently used against penicillin-resistant organisms, but because of the development of the newer β-lactam antibiotics (see Section 13.2), the macrolides are no longer of as great medical significance. The aglycone and sugar moieties are both necessary for anti-

Desosamine

R_2 : CH_3 = Cladinose
R_2 : H = Mycarose

Erythronolide

		R_1	R_2	
Erythromycin	A	OH	CH_3	(Main components)
	B	H	CH_3	
	C	OH	H	
	D	H	H	

Oleandrose

Oleandolide

Oleandomycin

Figure 13.27 Structure of erythromycin and oleandomycin. Straight arrow, acetate residue; bent arrow, propionate residue

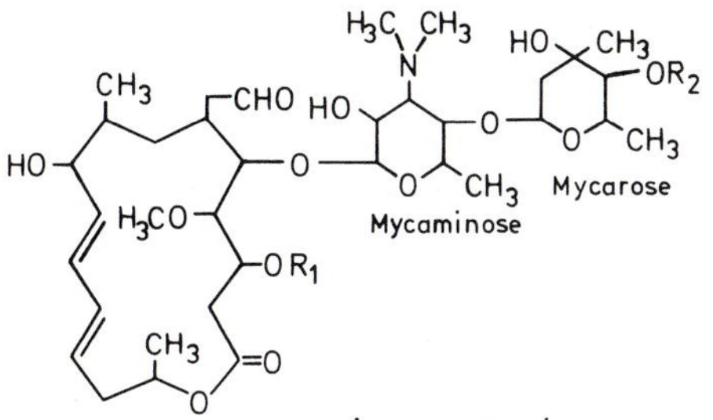

Leucomycin

	R_1	R_2
Leucomycin A_1	H	$COCH_2CH(CH_3)_2$
A_3 [Josamycin]	$COCH_3$	$COCH_2CH(CH_3)_2$
A_4	$COCH_3$	$COCH_2CH_2CH_3$
A_5	H	$COCH_2CH_2CH_3$
A_6	$COCH_3$	$COCH_2CH_3$
A_7	H	$COCH_2CH_3$
A_8	$COCH_3$	$COCH_3$

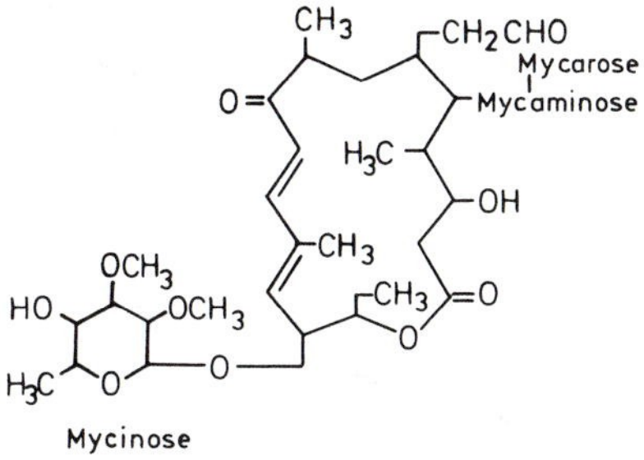

Tylosin

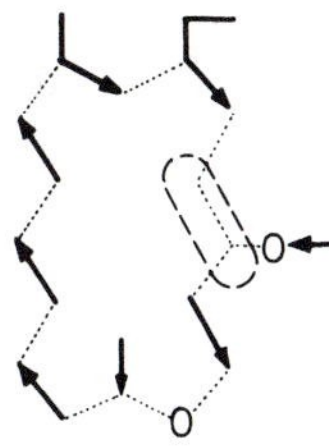

Leuconolide

Figure 13.28 Structure of leucomycin and tylosin. Straight arrow, acetate residue; bent arrow, propionate residue; upper right bent arrow, butyrate residue; dotted region, position of C-3, 4 unknown

Table 13.11 Commercially produced macrolide antibiotics

Macrolide antibiotic	Organism producing	Activity	Application
Carbomycin A	*Streptomyces halstedii*	G^+	Was marketed for several years but discontinued due to ineffectiveness
Erythromycin	*S. erythreus*	G^+	Human therapy, feed additive
Josamycin (=Leucomycin A_3)	*S. narbonensis* var. *josamyceticus*	G^+	Human therapy
Leucomycins A_1–A_9 (= Kitasamycins)	*Streptoverticillium kitasatoensis*	G^+	Human therapy
Maridomycin	*S. hygroscopicus*	G^+	
Midekamycin	*S. mycarofaciens*	G^+	
Oleandomycin	*S. antibioticus* *S. olivochromogenes*	G^+	Human therapy
Spiramycins I, II, III	*S. ambofaciens*	G^+	Human therapy
Tylosin	*S. fradiae* *S. hygroscopicus*	G^+, Mycoplasmas	Feed additive, veterinary medicine

bacterial activity. Therapy with macrolides results in a rapid buildup of resistance and there is also cross-resistance between the individual antibiotics. Because of the low toxicity, efforts are being made to isolate new compounds of this class, to chemically modify known macrolides, and to modify known macrolides through biotransformation.

All macrolide antibiotics are inhibitors of protein synthesis, binding to the 50 S subunit of bacterial ribosomes. Erythromycin is known to inhibit the elongation-factor G-dependent release of deacylated tRNA from the P site of the ribosome.

Biosynthesis and regulation The aglycone moiety of macrolides is built either of propionate subunits or of a combination of acetate and propionate subunits, sometimes with the participation of butyrate or related compounds. One of the first intensively studied macrolides was erythromycin; the biosynthesis of this antibiotic was studied by labeling experiments and from the behavior of mutants of *Streptomyces erythreus* blocked in antibiotic biosynthesis. Analogous to fatty acid synthesis, erythronolide production proceeds from propionyl-CoA as a starting material, onto which 6 molecules of 2-methylmalonyl-CoA condense. This polyketide synthesis takes place on a multi-enzyme complex called *lactone synthase,* which is very similar to the fatty acid synthetase of *S. erythreus.* After cyclization to the aglycone, the first free compound, 6-deoxyerythronolide B, can be isolated. The remaining glycosylation steps are diagrammed in Figure 13.29.

A number of observations have been made about the regulation of macrolide biosynthesis. Biosynthesis of the macrolide tylosin is stimulated by long-chain fatty acids; glucose inhibits the metabolism of fatty acids and tylosin production, probably by inducing an acetyl-CoA deficiency. Phosphate regulation is also observed with many macrolide antibiotics, as well as nitrogen (NH_4^+) catabolite regulation.

Erythromycin biosynthesis is regulated at various levels. For instance, erythronolide B inhibits its own production, the synthesis of 6-deoxyerythronolide B, and presumably also the activity of the lactone synthase. End-product inhibition of the transmethylase by erythromycin A has been observed. This enzyme catalyzes the methylation of mycarose to cladinose as a last step of the biosynthetic pathway. In high-yielding strains used for industrial production, this regulation is eliminated.

When propanol (0.2–0.5%) is added to the fermentation broth of *S. erythreus,* it induces the synthesis of acetyl-CoA-carboxylase and causes a 100% increase in erythromycin production.

Production method Macrolide antibiotics are produced in aerobic submerged fermentations. Yields in large-scale industrial processes are around 20 g/l. A typical medium for *S. erythreus* contains glucose 50 g, soy meal 30 g, $(NH_4)_2SO_4$ 3 g, NaCl 5 g, $CaCO_3$ 6 g, tap water 1 l, pH 7.0. Complex culture medium ingredients may also be used, such as starch, corn steep liquor, yeast extract, and oils (e.g., soy bean oil). Fermentation temperatures for most macrolides are 25–28°C and for erythromycin 33°C. The length of the fermentation is 3–7 days.

The macrolide tylosin can be produced in either batch or continuous fermentation. The specific production rate is increased to 1.1 mg/g biomass·h if glycerol is used instead of glucose and if the feeding rate of sodium glutamate is optimized at a low specific growth rate of 0.03/h (27% of the μ_{max}).

Polyene macrolide antibiotics

Polyenes consist of 26- to 38-membered lactone rings with 3–7 conjugated double bonds. Depending on the length of the chromophore, polyenes are classified as trienes, tetraenes, pentaenes, hexaenes, or heptaenes. Many polyenes also contain a glycosidically-bound amino sugar (mycosamine or perosamine). Some heptaenes carry p-aminoacetophenone or N-methyl-p-ami-

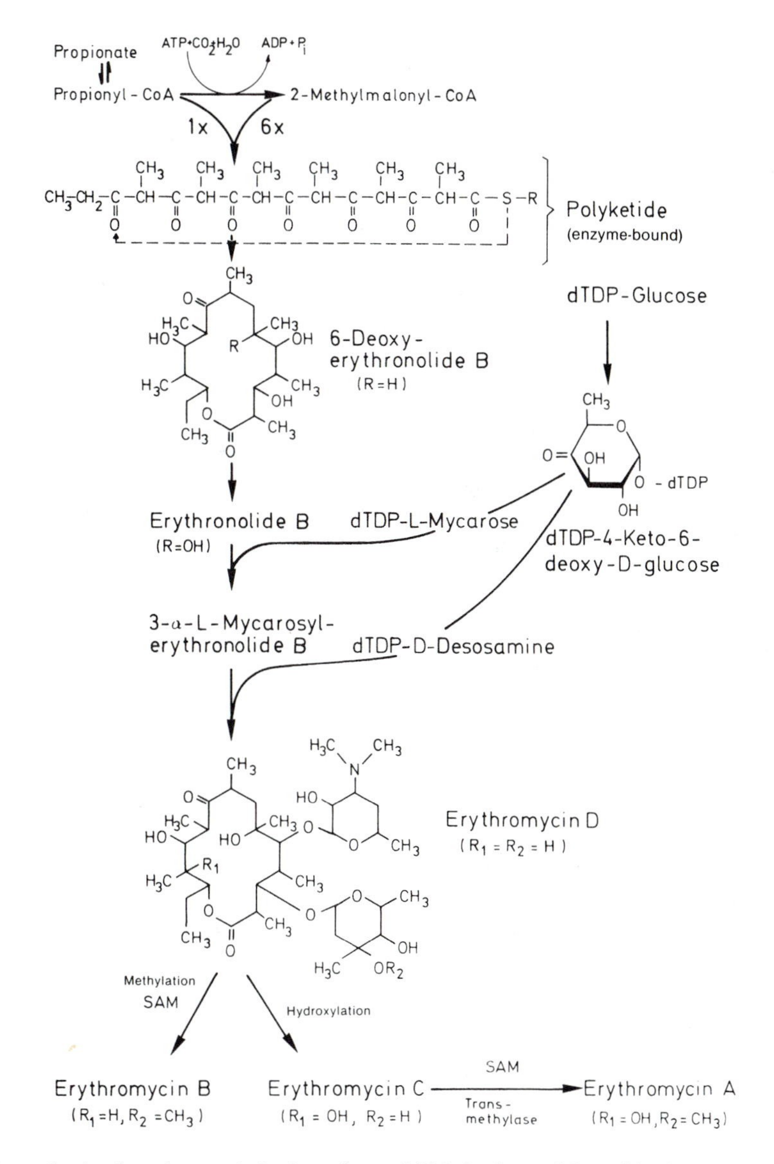

Figure 13.29 Biosynthesis of erythromycin in *S. erythreus*. SAM: S-adenosyl-L-methionine

noacetophenone as an aromatic moiety on the carbon atom which is involved in lactone formation. The structures of two polyenes are given in Figure 13.30.

More than 90 polyenes have been characterized, primarily from streptomycetes. These antibiotics are active against fungi and yeast but they are ineffective against bacteria. Polyenes affect membrane sterols of sensitive eucaryotes, causing a change in membrane permeability. Some polyenes are relatively toxic (nephro- or hepatotoxic) but they are widely used commercially for topical application because no better antifungal antibiotics are available.

Polyenes are not suitable for plant-disease control because they are sensitive to light and

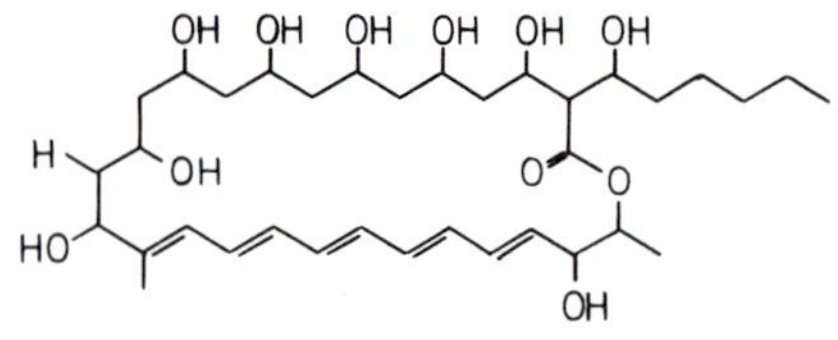

Filipin

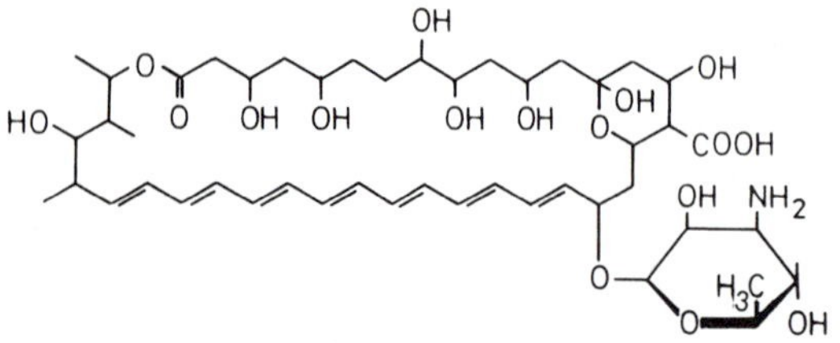

Amphotericin B

Figure 13.30 The polyene antibiotics filipin (pentaene) and amphotericin B (nonaromatic heptaene)

heat. Table 13.12 lists the commercially produced polyenes.

Biosynthesis and regulation Like macrolides, polyene antibiotics are produced from C_2 and C_3 units via the polyketide biosynthetic pathway. Among the aromatic polyenes, p-aminobenzoyl-CoA, the precursor of p-aminoacetophenone and N-methyl-p-aminoacetophenone, seems to function as the starting point for polyketide polymerization. The chromophore consists solely of acetate units. The amino sugar is derived from glucose, with thymidine diphosphate-4-keto-6-deoxy-D-glucose as an intermediate. Polyene biosynthesis shows phosphate regulation, catabolite repression, and feedback inhibition.

Production methods Glucose is usually the best carbon source for polyene production. Most fermentations using glucose are continuous, because catabolite repression occurs at the required glucose concentration (for candicidin 7–9.6%). The candicidin production phase can be extended to 300 h by glucose feeding. The addition of acetate, propionate, malonate, or n-propanol stimulates polyene production, and soy meal or soy peptone can be used as nitrogen sources.

For maximal polyene production, careful control of aeration is essential. During the growth phase, the required oxygen transfer rate is around 0.8 mmole O_2/l·min, but during the production phase, O_2 concentration should not exceed 40% of saturation. Figure 13.31 shows the relation of candicidin production to various fermentation parameters.

Since polyenes are not water-soluble, they accumulate as crystals or microparticles at the surface of the cell or in the medium, from which they can be extracted with appropriate solvents.

Table 13.12 Commercially produced polyene macrolide antibiotics

Antibiotic	Organism producing	Polyene type	Application
Amphotericin B	*Streptomyces nodosus*	Heptaene, M	F, Histoplasmosis
Candicidin B	*S. griseus*	Heptaene, M, AAP	F, Prostate hypertrophy
Candidin	*S. viridoflavus*	Heptaene, M	F
Filipin	*S. filipinensis*	Methylpentaene	F
Fungimycin (=Perimycin)	*S. coelicolor* var. *aminophilus*	Heptaene, P, MAAP	F
Hamycin	*S. primprina*	Heptaene, M, AAP	F
Levorins	*S. levoris*	Heptaene, M, AAP	F, Prostate hypertrophy
Mycoheptin	*Streptoverticillium mycoheptinicum*, *S. netropsis*	Heptaene, M	F
Nystatin A_1, A_2, A_3	*S. noursei*	Tetraene, M	F, *Candida albicans*; slightly toxic
Pimaricin	*S. natalensis*	Tetraene, M	F, Food preservative
Trichomycin A, B	*S. hachijoensis*	Heptaene, M, AAP	F, *Trichomonas vaginalis*

M=Mycosamine; AAP=p-Aminoacetophenone; P=Perosamine; MAAP=N-Methyl-p-aminoacetophenone; F=Antifungal activity.

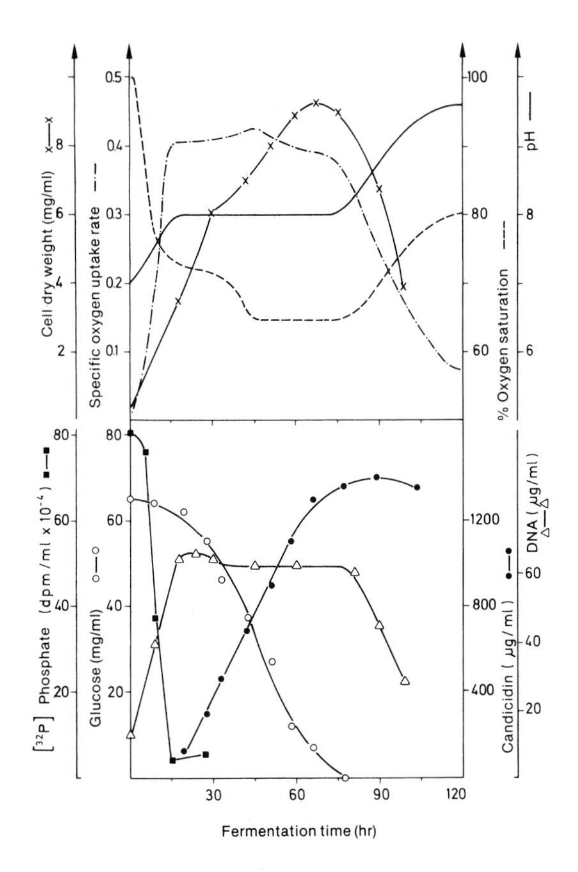

Figure 13.31 Relationship of candicidin formation to different fermentation parameters

Nonactin	*S. werraensis*	$R_1 = R_2 = R_3 = R_4 = CH_3$
Monactin	*S.* sp. ETH 23112	$R_1 = R_2 = R_3 = CH_3 \quad R_4 = C_2H_5$
Dinactin	*S.* sp. ETH 23112	$R_1 = R_2 = CH_3 \quad R_3 = R_4 = C_2H_5$
Trinactin	*S.* sp. ETH 23112	$R_1 = CH_3 \quad R_2 = R_3 = R_4 = C_2H_5$
Tetranactin	*S. flaveolus*	$R_1 = R_2 = R_3 = R_4 = C_2H_5$

Figure 13.32 Macrotetrolides

Avermectin	R_1	R_2	R_3
A_{1a}		C_2H_5	CH_3
A_{1b}		CH_3	CH_3
A_{2a}	OH	C_2H_5	CH_3
A_{2b}	OH	CH_3	CH_3
B_{1a}		C_2H_5	H
B_{1b}		CH_3	H
B_{2a}	OH	C_2H_5	H
B_{2b}	OH	CH_3	H

In those avermectins without an R_1 group, a double bond is present.

Figure 13.33 Structures of the avermectins

Macrotetrolides

Macrotetrolides are cyclic polylactones which function as ionophores. These sugar-free compounds are synthesized from four tetrhydrofuranyl hydroxy acids (Figure 13.32). Tetranactin, produced by *S. flaveolus*, is used as an insecticide.

Atypical macrolides

This is a heterogeneous group, classified together because all structures contain the lactone ring. Examples include avermectin, milbemycin, venturicidin, bafilomycin, and irumamycin. Avermectin, whose structure is shown in Figure 13.33, is produced by *S. avermitilis*. It has found use in veterinary medicine because of its high activity against nematodes and arthropods. Irumamycin is undergoing field trials in agriculture as a fungicide.

Ansamycins

Ansamycins are macrolactam antibiotics with an aromatic chromophore to which an aliphatic polyketide derivative is bound at both ends, via amide, ether, and/or carbon-carbon bonds, the so-called "ansa chain" (Figure 13.34). Ansamycins are distinguished by their chromophore. Those compounds with a benzene ring, such as geldanamycin or ansamitocin, act mainly on eucaryotes, whereas those with a naphthalene structure are antibacterially effective. In this latter class are the rifamycins, halomycins, and streptovaricins.

Rifamycins are important in chemotherapy. They are very effective against Gram-positive bacteria and mycobacteria, but less effective against Gram-negative bacteria; they show no cross resistance with other antibiotics. They are produced by *Nocardia mediterranei.* Approximately 20 different rifamycins have been isolated from culture solutions of *Nocardia* mutants, and genetic studies have shown that the genes for the last steps in the biosynthetic pathway lie in a single cluster. The wild strain produces rifamycin B as a major component and smaller amounts of the so-called rifamycin complex (A, C, D, E). The production of this complex can be largely suppressed by the addition of sodium diethylbarbiturate. Rifamycin B is completely inactive but during its breakdown or via chemical modification, the biologically active rifamycins O, S, and SV are formed. Rifamycin S and SV are biosynthetic precursors to rifamycin B. Mutants have been isolated which excrete rifamycin SV due to a block in the subsequent steps. The ansa chain of rifamycin is formed from two malonyl-CoA and eight methylmalonyl-CoA units and methionine. The polyketide condensation process proceeds from an aromatic C_7 unit. The biosynthesis of the naphthohydroquinone chromophore branches off from the aromatic biosynthetic pathway between 3-deoxy-D-arabinoheptulose-7-P and shikimate.

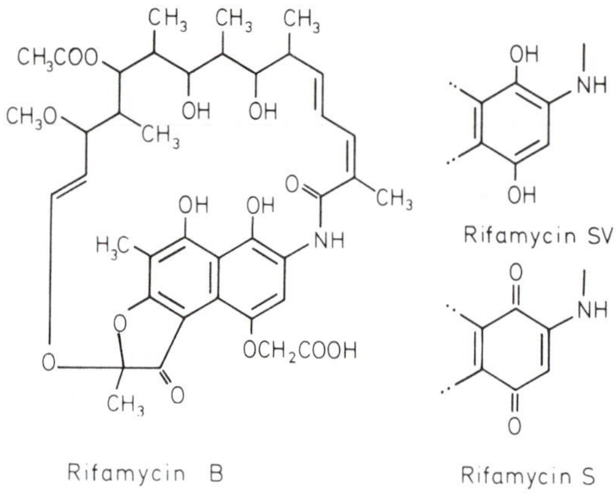

Rifamycin B

Rifamycin SV

Rifamycin S

Figure 13.34 The rifamycins B, S and SV

Rifampicin, a semisynthetic compound derived from rifamycin SV, is widely used in the treatment of tuberculosis. It is a specific inhibitor of bacterial DNA-dependent RNA polymerase. It binds with the β-subunit of the polymerase, and RNA polymerases of higher organisms are not affected.

13.6 TETRACYCLINES AND ANTHRACYCLINES

Tetracyclines

The tetracyclines are important antibiotics with widespread medical use. **Chlortetracycline**, the first antibiotic of this group, was isolated from cultures of *Streptomyces aureofaciens* in 1945. Tetracycline itself was first prepared by dehalogenation of chlortetracycline but was subsequently found to be produced by cultures of *S. viridifaciens* as well. Presently about 20 streptomycetes are known which produce a mixture of different tetracyclines. Several strains are listed in Table 13.13.

The basic structure of the tetracyclines consists of a naphthacene ring system. Figure 13.35 shows the structures of several therapeutically important compounds. There are also semisynthetic derivatives on the market as well as the fermentatively produced tetracyclines.

Table 13.13 Streptomycetes producing tetracycline

Species	Products[a]
S. alboflavus	CTC, TC, OTC, actinomycin
S. antibioticus	OTC, TC
S. aureofaciens	CTC, TC
S. aureus	TC
S. californicus	TC, actinomycin
S. cellulosae	OTC, actinomycin
S. feofaciens	TC
S. flaveolus	OTC, TC, actinomycin
S. flavus	CTC, TC, OTC, actinomycin
S. lusitanus	CTC, TC
S. parvus	OTC, TC, actinomycin
S. platensis	OTC
S. rimosus	OTC, TC, rimocidin
S. sayamaensis	TC, CTC
S. vendargensis	OTC, vengacide
S. viridifaciens	CTC, TC

[a] CTC 7-chlortetracycline; TC tetracycline; OTC oxytetracycline
(Perlman, 1979)

Tetracyclines	R_1	R_2	R_3	R_4	Production strain
Tetracycline	H	OH	CH_3	H	S. aureofaciens (in chloride-free medium) or through chemical modification of chlortetracycline
7-Chlortetracycline (Aureomycin)	H	OH	CH_3	Cl	S. aureofaciens
5-Oxytetracycline (Terramycin)	OH	OH	CH_3	H	S. rimosus
6-Demethyl-7-chlortetracycline (Declomycin)	H	OH	H	Cl	S. aureofaciens (+ inhibitor)
6-Deoxy-5-hydroxy-tetracycline (Doxycylcine)	OH	H	CH_3	H	Semisynthetic
7-Dimethylamino-6-demethyltetracycline (Minocyclin)	H	H	H	$N(CH_3)_2$	Semisynthetic
6-Deoxy-6-demethyl-6-demethyl-6-methylene 5-hydroxytetracycline (Methacycline)	OH	$=CH_2$		H	Semisynthetic

Figure 13.35 Structure of clinically important tetracyclines

Tetracyclines are broad-spectrum antibiotics, which are effective against Gram-positive and Gram-negative bacteria, as well as rickettsias, mycoplasmas, leptospiras, spirochetes, and chlamydias. There is a marked cross-resistance between different tetracyclines. After cephalosporin and penicillin, the tetracylcines are the most widely used antibiotics. Chlortetracycline and oxytetracycline are commonly used in human and veterinary medicine. In many countries they are also widely used as nutritional supplements in poultry and swine production and in some countries they are also used to preserve fish, meat and poultry.

Tetracyclines act as **inhibitors of protein synthesis**. Their site of action is the 30S ribosome, where binding of aminoacyl-tRNA to the ribosomal A-site is inhibited.

Biosynthesis and regulation Chlortetracycline and oxytetracycline are the tetracyclines most frequently produced by the streptomycetes, tetracycline itself usually being only a minor component. However, *S. aureofaciens* mutants with a block in the chlorination reaction excrete tetracycline as the major product. A simplified version of chlortetracycline biosynthesis is illustrated in Figure 13.36. The biosynthesis can be subdivided into three sections:

- Glucose is converted into acetyl-CoA.
- Malonyl-CoA is produced in the transformation of acetyl-CoA by means of acetyl-CoA-carboxylase. After transamination to malonamoyl-CoA, which is bound to an enzyme complex ("anthracene synthase"), this compound condenses with 8 molecules of malonyl-CoA. The polyketide presumably formed from this condensation has not been isolated. An alternative pathway to malonyl-CoA is via the production of oxalacetate catalyzed by phosphoenol pyruvate carboxylase, oxalacetate being converted to malonyl-CoA by oxidative decarboxylation. Cyclization into the tricyclic intermediate also takes place on the enzyme anthracene synthase.
- The step-by-step transformation of the assumed tricyclic intermediate, which has not yet been isolated, into the variety of known tetracyclines.

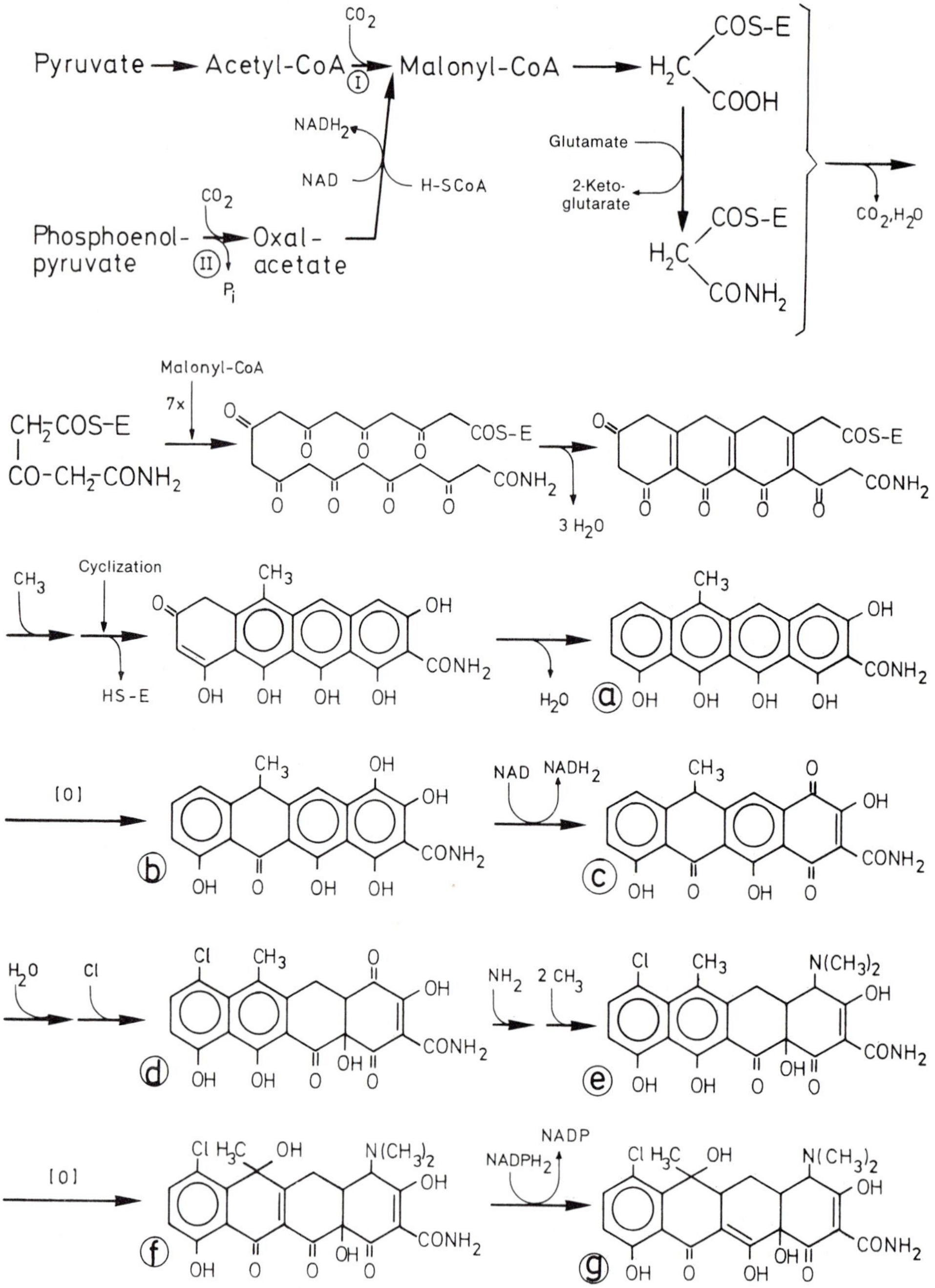

Figure 13.36 Biosynthesis of chlortetracycline by *S. aureofaciens*. a, 6-methylpretetramide; b, 4-Hydroxy-6-methylpretetramide; c, 6-Deoxytetramide green; d, 4-Oxyanhydrochlortetracycline; e, Anhydrochlortetracycline; f, Dehydrochlortetracycline; g, Chlortetracycline; I, Acetyl-CoA carboxylase; II, Phosphoenolpyruvate carboxylase (From Vanek et al., 1971; Hoŝtalek et al., 1979)

The scheme of chlortetracycline biosynthesis, still mainly hypothetical, has been deduced from studies of mutants blocked in tetracycline biosynthesis and from cosynthesis studies with *S. aureofaciens*. There are 72 intermediate products; 27 substances have been characterized but 45 have not yet been isolated and the chlorination reaction itself is not yet understood.

There is a clear correlation between tetracycline production and carbohydrate catabolism. High-yielding strains have a lower glycolysis rate than strains with low tetracycline production. The addition of 0.5–3 μg/ml benzylthiocyanate, an inhibitor of sugar uptake, causes a 50% increase in chlortetracycline production; simultaneously, the activity of the pentose-phosphate cycle increases. In the case of tetracycline biosynthesis, phosphate regulation is well understood. The addition of P_i to the producing culture decreases the rate of tetracyline formation by inhibition of ATC oxygenase. The rate of tetracycline synthesis is directly proportional to the activity of this enzyme, which is only detectable in the culture after phosphate has been completely depleted.

Strain development Studies on the genetics of *S. aureofaciens* have shown that over 300 genes are involved in the biosynthesis of chlortetracycline. Considering also the dearth of knowledge of biosynthesis and regulation in this pathway, it is understandable that a rational approach to strain improvement has not been possible. So far, all developments of higher-yielding strains have come from empirical mutagenesis studies.

Ultraviolet radiation, alone or in combination with other mutagens, has brought about the greatest number of higher-yielding strains using *S. aureofaciens* and *S. rimosus*. Table 13.14 shows the effect of different mutagens on the variability of chlortetracycline production, and Figure 13.37 shows the increase in chlortetracyline production with each step in the course of strain development. Strains with increased tetracycline production have been isolated by selecting revertants of mutants blocked in antibiotic synthesis and from prototrophic revertants of auxotrophs. They may also be isolated from mutants showing increased resistance to their own product. The highest published yield for tetracycline is 25 g/liter.

Hybridization has not resulted in strains with increased product formation, but in combination with mutation it has caused an increase in strain variability. Protoplast fusion has been used to increase yields.

A genetic map has been compiled for the wild type *S. rimosus* (ATCC 10970), as illustrated in

Table 13.14 Spontaneous and induced variability of chlortetracycline production in *S. aureofaciens*

Mutagen[a]	Distribution of productivity (% of control)[b]		Frequency of strains with improved productivity (%)[c]
	Total range	Most frequent class	
–	40–110	80–100	5.3
Ultraviolet radiation	0–130	90–110	17.4
X-rays	0–110	70– 90	7.1
γ-Rays	0–110	0– 10	1.8
Nitrosoguanidine	0–120	40– 80	9.4
N-Mustard	0–110	80–100	5.8

[a] Research was conducted with conidia at death rates >99%. Ultraviolet 50 sec; X-rays and γ-rays 100 kR; N-methyl-N′-nitro-N-nitrosoguanidine 1 mg/ml, 60 min, 0.02 M phosphate buffer, pH 6.7; N-mustard 0.01 M, 30 min, 0.15 M phosphate buffer, pH 8.0.
[b] Productivity of initial strain=100%.
[c] Percent of the tested isolates.
(Hoŝtalek et al., 1974)

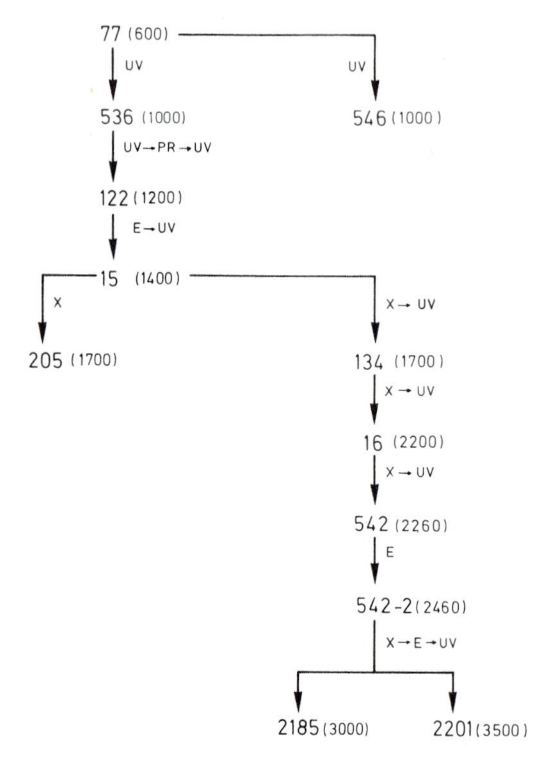

Figure 13.37 Genealogy of the chlortetracycline producer *S. aureofaciens* . UV, ultraviolet radiation; PR, photoreactivation; E, ethyleneimine treatment; X, X-rays. In (), Chlortetracycline titers in μg/ml (From Hoŝtalek et al., 1974)

Figure 13.38. This map has some similarities to that of *S. coelicolor*, the streptomycete best known genetically. The genes for the first steps of oxytetracycline biosynthesis are located between *ribB* and *cysD* in a narrow area of the *S. rimosus* chromosome. The genes for the steps from anhydrotetracycline to oxytetracycline are located in a second cluster between *proA* and *adeA*. Location of the genes in clusters facilitates genetic engineering research, and recombinant DNA methodology has been developed for oxytetracycline-producing strains of *S. rimosus*. Although plasmids of *S. rimosus* have been detected which function as fertility factors and which code for self-resistance, the role of extrachromosomal DNA in the regulation of tetracycline biosynthesis is not yet clear.

Production methods The production of tetracyclines is carried out in stirred fermenters with volumes up to 150,000 l. Figure 13.39 illustrates a typical process for **chlortetracycline** production. If glucose is used, continuous feeding is necessary; starch can be used as an additional carbon source. Because of phosphate repression, tetracycline fermentations are run with phosphate limitation.

For maximal yield, an optimal oxygen supply is important, particularly in the first hours (6–12 h) of the fermentation. Aeration rates of 1 vvm are customary; yield increases result when oxygen-enriched air is used.

There are several routes to **tetracycline**:

- By a chemical process using chlortetracycline.
- By fermentation in chloride-free culture medium, which can be made by pretreatment with ion exchangers.
- By adding chlorination inhibitors to the me-

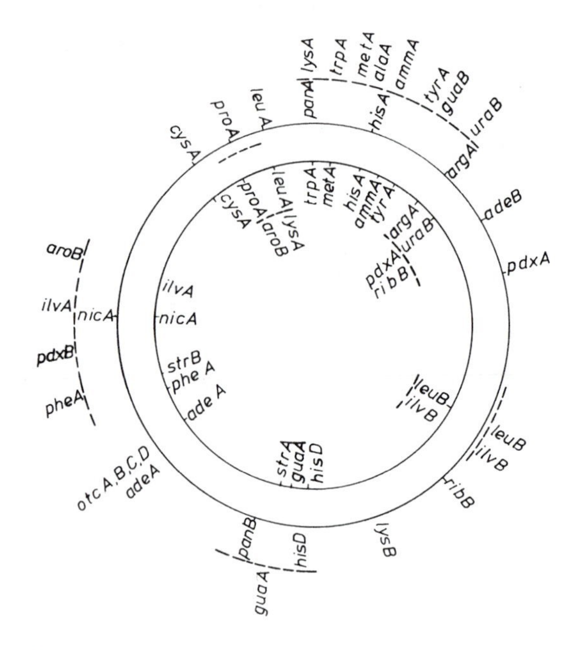

Figure 13.38 Outer circle, genetic map of *S. rimosus* R 7 (ATCC 10970). Inner circle, genetic map of *S. coelicolor*. The correct orders for the sites outside the dotted lines have not yet been determined (From Alaĉeviĉ, 1976)

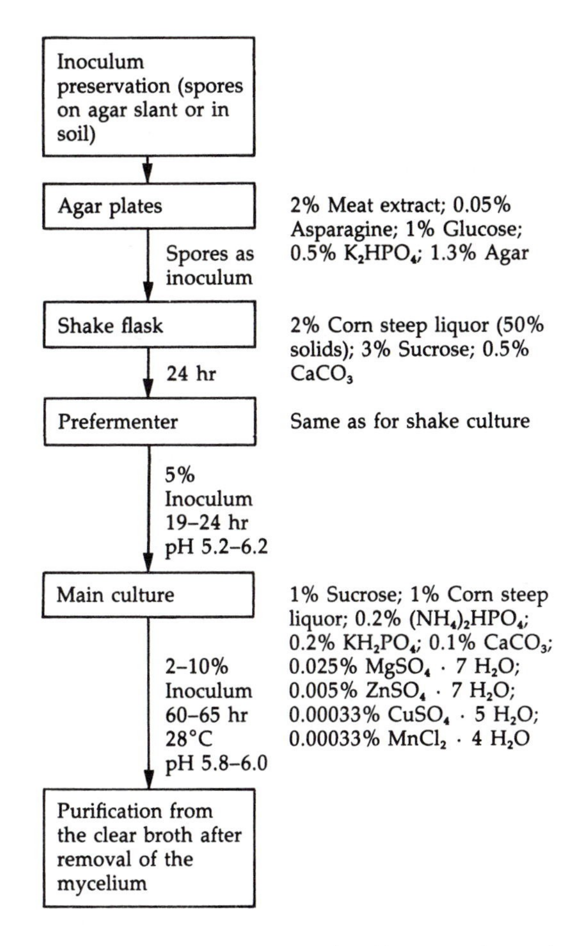

Figure 13.39 Production chart for chlortetracycline with *S. aureofaciens* (Perlman, 1979)

dium, such as mercaptobenzothiazole, 2-thiouracil, or thiourea. Bromide inhibits chlorination in some cultures, but causes formation of 7-bromtetracycline in other strains.

- With mutants which are blocked in the chlorination reaction.

Anthracyclines

Anthracyclines are glycosides which consist of 7,8,9,10-tetrahydro-5,12-naphthacenequinone as an aglycone combined with 1–3 sugar residues. Anthracyclines are very **toxic compounds**. They intercalate into DNA and inhibit DNA-dependent DNA polymerase in DNA replication.

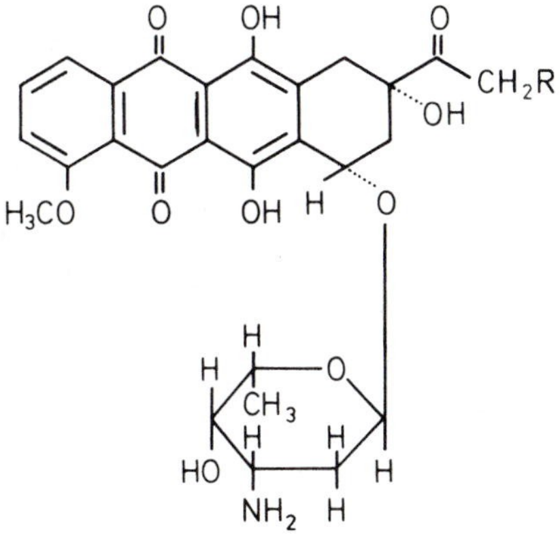

Daunomycin R = H

Adriamycin R = OH

Figure 13.40 Anthracyclines. Structure of adriamycin and daunomycin

Daunomycin (daunorubicin) and **adriamycin** (doxorubicin), produced by *Streptomyces peucetius*, are used clinically in cancer treatment. The structure of both compounds is diagramed in Figure 13.40. Daunomycin acts against certain forms of leukemia; adriamycin, which has low toxicity, is effective in treating various tumors. However, these compounds have significant side effects and research is under way, using mutasynthesis and protoplast fusion as well as by the formation of chemical derivatives, to produce compounds with decreased problems.

13.7 NUCLEOSIDE ANTIBIOTICS

About 200 antibiotics with nucleoside-like structures are known. Because of the diversity of structures, marked differences in biological activity are seen. Aminoacyl nucleosides used as fungicide antibiotics in plant protection include **blasticidin S** from *Streptomyces griseochromogenes*, **mildiomycin,** from *Streptoverticillium rimofaciens*, and the **polyoxins** from *S. cacaoi* var. *asoensis*. Polyoxins (Figure 13.41 A) inhibit the

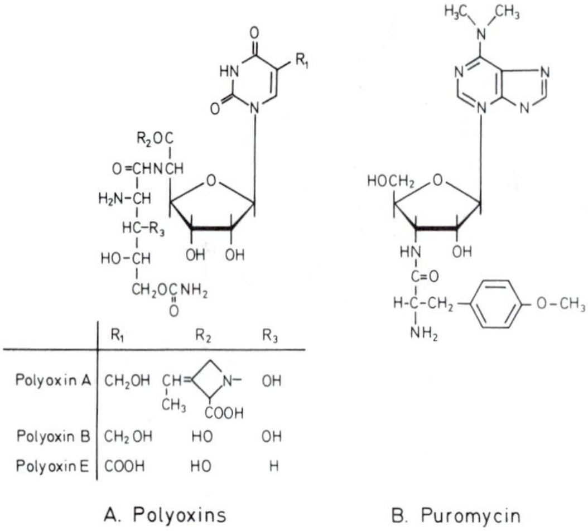

	R_1	R_2	R_3
Polyoxin A	CH_2OH	$CH(CH_3)=$ [ring] $N-$ COOH	OH
Polyoxin B	CH_2OH	HO	OH
Polyoxin E	COOH	HO	H

A. Polyoxins

B. Puromycin

Figure 13.41 Nucleoside antibiotics

incorporation of N-acetyl-D-glucosamine into the chitin moiety of fungal cell walls by competitive inhibition of chitin synthase.

The dapiramines as well as nikkomycin, which also act as chitinase inhibitors, are undergoing field tests as possible herbicides. Several compounds with cytostatic or antiviral activity are under development; for example, neplanosin A from cultures of *Ampullariella regularis.*

Puromycin (Figure 13.39 B) from *S. alboniger* has found wide use as a tool in studying ribosomal function in protein biosynthesis but has no clinical significance. Puromycin resembles the aminoacyl end of transfer RNA. In both procaryotes and eucaryotes it binds to the A-site of the ribosome and is transferred to the growing polypeptide chain, thus causing a break in the polymerization process.

13.8 AROMATIC ANTIBIOTICS

This is a heterogenous group of antibiotics with aromatic rings in the molecule. The biosynthesis of individual compounds differs considerably. Three antibiotics of this group are produced commercially.

Chloramphenicol

Chloramphenicol (Figure 13.42) is produced by *Streptomyces venezuelae, S. phaeochromogenes* var. *chloromyceticus, S. omiyaensis,* and other streptomycetes. Chloramphenicol is a broad-spectrum antibiotic which acts on Gram-positive and Gram-negative bacteria, actinomycetes, rickettsias, and chlamydias. It causes significant **side effects**, particularly bone-marrow damage, but risk of toxicity can be reduced if therapy is conducted carefully. Because of its potential toxicity, chloramphenicol has not been as widely used as would be expected from its antimicrobial spectrum and it is presently considered a reserve antibiotic. Nevertheless, chloramphenicol is indispensible in the treatment of persistent *Salmonella* infections.

Chloramphenicol binds specifically to the 50S subunit of 70S ribosomes and blocks the peptidyl transferase reaction, causing a premature chain break. In molecular biology, chloramphenicol is used as a research tool to inhibit the translation process in procaryotes without affecting the synthesis of nucleic acids.

The pathway for chloramphenicol biosynthesis is via the aromatic pathway from chorismic acid. However, chloramphenicol has been produced by chemical synthesis since about 1950 and even improved fermentation processes have not succeeded in replacing this chemical process.

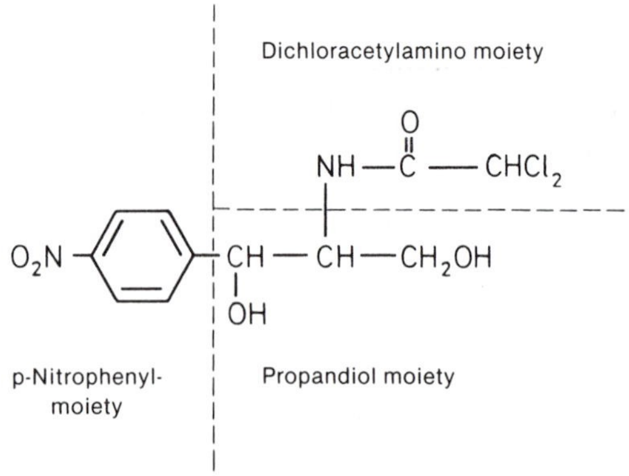

Dichloracetylamino moiety

p-Nitrophenyl-moiety

Propandiol moiety

Figure 13.42 Chloramphenicol

Griseofulvin

Griseofulvin, a benzofuran derivative (Figure 13.43), is produced by *Penicillium griseofulvum*, *P. patulum*, and other species of *Penicillium*. Griseofulvin is a fungistatic antibiotic but acts only on fungi with chitinous cell walls. In humans, griseofulvin is administered orally for the treatment of fungal skin infections (dermatomycoses). Although the antibiotic has been used in plant pathology in the treatment of leaf diseases due to *Botrytis*, *Alternaria solani*, and species of mildew-causing fungi, it is currently not in wide use here because of the high cost (due to the high production costs). Griseofulvin affects the morphogenesis of many fungi and in concentrations of 10–1000 ppm causes a characteristic "curling" effect of the hyphae. The exact mechanism of action has not yet been determined, but mitosis and chitin biosynthesis are presumed to be affected.

The biosynthesis of griseofulvin occurs on a multi-enzyme complex. The antibiotic is produced from acetate, 6 molecules of malonate, and 2 methyl groups derived from methionine. An intermediate stage in the biosynthetic process is a 14-member polyketide, from which the antibiotic is produced via cyclization, chlorination, and methylation reactions.

The chemical synthesis of griseofulvin has been achieved, but production by fermentation with *P. patulum* is more cost-effective. Griseofulvin production is carried out as an aerobic submerged process in a glucose-rich medium. In some processes, glucose feeding is used. Starch, lactose, or sucrose can also be used as carbon sources and usually $NaNO_3$ is used as a nitrogen source. The addition of potassium chloride (0.1–0.3%) stimulates production. The fermentation take place at 25 °C for 7–9 days within an optimal pH range between 6.8–7.2. High aeration (1 vvm O_2) is required when media with high carbon or nitrogen concentrations are used. The entire culture must be processed for recovery, since griseofulvin is only partially mycelium-bound.

Novobiocin

Novobiocin is a coumarin glycoside (Figure 13.44), which can be isolated from culture filtrates of *Streptomyces spheroides*, *S. niveus*, *S. griseus*, or *S. griseoflavus*. Novobiocin is effective against Gram-positive bacteria, particularly staphylococci, against Gram-negative meningococci and gonococci, and also against *Haemophilus* and some *Proteus* strains. Novobiocin shows no cross resistance with other antibiotics, but organisms do develop rapid resistance, a marked disadvantage for its medical use. Novobicin acts as a specific inhibitor of DNA gyrase during DNA replication.

The biosynthesis of the coumarin moiety is via aromatic metabolism. L-Tyrosine is incorporated into ring A and B; ring C is derived from glucose. The fermentation is carried out in a medium with glucose or sucrose as a carbon source. Organic acids, such fumaric acid, can be added, and peptone and meat extract are favorable nitrogen sources. p-Aminosalicylic acid can be used as a precursor and addition of tyrosine and phenylalanine stimulate the formation of the antibiotic. Fermentations take place over about 90 h at 25–28 °C and at a slightly alkaline pH range, with aeration rates of up to 1.5 vvm.

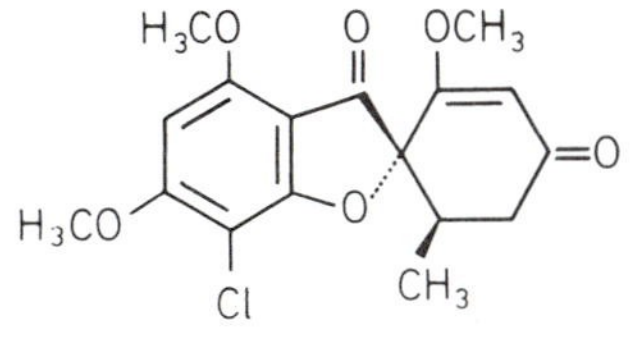

Figure 13.43 Griseofulvin

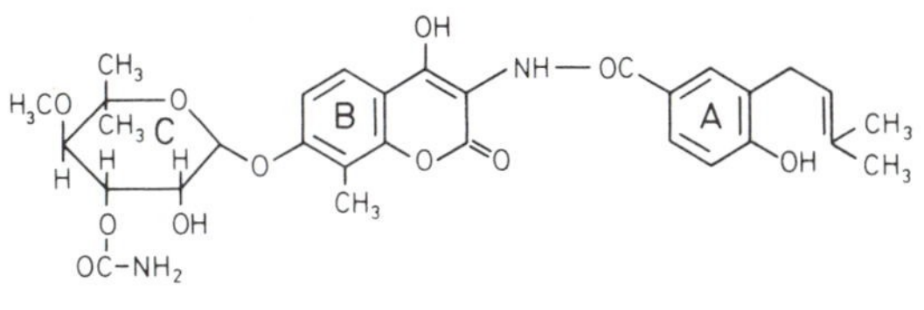

Figure 13.44 Novobiocin

13.9 OTHER COMMERCIALLY PRODUCED ANTIBIOTICS

Some antibiotics are discussed here which are used in therapy or have other uses. Their structural affiliation with certain classes of antibiotics was given in Table 13.2.

Fusidic acid

Among the steroid antibiotics, only fusidic acid (Figure 13.45) is therapeutically important. This compound, isolated from culture filtrates of *Fusidium coccineum*, inhibits the ribosomal translocation process during protein synthesis in both procaryotes and eucaryotes. Fusidic acid finds special use for the treatment of staphylococcal infections, especially wound infections and those of the skin.

Polyether antibiotics

Due to the numerous tetrahydropyran and tetrafuran moieties in these antibiotics, they are called polyethers. Over 150 different compounds are known. Their biosynthesis is derived from a polyketide derivative which is produced on a multienzyme complex. **Monensin** (Figure 13.46), commercially the most important compound of this group, consists of 5 acetate, 4 propionate and 3 butyrate units. Because of their ability to transport mono- or divalent cations through biological or artificial membranes, these antibiotics are classified with the **ionophores**.

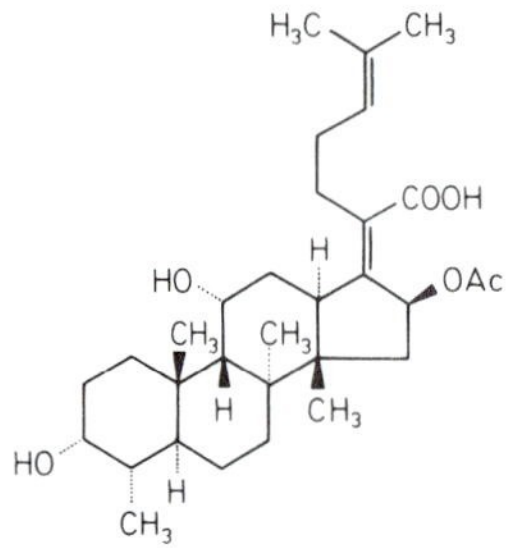

Figure 13.45 Fusidic acid

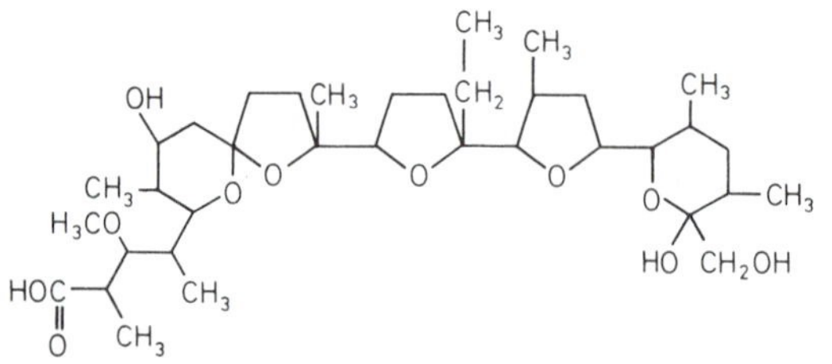

Figure 13.46 Monensin

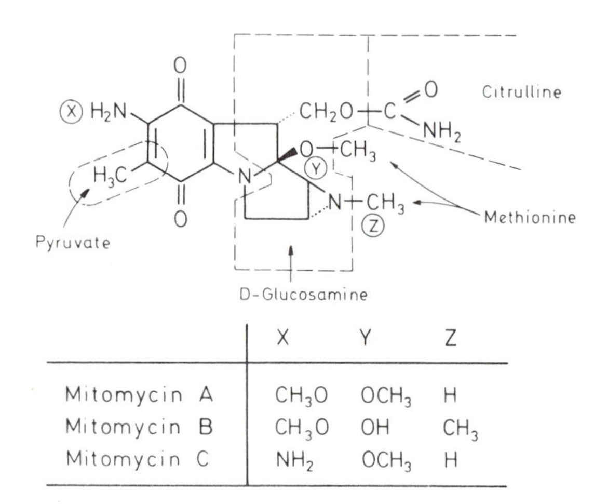

	X	Y	Z
Mitomycin A	CH_3O	OCH_3	H
Mitomycin B	CH_3O	OH	CH_3
Mitomycin C	NH_2	OCH_3	H

Figure 13.47 Structure of mitomycins A–C and biochemical precursors of mitomycin C

Antibiotics in this group which are commercially prepared include monensin, produced by *Streptomyces cinnamonensis*, **lasalocide,** produced by *S. lasaliensis,* and **salinomycin,** produced by *S. albus*. Salinomycin yields of 60 g/l are attained on media containing a fatty acid. Polyethers are active against coccidia (protozoa), as well as Gram-positive bacteria, mycobacteria, and fungi. They are added to poultry feed (about 100 ppm) to prevent coccidioses. Monensin, which was first introduced in 1971, has about 80% of the market; current annual gross world sales are estimated at $100 million.

Mitomycins

The mitomycins, produced by *Streptomyces caespitosus*, are benzoquinones; Figure 13.47 gives their structure and biochemical origin. Erythrose and pyruvate have been detected as precursors of the methylbenzoquinone. Mitomycin C is very effective against tumors, but it is relatively toxic. Although this compound was first described in the late 1950's, it was not marketed until 1974. Mitomycin reacts with double-stranded DNA, producing cross-linkages which inhibit DNA replication. Although various derivatives have been produced with the aim of achieving reduced toxicity, success along these lines has not yet been achieved.

REFERENCES

Abraham, E.P. 1986. Enzymes involved in penicillin and cephalosporin formation, pp. 115–132. In: Kleinkauf, H., H. von Döhren, H. Dornauer, and G. Nesemann (eds.). Regulation of secondary metabolite formation. Proceedings of the 16th workshop conference Hoechst, 1985. VCH Verlagsgesellschaft mbH, Weinheim.

Alačevič, M. 1976. Recent advances in *Streptomyces rimosus* genetics, pp. 513–519. In: Macdonald, K.D. (ed.), 2nd Int. Symp. on the Genetics of Industrial Microorganisms. (GIM 74). Academic Press, London.

Allsop, A., P.J. Normansell, J.F. Makins, and G. Holt. 1981. Biochemical genetics of β-lactam production. 2nd Europ. Congr. Biotechnol., Eastbourne, England. Abstracts of communications, p. 232.

Backus, M.P. and J.F. Stauffer. 1955. The production and selection of a family of strains in *Penicillium chrysogenum*. Mycologia 47: 429–463.

Ball, C. 1978. Genetics in the development of the penicillin process, pp. 165–176. In: Hütter, R., T. Leisinger, J. Nüesch, and W. Wehrli (eds.). Antibiotics and other secondary metabolites. Federation Europ. Microbiol. Soc. Symp. No. 5. Academic Press, London.

Barabas, G. 1980. Investigations of the biological role of antibiotics (some problems and ideas). Folia Microbiol. 25: 270–277.

Berdy, J. 1985. Screening, classification, and identification of microbial products. pp. 9–31. In: Verrall, M.S. (ed.). Discovery and isolation of microbial products. Ellis Horwood Publishers, Chichester.

Berdy, J. 1986. Further antibiotics with practical application, pp. 465–507. In: Rehm, H.J. and G. Reed (ed.). Biotechnology, Vol. 4. VCH Verlagsgesellschafft mbH, Weinheim.

Biot, A.M. 1984. Virginiamycin: Properties, biosynthesis, and fermentation, pp. 695–720. In: Vandamme, E.J. (ed.) Biotechnology of industrial antibiotics. Marcel Dekker, Inc., New York.

Brockman, H. 1960. Structural differences of the actinomycins and their derivatives. Ann. N.Y. Acad. Sci. 89: 323–335.

Buckland, B.C., D.R. Omstead, and V. Santamaria. 1985. Novel β-lactam antibiotics, pp. 49–68. In: Moo-Young, M. (ed.). Comprehensive biotechnology, Vol. 3. Pergamon Press, Oxford.

Burg, R.W., B.M. Miller, E.E. Baker, J. Birnbaum, S.A. Currie, R. Hartman, Y.L. Kong, R.L. Monaghan, G. Olsen, I. Putter, J.B. Tunac, H. Wallick, E.O. Stapley, R. Oiwa, and S. Omura. 1979. Avermectine, new family of potent antihelminthic agents. Antimicrob. Agents Chemother. 15:361–367.

Ciba Geigy AG. 1977. Switzerland Patent CH-006494.

Claridge, C.A. 1979. Aminoglycoside antibiotics, pp. 151–238. In: Rose, A.H. (ed.), Economic microbiol., vol. 3, Secondary products of metabolism. Academic Press, London.

Crameri, R. and J.E. Davies. 1986. Increased production of aminoglycosides associated with amplified antibiotic resistance genes. J. Antibiot. 39:128–135.

Crueger, A. 1983. Protoplast fusion in strain improvement of *Micromonospora inyoensis*, a sisomicin-producer, pp. 348–349. In: Potrykus, I., C.T. Harms, A. Hinnen, R. Hütter, P.J. King, and R.D. Shillito (eds.). Prototplasts 1983. Poster Proceedings. 6th Int. Protoplast Symposium, Basel. Birkhäuser Verlag, Basel.

Dalhoff, A. 1983. Studies on the action of aminoglycosides on bacterial membranes. Zbl. Bakteriol. 1. Abt. Orig. A, 253:427.

Dürckheimer, W., J. Blumbach, R. Lattrell, and K.H. Scheunemann. 1985. Neuere Entwicklungen auf dem Gebiet der β-Lactam Antibiotica. (New developments with β-lactam antibiotics.) Agnew. Chem. 97:183–205.

Elander, R.P. 1979. Mutations affecting antibiotic synthesis in fungi producing β-lactam antibiotics, pp. 21–35. In: Sebek, O.K. and A.I. Laskin (eds.), Genetics of industrial microorganisms. (GIM 78). Amer. Soc. Microbiol., Washington, D.C.

Flickinger, M.C. 1985. Anticancer agents, pp. 231–273. In: Moo-Young, M. (ed.) Comprehensive biotechnology, Vol. 3. Pergamon Press, Oxford.

Florent, J. 1985. Streptomycin and commercially important aminoglycoside antibiotics, pp. 137–162. In: Moo-Young, M. (ed.). Comprehensive biotechnology, Vol. 3. Pergamon Press, Oxford.

Frøyshov Ø. 1984. The bacitracins: Properties, biosynthesis, and fermentation, pp. 665–694. In: Vandamme, E.J. (ed.). Biotechnology of industrial antibiotics. Marcel Dekker, Inc., New York.

Fujiwara, T., Y. Takahashi, K. Matsumoto, and E. Kondo. 1980. Isolation of an intermediate of 2-deoxystreptamine biosynthesis from a mutant of *Bacillus circulans*. J. Antibiot. 33: 824–829.

Ghisalba, O. and J. Nüesch. 1978. A genetic approach to the biosynthesis of the rifamycin-chromophore in *Nocardia mediterranei*. I. Isolation and characterization of a shikimate-excreting auxotrophic mutant of *N. mediterranei* with normal rifamycin production. J. Antibiot. 31: 215–225.

Gil, J.A., P. Liras, and J. F. Martin. 1984. The polyenes: Properties, biosynthesis, and fermentation. In: Vandamme, E.J. (ed.). Biotechnology of industrial antibiotics. Marcel Dekker, Inc., New York.

Gray, P.P. and S. Bhuwapathanapun. 1980. Production of the macrolide tylosin in batch and chemostat cultures. Biotechnol. Bioeng. 22: 1785–1804.

Hamlyn, P.F. and C. Ball. 1979. Recombination studies with *Cephalosporium acremonium*. In: Sebek, O.K. and A.I. Laskin (eds.), Genetics of industrial microbiol. (GIM 78). Amer. Soc. Microbiol., Washington, D.C.

Higashide, E. 1984. The macrolides: Properties, biosynthesis, and fermentation, pp. 451–509. In: Vandamme, E.J. (ed.). Biotechnology of industrial antibiotics. Marcel Dekker, Inc., New York.

Holt, G., M.E. Rogers, and G. Saunders. 1986. Genetics of *Penicillium chrysogenum* and biosynthesis of β-lactam antibiotics, pp. 103–114. In: Kleinkauf, H., H. von Döhren, H. Dornauer, and G. Nesemann (eds.). Regulation of secondary metabolite formation. Proceedings of the 16th workshop conference Hoechst, 1985. VCH Verlagsgesellschaft mbH, Weinheim.

Horinouchi, S. and T. Beppu. 1987. A-Factor and regulatory network that links secondary metabolism with cell differentiation in *Streptomyces*, pp. 41–48. In: Alacevic, M., D. Hranueli, and Z. Toman (eds.). Genetics of Industrial Microorganisms, Zagreb.

Hoštalek, Z., M. Blumauerová, and Z. Vaněk. 1974. Genetic problems of the biosynthesis of tetracycline antibiotics, pp. 13–67. In: Ghose, T.K., A. Fiechter, and N. Blakebrough (eds.), Advances in biochemical engineering, vol. 3. Springer-Verlag, Berlin.

Hoštalek, Z., Blumauerová, and Z. Vaněk. 1979. Tetracycline antibiotics, pp. 293–354. In: Rose, A.H. (ed.), Economic microbiology, vol. 3, Secondary products of metabolism. Academic Press, London.

Howells, J.D., L.E. Anderson, G.L. Coffey, G.D. Senos, M.A. Untershill, D.L. Vogler, and J. Ehrlich. 1972. Butirosin, a new aminoglycoside antibiotic complex: Bacterial origin and some microbiological studies. Antimicrob. Agents Chemother. 2: 79–83.

Humphrey, A.E. 1980. Continuous culture—its influence on engineering practice, p. 68. VIth Int. Ferment. Symp. and VIth Int. Symp. on Yeasts. London, Ontario, Canada.

Ichikawa, T., M. Date, T. Ishikura, and A. Ozaki. 1971. Improvement of kasugamycin-producing strain by the agar piece method and the prototroph method. Folia Microbiol. 16: 218–224.

Jensen, E.B., R. Nielsen, and C. Emborg. 1981. The influence of acetic acid on penicillin production. Europ. J. Appl. Microbiol. Biotechnol. 13: 29–33.

Kleinkauf, H. and H. von Döhren. 1985. Peptide antibiotics, pp. 95–135. In: Moo-Young, M. (ed.). Comprehensive biotechnology, Vol. 3. Pergamon Press, Oxford.

Kleinkauf, H. and H. von Döhren. 1985. Enzyme systems synthesizing peptide antibiotics, pp. 173–207. In: Kleinkauf, H., H. von Döhren, H. Dornauer, and G. Nesemann (eds.). Regulation of secondary metabolite formation. Proceedings of the 16th workshop conference Hoechst, 1985. VCH Verlagsgesellschaft mbH, Weinheim.

Lancini, G. 1986. Ansamycins, pp. 431–463. In: Rehm, H.J. and G. Reed (eds.). Biotechnology, Vol. 4. VCH Verlagsgesellschaft, Weinheim.

Liu, C.M., L.E. McDaniel, and C.P. Schaffner. 1975. Factors affecting the production of candicidin. Antimicrob. Agents Chemother. 7: 196–202.

Malik, V.S. 1979. Regulation of chorismate-derived antibiotic production. Adv. Appl. Microbiol. 25: 75–93.

Martin, J.F. and P. Liras. 1985. Biosynthesis of β-lactam antibiotics: Design and construction of overproducing strains. Trends Biotechnol. 3:39–44.

Martin, J.F., B. Diez, E. Alvarez, J.L. Barredo, and M.J. Cantoral. 1987. Development of a fermentation system in *Penicillium chrysogenum*: Cloning of genes involved in penicillin biosynthesis, pp. 297–308. In: Alacevic, M., D. Hranueli, and Z. Toman (eds.). Proceedings of the 5th Int. Symp. on the Genetics of Industrial Microorganisms, Zagreb.

McClure, W.R. and C.L. Cech. 1978. On the mechanism of rifampicin inhibition of RNA synthesis. J. Biol. Chem. 253: 8949–8956.

Mehta, R.J., J.L. Speth, and C.H. Nash. 1979. Lysine stimulation of cephalosporin C synthesis in *Cephalosporium acremonium*. Europ. J. Appl. Microbiol. Biotechnol. 8: 177–182.

Messenger, A.J.M. and C. Ratledge. 1985. Siderophores, pp. 275–295. In: Moo-Young, M. (ed.). Comprehensive biotechnology, Vol. 3. Pergamon Press, Oxford.

Morikawa, Y., J. Karube and S. Suzuki. 1980. Continuous production of bacitracin by immobilized living whole cells of *Bacillus* sp. Biotechnol. Bioeng. 22: 1015–1024.

Neumann, M. 1981. Antibiotika-Kompendium (Handbook of antibiotics). Verlag Hans Huber, Bern.

Okachi, R. and T. Nara. 1984. The aminoglycosides: Properties, biosynthesis, and fermentation, pp. 329–365. In: Vandamme, E.J. (ed.). Biotechnology of industrial antibiotics, Marcel Dekker, Inc., New York.

Okita, W.B. and D.J. Kirwan. 1986. Simulation of secondary metabolite production by immobilized living cells: Penicillin production. Biotechnol. Progress 2: 83–90.

Okumura, Y. 1983. Peptidolactones, pp. 147–178. In: Vining, L.C. (ed.). Biochemistry and genetic regulation of commercially important antibiotics. Addison-Wesley Publishing Company, London.

Ollis, W.D., C. Smith, and D.E. Wright. 1979. The orthosomycin family of antibiotics I. The constitution of flambamycin. Tetrahedron 35: 105–127.

Omura, S. and Y. Tanaka. 1986. Macrolide antibiotics, pp. 359–391. In: Rehm, H.J., and G. Reed (eds.). Biotechnology, Vol. 4. VCH Verlagsgesellschaft mbH, Weinheim.

O'Sullivan, J. and C. Ball. 1983. β-Lactams, pp. 73–94. In: Vining, L.C. (ed.). Biochemistry and genetic regulation of commercially important antibiotics. Addison-Wesley Publishing Co., London.

O'Sullivan, J. and R.B. Sykes. 1986. α-Lactam antibiotics. pp. 247–281. In: Rehm, H.J., and G. Reed (eds.) Biotechnology, Vol. 4. VCH Verlagsgesellschaft mbH, Weinheim.

Penco, S. 1980. Antitumor anthracyclines: New developments. Proc. Biochem. June/July: 12–16.

Peperdy, J.F., P.F. Hamlyn, G. Perez-Martinez, and E. Nesemann. 1987. Parasexuality and the genetics of *Ceph-*

alosporium acremonium, pp. 139–148. In: Alacevic, M., D. Hranueli, and Z. Toman (eds.).Proceedings of the 5th Int. Symp. on the Genetics of Industrial Microorganisms, Zagreb.

Perlman, D. 1979. Microbial production of antibiotics, pp. 241–280. In: Peppler, H.J. and D. Perlman (eds.), Microbial technology, vol. I, 2nd edition. Academic Press, New York.

Podojil, M., M. Blumauerova, Z. Vanek, and K. Culik. 1984. The tetracyclines: Properties, biosynthesis, and fermentation, pp. 259–279. In: Vandamme, E.J. (ed.). Biotechnology of industrial antibiotics. Marcel Dekker, Inc., New York.

Pressman, B.C. 1976. Biological applications of ionophores. Ann. Rev. Biochem. 45: 501–530.

Queener, S. and R. Schwartz. 1979. Penicillins: Biosynthetic and semisynthetic, pp. 35–122. In: Rose, A.H. (ed.), Economic microbiology, vol. 3. Secondary products of metabolism. Academic Press, London.

Rhodes, P.M., N. Winskill, E.J. Friend, and M.W. Warren. 1981. Biochemical genetics of oxytetracycline biosynthesis. 2nd Europ. Congr. Biotechnol., Eastbourne, England. Abstracts of communications, p. 229.

Rhodes, P.M., I.S. Hunter, E.J. Friend, and M. Warren. 1984. Recombinant DNA methods for the oxytetracycline producer *Streptomyces rimosus*. Biochem. Soc. Transactions 12: 586–587.

USSR Patent SU 497,333. 1974. *Bac. lichenformis* 1001 strain-producing antibiotic bacitracin, used in cattle breeding.

Samson, S.M., R. Belagaje, D.T. Blankenship, J.L. Chapman, D. Perry, P.L. Skatrud, R.M. Van Frank, E.P. Abraham, J.E. Baldwin, S.W. Queener, and T.D. Ignolia. 1985. Isolation, sequence determination and expression in *Escherichia coli* of the isopenicillin N synthetase gene from *Cephalosporium acremonium*. Nature 318:191–194.

Skatrud, P.L., D.L. Fisher, T.D. Ignolia, and S.W. Queener. 1987. Improved transformation of *Cephalosporium acremonium*, pp. 111–119. In: Alacevic, M., D. Hranueli, and Z. Toman (eds.). Proceedings of the 5th Int. Symp. on the Genetics of Industrial Microorganisms, Zagreb.

Suami, T., S. Ogawa, and N. Chida. 1980. The revised structure of validamycin A. J. Antibiot. 33: 98–99.

Swartz, R.W. 1979. The use of economic analysis of penicillin G manufacturing costs in establishing priorities for fermentation process improvement. Ann. Rep. Ferm. Proc. 3: 75–110.

Swartz, R.W. 1985. Penicillins, pp. 7–47. In: Moo-Young, M. (ed.). Comprehensive biotechnology, Vol. 3. Pergamon Press, Oxford.

Takita, T. 1984. The bleomycins: properties, biosynthesis, and fermentation, pp. 595–603. In: Vandamme, E.J. (ed.). Biotechnology of industrial antibiotics. Marcel Dekker, Inc., New York.

Treichler, H.J., M. Liersch, and J. Nüesch. 1978. Genetics and biochemistry of cephalosporin biosynthesis, pp. 177–199. In: Hütter, R., T. Leisinger, J. Nüesch, and W. Wehrli (eds.), Antibiotics and other secondary metabolites. Federation Europ. Microbiol. Soc. Symp. No. 5. Academic Press, London.

Umezawa, S., S. Kondo, and Y. Ito. 1986. Aminoglycoside antibiotics, pp. 309–357. In: Rehm, H.J. and G. Reed (eds.). Biotechnology, Vol. 4. VCH Verlagsgesellschaft mbH, Weinheim.

Valin, C., R. Rodriguez, A. Ramos, E. Alonso, and S. Biro. 1986. Increased oxytetracycline production in *Streptomyces rimosus* after protoplast fusion. Biotechnol. Letters 8:343–344.

Vandamme, E.J. 1984. Biotechnology of industrial antibiotics. Marcel Dekker, Inc., New York.

Vaněk, Z., J. Cudlin, M. Blumauerová, and Z. Hoštalek. 1971. How many genes are required for the synthesis of chlortetracycline? Folia Microbiol. 16: 225–240.

Vining, L.C. (ed.). 1983. Biochemistry and genetic regulation of commercially important antibiotics. Addison-Wesley Publishing Company, London.

Welzel, P., F.J. Witteler, D. Müller, and W. Reimer. 1981. Struktur des Antibioticums Moenomycin A (Structure of the antibiotic moenomycin A). Angew. Chem. 93: 130–131.

Zähner, H. 1965. Biologie der Antibiotica (Biology of antibiotics). Springer-Verlag, Berlin.

Zähner, H. 1978. The search for new secondary metabolites, pp. 1–17. In: Hütter, R., T. Leisinger, J. Nüesch, and W. Wehrli (eds.), Antibiotics and other secondary metabolites. Federation Europ. Microbiol. Soc. Symp. No. 5. Academic Press, London.

Zähner, H. 1982. Antibiotica und andere sekundäre Metabolite (Antibiotics and other secondary metabolites). In: Präve et al. (eds.), Handbuch der Biotechnologie. Akademische Verlagsgesellschaft, Wiesbaden.

Zimmer, T.L., Ø. Frøyshov, and S.G. Laland. 1979. Peptide antibiotics, pp. 123–150. In: Rose, A.H. (ed.), Economic microbiology, vol. 3. Secondary products of metabolism. Academic Press, London.

14

Ergot alkaloids

14.1 OCCURRENCE AND SIGNIFICANCE

Alkaloids, some of which have therapeutic applications, are generally derived from plants, but the ergot alkaloids are produced by fungi. Ergot alkaloids were first obtained from the sclerotium of the parasitic ascomycete *Claviceps purpurea*, which develops on rye and other grasses. The term *ergot* is used to refer both to the fungus developing in the rye plant and to the alkaloids which are produced by the fungus. Two related groups of ergot alkaloids are known, those based on lysergic acid and those based on clavin (see later). Presently there are over 40 known ergot alkaloids produced by various *Claviceps* strains. Although tests have been conducted on over 1000 different species of various filamentous fungi, lysergic acid alkaloids have been found only in species of the genus *Claviceps*. However, clavine alkaloids were found in species of the genera *Aspergillus, Penicillium,* and *Rhizobium,* as well as in *Claviceps*. Several of the most important alkaloid producers are listed in Table 14.1. Ergot alkaloids have also been found in higher plants, but they seem to be limited to several species of the plant family Convolvulaceae.

Some **pharmacological properties** of ergot alkaloids include the stimulation of the sympathetic nervous system, an effect on the smooth muscles (uterus, artery walls), and antagonism to adrenalin and serotonin. Chronic ergot poisoning (ergotism), which used to occur frequently due to consumption of bread made from grain containing ergot alkaloid, results in cramps, vomiting, and gangrene, and can be fatal in severe cases. In the Middle Ages, ergotism was common in northern Europe, where it was called St. Anthony's Fire, but it has been eliminated by modern methods of grain cleaning.

Some lysergic acid alkaloids are used therapeutically in obstetrics, including ergotamine, ergobasine, and the semisynthetic methylergobasine, all of which act to stimulate labor and cause the uterus to contract. They are also employed during the period after birth to contract the post-

Table 14.1 Several producers of ergot alkaloids and the alkaloids produced by them

Organism	Plant host	Alkaloid content (%)	Alkaloids produced (main components)
Claviceps purpurea	Rye, wheat, barley and others	0.1 –0.5	Ergotoxine group: Ergocornine Ergocryptine Ergocristine Ergotamine group: Ergotamine Ergosine
C. fusiformis	*Pennisetum typhoides*	0.3	Ergometrine Agroclavine Elymoclavine Chanoclavine
C. gigantea	*Zea mais*	0.03	Festuclavine Pyroclavine Dihydroelymoclavine Chanoclavine
C. paspali	*Paspalum* spp.	0.003	D-Lysergic acid α-Hydroxyethyl-lysergamide

(Mantle, 1975)

partum uterus. Hydrogenation of the $\Delta^{9,10}$-double bond in the D-ring reduces the contracting action on the smooth muscles; dihydroergotamine is thus used to treat migraine headaches. Also, a mixture of dihydro compounds from ergocristine, ergocryptine, and ergocornine is used for treatment of disturbances of the peripheral and central circulation system. Finally, a few naturally occurring alkaloids, particularly the semisynthetic compound lysergic acid diethylamide (LSD), are very effective **hallucinogens**.

14.2 DEVELOPMENTAL CYCLE OF *CLAVICEPS*

The *Claviceps* fungus lives as a parasite on grasses. During flowering of the host plant, the infection cycle begins. Hyphae from germinating ascospores or conidia grow into the plant ovary and within a week the ovary tissue is penetrated by a filamentous mycelium (sphacelia), which produces mononuclear conidia for asexual reproduction. No alkaloids are produced during this stage. About two weeks after infection, production of conidia stops, the filamentous mycelium is replaced with a plectenchymous, nonsporulating tissue, and the sclerotium appears. It is during this latter developmental phase that alkaloids are produced, usually as part of a mixture of different substances. The sclerotia survive until the host plant blooms again; hyphae grow out from the sclerotia and long-stalked stromata grow on the ends of the hyphae. The stromata contain perithecia with asci; the products of this sexual stage are 8 needle-shaped ascospores per ascus. The ascospores germinate on the plant and the cycle begins again, as illustrated in Figure 14.1.

14.3 STRUCTURE

The active ingredients of ergot were characterized in 1935 as derivatives of lysergic acid and the structural formulas were confirmed through synthesis in 1954. Ergot alkaloids are classified among the **indole alkaloids** and are derived from the tetracyclic ergoline ring system (Figure 14.2).

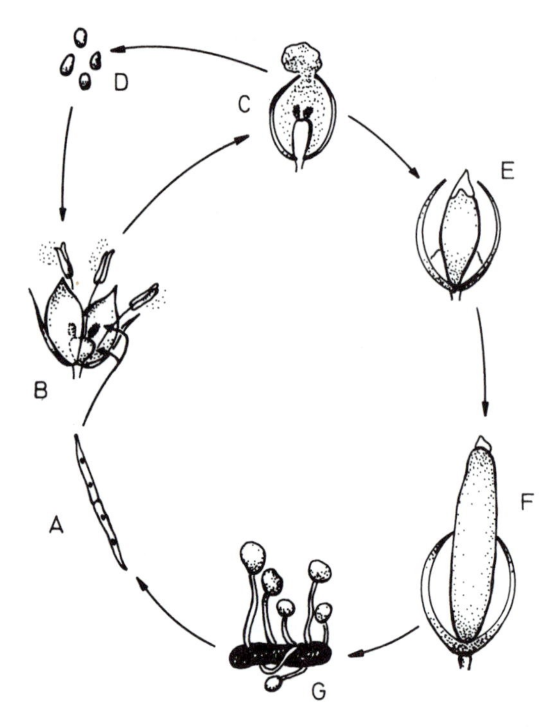

Figure 14.1 Development cycle of *Claviceps purpurea*. A, ascospore (1×100 μm); B, infection of grass flower; C, growth of filamentous mycelium in the ovary; D, asexual spores excreted with honey dew cause additional infections but no alkaloid formation; E, filamentous mycelium is replaced by plectenchymatic tissue; alkaloid formation is initiated; F, the sclerotium ripens within 2 months; G, sexual stage with perithecium formation (From Mantle, 1975)

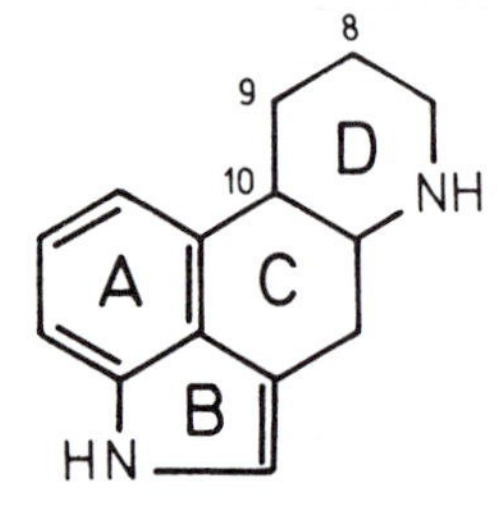

Figure 14.2 Structure of the ergoline ring system

Two main classes of ergot alkaloids have been recognized. The **clavine alkaloids** or **clavines** have ergoline as a basic structure, but contain no peptide components. Some substances of this group are intermediates in lysergic acid biosynthesis (see Figure 14.5). Various *Claviceps* species produce clavines, particularly in culture away from the plant, but traces of different clavines mixed with lysergic acid alkaloids have also been found in parasitic growth. Despite their pharmacological effect, clavines have not been used therapeutically.

The other class of alkaloids are the **lysergic acid alkaloids** or "Ergo" alkaloids. These are the classical alkaloids which are isolated from sclerotia. In these compounds, D-lysergic acid (or its stereoisomer D-isolysergic acid) is linked with a tricyclic peptide or an amino alcohol by an amide bond. The best known member of the latter group is ergometrin, which has 2-aminopropanol-(1) as a side chain. The structure of this compound, as well as the structures of several other of the simpler lysergic acid alkaloids, are shown in Figure 14.3.

Two groups of peptide alkaloids have been isolated, the classical **ergopeptins** and the more recently discovered **ergopeptams**. The ergopeptins consist of a lysergic acid molecule connected by a peptide bond to a tricyclic peptide which consists of an α-hydroxy-γ-amino acid derived from L-valine or L-alanine, with another amino acid and L-proline (Table 14.2; Figure 14.4). The first ergopeptams were only isolated in 1982. They are ergocristine, α-ergocryptine, and ergocornine. In these compounds, the tripeptide moiety, an acyclic lactam, is composed of two amino acids (which can be identical to the amino acids of the ergopeptins) and D-proline.

Lysergic acid and its derivatives epimerize easily at C-8, leading to the formation of the pharmacologically ineffective D-isolysergic acid compounds. They are distinguished by the ending "-inine", so that ergocristinine is the C-8 epimer of ergocristine.

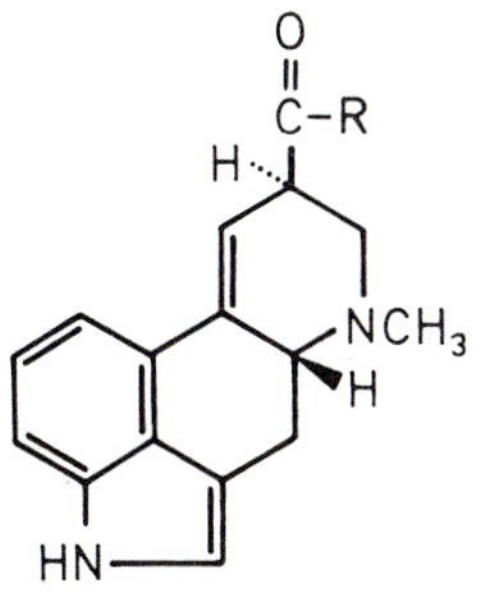

Name	R
Ergometrine (Ergobasine)	$-HN-C(CH_3)(H)-CH_2OH$
α-Hydroxyethyl-lysergamide	$-HN-CH(CH_3)-OH$
Lysergic acid	$-OH$
$\Delta^{8,9}$-Lysergic acid^{x}	$-OH$

Figure 14.3 Structure of several simply constructed lysergic acid alkaloids (x = double bond in the $\Delta^{8,9}$ position)

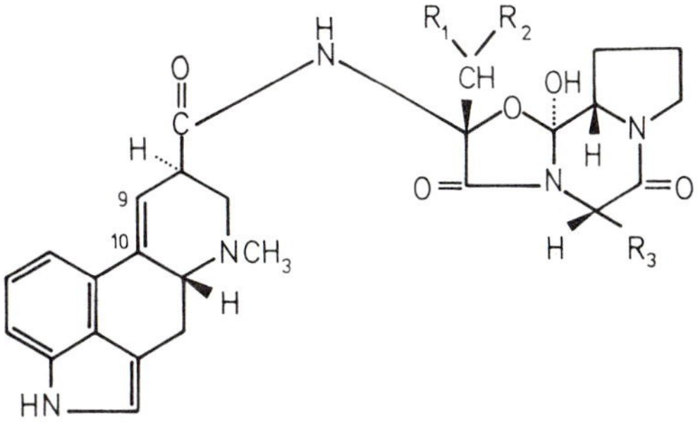

Name	R_1	R_2	R_3
Ergotamine	H	H	$CH_2-C_6H_5$
Ergosine	H	H	$CH_2CH(CH_3)_2$
Ergocristine	CH_3	CH_3	$CH_2-C_6H_5$
α-Ergocryptine	CH_3	CH_3	$CH_2CH(CH_3)_2$
β-Ergocryptine	CH_3	CH_3	$CH(CH_3)CH_2CH_3$
Ergocornine	CH_3	CH_3	$CH(CH_3)_2$
Ergostine	H	CH_3	$CH_2-C_6H_5$

Figure 14.4 Naturally occurring ergot alkaloids of the peptide type (ergopeptins)

Table 14.2 Amino acid sequence of the tripeptide moiety in ergotoxine and ergotamine alkaloids (ergopeptins)

Alkaloid	Amino acid sequence of the tripeptide
Ergotoxine-alkaloids:	
Ergocristine	—L-Valine—L-Phenylalanine—L-Proline
Ergocornine	—L-Valine—L-Valine—L-Proline
α-Ergocryptine	—L-Valine—L-Leucine—L-Proline
β-Ergocryptine	—L-Valine—L-Isoleucine—L-Proline
Ergotamine-alkaloids:	
Ergotamine	—L-Alanine—L-Phenylalanine—L-Proline
Ergosine	—L-Alanine—L-Leucine—L-Proline

14.4 BIOSYNTHESIS

Alkaloid synthesis occurs upon the endoplasmic reticulum in *Claviceps* when it is grown in submerged culture. About 95% of the clavines and the simple lysergic acid derivatives are excreted into the medium; the peptide alkaloids remain mostly mycelium-bound.

The formation of the ergoline ring proceeds from mevalonic acid and tryptophan, as outlined in Figure 14.5. In a series of reactions analogous to those involved in isoprenoid synthesis, mevalonic acid is transformed via mevalonic acid pyrophosphate and isopentenyl pyrophosphate

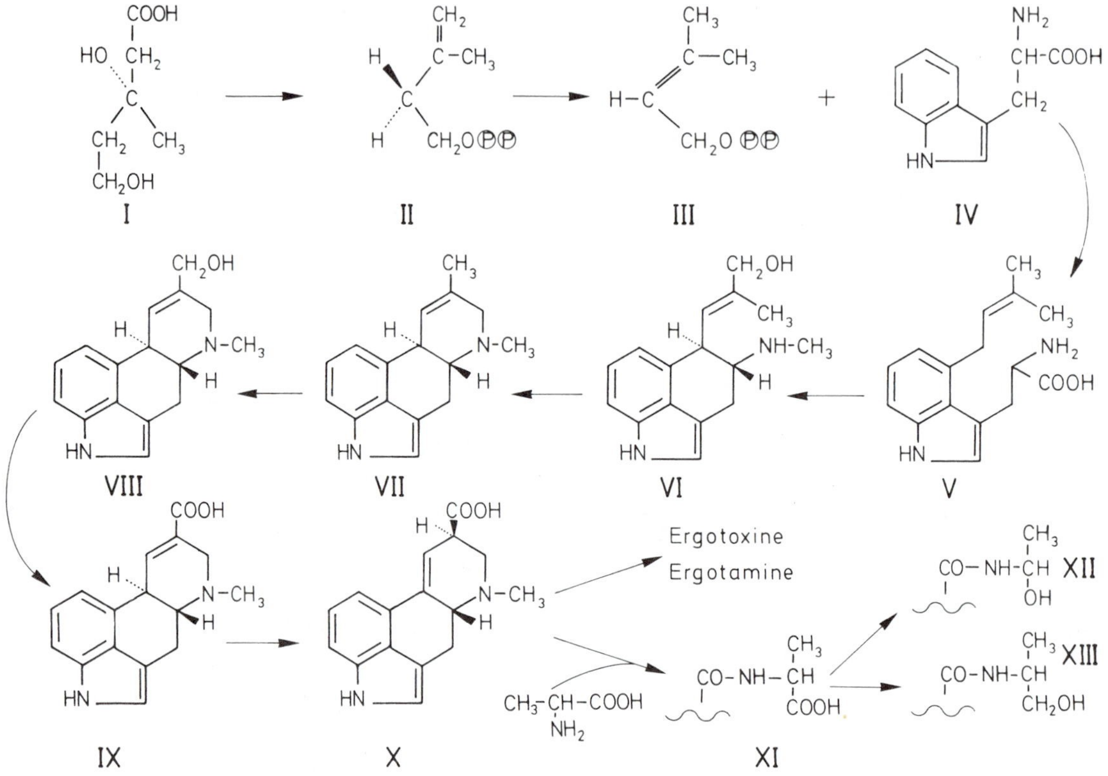

Figure 14.5 Biosynthesis of the ergoline ring system. I, Mevalonate; II, Isopentenylpyrophosphate; III, Dimethylallylpyrophosphate; IV, Tryptophan; V, 4-Dimethylallyltryptophan; VI, Chanoclavin-I; VII, Agroclavine; VIII, Elymoclavine; IX, $\Delta^{8,9}$-Lysergic acid; X, Lysergic acid; XI, Lysergylalanine; XII, α-Hydroxyethyllysergamide; XIII, Ergometrine

into dimethylallyl pyrophosphate. The first reaction specifically for alkaloid synthesis is the linkage of dimethylallyl pyrophosphate with L-tryptophan (in position C-4), tryptophan being converted into 4-dimethylallyl tryptophan. In the next intermediate step, chanoclavin-I, a clavine with an open D-ring, is produced. The methyl group on the nitrogen in position 6 is derived from methionine. The ring closes after oxidation to chanoclavin-I-aldehyde and the tetracyclic compound agroclavine is formed. Hydroxylation of the methyl group of agroclavine to form elymoclavine is carried out by a peroxidase or oxygenase in the presence of cytochrome P-450. The further transformation from elymoclavine into derivatives of lysergic acid occurs via $\Delta^{8,9}$-lysergic acid, which in contrast to lysergic acid occurs freely in relatively high concentrations in the medium.

Biosynthesis of tripeptide moieties of the peptide alkaloids is not fully understood. Attempts to incorporate ^{14}C-labeled precursors showed that lysergic acid and various amino acids are incorporated as single compounds. From results of experiments with inhibitors, ergopeptin formation is postulated to take place on a multienzyme complex, analogous to the biosynthesis of peptide antibiotics (see Section 13.3). The proposed biosynthetic pathway, which begins with the linkage of L-proline to the enzyme, is illustrated in Figure 14.6. It has been possible to obtain incorporation of the appropriate amino acids

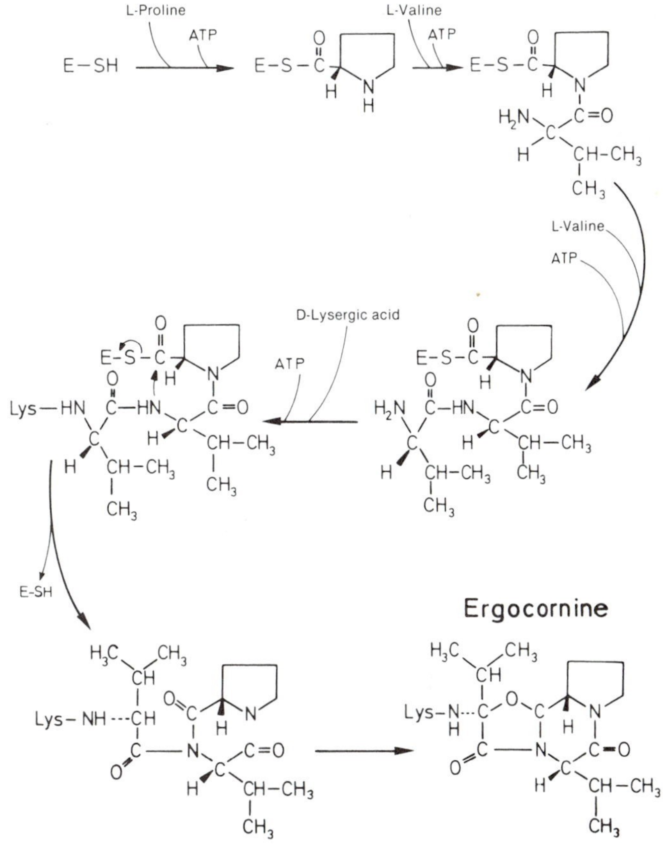

Figure 14.6 Hypothetical scheme for the formation of the peptide side chain of ergopeptines on a multi-enzyme complex (From Floss, 1976)

into the respective ergopeptin in a cell-free system in ergosine or ergotamine-producing strains of *Claviceps purpurea*, and experiments are underway to characterize the "cyclol-synthetase complex" which is involved. It is likely that ergopeptams are formed by irreversible epimerization as intermediate products before ring closure.

14.5 PRODUCTION OF ERGOT ALKALOIDS

Annual world production of peptide alkaloids was 4000 kg in 1976, and lysergic acid production was 12,000 kg.

There are a number of methods of obtaining ergot alkaloids: by **chemical synthesis**, by **culture** of *Claviceps* strains on the respective host plants, by **microbial fermentation**, either in surface culture or with immobilized cells. Culture on the host plant or in submerged culture are the processes most used commercially.

Chemical synthesis

Total chemical synthesis of ergot alkaloids is possible, but currently this is not cost-effective. Lysergic acid which is produced fermentatively can be chemically transformed into the desired alkaloids. Ergoline derivatives used medically are produced commercially using this method, with the exception of peptide alkaloids.

Obtaining alkaloids from sclerotia

In 1976, more than 95% of the peptide alkaloids were prepared by extraction from sclerotia, chiefly in Switzerland and eastern European countries. In this process, rye flowers are mechanically inoculated with conidia. The fungus grows in the ovary tissue and forms conidia which then serve as the inoculum for an extensive plant infection. About 200–500 kg of sclerotia can be harvested from 1 hectare (2.47 acres) of rye. Disadvantages of this process are that the harvest is only possible once a year and the extent of infection is extremely dependent on the weather. Since 1978, this production method has been replaced in Switzerland with the fermentation method.

Production by fermentation

Three species of *Claviceps* are currently used in the production of alkaloids by fermentation:

Claviceps paspali — α-Hydroxyethyl lysergamide, $\Delta^{8,9}$-Lysergic acid, ergometrine
C. fusiformis — Clavines (chiefly agroclavine)
C. purpurea — Ergotamine, ergosine, ergocornine, ergocristine, ergocryptines

Research with *C. paspali* at the Italian pharmaceutical company Farmitalia led to the initial breakthrough in fermentative production of ergot alkaloids. Whereas the initial isolate grown in submerged culture produced about 20 mg alkaloid/l, strain development and culture medium optimization resulted in a commercial process which had alkaloid titers of 5 g/l, with α-hydroxyethyl lysergic amide as the main product.

The stages of the fermentation process are outlined in Figure 14.7. After hydrolytic splitting, D-lysergic acid can be used in the production of semisynthetic alkaloids. Chemical conversion to the peptide alkaloids has also been accomplished.

A second simplified process for the production of lysergic acid has been developed by the Swiss company, Sandoz. An isolate of *C. paspali* produces high yields of $\Delta^{8,9}$-lysergic acid which can be rearranged by isomerization into D-lysergic acid (with the double bond in the $\Delta^{9,10}$ position). Production of the alkaloid agroclavine is possible with *C. fusiformis*, which in semicontinuous culture can achieve yields of 6g/l.

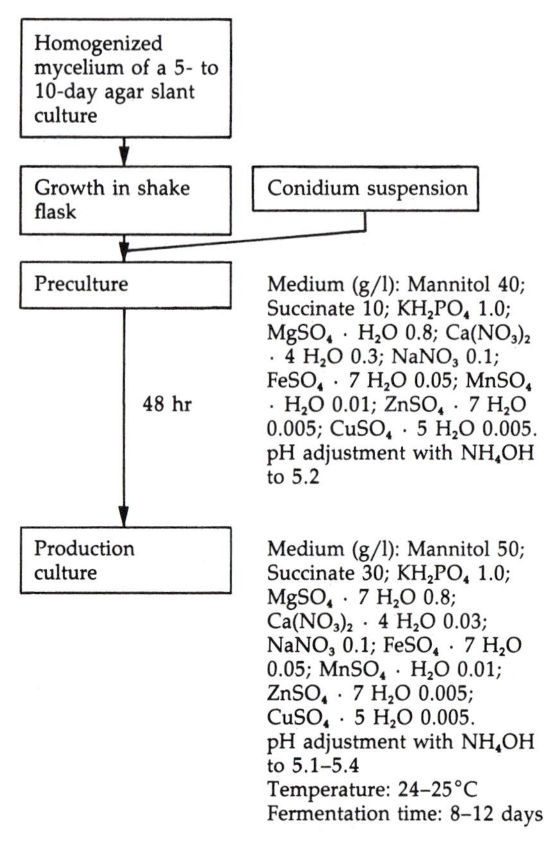

Figure 14.7 Flow scheme for the production of α-hydroxyethyl lysergamide with *C. paspali* (From Kelleher, 1970)

Processes for the production of therapeutically important peptide alkaloids have also been developed. Table 14.3 lists commercially produced compounds.

Alkaloid composition is affected by the strain used and the culture medium, as shown in Table 14.4. In *C. purpurea* the relative proportions of ergocornine and ergocryptine can be altered from 2:1 to the more desirable 1:1 by addition of L-valine to the fermentation medium. Good alkaloid formation seems to be linked to the ability to metabolize high sucrose and citrate concentrations, but at the same time, the medium must not contain any phosphate. Synthesis proceeds in parallel with lipid and sterol biosynthesis. The progress of a typical fermentation is illustrated in Figure 14.8 for *C. purpurea*.

Table 14.3 Production of ergot alkaloids by fermentation

Alkaloid	Manufacturers 1	2	3	4
Ergocornine	+	+	+	+
Ergocristine	+	+		+
Ergocryptine	+	+	+	+
Ergotamine		+	+	
Ergometrine			+	
Lysergic acid		+		
$\Delta^{8,9}$-Lysergic acid	+			+

Manufacturers: 1. Biochemie GmbH, Kundl (Austria); 2. Farmitalia S.p.A. (Italy); 3. Gedeon Richter (Hungary); 4. Sandoz-Wander AG (Switzerland).
(Rehacek, 1980)

Scale-up has been a problem, since the production strains are sensitive to the shearing action of the impeller but at the same time exhibit a high oxygen requirement. Maintenance of sterility is also difficult for long fermentation periods. Very slight alterations in the composition of the medium or the addition of antifoam agents usually cause lower yields. In addition, the degeneration of high-yielding strains frequently occurs, so that a careful strain maintenance program is necessary.

Surface culture

A process has been described for the large-scale production of ergot alkaloids using surface cultures of *C. purpurea* grown under sterile conditions. The advantage of this process is that a higher proportion of the more desirable ergotamine and ergotoxine alkaloids are formed.

Semicontinuous transformation with immobilized mycelium

A process has been developed for semicontinuous alkaloid production using immobilized mycelium of *Claviceps purpurea* and *C. fusiformis.* In *C. purpurea* CBS 164.59, which forms ergometrin and a mixture of chanoclavin, agroclavin and elymoclavin, the best results have been obtained by immobilization in 4% calcium alginate gel. At concentrations of alginate up to 8%, the overall yield of alkaloids was increased by 35%, but because of the reduced O_2 diffusion into the alginate beads, the main product was agroclavin. To prevent contamination problems during the long incubation period with the immobilized mycelium (200–400 days), an antibiotic such as chloramphenicol was added.

14.6 REGULATION OF ALKALOID PRODUCTION IN CULTURES

To produce alkaloids, the medium must always contain an organic acid of the TCA cycle or a related compound as well as a carbohydrate. Mannitol and succinate serve for the production of lysergic acid, and sucrose and citrate for the

Table 14.4 Ergot alkaloid production by various strains of *C. purpurea*

Characteristics	275 FI	FI 32/17	FI 43/14	FI S 40
Alkaloid produced	Ergotamine	Ergocryptine Ergotamine	Ergocornine Ergosine	Ergocristine
Alkaloid yield (mg/l)	1150	1800	1000–1100	1000–1100
Culture medium	T 25[a]	T 25[a]	T 25[a]	TS[b]
Sucrose consumed (g/l)	160	75	145	100
Citric acid consumed (g/l)	14	15	6	
Lipid production (g/l)	14	24	47	18.5
Sterol production (mg/l)	316	444	393	510
Fermentation time (days)	14	16	12–14	8–10

[a] Medium T 25 (g/l): Sucrose 300; citric acid 15; KH_2PO_4 0.5; $MgSO_4 \cdot 7\ H_2O$ 0.5; KCl 0.12; $FeSO_4 \cdot 7\ H_2O$ 0.007; $ZnSO_4 \cdot 7\ H_2O$ 0.006; yeast extract 0.1. Tap water; pH adjusted to 5.2 with NH_4OH. Sterilization at 120°C, 20 min.
[b] Medium TS (g/l): Sucrose 100; asparagine 10; KH_2PO_4 0.5; $MgSO_4 \cdot 7\ H_2O$ 0.3; $FeSO_4 \cdot 7\ H_2O$ 0.007; $ZnSO_4 \cdot 7\ H_2O$ 0.006; yeast extract 0.1. Distilled H_2O; pH adjusted to 5.2 with NaOH. Sterilization at 110°C, 20 min.
(Amici et al., 1969)

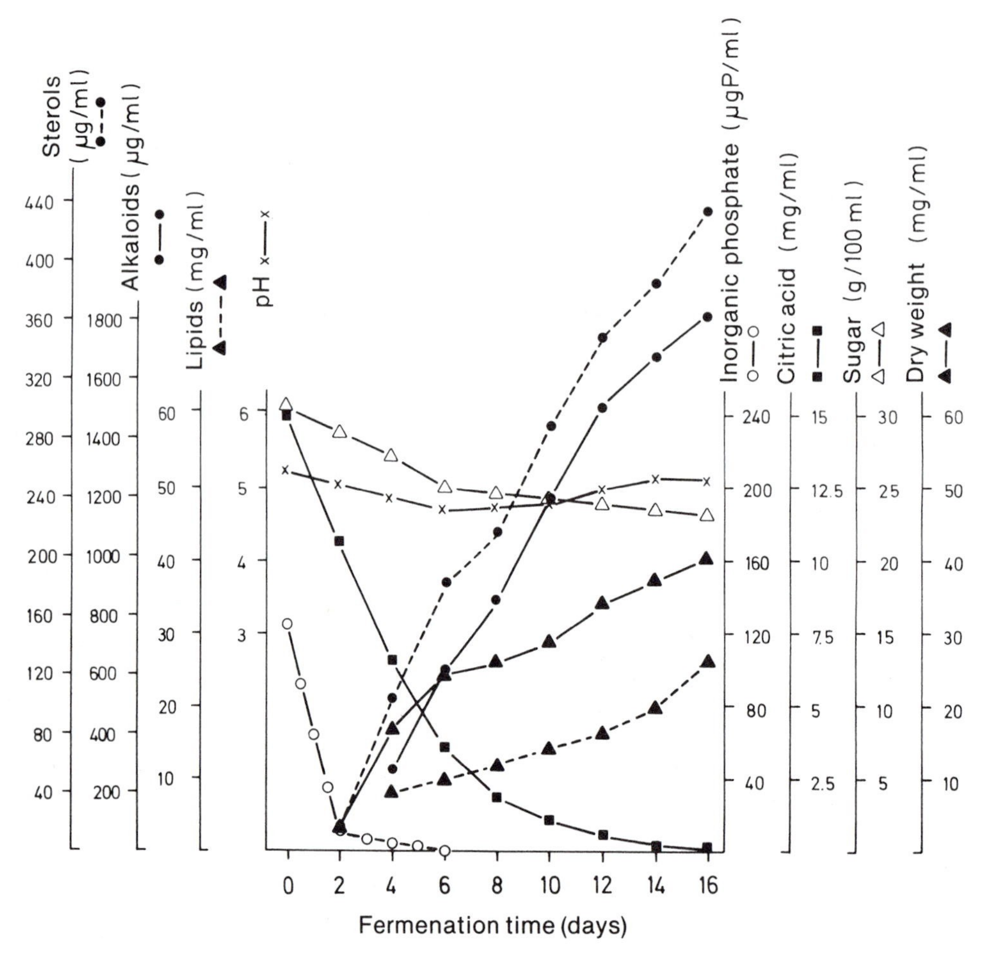

Figure 14.8 Fermentation kinetics of the ergocryptine/ergotamine producer *C. purpurea* FI 32/17 in medium T 25 (composition in Table 14.4)

production of ergotamine or ergocristine. Both carbon sources are metabolized simultaneously without any evidence of a diauxy effect, although glucose itself represses alkaloid production.

Ergot alkaloid production exhibits a typical **phosphate regulation**, as shown in Figure 14.9. In the trophophase (3–4 days), after free phosphate has been used up growth ceases and the culture enters the idiophase. Tryptophan, which induces alkaloid synthesis and simultaneously serves as a precursor, accumulates. At the same time, the inducible enzymes of secondary metabolism appear, such as 4-dimethylallyl tryptophan synthase, which is induced by tryptophan or trytophan analogs (such as 5-methyltryptophan). The phosphate inhibition mentioned above can be counteracted by the addition of tryptophan. During the transition from the trophophase to the idiophase, another key enzyme of secondary metabolism, chanoclavin-I-cyclase, appears.

Measurements of the enzyme activities of intermediary metabolism have indicated the predominant role of the TCA cycle at the beginning

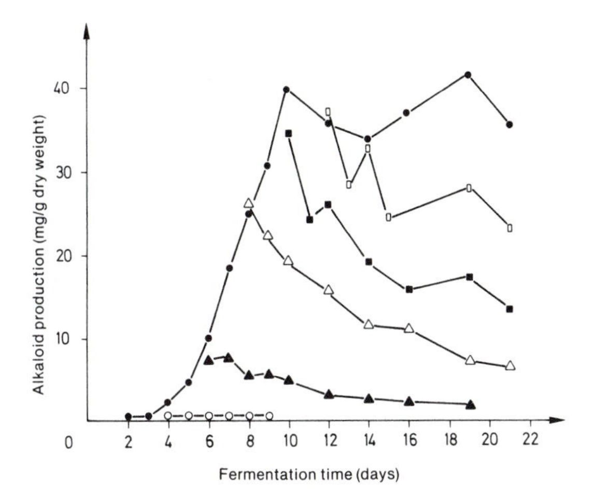

Figure 14.9 Clavine formation with *Claviceps* SD 58 after a single addition of phosphate (concentration in the medium: 1.1 g/l KH_2PO_4) at different times: • — • control; o — o addition after 3 days; ▲ — ▲ addition after 5 days; △ — △ addition after 7 days; ■ — ■ addition after nine days; and □ — □ addition after 11 days (From Robbers et al., 1978)

of the submerged fermentation process. In contrast, the glyoxylate cycle prevails during the alkaloid production phase. Acetyl-CoA can be produced via an anaplerotic sequence (glycerate pathway) of the dicarboxylic acid cycle. The activity of phosphofructokinase, a key enzyme of glycolysis, is reduced. The changeover from the TCA to the glyoxylate cycle is stimulated by the addition of citrate, succinate, or malate. Moreover, excess citrate causes inhibition of citrate synthase, making acetyl-CoA available for conversion to mevalonic acid, a key precursor of the ergoline ring system. Increased amounts of acetyl-CoA flow to fatty acid biosynthesis during activation of acetyl-CoA carboxylase, thus explaining the parallelism of lipid and alkaloid production. Since the enzymes of the pentose phosphate cycle are derepressed, a sufficient quantity of $NADPH_2$ is produced, which is necessary for lipid synthesis.

In submerged culture, synthesis of the alkaloids is increased considerably when asparagine is used as a **nitrogen source** instead of ammonium salts. Measurements during the growth phase showed there was an increase in asparaginase activity at the same time that an accumulation of ammonium ions occurred. The initiation of alkaloid formation is coupled with a significant decrease in the intracellular ammonium concentration brought about by ammonium assimilation via glutamine synthetase. Since glutamine is the amino group donor, this lead to an increase in the intracellular concentration of tryptophan, an ergolin precursor.

14.7 STRAIN DEVELOPMENT

In nature the production of ergot alkaloids is linked exclusively to the sexual phase of the developmental cycle, which does not occur in vitro. In submerged culture, the alkaloid-producing mycelium has morphological similarities with the hyphae in sclerotia. These thick-walled, isodiametric cells have a high fatty acid content. The composition of the culture medium affects the differentiation into sclerotic cell types. Increasing the sucrose concentration from 10 to 30% or adding citrate, $(NH_4)_2SO_4$, or high Ca^{2+} concentrations promotes the formation of the sclerotial cell types. Most alkaloid-producing strains are homocaryotic. In several strains that produce primarily peptide alkaloids, heterocaryosis leads to an increase in product formation. **Plasmids** have been demonstrated so far only in wild type strains. These plasmids are linear genetic elements (5.6–6.3 kb) which are present in free form in the mitochondria.

A frequently occurring problem in strain preservation is the **degeneration** of high-yielding strains. The proportion of unproductive variants after a single transfer was 3% in an ergotoxine-producing *C. purpurea* strain, but after 7 passages it was 78%. Thus, strains must be carefully maintained in storage (mycelium fragments or conidia in nutrient solution containing 10–20% glycerol can be stored in liquid nitrogen). Occasional reisolation of high-yielding variants may also be necessary in order to guarantee alkaloid production.

Yield increases have been achieved in genetic research primarily by use of mutation and selection. Conidia, mycelium fragments, or protoplasts have been treated with mutagens, such as ethyl methanesulfonate, nitrosoguanidine, nitrous acid, ethylenimine, or ultraviolet radiation. Such studies have led in every case to increased yields: for example, a 160-fold increase in an ergocristine producer. Mutagenesis has resulted primarily in changes in the amount of alkaloids produced rather than in the proportion of the various products. However, in *C. purpurea* shifts in the proportions of ergotamine to ergotoxine, ergocornine/ergocryptine to ergocrystine, or ergotamine to ergosine. Due to the absence of the sexual phase in vitro, for a long time there was no means of performing genetic analysis or using genetic recombination for strain development, but **protoplast fusion** has overcome this difficulty. Fusion of an ergocristine producer with an ergocornine/ergocryptine producer resulted in strains which appeared to be hybrids, as judged by their nutrient requirements (citrate or succinate) and by the amount of alkaloids produced, which ranged between the values of the parent strains.

Experiments with **mutasynthesis** (see Section 3.6) using *C. purpurea* show promise of obtaining new pharmacologically useful alkaloids. Auxotrophs of an ergocristine-producing strain (Phe^-) and an ergocornine/ergocryptine-producing strain (Leu^-) produced the appropriate alkaloid analogs after the addition of phenylalanine analogs (p-chlorophenylalanine, p-fluorophenylalanine) or leucine analogs (norleucine, norvaline). By feeding synthetic amino acid analogs it has been possible to exchange both of the other amino acids of the tripeptide moiety.

For the development of a **cloning system**, mitochondrial (mt) DNA from *C. purpurea* has been used as a vector. Chromosomal DNA of *C. purpurea* containing the gene for the phosphoribosylanthranilate isomerase, has been cloned in *E. coli*. This raises the possibility of using protein engineering to increase the action of the rate-limiting enzymes in the biosynthetic sequence. It can be anticipated that these new genetic techniques may lead not only to further increases in yield but to the discovery of new structural variants of the ergot alkaloids.

REFERENCES

Amici, A.M., A. Minghetti, T. Scotti, C. Spalla, and L. Tognoli. 1969. Production of peptide ergot alkaloids in submerged culture by three isolates of *Claviceps purpurea*. Appl. Microbiol. 18: 464–468.

Arcamone, F., E.B. Chain, A. Ferretti, A. Minghetti, P. Pennella, A. Tonolo, and L. Vero. 1961. Production of a new lysergic acid derivative in submerged culture by a strain of *Claviceps paspali*. Proc. Roy. Soc. (London) Ser. B 155: 26–54.

Desai, J.D., H.C. Patel, and A.J. Desai. 1986. Alkaloid production during the cultivation with shaking of *Claviceps* sp.: Effects of asparagine. J. Ferment. Technol. 64:339–342.

Floss, H.G. 1976. Biosynthesis of ergot alkaloids and related compounds. Tetrahedron 32: 873–912.

Hofmann, A. 1964. Die Mutterkornalkaloide (The ergot alkaloids). Enke Verlag, Stuttgart.

Hoffmann, A. and H. Tscherter. 1960. Isolierung von Lysergsäure-Alkaloiden aus der mexikanischen Zauberdroge Ololiuqui (*Rivea corymbosa*). (Isolation of lysergic acid alkaloids from the Mexican magic drug plant). Experientia 16: 414.

Kelleher, W.J. 1970. Ergot alkaloid fermentations. Adv. Appl. Microbiol. 11: 211–244.

Kobel, H. and J.J. Sanglier. 1986. Ergot alkaloids, pp. 569–609. In: Rehm, H.J. and G. Reed (eds.). Biotechnology, vol. 4. VCH-Verlagsgesellschaft mbH, Weinheim.

Kobel, H., E. Schreier, and J. Rutschmann. 1964. 6-Methyl $\Delta^{8,9}$-ergolen-8-carbonsäure, ein neues Ergolinderivat aus Kulturen eines Stammes von *Claviceps paspali*. (6-Methyl-$\Delta^{8,9}$-ergolen-8-carboxylic acid, a new ergoline derivative from cultures of strains of *Claviceps paspali*). Helv. Chim. Acta 47: 1052–1064.

Kopp, B. and H.J. Rehm. 1984. Semicontinuous cultivation of immobilized *Claviceps purpurea*. Appl. Microbiol. Biotechnol. 19: 141–145.

Kren, V., S. Chomatova, J. Bremek, P. Pilat, and Z. Rehacek. 1986. Effect of some broad-spectrum antibiotics on the high-production strain *Claviceps fusiformis* W1. Biotechnol. Letters 8:327–332.

Kybal, J. and B. Sikyta. 1986. Renaissance of surface culture: Production of ergot alkaloids and spore formation. Acta Biotechnol. 6:245–351.

Kybal, J., E. Svoboda, K. Strnadova, and M. Kejzlar. 1981. Role of organic acid metabolism in the biosynthesis of peptide ergot alkaloids. Folia Microbiol. 26: 112–119.

Maier, W., D. Erge, and D. Gröger. 1981. Studies on the cell-free biosynthesis of ergopeptines in *Claviceps purpurea*. Federation Europ. Microbiol. Soc. Microbiol. Lett. 12: 143–146.

Mantle, P.G. 1975. Industrial exploitation of ergot fungi, pp. 281–300. In: Smith, J.E. and D.R. Berry (eds.), The

filamentous fungi, vol. 1, Industrial mycology. Edward Arnold, London.
Puc, A., S. Milicic, M. Kremser, and H. Socic. 1987. Regulation of ergotoxine biosynthesis in *Claviceps purpurea* submerged fermentation. Appl. Microbiol. Biotechnol. 25:449–452.
Rehacek, Z. 1980. Ergot alkaloids and their biosynthesis. Adv. Biochem. Eng. 14: 33–60.
Rehacek, Z. and P. Sajdl. 1979. Changes in activity of Krebs and glyoxylate cycles during biosynthesis of agroclavine and elymoclavine. Biotechnol. Lett. 1: 53–57.
Robbers, J.E., W.W. Eggert, and H.G. Floss. 1978. Physiological studies on ergot: Time factor influence on the inhibitory effect of phosphate and the induction effect of tryptophan on alkaloid production. Lloydia 41: 120–129.

15

Microbial transformations

15.1 INTRODUCTION

Microorganisms have the ability to chemically modify a wide variety of organic compounds. Such changes are called **biological** or **microbial transformations**, or more generally **bioconversions**. In these enzymatic reactions, the substrate may be metabolized, but in some cases the conversion may take place without energy gain (cometabolism). In general, an industrial process can be implemented either by chemical synthesis or by bioconversion, but bioconversion is often preferable for the following reasons:

- Substrate specificity: Only one specific reaction step is normally catalyzed by an enzyme.
- Site specificity (regiospecificity): If several functional groups of one type are present in the molecule, only one specific position may be affected.
- Stereoselectivity: If a racemic mixture is used as starting material, only one specific enantiomer is converted. If a center of asymmetry appears as a result of the enzyme reaction, the reaction product is normally optically active.
- Reaction conditions: Enzymatic reactions do not cause destruction of sensitive substrates, due to the mild conditions of conversion. Several reactions can be combined, either in one fermentation step using an organism with suitable enzyme systems, or by step-wise conversions using different microorganisms. The reactions cause less environmental hazard, as they take place chiefly in water.

15.2 TYPES OF BIOCONVERSION REACTIONS

The most important microbial transformation reactions are outlined in Table 15.1. Chemicallly, these transformations can be grouped under the following categories: oxidation, reduction, hydrolysis, condensation, isomerization, formation of new C—C bonds, introduction of hetero functions. Oxidation reactions are particularly useful

Table 15.1 Bioconversion reactions of various types

Type of reaction	Example	Conversion efficiency (%)
OXIDATIONS		
Hydroxylation	Tryptophan → (*B. subtilis*) → 5-Hydroxy-tryptophan	100
Epoxidation	1,7-Octadiene → (*Pseudomonas oleovorans*) → 7,8-Epoxy-1-octene	~25
Dehydration of –CH–CH–	Glaucine → (*Fusarium solani*) → Didehydro-glaucine	60
Oxidation of aliphatic side chains with the formation of aldehydes, ketones or carboxyl functions	n-Dodecyl-benzene ($C_{12}H_{25}$) → (*Nocardia* sp.) → Phenyl-acetic acid (CH_2–COOH)	80
Oxidative breakdown of alkyl side chains	1-Phenyl-dodecane ($(CH_2)_{11}$–CH_3) → (*Nocardia opaca* Stamm T16) → 2-Hydroxy-phenylacetic acid	?
	1-Phenyl-dodecane → (*Nocardia opaca* P2) → Phenylacetic acid	67–85

Table 15.1 (continued)

Type of reaction	Example	Conversion efficiency (%)
Oxidative splitting of aromatic rings	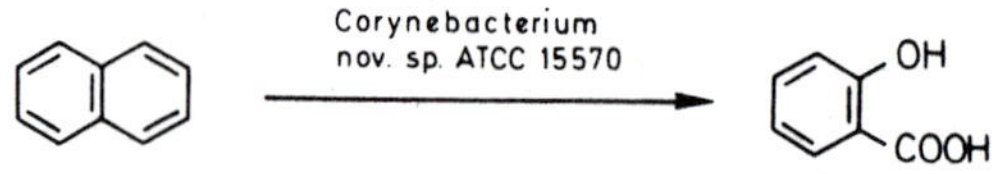 Napthalene → (Corynebacterium nov. sp. ATCC 15570) → Salicylic acid	70
Oxidative splitting of substituents (oxidative deamination, N–CH_3-demethylation, O–CH_3-demethylation)	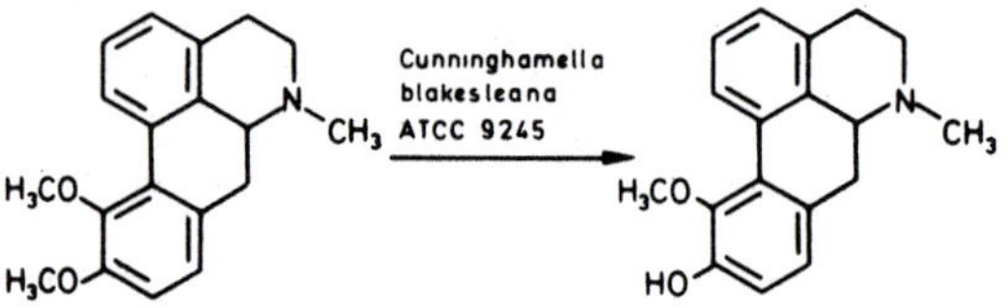 10,11-Dimethoxy-aporphine → (Cunninghamella blakesleana ATCC 9245) → Isoapocodeine	100
Oxidation of heterofunctions (amino groups to nitro groups; formation of N-oxides and sulfoxides)	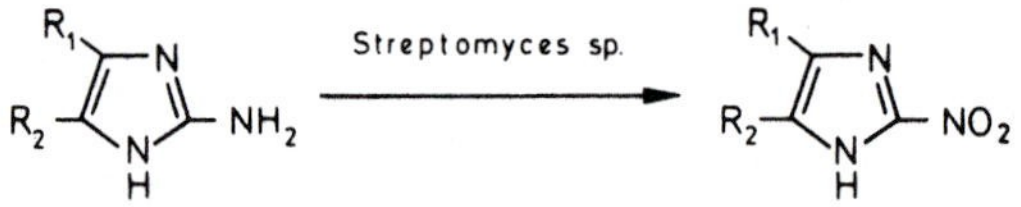 2-Amino-4-alkyl-imidazole → (Streptomyces sp.) → 2 Nitro-4-alkyl-imidazole	
	R_1 = H; R_2 = H	50
	R_1 = $-CH_3$; R_2 = H	25
	R_1 = $-C_2H_5$; R_2 = H	36
REDUCTIONS		
Reduction of carboxyl functions	 Benzaldehyde → (Saccharomyces cerevisiae) → Benzyl alcohol	50
Reduction of heterofunctions (particularly $-NO_2$)	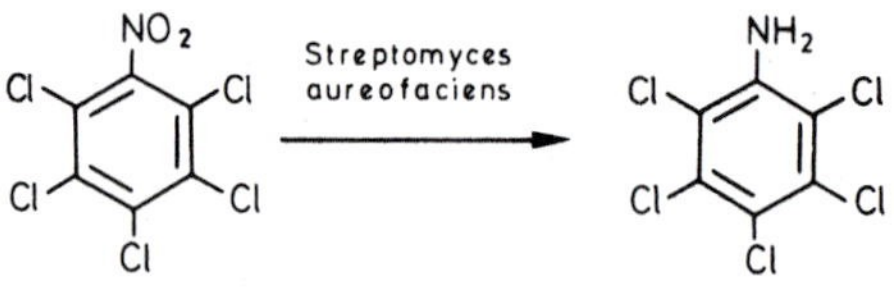 Nitropenta-chlorbenzol → (Streptomyces aureofaciens) → Pentachlor-aniline	?

Table 15.1 (continued)

Type of reaction	Example	Conversion efficiency (%)
Hydration of carbon-carbon double bonds	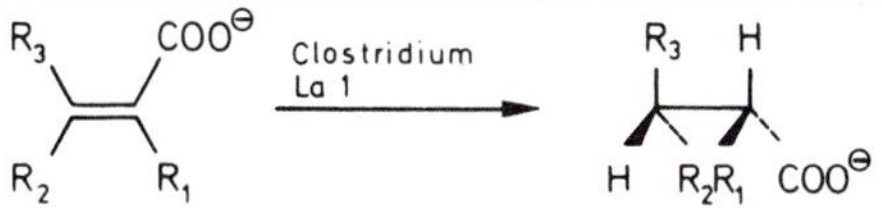 α,β-unsaturated carboxylic acids —(*Clostridium* La 1)→ (structure)	?
HYDROLYTIC REACTIONS		
Hydration of carbon-carbon double bonds	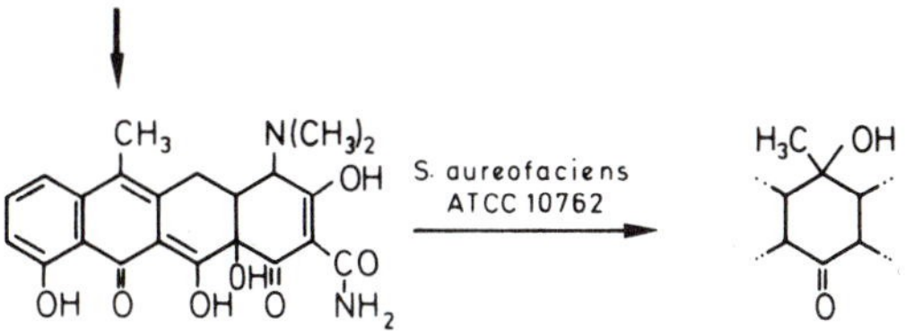 Anhydrotetracycline —(*S. aureofaciens* ATCC 10762)→ Tetracycline	55
Hydrolysis of carboxylic acid esters	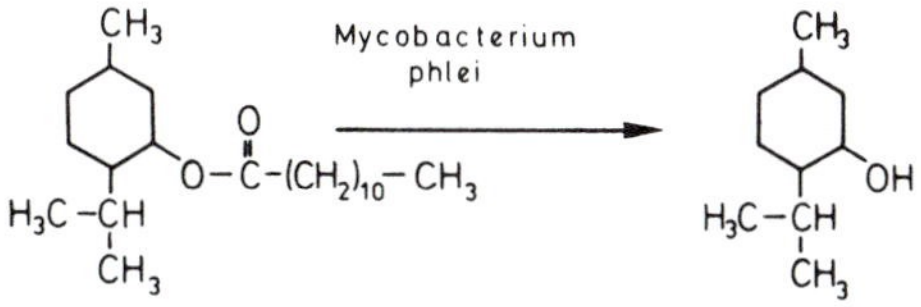 d,l-Menthyl laureate —(*Mycobacterium phlei*)→ l-Menthol	?
Hydrolysis of N-derivatives	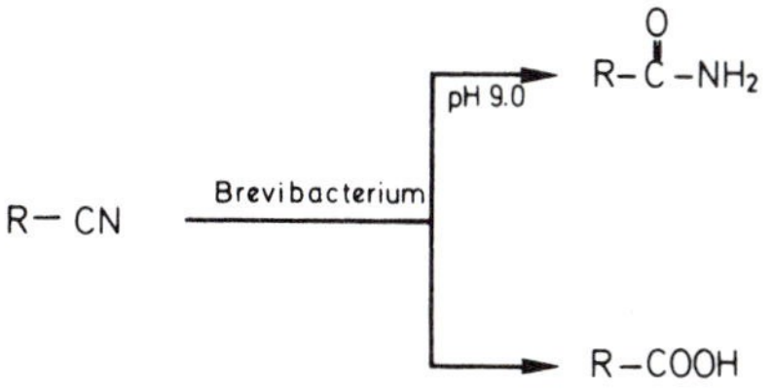 R–CN —(*Brevibacterium*, pH 9.0)→ $R-C(=O)-NH_2$	?
	R–CN —(*Brevibacterium*)→ R–COOH	?
CONDENSATIONS		
Phosphorylation	Streptomycin —(*S. griseus*)→ Streptomycin-P	?

Table 15.1 (continued)

Type of reaction	Example	Conversion efficiency (%)
N-Glycosidation	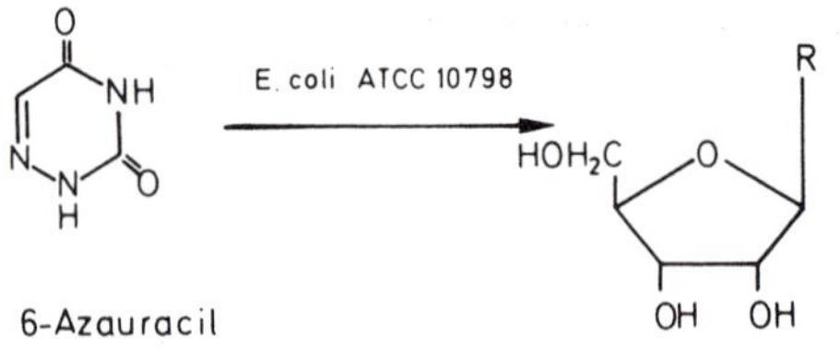 6-Azauracil —E. coli ATCC 10798→ 6-Azauracil-riboside	16
O-Glycosidation	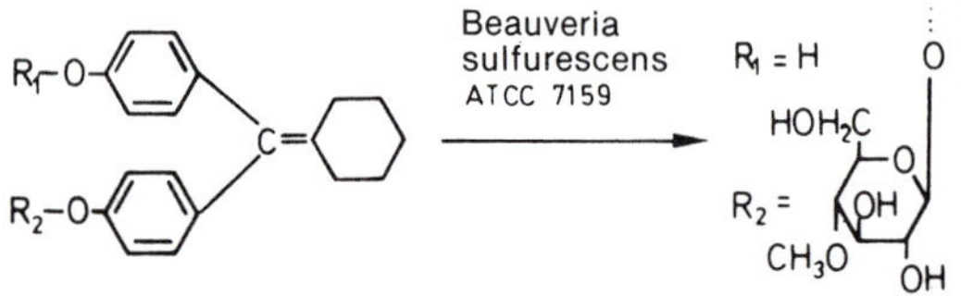 Cyclofenil ($R_1 = R_2 = CH_3$-CO-) —Beauveria sulfurescens ATCC 7159→ Methylglucopyranoside derivative (R_1 = H)	60
N-Acylation	6-APS + D-Phenylglycine methyl ester —Kluyvera citrophila→ Ampicillin	63

? = data unavailable or substances in mixture not isolated.

in industrial production. To a lesser extent, isomerization, reduction, hydrolysis, and condensation also have industrial application.

15.3 PROCEDURES FOR BIOTRANSFORMATION

Spores, growing cultures, resting cells, enzymes, immobilized cells, or immobilized enzyme systems can be used in the microbial conversion of organic compounds.

In processes with **growing cultures**, the strain used is cultivated in a suitable medium and a concentrated substrate solution is added after suitable growth of the culture (6–24 h). A variant of this procedure is to use a very large inoculum and to add the concentrated substrate immediately without allowing for a growth period. Emulsifiers such as Tween or water-miscible solvents with low toxicity (ethanol, acetone, dimethyl formamide, dimethyl sulfoxide) may be used to help solubilize poorly soluble compounds. Steroid conversions, in which the low solubility limits the amount of substrate which can be added, are commonly carried out at substrate concentrations between 0.1–10 g/l medium, although in some cases, up to 30 g/l can be converted. Solvent concentrations between 5 and 15 ml per liter of medium can be used. In some steroid transformations, the substrate is added and converted in fine crystalline form. These so-called pseudo-crystalline fermentations can be carried out with relatively high concentrations of substrate (for example, 15–50 g/l with progesterone).

For the biotransformation of lipophilic materials it is possible to employ a polyphase system. The aqueous phase containing the cell material or the enzyme is overlayed with a water-immiscible fluid phase in which the substrate has been dissolved. The substrate passes slowly into the aqueous phase and as the transformation reaction proceeds, the product passes back into the solvent phase. In some cases, the actual transformation occurs at the interface of the aqueous and solvent phases.

Conversion time is related to the type of reaction, the substrate concentration, and the microorganism used. Oxidation and dehydration reactions using bacteria are often completed in a few hours; conversions with yeast and especially fungi can take several days. Hydrolysis reactions with most kinds of microorganisms can be accomplished in a few hours.

Transformation reactions in large-scale equipment are carried out under sterile conditions in aerated and stirred fermenters, the conversion process being monitored chromatographically or spectroscopically. The process is terminated when a maximal titer is reached. Sterility is necessary because contamination can suppress the desired reaction, induce the formation of faulty conversion products, or cause total substrate breakdown.

If enzyme induction by the added substrate is not necessary, **resting cells** may be used. This has the considerable advantage that growth inhibition by the substrate is eliminated. High cell densities, which promote increased productivity, may be used; at the same time, risk of contamination is reduced. Since the transformation reaction occurs predominantly in the buffer solution, the recovery of the product is relatively easy. A number of transformation processes employ **immobilized cells**, offering the advantage that the process can be carried out continuously and the cells can be used over and over again. Immobilized bacterial cells, which catalyze one-stage or multi-stage reactions, are presently used commercially in the production of aspartic acid, L-alanine, and malic acid.

Cell-free enzyme extracts, generally inolving the use of **immobilized enzymes** (see Section 11.11), are usually employed in biotransformation reactions when undesirable side reactions or further breakdown of the reaction products must be avoided or when the rate of the reaction is hindered by transport of substrate or product through the cell membrane. The use of carrier-bound enzymes also makes possible the development of continuous processes. We discussed in Chapter 11 the use of immobilized enzymes for several processes which could be considered biotransformations, for example penicillin acylase, glucoamylase, and glucose isomerase. A pilot-scale process has been developed for the synthesis of amino acids by the reductive amination of α-keto acids by amino acid dehydrogenases. This system permits the regeneration of the required NADH cofactor.

The end products of transformation reactions are usually extracellular and may occur in either dissolved or suspended form. For further processing, bacteria and yeasts are generally not separated, whereas fungal mycelium is usually removed by filtration. In all cases, separated cell material must be washed repeatedly with water or organic solvents since a significant amount of the reaction product can be adsorbed to the cells. Depending on the solubility of the product, recovery is performed by precipitation as the calcium salt, by adsorption to ion exchangers, by extraction with appropriate solvents, or, for volatile substances, by direct distillation from the medium.

15.4 APPLICATIONS OF BIOCONVERSIONS

Although a vast array of biotransformations have been described, only a few of these processes have found industrial application. Some processes have insufficient yields and for others the market is too limited. In the future, a wide range of applications is expected to arise in connection with new technology, such as the more cost-effective processes using immobilized cells or enzymes. Other improvements are expected with the use of strains which have been genetically optimized for specific processes.

15.5 TRANSFORMATION OF STEROIDS AND STEROLS

Naturally occurring steroids have **hormone properties**. Examples are the adrenal cortex hormones (glucocorticoids and mineral corticoids), androgens, estrogens, and hormones active during pregnancy, such as progesterone. All steroids have the same basic structure, a cyclopentanoperhydrophenanthrene. Figure 15.1 shows the structure of the most important types. Estrogens, progesterone, and androgens are used therapeutically; derivatives of progesterone and estrogens are also used as contraceptives. In addition, steroids are used as sedatives, in antitumor therapy, and as veterinary products. The glucocorticoids are valuable compounds with wide therapeutic uses. Cortisone is especially useful because of its anti-inflammatory action in such conditions as rheumatoid arthritis and skin diseases. By altering the structure, specifically by introducing a 1,2 double bond in ring A of the cortisol or cortisone molecule to produce prednisolone or prednisone, substances can be produced with markedly increased anti-inflammatory effect. Addition of fluorine and methyl groups leads to the formation of compounds with reduced mineral corticoid activity (Na^+ retention, K^+ excretion), such as 16α-hydroxy-9α-fluoroprednisolone (triamcinolone) or 6α-methylprednisolone (medrol). By means of transformation, anabolic steroids with reduced androgenic effects have been developed, such as 1-methyl-Δ^1-androstenolone.

At the present, the available steroids serve most of the medical requirements quite well so that further development of new steroid transformation processes is limited. Research is now primarily directed at the optimization of the existing processes, primarily by use of immobilized cells or enzymes or by the optimization of the reaction system, such as by the use of polyphase systems. Further optimization work is in the direction of finding better starting materials or reducing the degradative side reactions. Genetic engineering research on the microorganisms used for steroid transformations is also under investigation. Additionally, studies are under way on the use of plant cell cultures for steroid transformations.

Types of transformations

Because the steroid molecule contains several asymmetric centers, **total synthesis** is very difficult. The original chemical process involved 31 separate reaction steps and yielded 1 g cortisone acetate from 615 g deoxycholic acid. More recent chemical processes are simpler and deoxycholic acid (from ox bile) is presently used as substrate in several production processes.

Preliminary research on the 11α-hydroxylation of progesterone pointed to the possibility of the microbial introduction of oxygen into the steroid nucleus in a site-specific and stereospecific manner without prior activation. These reactions worked well, and cost-effective production of cortisone became possible. The oxygen atom at C-11 is essential for the anti-inflammatory effect of cortisone. In 1949, 1 g of cortisone cost $200 to produce, but as a result of the introduction of the microbial process for the 11-α-hydroxylation of progesterone, the cost had decreased to under $1 by 1979. Currently, almost all positions of the steroid molecule can be specifically hydroxylated by different microorganisms and the number of transformation reactions is larger than the number carried out by animal tissue (Figure 15.2). The microbial hydroxylation reactions are carried out by highly specific monooxygenases; some examples are an 11α- or 11β-hydroxylase, a 17α-hydroxylase and a 21-hydroxylase.

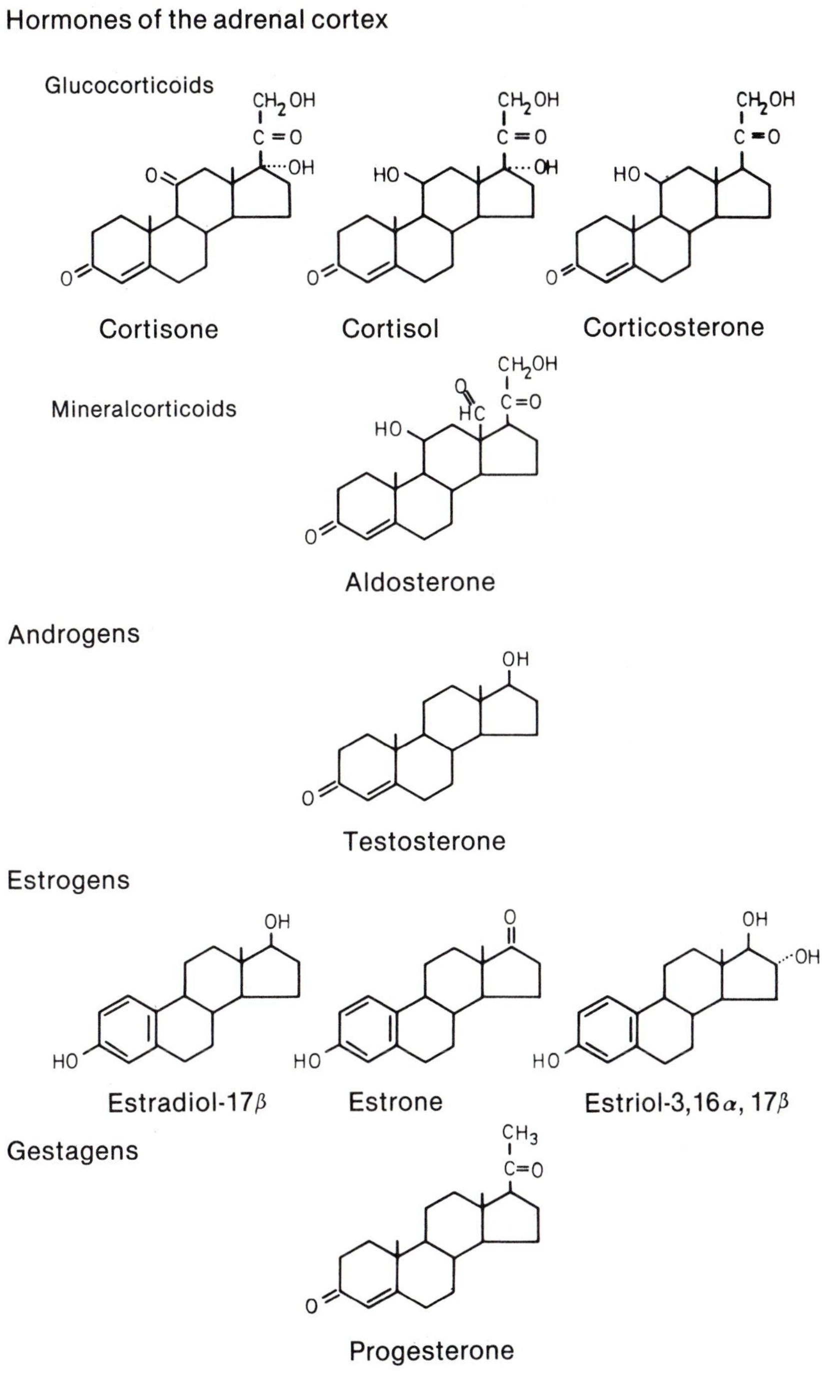

Hormones of the adrenal cortex

Glucocorticoids

Cortisone Cortisol Corticosterone

Mineralcorticoids

Aldosterone

Androgens

Testosterone

Estrogens

Estradiol-17β Estrone Estriol-3,16α, 17β

Gestagens

Progesterone

Figure 15.1 Structure of several naturally occurring steroid hormones

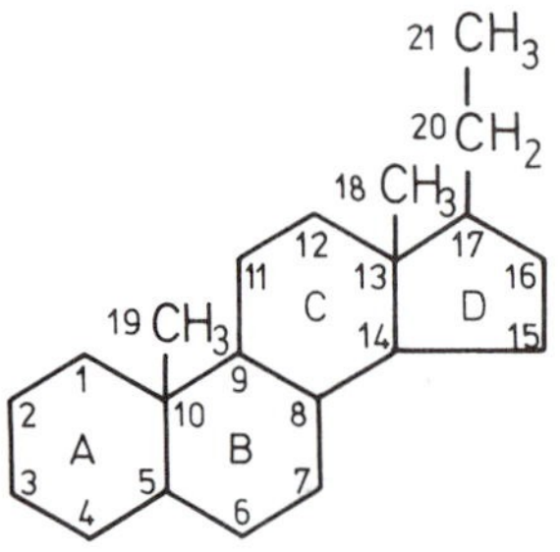

5β-Pregnane

Microorganisms			Animal tissue	
1α	7α	15α	1α	15α
1β	7β	15β	2α	16α
2α	9α	16α	2β	17α
2β	10β	16β	6α	18
3β	11α	17α	6β	19
5α	11β	18	7α	21
5β	12α	19	11β	
6β	12β	21	12α	
	14α			

Figure 15.2 Steroid hydroxylation by microorganisms as contrasted with the process in animal tissue

As outlined in Table 15.2, microorganisms are capable of carrying out a wide variety of other steroid transformation reactions besides hydroxylation.

Table 15.2 Reaction types of microbial transformations from steroids

A. Oxidations

- Conversion of secondary alcohols to ketones
- Introduction of primary hydroxyl groups into the steroid side chain
- Introduction of secondary hydroxyl groups into the basic steroid framework
- Introduction of tertiary hydroxyl groups into the basic steroid framework
- Dehydration of ring A in positions 1(2) and 4(5)
- Aromatization of ring A
- Oxidation of the methylene group to the keto group
- Splitting of the side chain of pregnane at C-17 during production of a ketone
- Splitting of the side chain of pregnane at C-17 and opening of ring D during production of testololactone
- Splitting of the steroid side chain during production of a carboxyl group
- Splitting of the side chain of pregnane at C-17 during production of a secondary alcohol
- Production of epoxides
- Decarboxylation of acids

B. Reductions

- Reduction of ketones to secondary alcohols
- Reduction of aldehydes to primary alcohols
- Hydration of double bonds in position 1(2) in ring A
- Hydration of double bonds in position 4(5) in ring A and 5(6) in ring B
- Elimination of secondary alcohols

C. Hydrolysis

- Saponification of steroid esters

D. Ester production

- Acetylation

(Sebek and Perlman, 1979)

Economically important transformations

In recent decades, thousands of modified steroids produced by a combination of chemical and microbial reaction steps have been tested for their therapeutic effectiveness. A typical example of such a combined synthesis is the production of cortisone and its 1-dehydro-derivatives from diosgenin via Reichstein's Substance S (11-deoxycortisol), as shown in Figure 15.3.

The microbial reaction steps listed in Table 15.3 are of great economic significance. Moreover, progesterone transformation to a C_{19}-steroid is used industrially in testosterone and estrogen production and the microbial dehydration of ring A is used in estrogen production. The conditions for several bioconversion reactions of this type are given in Table 15.4 (examples 1–6). Steroid transformations have thus far been conducted as batch fermentations, but progress is being made with the use of immobilized cells or enzymes. Advantages of the latter include reduced risk of contamination, simplified product recovery, shorter conversion times, and increased substrate concentrations. In several processes, more than one biochemical step can be combined. For instance, immobilized mycelium of *Curvularia lunata* or immobilized cells of *Arthrobacter simplex* have been used to carry out the

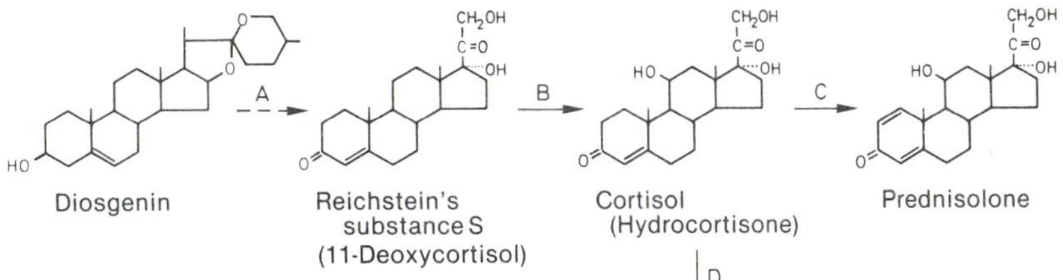

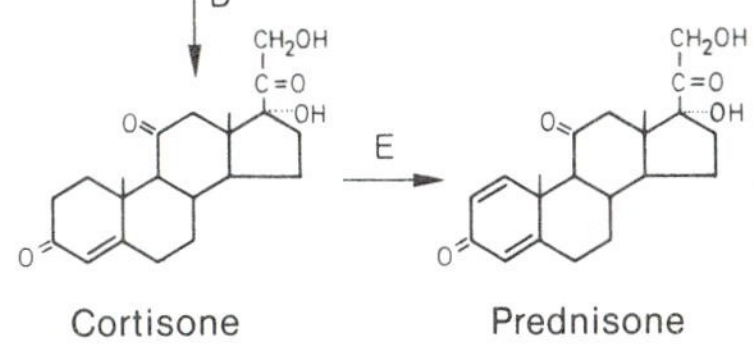

Figure 15.3 Production of cortisol, cortisone and the 1-dehydro compounds from diosgenin via Reichstein's substance S (A, several steps; B, 11β-hydroxylation with *Curvularia lunata*; C, 1-dehydration with *Corynebacterium simplex*; D, chemical oxidation; E, 1-dehydration with *Corynebacterium simplex*)

Table 15.3 Examples of commercial steroid processes

Reaction	Substrate → Product	Microorganism	Manufacturer
11α-Hydroxylation	Progesterone → 11α-Hydroxyprogesterone	*Rhizopus nigricans*	Upjohn Company
11β-Hydroxylation	Component S → Cortisol	*Curvularia lunata*	Pfizer Inc., Gist-Brocades
16α-Hydroxylation	9α-Fluorocortisol → 9α-Fluoro-16α-hydroxycortisol	*Streptomyces roseochromogenes*	E. R. Squibb and Sons, Lederle Laboratories
1-Dehydrogenation	Cortisol → Prednisolone Diendiol[a] → Triendiol[b]	*Arthrobacter simplex* *Septomyxa affinis*	Schering Corp. Upjohn Company
1-Dehydration, side chain splitting	Progesterone → 1-Dehydrotestololactone	*Cylindrocarpon radicicola*	E. R. Squibb and Sons
Side chain splitting	β-Sitosterol → Androstadiendione, and/or 9α-hydroxyandrostendione (see Fig. 15.5 for structure)	*Mycobacterium* sp., *M. fortuitum* mutants	G. D. Searle and Co. Upjohn Company

[a] Diendiol = 11β,21-dihydroxy-4,17(20)-pregnadiene-3-one
[b] Triendiol = 11β,21-dihydroxy-1,4,17(20)-pregnatriene-3-one (precursors in the production of 6α-methylprednisolone).
(Sebek and Perlman, 1979)

two-step reaction from Reichstein's component S to prednisolone. In several processes, fungal spores are being used directly to catalyze the transformation. Since most steroid substrates are not very soluble, transformation conditions have been developed for some steroids in a solvent system which is water-immiscible, e.g., for testosterone with immobilized cells of *Nocardia rhodochrous*. Since the organic solvent is often toxic to the cells or enzyme, an alternative is the use of an aqueous two-phase system. For example, the 1-dehydration of cortisol to prednisolone can be carried out by cells of *Arthrobacter simplex* in a system consisting of 25% (w/w) polyethylene glycol (PEG) 8000 and 6 (w/w) dextran T40.

Microbial breakdown of sterol side chains

The growing demand for steroids caused a shortage of steroid precursors for bioconversion, such as the compound diosgenin, which is obtained from the Mexican yam root (*Dioscorea composita*) or the South African plant *Testudinaria sylvatica*. Intensive studies were conducted on the use of low-cost sterols of animal origin, such as cholesterol, or of plant origin, such as β-sitosterol

Table 15.4 Conditions for operation of several steroid and sterol transformations

No.	Substrate	Product	Yield (weight %)	Microorganism	Medium	Conditions used
1	Progesterone	1-Dehydrotestololactone	50	*Cylindrocarpon radicicola*	a	72 h, 25°C
2	Progesterone	1,4-Androstadiene-3,17-dione	85	*Fusarium solani*	b	96 h, 25°C
3	Progesterone	15α-Hydroxy-4-pregnene-3,20-dione	11	*Streptomyces aureus*	c	72 h, 25°C
4	4-Androstene-3,17-dione	11α-Hydroxy-4-androstene-3,17-dione	25	*Rhizopus arrhizus*	d	96 h, 28°C
5	Progesterone	11-α-Hydroxy-progesterone	91	*Aspergillus ochraceus*	e	120 h, 28°C
6	Hydrocortisone	Prednisolone	93	*Arthrobacter simplex*	f	120 h, 28°C Pseudo-crystal fermentation
7	Cholesterol	1,4-Androstadiene-3,17-dione	90	*Arthrobacter simplex*	g	44 h, 30°C + chelating agents
8	β-Sitosterol Cholesterol Stigmasterol Campesterol	9α-Hydroxy-4-androstene-3,17-dione	Data unavailable	*Mycobacterium fortuitum*	h	336 h, 30°C mutant with block in breakdown of steroid nucleus

Media in Table 15.4

a. 3 g Corn steep liquor (dry weight), 3 g $NH_4H_2PO_4$, 2.5 g $CaCO_3$, 2.2 g soy bean oil, 0.5 g progesterone, distilled H_2O to 1 l, pH 7.0.
b. 15 g Peptone, 6 ml corn steep liquor, 50 g glucose, distilled H_2O to 1 l, pH 6.0; 0.25 g progesterone, added after 48 h.
c. 2.2 g Soy bean oil, 15 g soy meal, 10 g glucose, 2.5 g $CaCO_3$, 0.25 g progesterone, distilled H_2O to 1 l.
d. 20 g Peptone, 5 ml corn steep liquor, 50 g glucose, tap water to 1 l, pH 5.5–5.9; 0.25 g androstendione, added after 27 h.
e. 10 g peptone, 5 g yeast extract, 30 g sucrose, tap water to 1 liter; pH 6.5; 40 g progesterone dissolved in acetone added to the growing culture after 24 h.
f. 5g Peptone, 5 g corn steep liquor, 5 g glucose; distilled H_2O 1 l; pH 7.0; 1–50% finely ground hydrocortisone, suspended in ethanol, added to a 24 hour old culture.
g. 10 g Corn steep liquor, 2 g meat extract, 5 g glucose, 0.5 g K_2HPO_4; distilled H_2O 1 l; pH 7.0; cholesterol (1 g) added after 20 h (dispersed in water); 0.8 mM α,α′-dipyridyl (an iron chelating agent) in ethanol added after 26 h.
h. 10 g Glycerol, 8.4 g Na_2HPO_4, 4.5 g KH_2PO_4, 2 g NH_4Cl, trace elements; distilled H_2O 1 l; 1 g soy meal and 10 g sitosterol are added.

(Sebek and Perlman, 1979)

and stigmasterol (from soy beans) or campesterol (produced in great amounts as a byproduct of paper manufacture).

The objective of these studies was the selective removal of the aliphatic side chain without further breakdown of the steroid nucleus. However, a screening procedure with cholesterol as the substrate led only to strains carrying out a total breakdown of sterol to CO_2 and H_2O. The breakdown of the side chain to yield a C-17 ketocompound, shown in Figure 15.4, involves a mechanism which is similar to that of the β-oxidation of fatty acids. A C-1(2)-dehydration and 9α-hydroxylation are mandatory for further

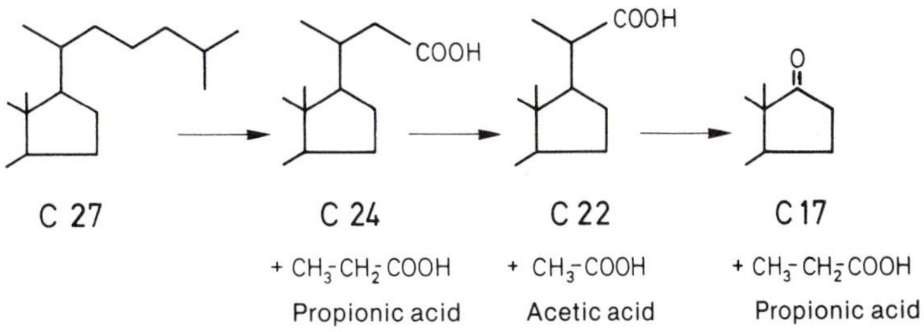

C 27 → C 24 → C 22 → C 17

+ CH_3-CH_2-COOH Propionic acid; + CH_3-COOH Acetic acid; + CH_3-CH_2-COOH Propionic acid

Figure 15.4 Side chain breakdown of sterols

breakdown of the steroid ring. As shown in Figure 15.5, the breakdown product 3-hydroxy-9,10-secoandrostatriene-9,17-dione is produced from cholesterol via an opening of the B ring, with the production of two useful intermediate products, androstendione and androstadiendione. In order to modify the steroid nucleus without breaking any of the rings, methods for selectively blocking attack on the ring are needed. Several methods are available to selectively block the breakdown of the steroid nucleus:

- The breakdown reaction can be blocked by chemical modification of the substrate.
- The conversion can take place in the presence of inhibitors which prevent C-1(2)-dehydration or 9α-hydroxylation, such as compounds which chelate Fe^{2+} or Cu^{2+}, bivalent ions which replace Fe^{2+}, or substances which block sulfhydryl functions, such as Ni^{2+}, Co^{2+}, and Pb^{2+}.
- Mutants with inactive C-1(2)-dehydrogenase and/or 9α-hydroxylase can be used. *Mycobacterium* mutants which transform cholesterol, stigmasterol, and sitosterol into the main product androstenedione and the secondary product androstadiendione have been isolated. With these mutants, there is no further breakdown of these latter compounds.

Some of the processes just mentioned are used commercially (see Table 15.4, examples 7, 8).

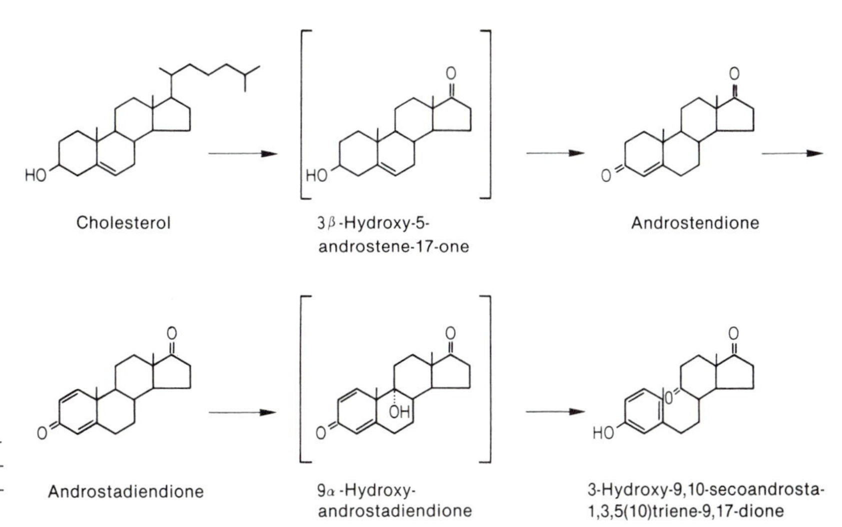

Figure 15.5 Biotransformation of cholesterol to androstadiendione and androstatriendione by mycobacteria

15.6 TRANSFORMATION OF NONSTEROID COMPOUNDS

In addition to the steroid transformations, microbial transformations of alkanes, of alicyclic, aromatic, and heterocyclic compounds, of terpenes, and of alkaloids have been described. In some cases, microbial transformations are difficult to control. For example, in experiments with gibberellins, carotenoids, and camphor, breakdown reactions predominate and there is only slight accumulation of utilizable intermediary products; attempts to isolate mutants blocked in the breakdown reactions have been unsuccessful. In other cases, as in alkaloid biotransformations, reaction rates have been too low for commercial processes. However, some transformation reactions are economically important; some of these have already been covered earlier (e.g., gluconic acid, kojic acid, in Chapter 8; manufacture of L-amino acids through racemic separation or hydantoin cleavage, in Chapter 9; penicillin acylase, glucose isomerase, in Chapter 11). A few more commercially significant bioconversions among the many known reactions are discussed here.

L-Ascorbic acid (Vitamin C)

The established process for producing L-ascorbic acid is the so-called Reichstein-Grüssner synthesis, shown in Figure 15.6. This process consists of several chemical steps and one microbial conversion. L-ascorbic acid is used in vitamin preparation or as an antioxidant in food manufacture and world production by this process is about 40,000 tons per year. The oxidation stage from D-sorbitol to L-sorbose is carried out by *Acetobacter suboxydans* in a submerged process at 30–35°C with vigorous stirring and aeration. Sorbitol is added at an initial concentration of 20% to a nutrient solution consisting of 0.5% yeast extract or corn steep liquor, and $CaCO_3$. A quantitative conversion is completed after about 24 hours; higher sorbitol concentrations prolong the conversion time. Today this process is carried out continuously in two stages; there are even some installations in which polyacrylamide-immobi-

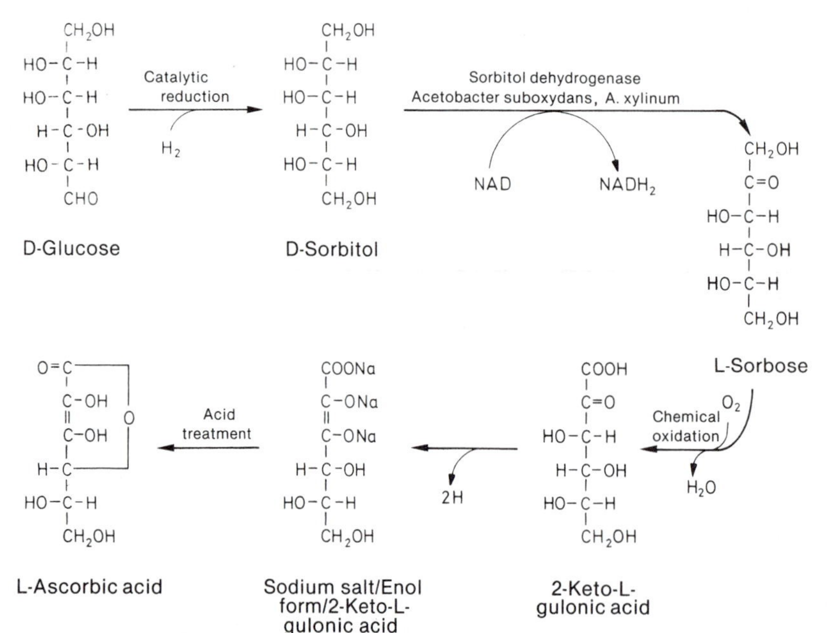

Figure 15.6 Microbial dehydration of D-sorbitol to L-sorbose in the production of L-ascorbic acid (Reichstein-Grüssner synthesis)

lized cells are used. In the overall process approximately 1 kg of L-ascorbic acid is produced from 2 kg of glucose.

In addition to the Reichstein-Grüssner synthesis shown in Figure 15.6, a two-step fermentation process has been developed and carried to commercial production. As shown in Figure 15.7, the first step involves the oxidation of glucose by an *Erwinia* species to 2,5-diketo-D-gluconic acid (2,5-DKG), via D-gluconic acid and 2-keto-D-gluconic acid. During a 26 hour incubation, 328 g/l calcium 2,5-DKG is formed with a 94% efficiency. The second step, a reduction of 2,5-DKG to 2-keto-1-gulonic acid, is catalyzed by a *Corynebacterium* sp. In this process, after the *Corynebacterium* has grown for 16 hours, it is fed with a sterilized *Erwinia* culture. After 66 hours incubation, 106 g of calcium 2-keto-L-gulonate is formed, with an efficiency of 92%. The latter product is easily transformed chemically into L-ascorbic acid and the overall balance, based on glucose consumed, is 86%.

Scientists at the Genentech Company have succeeded in cloning and expressing the gene for the 2,5-DKG reductase of *Corynebacterium* into *Erwinia herbicola*, opening up the possibility of a single-step process from glucose to 2-keto-L-gulonic acid. However, the hybrid strain produced exhibited a low tolerance to high concentrations of glucose and a yield of only 0.6 g/l of 2-keto-L-gulonic acid was obtained from 20 g/l glucose after 57 hours incubation.

Dihydroxyacetone from glycerol

The microbial conversion of glycerol to dihydroxyacetone (used in suntan lotions and cosmetics), is of some significance (Figure 15.8). Various acetic acid bacteria convert 10% glycerol in a suitable nutrient solution (0.5% yeast extract, 0.5% KH_2PO_4, 2% $CaCO_3$) at 28°C and at a pH below 6.0. Conversion time is 72–96 hours but can be reduced to 33 hours with *Gluconobacter melanogenus* IFO 3293 by using O_2-enriched air (oxygen partial pressure 0.05 bar). Growth is not affected, but the amount of dihydroxyacetone produced is raised from 12.2 g/g biomass in normally aerated cultures to 35.8 g/g in cultures with improved oxygen supply.

Prostaglandins

Prostaglandins are unsaturated C_{20} fatty acids that function as tissue hormones. They are of increasing medical significance because of their varied physiological activities. Currently marketed are PGE_2 as a contraceptive, PEG_2 for the alleviation of pain of childbirth, PEG_1 for the treatment of congenital heart failure, and a methyl derivative of PEG_1 for the treatment of digestive diseases. The prostaglandins PGE_1, PGE_2, PGF_1, and PGF_2 can be produced from unsaturated fatty acids (particularly arachidonic acid) (Figure 15.9) by microbial transformation with fungi. The isolation of compounds with im-

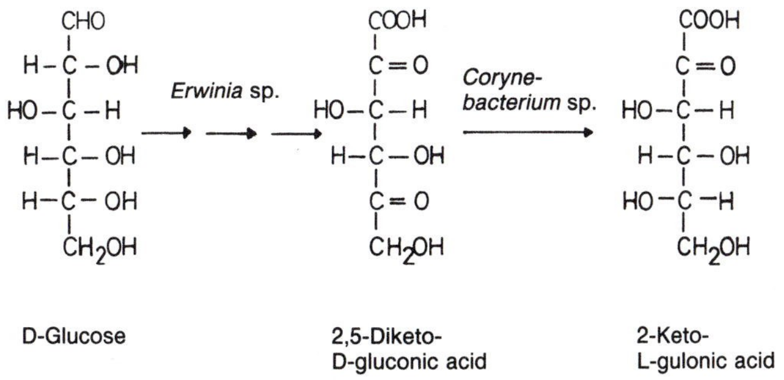

D-Glucose —*Erwinia* sp.→ 2,5-Diketo-D-gluconic acid —*Corynebacterium* sp.→ 2-Keto-L-gulonic acid

Figure 15.7 Two-stage fermentation for the production of 2-keto-L-gulonic acid from D-glucose

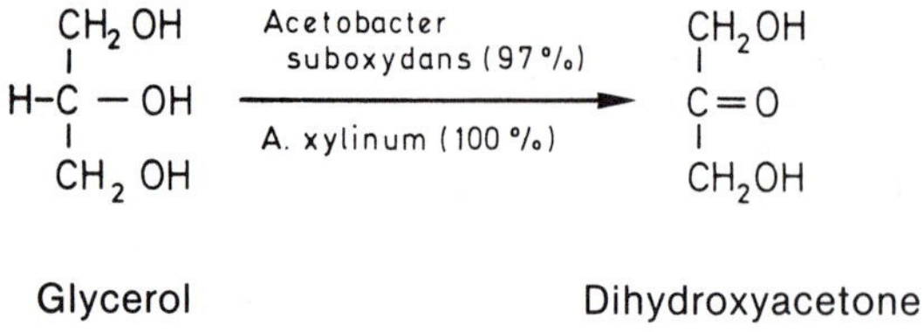

Glycerol → Dihydroxyacetone (*Acetobacter suboxydans* (97%); *A. xylinum* (100%))

Figure 15.8 Conversion of glycerol to dihydroxyacetone

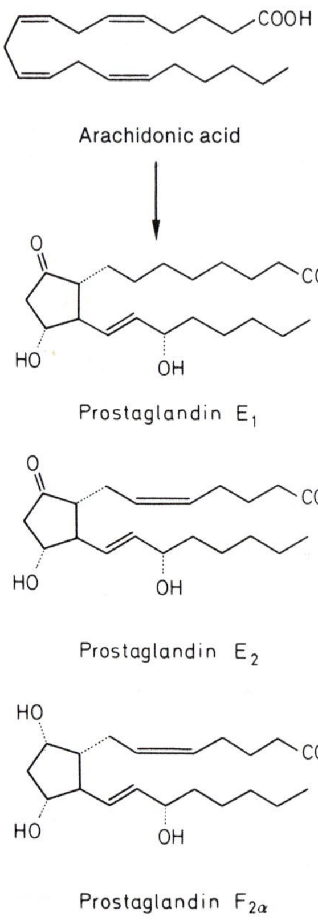

Arachidonic acid → Prostaglandin E_1, Prostaglandin E_2, Prostaglandin $F_{2\alpha}$

Figure 15.9 Biotransformation of arachidonic acid to prostaglandins

proved effectiveness can be expected in light of new reports on successful biotransformations of this molecule.

15.7 TRANSFORMATION OF ANTIBIOTICS

In addition to screening procedures, studies in recent years have also been directed at the production of new and improved antibiotics by the microbial transformation of existing compounds. The objective here is the development of modified antibiotics with improved effectiveness, reduced toxicity, wider antimicrobial spectrum, improved oral absorption, decreased development of resistance, or lower allergic effects. In most cases, any transformation step causes a partial or complete inactivation of the antibiotic. Thus, many biotransformations must be attempted and their products assayed to find those rare cases where the modified product has desirable improved properties.

Several typical examples of the many possible reactions are given here.

Indirect transformation The addition of inhibitors or modified precursors to the medium may result in the synthesis of altered antibiotics via **controlled biosynthesis**. For instance, in the presence of cis-4-methylproline, *Streptomyces parvulus* produces two new actinomycins (K_{1c} and K_{2c}) in which proline is replaced by this proline analog. New compounds have been found when mutants blocked in the synthesis of a particular antibiotic were used. For instance, ribostamycin (Figure 15.10) accumulates as an intermediate product of neomycin biosynthesis with a mutant of the neomycin producer *S. fradiae*. A few improved antibiotics have been isolated after **mutational synthesis** (see Section 3.6), of which only 5-*epi*-sisomicin has proved of sufficient utility to undergo clinical trials. Several genetic techniques, mutagenesis, intra- and interspecific recombination by protoplast fusion, and recombinant DNA technology, have been used to isolate strains capable of producing modified versions of rifamycins, aminoglycosides, and tylosin.

Direct transformation Acylation reactions have been described for various antibiotics. Inactive compounds develop in most cases but in the case of lankacidin-C-14-butyrate, a bioconversion product formed from lankacidin C and methylbutyrate by *Bacillus megaterium* IFO 12108, improved antimicrobial activity with lower toxicity

R = -H Ribostamycin

R = Neomycin B

Figure 15.10 Ribostamycin formation by a mutant of *Streptomyces fradiae* blocked in neomycin production

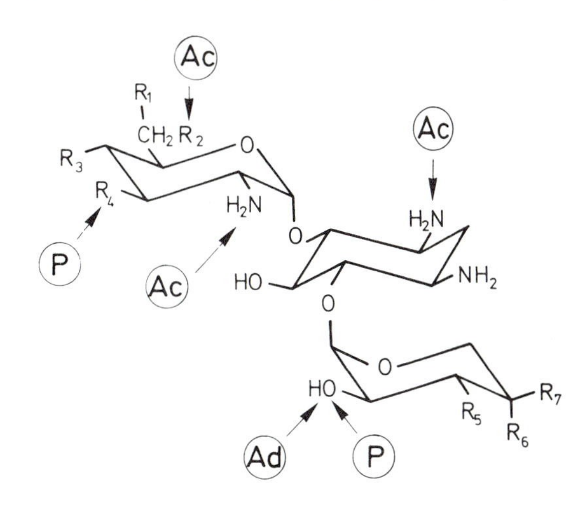

Gentamicin	R_1	R_2	R_3	R_4	R_5	R_6	R_7
Gentamicin A	H	OH	OH	OH	$NHCH_3$	H	OH
Gentamicin A_2	H	OH	OH	OH	OH	OH	H
Gentamicin X_2	H	OH	OH	OH	$NHCH_3$	OH	CH_3
Gentamicin C_{1a}	H	NH_2	H	H	$NHCH_3$	OH	CH_3
Gentamicin C_2	CH_3	NH_2	H	H	$NHCH_3$	OH	CH_3
Gentamicin C_1	CH_3	$NHCH_3$	H	H	$NHCH_3$	OH	CH_3

Figure 15.11 Structure of gentamicins with indications of the site of action of the inactivation enzymes

was obtained. Deacylation reactions have been described in particular for the macrolide antibiotics. The biologically less active deacylated products can then be used for the production of semisynthetic compounds. Phosphorylation via different phosphotransferases and adenylylation via adenyltransferases take place chiefly with aminoglycosides and lead to inactivation. Acetylation (Ac), phosphorylation (P), and adenylylation (Ad) are reactions that are primarily responsible for the development of bacterial resistance to aminoglycosides. The example of gentamicin in Figure 15.11 shows the target at which the three modifying enzymes act.

Hydrolysis reactions are especially significant in β-lactam antibiotics (Figure 15.12). Hydrolysis involves the splitting of the lactone ring of penicillins and cephalosporins by β-lactamases, leading to inactivation of the antibiotics. Bacterial resistance to β-lactam antibiotics is mainly caused by this reaction. On the other hand, the enzymatic splitting of penicillin by penicillin acylase into 6-aminopenicillanic acid (Section 11.9) is of great economic significance.

Hydroxylation is another frequent microbial transformation reaction in antibiotics (example in Figure 15.13).

15.8 TRANSFORMATION OF PESTICIDES

Until now, we have been discussing microbial transformations that lead to the production of commercially useful compounds. We now change our perspective and discuss some transformations that cause the *destruction* of a major class of commercially useful compounds, the pesticides. Agents for plant disease and pest control are necessary for the survival of the world's population. Presently one-third of the world's harvest is lost because of damaging organisms. It is estimated that another one-third would be lost without the use of chemical control agents. Contagious diseases such as malaria, chagas disease,

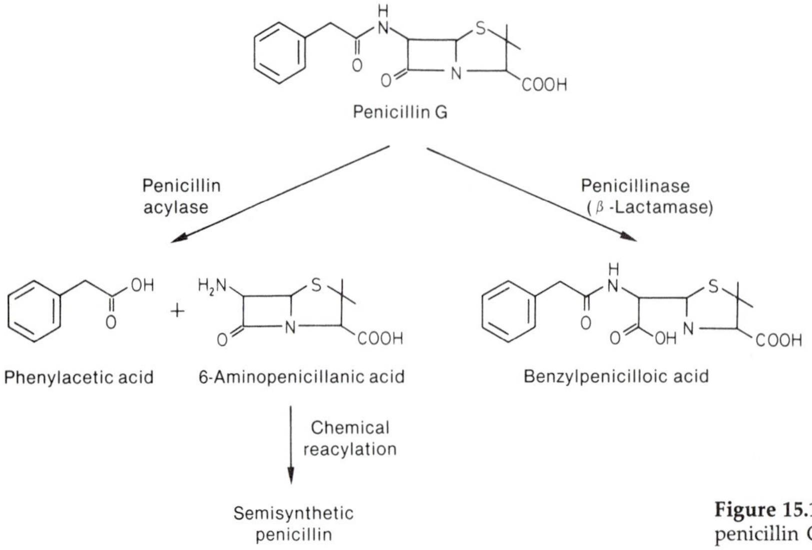

Figure 15.12 Microbial transformation of penicillin G

typhus, cholera, and spotted fever have been reduced in severity and in some regions completely eliminated through the control of disease-carrying insect vectors. High stability (persistence) of the compounds used is vital for these vector-control programs, but this stability has a negative effect on the environment.

This problem of environmental persistence is apparent with chlorinated hydrocarbon insecticides such as DDT, lindane, and dieldrin. A remarkable success in infectious disease control can be attributed to DDT, but due to its resistance to decomposition, the compound accumulates in microorganisms and thus enters into the food chain. This development, along with the restricted use of DDT since the early 1970's, has led to research for new control methods and toxicologically and environmentally safe preparations.

In this context, microbial transformation is of interest not for the production of new active agents, but for the greatest possible detoxification of the environment. This involves enzymatic conversions of so-called *xenobiotics,* substrates which do not normally occur in nature, such as halogenated hydrocarbons, aromatic nitro-compounds, and sulfonic acid derivatives. Many of the enzymes involved in these transformations must be induced and different organisms are frequently involved in the breakdown of a compound. Depending on chemical structure, some compounds cannot be easily converted; thus persistence times in the soil range from a few days to several years.

Removal of xenobiotics from ecosystems can be accomplished through various mechanisms.

Metabolism Xenobiotics can serve as substrates for microbial growth and energy production. Complete breakdown of some substances to CO_2 and H_2O has been described. Figure 15.14 shows the example of the herbicide dalapon, a chlorinated fatty acid, which is converted by *Arthrobacter* sp. into pyruvate after oxidative dehalogenation.

Cometabolism In cometabolism, microorganisms do not obtain energy from the transformation reaction and require another substrate for growth. Hence, cometabolism normally causes mere modification of molecules, which may result in either a decrease or an increase in toxicity. A further breakdown can be achieved through the combined action of different organisms.

Narbomycin

Streptomyces sp.

Picromycin

Figure 15.13 Hydroxylation of narbomycin

Some examples of cometabolism are given in Table 15.5.

Dehalogenation or oxidative dehalogenation reactions (Figure 15.15) are important cometabolism reactions which may make pesticide molecules accessible for further breakdown. Some compounds such as chlordecone, a hexachlorocyclopentadiene derivative with excellent insecticide and acaricide effects, are not easily attacked by microorganisms because of their complicated structure and high degree of halogenation.

Table 15.5 Cometabolism of pesticides

Substrate	Conversion product	Microorganism
Chlorbenzilate (ethyl-4,4′-dichlorobenzilate)	4,4′-Dichlorbenzophenone	*Rhodotorula gracilis*
Chloroneb (1,4-dichloro-2,5-dimethoxybenzene)	2,5-Dichlor-4-methoxyphenol	*Fusarium* sp.
DDT	p,p′-Dichlordiphenylmethane	*Aerobacter aerogenes*
p,p′-Dichlorodiphenylmethane	p-Chlorphenylacetate	*Hydrogenomonas* sp.
2,4,5-Trichlorophenoxyacetate	3,5-Dichlorcatechol	*Brevibacterium* sp.
3,5-Dichlorcatechol	3,5-Dichlor-2-hydroxymuconic-acid semialdehyde	*Achromobacter* sp.

(Bollag, 1974)

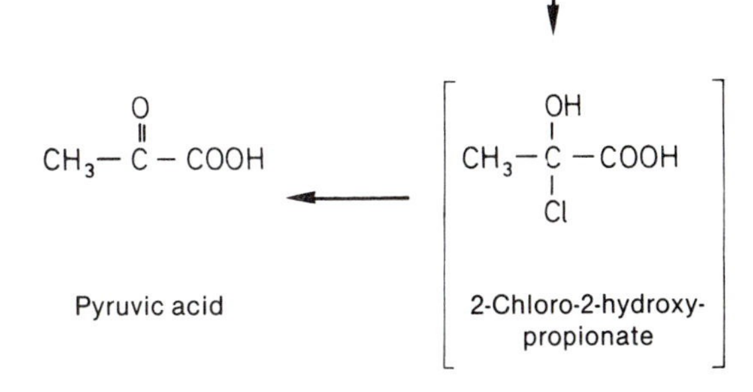

Figure 15.14 Microbial breakdown of dalapon

Conjugate formation Linkage of xenobiotics or decomposition products with naturally occurring compounds such as amino acids or carbohydrates results only in a temporary detoxification; the toxic compound can be released again at any time. Figure 15.16 shows the conversion of a dithiocarbamate fungicide.

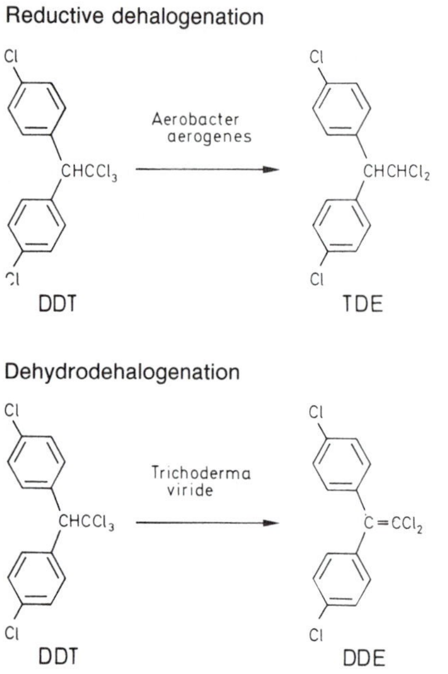

Reductive dehalogenation

DDT → (Aerobacter aerogenes) → TDE

Dehydrodehalogenation

DDT → (Trichoderma viride) → DDE

Figure 15.15 Transformation of the chlorinated hydrocarbon DDT (DDT=2,2-Bis-[4-chlor-phenyl]-1,1,1-trichlor-ethane; TDE=2,2-Bis-[4-chlor-phenyl]-1,1-dichlorethane; DDE=1,1-dichlor-2,2-Bis-[4-chlor-phenyl]-ethylene)

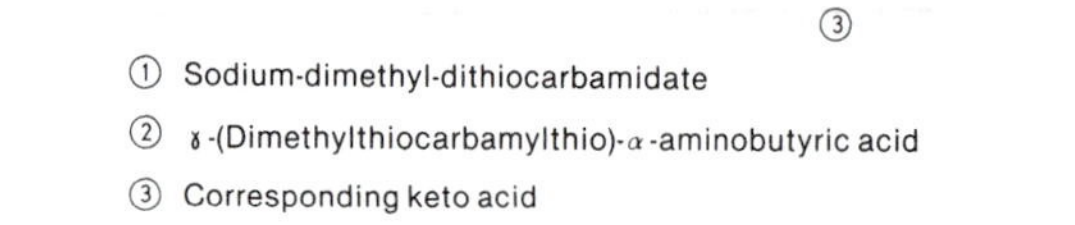

Hansenula anomala, Saccharomyces cerevisiae, E. coli, Aspergillus niger

① Sodium-dimethyl-dithiocarbamidate

② γ-(Dimethylthiocarbamylthio)-α-aminobutyric acid

③ Corresponding keto acid

Figure 15.16 Transformation of a thiocarbamate fungicide

Accumulation of xenobiotics When microorganisms absorb xenobiotics, only temporary detoxification of the environment occurs. It has been found that marine microorganisms and plankton absorb DDT and concentrate it by a factor of 100. These microorganisms are eaten by marine animals and the DDT is then stored in fat tissue, leading to an even greater concentration factor. The end result is an accumulation of the compound to high levels as it passes up the food chain.

Intensive research of recent years has shown a whole series of aerobic and anaerobic biodegradative reactions that can occur with environmentally significant compounds. Further progress can be anticipated through the following:

- The search for microorganisms capable of breaking down compounds of interest, by use of strong selective pressure in the chemostat for enrichment culture. If particular enzyme systems are being sought, special genetic probes can be used to quickly screen and identify a wide variety of natural isolates.
- The use of recombinant DNA technology to construct microorganisms that contain the complete biochemical sequence for the breakdown of a particular chemical. It should be possible to combine in one organism that processes that occur now only in co-metabolizing organisms. For example, a hybrid *Pseudomonas* has been constructed capable of breaking down chlorosalicylic acid.
- Immobilized cells should tolerate higher concentrations of xenobiotics and should also carry out the desired biochemical reactions more rapidly. This approach has already been used to develop a continuous process for the biodegradation of phenolic compounds.

REFERENCES

Anderson, S., C. Marks, R. Lazarus, J. Miller, K. Stafford, J. Seymour, D. Light, W. Rastetter, and D. Estell. 1985. Production of 2-keto-1-gulonate, an intermediate in L-ascorbate synthesis, by a genetically modified *Erwinia herbicola*. Science 230: 144–149.

Bettman, H. and H.J. Rehm. 1985. Continuous degradation of phenol(s) by *Pseudomonas putida* P8 entrapped in polyacrylamide hydrazide. Appl. Microbiol. Biotechnol. 22: 389–393.

Bollag, J.M. 1982. Microbial transformation of pesticides. Adv. Appl. Microbiol. 18:75–130.

Fukui, S., S.A. Ahmed, T. Omata, and A. Tanaka. 1980. Bioconversion of lipophilic compounds in non-aqueous solvent. Effect of gel hydrophobicity on diverse conversions of testosterone by gel-entrapped *Nocardia rhodochrous* cells. Europ. J. Appl. Microbiol. Biotechnol. 10: 289–301.

Kaul, R. and B. Mattiasson. 1986. Extractive bioconversion in aqueous two-phase systems. Production of prednisolone from hydrocortisone using *Arthrobacter simplex* as catalyst. Appl. Microbiol. Biotechnol. 24: 259–265.

Kieslich, K. 1978. Microbial transformations—type reactions, pp. 57–85. In: Hütter, R., T. Leisinger, J. Nüesch, and W. Wehrli (eds.), Antibiotics and other secondary metabolites. Federation Europ. Microbiol. Soc. Symp. No. 5. Academic Press, London.

Leuenberger, H.G.W. 1978. Microbial transformations—some applications in natural product chemistry, pp. 87–100. In: Hütter, R., T. Leisinger, J. Nüesch, and W. Wehrli (eds.), Antibiotics and other secondary metabolites. Federation Europ. Microbiol. Soc. Symp. No. 5. Academic Press, London.

Mazumder, T.K., K. Sonomoto, A. Tanaka, and S. Fukui. 1985. Sequential conversion of cortexolone to prednisolone by immobilized mycelia of *Curvularia lunata* and immobilized cells of *Arthrobacter simplex*. Appl. Microbiol. Biotechnol. 21: 154–161.

Miller, T.L. 1985. Steroid fermentations. pp. 297–318. In: Moo-Young, M. (editor), Comprehensive Biotechnology, Volume 3. Pergamon Press, Oxford.

Rehm, H.J. and G. Reed. 1984. Biotechnology, Volume 6a. Biotransformations. VCH Publishers, Deerfield Beach, FL.

Rubio, M.A., K.H. Engesser, and H.J. Knackmuss. 1986. Microbial metabolism of chlorosalicylates: accelerated evolution by natural genetic exchange. Arch. Microbiol. 145: 116–122.

Sariaslani, F.S. and J.P.N. Rosazza. 1984. Biocatalysis in natural products chemistry. Enzyme Microb. Technol. 6: 242–253.

Sebek, O.K. 1974. Microbial conversion of antibiotics. Lloydia 37: 115–133.

Sebek, O.K. and D. Perlman. 1979. Microbial transformation of steroids and sterols, pp. 483–496. In: Peppler, H.J. and D. Perlman (eds.), Microbial technology, vol. 1. Academic Press, New York.

Sonoyama, T., H. Tani, K. Matsuda, B. Kageyama, M. Tanimoto, K. Kobayashi, S. Yagi, H. Kyotani, and K. Mitsushima. 1982. Production of 2-keto-L-gulonic acid from D-glucose by two-stage fermentation. Appl. Environ. Microbiol. 43: 1064–1069.

Wichmann, R., C. Wandrey, A.F. Bückmann, and R.M. Kula. 1981. Continuous enzymatic transformation in an enzyme membrane reactor with simultaneous NAD(H) regeneration. Biotechnol. Bioeng. 23: 2789–2802.

16

Single-cell protein (SCP)

16.1 INTRODUCTION

The term "single-cell protein" (SCP) was coined at Massachusetts Institute of Technology in 1966 and is used today to refer to microbial biomass used as food and feed additives. Either the isolated cell protein or the total cell material may be called SCP.

The **human food chain** is diagrammed in Figure 16.1. Animals and plants have always provided the main food sources, but microorganisms are also eaten in small quantities in such products as cheese, vinegar, mushrooms, koji, yeast, and blue-green algae (the species *Spirulina* is eaten in parts of Africa).

In view of the insufficient world food supply and the high protein content of microbial cells (Table 16.1), the use of biomass produced in the fermenter would be an ideal supplement to conventional food supply. Single-cell protein is of great nutritional value because of its high protein, vitamin, and lipid content and the general presence of a complete array of all essential amino acids. In many countries, however, there are psychological barriers to the use of microorganisms as a major food source. The following points must also be considered with regard to safety:

- The high nucleic acid content (4–6% in algae, 10–16% in bacteria, 6–10% in yeast, and 2.5–6% in fungi) can be hazardous to health.
- Toxic or carcinogenic substances adsorbed from the microbial growth substrate (odd-numbered and branched-chain hydrocarbons, polycyclic aromatic compounds) may be present. There is also the possibility that microorganisms may produce highly toxic substances (aflatoxins are produced by some fungi, for example.).
- Slow digestion of microbial cells in the digestive tract may cause indigestion and allergic reactions.

The following substrates are presently being studied for SCP production: alkanes, methane,

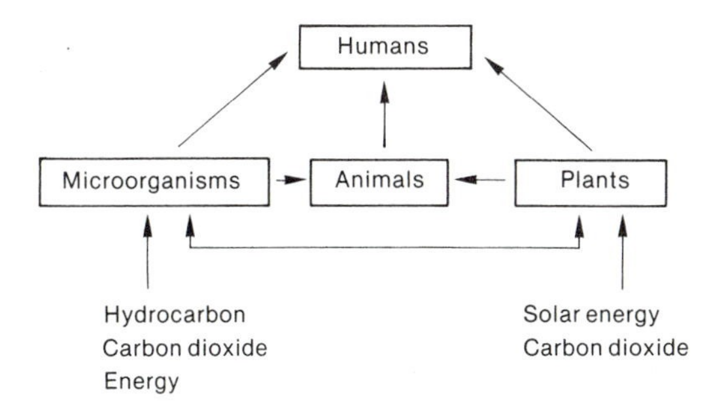

Figure 16.1 Food chain of humans

methanol, cellulose, carbohydrates, and waste materials.

Cost of production will also be a deciding factor in determining the ultimate place of SCP in the human or animal diet. It is clear that SCP must compete with natural protein sources such as soy meal or fish meal. The production costs will be strongly influenced by the nature of the raw starting material used. If a starting material is used that must otherwise be disposed of as a waste product, for instance, sulfite-waste liquor from the paper industry, then the raw material cost is actually negative. However, if a useful material such as natural gas is used, the cost of the starting material will play a major role in the final cost of the SCP. Another factor is that the price of SCP varies depending on how it is used. For instance, SCP for humans is about 10 times more expensive than SCP for animal feed because a more highly refined product must be used. Further, in less industrialized countries the price structure is more critical than in more industrialized countries.

16.2 PRODUCTION OF SINGLE-CELL PROTEIN FROM ALKANES

The use of longer-chained alkanes will be discussed in this section, and gaseous methane, the simplest alkane, will be discussed in the next section. Alkanes can be catabolized by many yeasts and by some fungi (for instance, fungi of the orders *Mucorales* and *Moniliales*) and by some bacteria (partially through cooxidation). The following yeast species have been intensively studied for SCP production: *Candida tropicalis, Candida oleophila*, and *Saccharomycopsis lipolytica*.

The disadvantage of the use of alkanes is that they are not easily soluble. During growth in bioreactors with impellers or in airlift fermenters, large alkane drops are formed which are 1–100 μm in size and which remain suspended. Considering the low water solubility of alkanes (solubility 10^{-4}–10^{-9} v/v), the observed high growth rates of microorganisms on alkanes cannot be explained merely by transport of alkane dissolved in water; other mechanisms of uptake must be present. It seems likely that the cells form emulsifying substances which convert the insoluble alkanes into droplets of 0.01–0.5 μm. Alkane molecules can then reach the cytoplasmic membrane through the cell wall via passive diffusion. Cells growing on alkanes are enriched in lipids and it seems likely that these lipids play a role in the transfer of alkanes through the cell membrane.

Table 16.1 Composition of crude single-cell protein (percent)

	Alkane yeast	Methanol bacterium	Protein isolate	Fungus	Alga	Soy meal	Milk powder
Raw protein	60.0	83	80	42	70	45.0	34.0
Fat	9.0	7.4	8–10	13	5	1.8	1.0
Nucleic acid	5.0	15	1–2	9.7	4		
Mineral salts	6.0	8.6	8–12	6.6	7	6.0	8.0
Amino acids	54	65				40	
Moisture	4.5	2.8	4.0	13.0	6	12.0	5.0

The values for soy meal and milk powder are given for comparison.

Catabolism of longer-chained alkanes

The first step in the utilization of alkanes is the introduction of molecular oxygen into the molecule. There are two pathways for oxygen introduction: terminal oxidation and subterminal oxidation (Figure 16.2). In **terminal oxidation**, the corresponding monocarboxylic acid is produced via the intermediate stages of the primary alcohol and aldehyde. After this terminal oxidation, breakdown generally proceeds to acetyl-CoA units by means of β-oxidation. Terminal oxidation is the chief pathway of metabolism for bacteria and yeast. In some cases, terminal oxidation at both ends of the molecule occurs by means of ω-oxidation, leading to the production of the corresponding dicarboxylic acid, which is further broken down into acetate units and succinate by means of β-oxidation. The enzymatic mechanisms of terminal oxidation are not yet fully understood.

In **subterminal oxidation**, the appropriate ketone is first produced via a secondary alcohol. This can either happen in the C_2 position or in the interior of the molecule, e.g. at C_3, C_4, C_5 or C_6. Internal oxidation refers to subterminal oxidation which does not take place in position C_2. Organisms show specificity in which carbon atom is attacked in the subterminal oxidation process. Further catabolism of the secondary alcohol is variable. α-Oxidation with subsequent decarboxylation and β-oxidation has been found in *Candida*. It is also possible for the molecule to be converted to an ester which is then split with an esterase. The acid and the alcohol, which are products of the esterase action, are then metabolized further.

Large-scale processes using yeast

Two petroleum products have been used as starting material:

- *Gas oil*, also known as fuel oil or diesel oil, is a fraction derived from crude oil and contains 10–25% C_{15}–C_{30} alkanes.
- C_{10}–C_{13} alkanes or C_{13}–C_{17} alkanes are separated from gas oil with molecular sieves.

British Petroleum Co. has developed processes using both of these substrates.

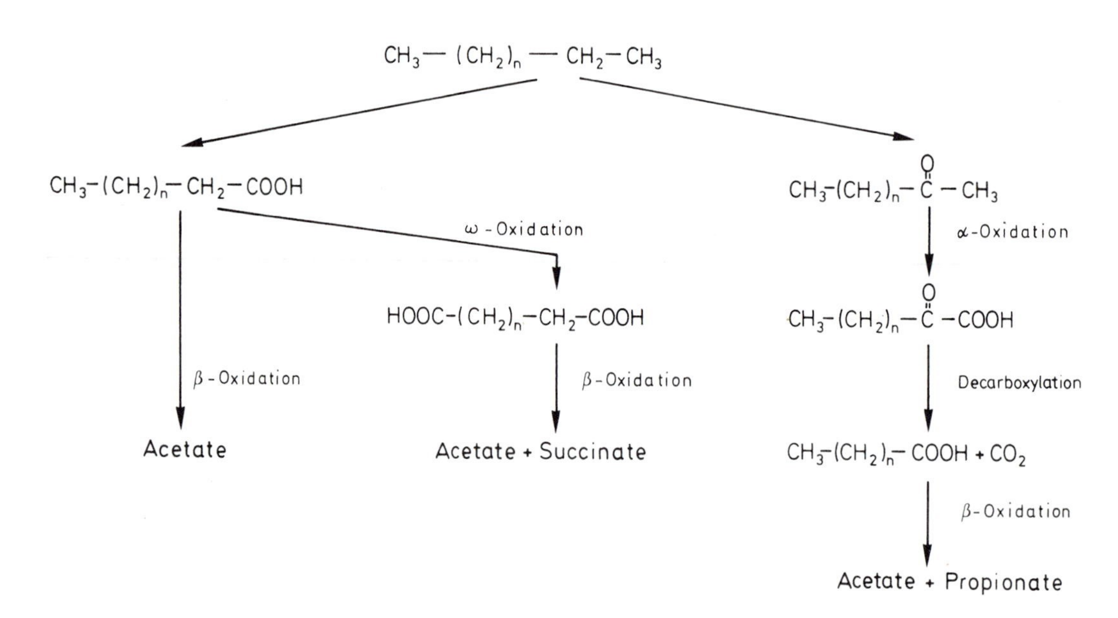

Figure 16.2 Alkane oxidation in Candida

Gas-oil process *Candida tropicalis* was tested in the gas-oil process in a nonsterile, continuous system (16,000 tons/year) in Cap Lavera, France from 1973–1975. In Grangemouth (U.K.), *Saccharomycopsis lipolytica* (previously called *Candida lipolytica*) was grown aseptically with n-alkanes in a 300 m^3 bioreactor in a system which produced 4000 tons SCP/year over a period of several years. A complete factory of 100,000 tons/year with three 1000 m^3 fermenters was built in Sardinia for the alkane procedure, but was subsequently not put into operation for political reasons.

The gas-oil fermentation was run in an airlift bioreactor with an increased aeration rate. Since alkanes make up only a small proportion of gas oil, insoluble gas oil had to be fed repeatedly to the growth medium. This resulted in poor mixing, poor oxygen transfer, and low yields.

Alkane process In comparison to glucose, *Candida tropicalis* grows much more poorly on alkanes (μ_{max} = 0.28 as compared with 0.62 on glucose), but the yield is better (0.98 g cells/g alkane versus 0.51 g cells/g glucose) and the oxygen uptake rate is similar (14 $\mu M\ O_2/g \cdot h$ on alkane versus 11 $\mu M\ O_2/g \cdot h$ on glucose).

16.3 BACTERIA WHICH UTILIZE METHANE

There is an excess of methane, the chief component of natural gas, in some parts of the world, making this a desirable energy source for SCP production. Methane can be obtained as a very pure gas. However, in contrast to higher hydrocarbons, methane cannot easily be liquefied, making long-distance transport difficult and expensive. Also, considerable security measures must be taken when handling methane, due to the risk of explosion.

Methane-oxidizing bacteria are classified among the obligate methylotrophs. This group grows only on C_1 substrates (methane, methanol, methylamine, formaldehyde, or formate). Yeasts which assimilate methane have not yet been isolated, and thus far only relatively few methane-utilizing bacteria have been identified. Among the bacteria are *Methylomonas methanica, Methylococcus capsulatus, Methylovibrio soehngenii, Methanomonas margaritae,* and some unclassified organisms.

The enzyme methane oxygenase oxidizes methane to methanol, which is further channeled into the primary metabolism (see Section 16.4).

$$CH_4 + O_2 + XH_2 \rightarrow CH_3OH + H_2O + X$$

Methanol accumulates as a result of the oxidation process and inhibits the growth of bacteria. Primary metabolites, such as amino acids, sugars, or acetate, can also inhibit growth and methane oxidation.

Using *Methylococcus capsulatus,* 0.4 g/l dry weight was obtained with a yield of 1.00–1.03 g dry weight/g methane.

Since methanol is much easier to handle and can be produced chemically from methane, methanol is the preferred starting material for all systems using C_1 substrate (see below).

16.4 METHANOL FERMENTATIONS

Methanol was at one time the most important substrate for single-cell protein production and extensive research on methanol-utilizing organisms was carried out. Although cost factors currently rule out methanol as a starting material, it still has many advantages as a substrate for SCP production and changing economics could easily bring it back.

Methanol may be obtained from synthesis gas, natural gas, methane, oil, or coal. Wood could theoretically also be used as a starting material for methanol production. Bacteria, yeasts, and fungi may all be considered for the production of SCP from methanol (Table 16.2). Besides the obligate methylotrophic bacteria which only grow on C_1 compounds, facultative methylotrophic bacteria, yeasts, and fungi which metabolize longer-chained hydrocarbons as well are

Table 16.2 Microorganisms that grow on methanol

1. Obligate methylotrophic bacteria	
Methylobacter	*Methylocystis*
Methylococcus	*Methylosinus*
Methylomonas	
2. Facultative methylotrophic organisms	
a. Bacteria	
Arthrobacter	*Protaminobacter*
Bacillus	*Pseudomonas*
Hyphomicrobium	*Rhodopseudomonas*
Klebsiella	*Streptomyces*
Micrococcus	*Vibrio*
b. Yeast	
Candida boidinii	*Pichia haplophila*
Candida parapsilosis	*Pichia lindnerii*
Hansenula capsulata	*Pichia pastoris*
Hansenula henricii	*Torulopsis glabrata*
Hansenula minuta	*Torulopsis methanolovescens*
Hansenula nonfermentans	*Torulopsis methanosorbosa*
Hansenula wickerhamii	*Torulopsis molischiana*
	Torulopsis memodendra
c. Fungi	
Gliocladium delinquescens	
Paecilomyces varioti	
Trichoderma lignorum	

(Sahm, 1979)

included in this list. In contrast to many bacteria, yeasts are unable to use any C_1 compounds other than methanol.

In **methanol fermentation** for SCP production, bacteria rather than yeasts are employed in essentially all existing production processes for the following reasons: rapid growth (Table 16.3), higher protein content, better yields, and simpler culture medium requirements.

Methanol oxidation

Methanol is oxidized to CO_2 by bacteria via the following intermediate steps:

$$CH_3OH \rightarrow HCHO \rightarrow HCOOH \rightarrow CO_2$$

The first step to formaldehyde requires an inducible nonspecific methanol dehydrogenase.

Yeasts oxidize methanol by means of a nonspecific, FAD-containing, inducible methanol oxidase. Methanol can also be oxidized by means of H_2O_2 with the peroxidase activity of an inducible catalase, the hydrogen peroxide being produced by methanol oxidase. Methanol oxidation to formaldehyde results in no energy gain for the yeast.

The next step is formaldehyde oxidation, which in bacteria can be carried out by several enzymes:

- Conversion of formaldehyde to formate with reduced glutathione (GSH) by means of an NAD-dependent formaldehyde dehydrogenase, a reaction which also occurs in yeast.

$$HCHO + NAD + H_2O \xrightarrow{GSH} HCOOH + NADH_2$$

- A dichlorophenol-indophenol (DCPIP)-dependent formaldehyde dehydrogenase.
- An unspecific methanol dehydrogenase.

Two to three ATP per mole of substrate are obtained at this oxidation level.

The last step, the oxidation of formate, is common to all the methanol-utilizing microorganisms. Formate oxidation involves a NAD-dependent formate dehydrogenase and yields 3 ATP per mole of substrate:

$$HCOOH + NAD \rightarrow CO_2 + NADH_2$$

Table 16.3 Maximal specific growth rates (μ_{max}) of various methanol utilizers

Microorganism	$\mu_{max}(h^{-1})$
Pseudomonas rosea (ICI)	0.38–0.50
Protaminobacter ruber	0.10
Pseudomonas extorquens	0.18
Methylomonas methanolica	0.53
Pseudomonas B45	0.198
Kloeckera sp. 2201	0.075
Hansenula polymorpha	0.22
Torulopsis glabrata	0.11

(Braunegg, 1975)

In some obligate methanol-utilizing bacteria (for example, *Methylomonas* M 15), formaldehyde dehydrogenase and formate dehydrogenase are not found. Instead, these organisms oxidize formaldehyde by means of the phosphogluconate pathway (Figure 16.3).

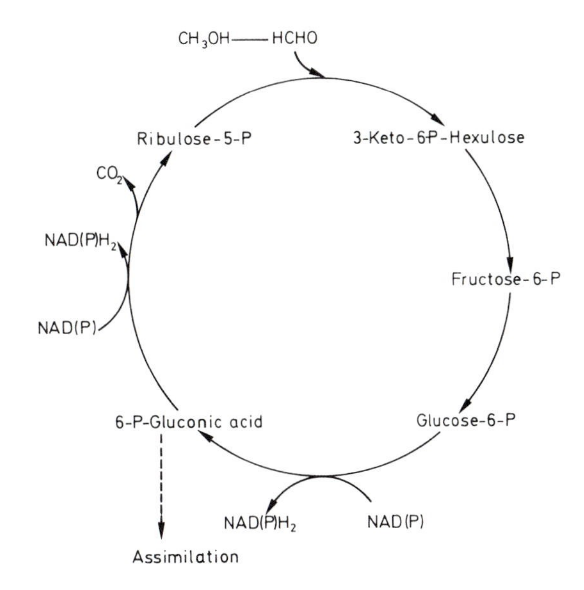

Figure 16.3 Phosphogluconate pathway

Carbon assimilation by methanol-oxidizing organisms

Bacteria growing on methanol must produce C_3 molecules which then feed into primary metabolism as pyruvate. Three distinct pathways for one-carbon assimilation have been recognized:

- CO_2 may be taken up in photosynthetic organisms via the **ribulose diphosphate (Calvin) cycle**.
- Formaldehyde may be condensed with ribulose-5-P in the **ribulose monophosphate cycle (Quayle cycle)**. The enzyme hexulose phosphate synthase (HUPS) carries out this reaction (Figure 16.4). After isomerization of 3-keto-6-P-hexulose to fructose-6-P, either dihydroxyacetone-P is produced through glycolysis or pyruvate is produced through the Entner-Doudoroff pathway. Ribulose-5-P is regenerated in the ribulose monophosphate cycle through transketolase and transaldolase reactions. HUPS, the key enzyme of the cycle, is partly constitutive, partly inducible by methanol, and is repressed by formaldehyde.
- In the **serine pathway**, condensation of for-

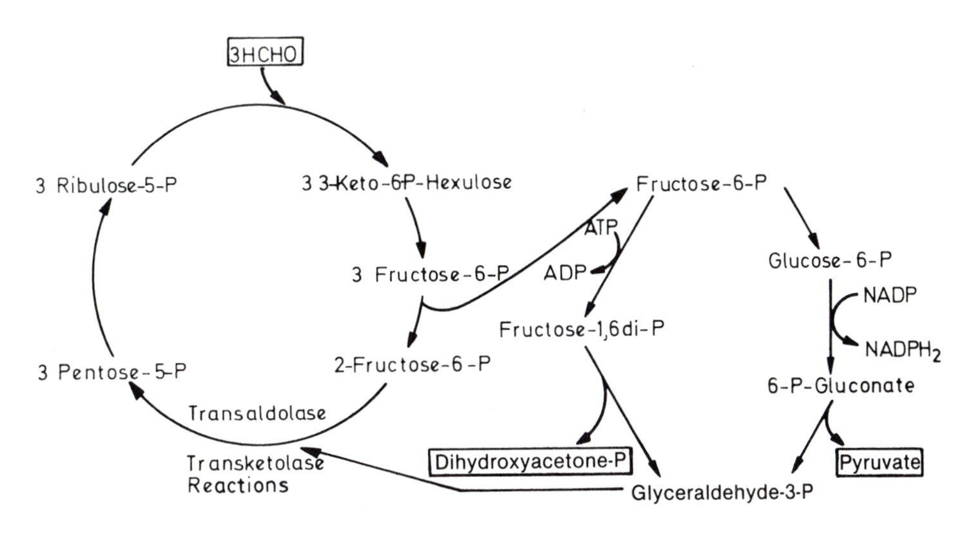

Figure 16.4 Ribulose monophosphate (Quayle) cycle

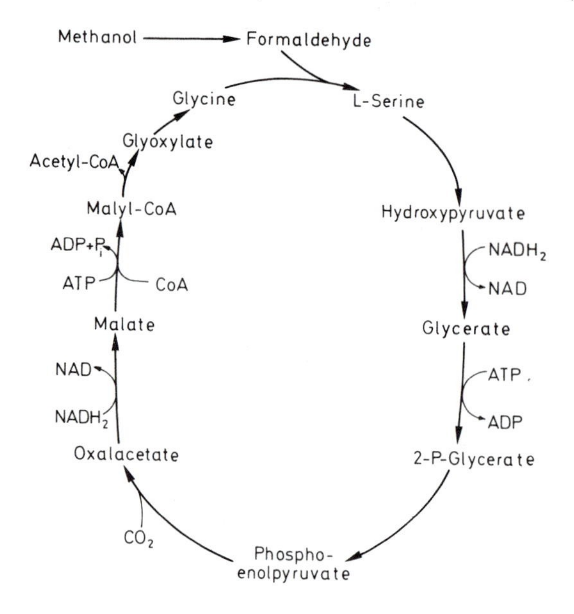

Figure 16.5 Formaldehyde fixation via the serine pathway

maldehyde and glycine takes place through the action of serine transhydroxymethylase (Figure 16.5).

A pathway very similar to the ribulose monophosphate cycle, the dihydroxyacetone cycle, has been found in yeasts (Figure 16.6).

Table 16.4 shows theoretical yield coefficients for bacteria and yeasts that use the various one-carbon pathways. It is clear from this table that bacteria using the ribulose monophosphate pathway are the best producers; therefore, all commercial processes are conducted with this group.

Production processes

Imperial Chemical Industries (ICI) was the first company to develop a continuous methanol fermentation for the commercial production of SCP. They studied the effect of O_2, CO_2, and methanol concentration on productivity and the effect of the pressure differential between the bottom and the surface of the bioreactor. The "ICI Pressure Cycle Fermenter", a combination of air lift and loop reactor, is illustrated in Figure 16.7. This 37 m^3, 30 m-high pilot fermenter consists of 3 units: air lift column (I), down-flow tube with heat removal (II), and gas release space (III). The pilot

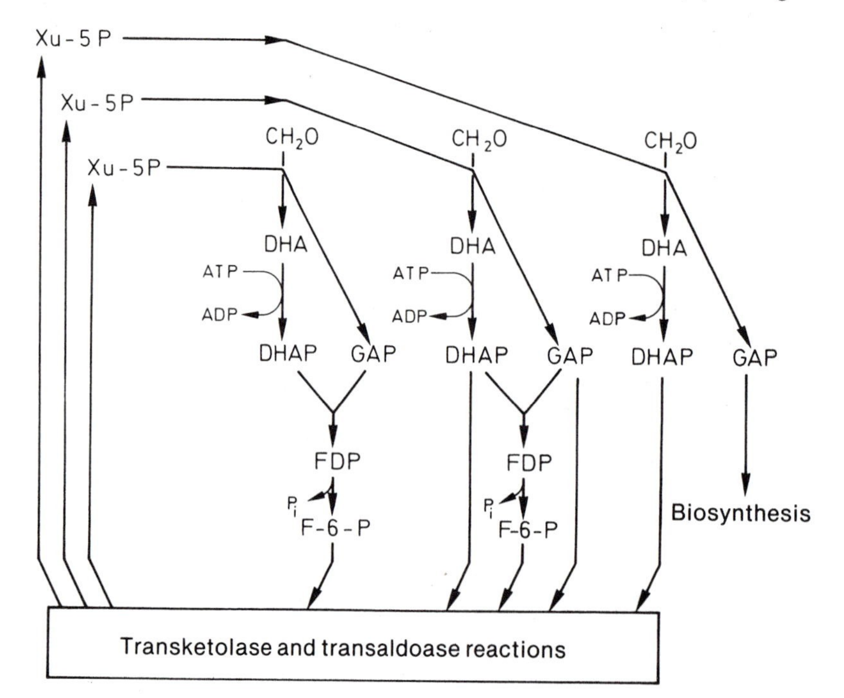

Figure 16.6 Formaldehyde fixation in yeast via the dihydroxyacetone pathway

Table 16.4 Yield coefficients of microorganisms growing on methanol

Microorganism	Pathway	Y_s (g cell dry weight/ g methanol)
a. Bacteria		
Pseudomonas C	RMP	0.54
P. methylotrophus	RMP	0.53
Methylomonas methalonica	RMP	0.49
Pseudomonas AM 1	SER	0.30
Pseudomonas M 27	SER	0.41
Pseudomonas rosea	SER	0.41
b. Yeasts		
Candida boidinii	DA	0.32
Hansenula polymorpha	DA	0.38

RMP Ribulose monophosphate cycle, SER Serine pathway, DA Dihydroxyactone cycle
(Sahm, 1979)

plant system, which had a capacity of 1000 tons/ year, operated at pH 6.5–6.9 and 34–37°C. The organism used for industrial SCP production is the methanol oxidizer *Pseudomonas methylotrophus*, which was isolated by ICI and then significantly improved through genetic and physiological development (Table 16.5).

In the product recovery process, partial cell lysis is first achieved via heat and acid treatment and the nutrient solution is then clarified by decanting. The water is then recycled back into the

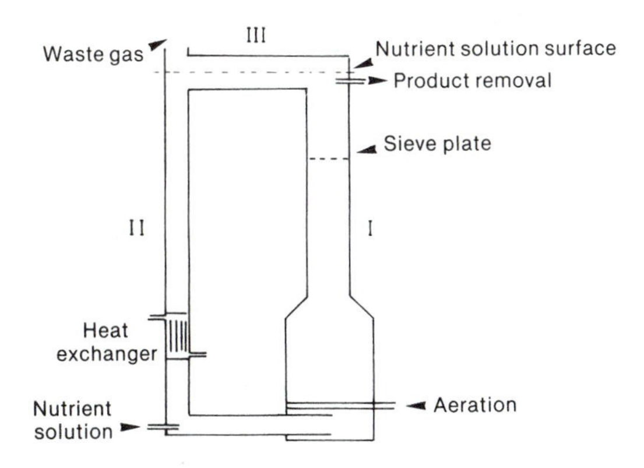

Figure 16.7 ICI pressure cycle fermenter

fermenter and the cells are spray-dried. Based on the results of this pilot study, ICI invested £40 million in 1979 to install a continuous culture system with a capacity of 50,000–70,000 tons/ year, which began operation in 1980. In this fermenter, which had a volume of 1000 m^3, cultivation of the inoculum could be carried out right in the fermenter.

Although the original chemostat process was methanol-limited, it was later operated as a nitrogen-limited system. With the bacterial strain originally used, NH_4-assimilation occurred via two inefficient enzyme systems:

Glutamine-Ketoacid Transaminase (GOGAT)
α -Ketoglurate + NAD(P)H + Glutamine $\rightarrow$ 2 Glutamate + NAD(P)

Glutamine synthetase
Glutamate + NH_3 + ATP $\rightarrow$ Glutamine + ADP + P_i

The more effective glutamate dehydrogenase system from *Escherichia coli* was therefore transferred to a $GOGAT^-$ mutant of *Pseudomonas methylotrophus* by means of genetic engineering, and such strains, which presumably grow more rapidly due to improved nitrogen assimilation, were placed in commercial production. However, the yield increase amounted to only 3%.

Because of difficulties in removing the cells from the liquid, the completely automatic large-scale process for SCP was never placed in full production due to economic reasons. For instance, in 1984 the price of soy meal was around \$125–190 per ton, whereas the SCP from ICI (going under the trade name Pruteen) was being sold at \$600 per ton!

Table 16.5 Results of the optimization of the ICI fermentation process for growing *Pseudomonas methylotrophus* on methanol

Raw protein	83%	→	85%
Pure protein	59%	→	64%
Cell dry weight	4	→	30 g/l
μ_{max}	0.38 h^{-1}	→	0.5 h^{-1}
Y		→	0.5 g/g

(Gow et al., 1975)

16.5 SINGLE-CELL PROTEIN FROM WOOD

At the present time cellulose from natural sources and waste wood is still an attractive starting material for single-cell protein production as well as a potential source for production of fermentation ethanol. Although there is an abundance of cellulose on earth, it is usually mixed with substances such as lignin, hemicellulose, starch, protein, and salts (Table 16.6). Therefore, cellulose sources must be pretreated physically and chemically in order to break down the cellulose into fermentable sugars. Pretreatment may either be enzymatic (cellulases) or chemical (acid hydrolysis). We discuss here the enzymatic processes for cellulose hydrolysis. Great advances are being made in the breakdown of lignin, and ligninase enzymes can now be produced by submerged fermentation.

Physiology

Only extracellular cellulases can be used commercially. Such cellulases are excreted by both bacteria (*Cellulomonas* and actinomycetes) and fungi (*Trichoderma, Penicillium, Thermoascus, Sporotrichum,* and *Humicola*). For pilot plant studies, the following fungi have been used: *Trichoderma reesei (T. viride), T. koningii,* and *Sporotrichum pulverulentum.*

Table 16.6 Composition of cellulosic substrates

	Cellulose %	Hemi-cellulose %	Lignin %
Wood (Angiosperms)	40–55	24–40	18–25
Wood (Gymnosperms)	45–50	25–35	25–35
Grasses	25–40	25–50	10–30
Leaves	15–20	80–85	—
Newspaper	40–55	25–40	18–30
Waste from paper manufacture	60–80	20–30	2–10

Cellulase is an enzyme complex consisting of at least three enzymes:

- Endo-β-1,4-glucanase, also called endocellulase, carboxymethyl cellulase (CMCase), or C_x cellulase.
- Exo-β-1,4-glucanase, also called cellobiohydrolase, avicelase, or C_1 cellulase. C_1 cellulase comprises 80% of the cellulase complex during fermentation with *T. viride.*
- β-1,4-glucosidase, or cellobiase. The extracellular concentration of these cellobiose-splitting enzymes is low, with the bulk of the enzyme, in *T. viride,* being intracellular.

Current status of cellulase research

Enzyme production has been optimized but cellulase yields are still low. The effects on *Trichoderma viride* of pH control, inoculum age, and culture medium additives have been tested in a basal medium containing KH_2PO_4 0.2%, $(NH_4)_2SO_4$ 0.14%, urea 0.03%, $MgSO_4 \cdot 7\ H_2O$ 0.03%, $CaCl_2$ 0.03%, $FeSO_4 \cdot 7\ H_2O$ 5.0 mg/l, $MnSO_4 \cdot H_2O$ 1.6 mg/l, $ZnSO_4$ 1.4 mg/l and $CoCl_2$ 2.0 mg/l. Maximal yields were obtained at 30°C after 140 hours' growth. As seen in Figure 16.8, yields are dependent on the cellulose concentration. Purification of the enzyme was carried out using ion exchangers.

Based on pilot plant experiments, computations can be made for the production of cellulase and for sugar production from cellulose with cellulases. Production costs have been quoted at $0.011/l for crude culture broth and $0.11/l for purified enzyme solution. Depending on the process and the extent of cellulose breakdown, the production cost to make fermentable sugar from cellulose ranges from $0.037–0.489/kg sugar. The higher price corresponds to the world market price of sucrose in 1980. Obviously, it would not be feasible to use the higher-priced sugar equivalents as substrate for the production of SCP which itself entails a considerable production cost. Before SCP can be economically produced from cellulose, cellulase yields must show a dras-

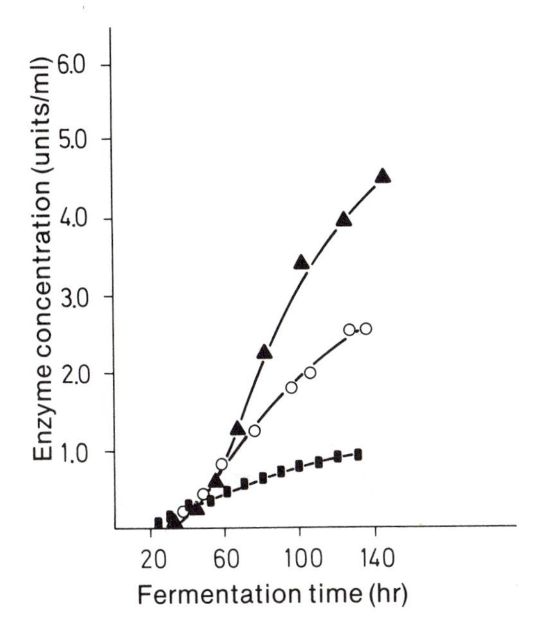

Figure 16.8 Effect of cellulose concentration on cellulase production in *Trichoderma viride* QM 9414. Squares, 0.94% cellulose; circles, 2.55% cellulose; triangles, 5.04% cellulose (Nystrom and DiLuca, 1978)

tic increase, less expensive pre-saccharification wood treatment methods must be developed, and energy-saving measures must be implemented throughout the process.

An alternate possibility is the direct fermentation of pretreated lignocellulose-containing materials with mixed cultures. The resulting biomass and the unmetabolized substrate can be used directly as animal feed.

16.6 SCP FROM CARBOHYDRATES

Large-scale cultivation of yeasts on molasses is widely used in the manufacture of baker's yeast, and in smaller-scale for food yeast and for the production of wine yeasts. In addition, several large-scale yeast processes employ whey (whose principal sugar is lactose) as a starting material.

A promising product is a mycoprotein which was introduced on the market in England in 1986. The fungus *Fusarium graminearum* is grown completely continuously on glucose with a dilution rate of 0.1 h^{-1}. Mineral salts (Na-, K-, Mg-, Mn-, Cu-, Ca-, Co-ions) and biotin are added, with ammonium ions serving as the nitrogen source. The pH is regulated at 6.0 and the incubation temperature is 30°C. From 1 kg of glucose 1 kg wet weight of fungal mycelium is obtained, with a protein content of 136 g. The yield is 3—3.5 kg/m^3·h. The culture taken directly from the fermenter is heated to 64°C to inactivate proteases and to activate endgenous RNases. After 30 min heating, the ribosomal RNA content is reduced from 9 to 1.1% and the breakdown products (principally 5′ nucleotides) diffuse from the cells. The dried protein is eventually formed into structures that resemble pork, chicken, or beef meat and an appropriate artificial flavor added. For test marketing, ICI has used fermenters of 2 × 1.3 m^3 size, with scale-up in pilot size of 30 m^3. The final product is intended to be produced in the same fermenters that ICI used for Pruteen production (see Section 16.4).

16.7 SCP FROM SEWAGE

Domestic sewage is at present not suitable for large-scale SCP production but may be more important for methane production in the future (Section 17.5). Thus far, industrial waste waters from cellulose processing, coffee production, starch production, and food processing have been used for SCP production. Sulfite waste liquors from paper and cellulose production have also been extensively used to produce SCP for the following reasons: Consistent availability of large quantities, low investment and production costs, and availability of organisms capable of high growth rates on the substrate.

Appropriate microorganisms for growth on sulfite waste liquor are *Candida utilis, C. tropicalis, Chaetomium cellulolyticum,* and *Paecilomyces varioti.*

A continuous process was used in a Finnish plant with two 360 m^3 fermenters. The SCP yields were 2.7–2.8 kg/m^3·h using *Paecilomyces* at dilution rates of 0.2 h^{-1}. With a solution containing 32 g/l reducing sugar, 55% of the sugar was converted into biomass. This installation is no longer in operation.

In Czechoslovakia a SCP factory producing 25,000 tons per year is operating using the effluent from a paper-manufacturing facility. The so-called System Paskov uses *Candida utilis* in 3 × 800 m^3 fermenters, operated continuously with a capacity of 3 × 1.5 tons/h. Product recovery involves two concentration steps in a separator (to 18% and 25%), followed by drying. In this installation, SCP production is an ancillary result of waste-water stabilization and purification.

REFERENCES

Braunegg, G. 1975. Methanol—eine neue, billige Kohlenstoffquelle in der Fermentationstechnik (Methanol—a new inexpensive carbon source for use in large-scale fermentation), pp. 218–234. 1. Arbeitstagung Biotechnologie in Österreich.

Busche, R.M. 1985. The business of biomass. Biotechnol. Progress 1: 165–180.

Fukui, S. and A. Tanaka. 1981. Metabolism of alkanes by yeasts. Adv. Biochem. Eng. 19: 217–237.

Gow, J.S., J.D. Littlehailes, S.R.L. Smith, and R.B. Walter. 1975. SCP-production from methanol: *Bacteria*, pp. 370–384. In: Tannenbaum, S.R. and D.I.C. Wang (eds.), Single cell protein, vol. II. MIT Press, Cambridge, MA.

Harwood, J.H. and S.J. Pirt. 1972. Quantitative aspects of growth of the methane oxidizing bacterium *Methylococcus capsulatus* on methane in shake flasks and continuous chemostat culture. J. Appl. Bacteriol. 35: 597–607.

Leisola, M., V. Thanei-Wyss, and A. Fiechter. 1985. Strategies for production of high ligninase activities by *Phanerochaete chrysosporium*. J. Biotechnol. 3: 97–107.

Litchfield, J.H. 1985. Bacterial biomass. pp. 463–481. In: Moo-Young, M. (ed.), Comprehensive Biotechnology, Volume 3, Pergamon Press, Oxford.

Magee, R.J. and N. Kosaric. 1985. Bioconversion of hemicellulosics. Adv. Biochem. Eng./Biotechnol. 32: 61–93.

Nystrom, J.M. and A.L. Allen. 1976. Pilot scale investigations and economics of cellulase production. Biotechnol. Bioeng. Symp. 6: 55–74.

Nystrom, J.M. and P.H. DiLuca. 1978. Enhanced production of *Trichoderma* cellulase on high levels of cellulose in submerged cultures. Proc. Bioconversion Symp. IIT Delhi, pp. 293–304.

Oura, E. 1983. Biomass from carbohydrates. pp. 3–41. In: Rehm, H.J. and G. Reed (editors), Biotechnology, Volume 3, VCH Publishers, Deerfield Beach, FL.

Rehm, H.J. and I. Reiff. 1981. Mechanisms and occurrence of microbial oxidation of long-chain alkanes. Adv. Biochem. Eng. 19: 175–215.

Romantschuk, H. and M. Lehtomäki. 1978. Operational experiences of first full scale Pekilo SCP-mill application. Proc. Biochem. 13: 16–17,29.

Sahm, H. 1979. Production of SCP by methanol utilizing microorganisms. Int. Microbiol. and Food Ind. Congr., Paris.

Smith, A.J. and D.S. Hoare. 1977. Specialist phototrophs, lithotrophs and methylotrophs: a unity among a diversity of procaryotes? Bacteriol. Rev. 41: 419–448.

Solomons, G.L. 1985. Production of biomass by filamentous fungi. pp. 483–505. In: Moo-Young, M. (editor), Comprehensive Biotechnology, Volume 3, Pergamon Press, Oxford.

Tanaka, M. and Matsuno, R. 1985. Conversion of lignocellulosic materials to single-cell protein (SCP): recent developments and problems. Enzyme Microb. Technol. 7: 197–206.

17

Newer approaches to sewage treatment

17.1 INTRODUCTION

In urbanized countries, vast amounts of industrial and domestic sewage are treated biologically. At present, sewage is treated primarily by the classic, aerobic, activated sludge process with surface aeration (Figure 17.1) or with oxygen supplied by forced air. In another type of aerobic treatment process (the trickling filter), a bed of stones or sand is used and the waste water is allowed to trickle down over the supporting medium, the organisms growing attached to the surfaces of the filter bed and oxidizing the organic compounds in the waste material. In another process, anaerobic digestion is used (Figure 17.2), occasionally directly on raw sewage or more frequently on the solid material (sludge) obtained from sedimentation in the aerobic treatment processes.

Aerobic systems have the following disadvantages: open construction which restricts process control, uncontrolled populations of organisms, environmental pollution due to odor emission, and fog formation.

The developments in sewage treatment over the past century have been carried out primarily in an empirical fashion, but aerobic and anaerobic sewage treatment processes are being increasingly studied scientifically, since it is felt that optimization will lead to the greatest success in improving these processes. Although studies on aerobic processes have been primarily directed to an understanding of carbon metabolism, in the future it is anticipated that research will be directed at sulfur, nitrogen, and phosphate metabolism, since elimination of these elements from sewage will become increasingly important. Further, accumulations of heavy metals in the stabilized sludge are being increasingly recognized to be of serious concern.

A further critical problem is the phenomenon of **bulking** in sludge, which inhibits the settling process in the clarification basins. Bulking is due to the development of filamentous microorga-

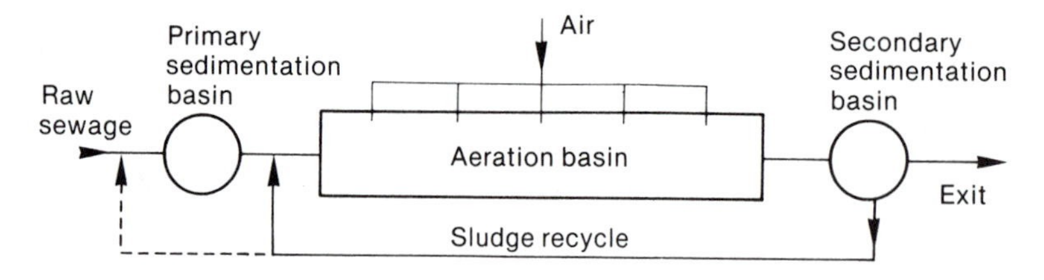

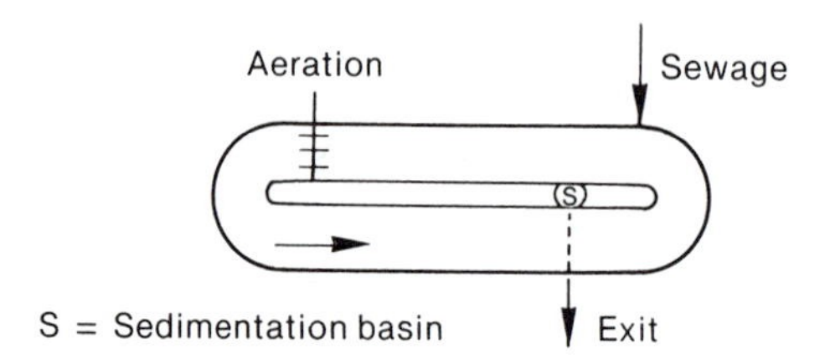

Figure 17.1 Activated sludge process: above, aeration basin; below, oxidation ditch

nisms that do not settle well. Although bulking can be controlled in some cases by introduction of flotation processes, in domestic sewage this technical solution is too expensive. Therefore, a microbiological solution to the bulking problem must be found so that a satisfactory clear effluent can be delivered from the final settling basin.

One of the most encouraging developments over the past several decades has been the remarkable decrease in organic material in industrial effluents. This has resulted from extensive application of research and development efforts and considerable expenditures for new treatment installations.

In this chapter four new processes will be discussed which have the potential for increasing the efficiency of sewage treatment.

Figure 17.2 Sludge digester

17.2 STARTER CULTURES FOR TREATMENT PROCESSES

Conventional sewage treatment involves the use of microorganisms which develop naturally within the sewage treatment system, no attempt being made to optimize the organisms involved. An approach which may have some potential for increasing the efficiency of the sewage treatment process is to inoculate the system wih microorganisms which have been specially selected for the particular sewage-treatment process. In analogy with their use in food fermentations, such organisms might be called ''starter cultures''. Although starter cultures might find use in the treatment of domestic sewage, it seems more likely that they will find use in the treatment of special or unusual industrial wastes or in the treatment of accidental spills of industrial chemicals. Such wastes often cannot be channeled into ordinary treatment plants because their toxicity and lack of biodegradability cause significant damage to the nonadapted organisms. Another way in which such starter cultures might be used is in shortening the start-up time that is generally required after a sewage treatment plant is shut down for one or another reason. Before the system can become fully operative once again, the optimal bacterial culture mixture must be re-established; this usually requires a week-long enrichment process. A starter culture could be expected to shorten this start-up time. In the United States, starter cultures resembling those of the dairy industry have been developed and are suitable for a variety of special applications in sewage treatment, such as tank cleaning, pipeline cleaning, start-up of city purification plants, and breakdown of special substances contained in sewage.

Bacteria from cold habitats have been isolated which degrade alkanes and aromatic compounds at 0–15 °C in saline habitats; these could be useful in the degradation of oil spills in the ocean. Mixed cultures which metabolize DDT, polychlorinated diphenols, and phenols or which possess high protease, lipase, or cellulase activity are also on the market. If special operating conditions are desired, such as growth at $pH \leq 5.0$ or ≥ 9.0, these requirements can also be met with selected enrichment cultures.

A patented process has been developed with a strain of *Pseudomonas putida* containing plasmids which code for the breakdown of octane, xylene, metaxylene, and camphor. Starter cultures have also been used to deodorize animal excrements.

In 1978, the starter-culture industry grossed $2–4 million in the United States, but the potential total value is estimated at $200 million. Twenty manufacturers produce single strains or mixed cultures under sterile conditions in vessels up to 10 m^3 capacity. The culture conditions used are determined by how the cultures are to be used, in order to ensure that the cells are fully induced. Research in this field is still in its infancy and scientific studies on the breakdown of individual compounds are underway, but practical applications are not yet widespread.

17.3 AEROBIC SEWAGE TREATMENT–AIRLIFT PROCESS

A disadvantage of customary aerobic sewage treatment by the activated sludge system is the low efficiency of oxygen transfer and the large amount of space required by the installation. Because these installations are open to the atmosphere, another problem is the odor which they emit. Tower reactors have been developed in which aeration and oxygen utilization are improved and the efficiency of the overall process increased. Both tube reactors and airlift fermenters have been used. These installations are favorable economically because of 30% less space required, 20% lower investment costs, and 20% less energy costs.

The British chemical company ICI uses a tubular loop reactor which is embedded 100 meters into the ground. Two German companies, Bayer (Biotower) and Uhde/Hoechst (Bio-high Reactor) have constructed bioreactors 30 m in height. In these systems, the circular settling basins are located around the top rim of the bioreactor (Figure 17.3). The dimensions of this type of activated sludge fermenter allow a considerably better oxygen-transfer efficiency. Table 17.1 shows performance data from various systems of the

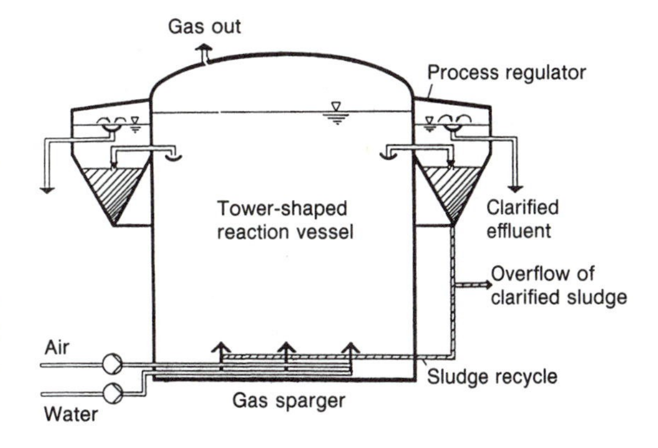

Figure 17.3 Bayer Biotower

Table 17.1 Data from an activated sludge system (Uhde/Hoechst)

Water depth (m)	4	8	20
O_2-Saturation (g O_2/m³)	13	17	28
Oxygen yield (%)	8–15	15 –30	40–60
Air required (Nm³/kg O_2)	42–23	23 –11	9– 6
Energy-specific O_2 input (kg O_2/kWh)	2– 2.5	2.5– 3	3– 3.5
Bioreactor diameter (m)	18	13	8
Reactor surface area for 1000 m³ volume (m²)	250	125	50

Table 17.2 Parameters for a Bayer Biotower in continuous operation

Process	Parameters
Pretreatment	Screening, neutralization, primary sedimentation
Bioreactors	4 parallel, rubber-lined tanks (26 × 30 m) with 13,600 m³ total volume
Aeration	22,000 m³/h
Sewage	
From industry	90,000 m³/d with 95 tons/d BOD
From domestic sources	70,000 m³/d with 14 tons/d BOD
Residence time	14.5 h
Treatment capacity	105 tons/d BOD
Oxygen requirements	120 tons/d O_2

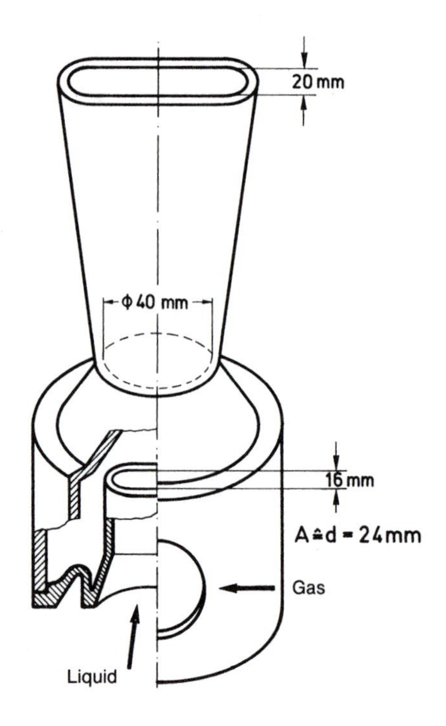

Figure 17.4 Bayer injector (Zlokarnik, 1985)

Uhde/Hoechst type and Table 17.2 shows data for the Bayer system.

The Hoechst system has baffles to facilitate mixing. In the Bayer Biotower process, two-component nozzles have been developed as injectors by means of which the kinetic energy of the liquid stream is utilized to disperse air in small gas bubbles in a manner similar to a water spray pump (Figure 17.4). These injectors are installed in groups of four (injector clusters) above the bottom of the tower reactor (1 injector/1–2 m² bottom surface).

17.4 AERATION WITH PURE OXYGEN

Aeration of activated sludge treatment facilities with pure oxygen has been tried in both the United States and the Federal Republic of Germany. The objective has been to increase the efficiency of oxygen transfer, since the loading capacity of a conventional purification plant and the size of the microbial population are limited by the low solubility of dissolved oxygen.

Pure oxygen has 4.8 times the partial pressure of oxygen from the air, so that aeration with pure oxygen markedly increases the oxygen content of the activated sludge system. Dissolved oxygen in such a system is 90–95% utilized, and the off gas (1% of that from conventional aeration) can subsequently be treated chemically or thermally in the closed system in order to prevent unpleasant odors.

High loading capacity in the oxygen process allows smaller aeration tanks and smaller settling basins. Due to the smaller amount of sludge produced, the costs of facilities for further treatment are also reduced. On the negative side is the higher capital investment, the more complicated installation, the necessity for careful control of the process parameters, and the need for highly trained personnel.

In both domestic and industrial installations systems are in use that employ pure oxygen. These systems are especially used in the paper, chemical, and food industries.

The Unox system is one such example. The first aeration stage consists of 2 channels, each of which is connected in a cascade arrangement to three bioreactors (Figure 17.5) with capacities of 1500, 750 and 730 m^3. In the first bioreactor (activated sludge basin), pure oxygen from an air-separation plant is forced in via injectors. In this stage, 90–93% of the organic, decomposable material is eliminated. The CO_2 produced is collected along with the residual oxygen and treated at 1000°C. In the second treatment stage (settling basin), the accumulating sludge is pumped back into the activated sludge basin. The waste water flows through this second treatment stage to remove material which is not easily broken down (Figure 17.6). After a third treatment stage, the clarified waste water proceeds to the receiving stream. Table 17.3 gives performance characteristics using the Unox process.

17.5 METHANE PRODUCTION

Methane, or natural gas, is a microbial product of the anaerobic decomposition of organic matter. Methanogenesis is a widely used process in organic waste disposal, primarily because the methane gas, being insoluble, is readily removed from the treatment system. Since it is a fuel, the methane produced in the treatment system can also be used as a source of energy. In some agricultural situations, the methane formed is a significant energy source (biogas) and the main goal of the treatment process is the maximization of

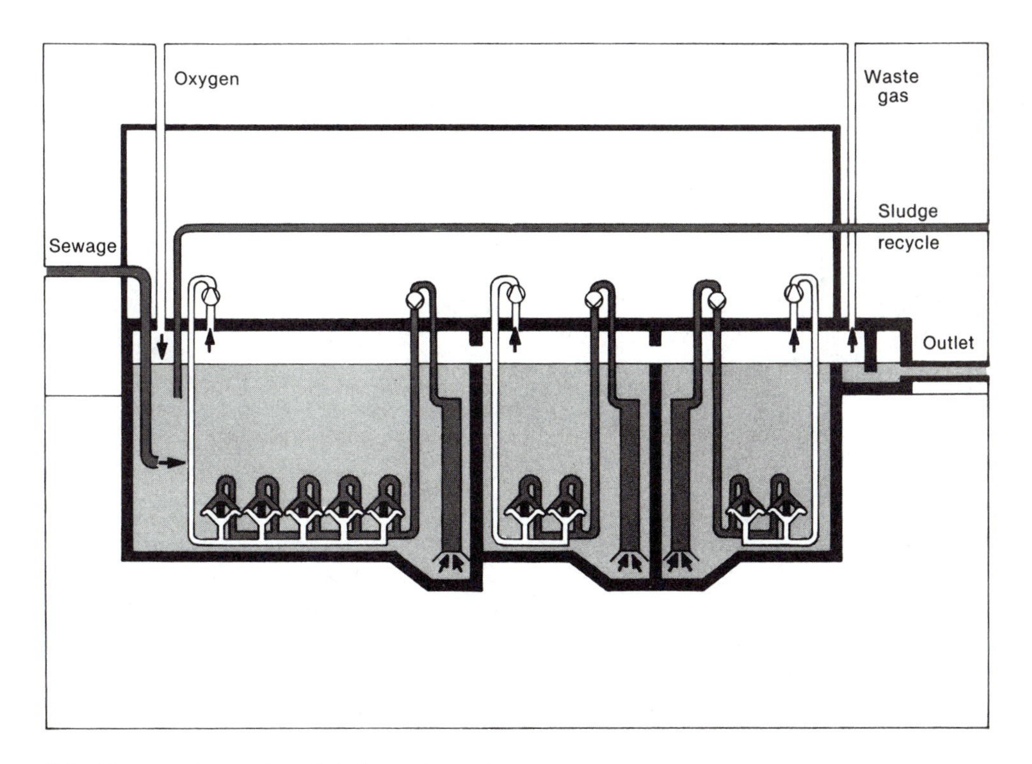

Figure 17.5 Diagram of an activated sludge system using pure oxygen

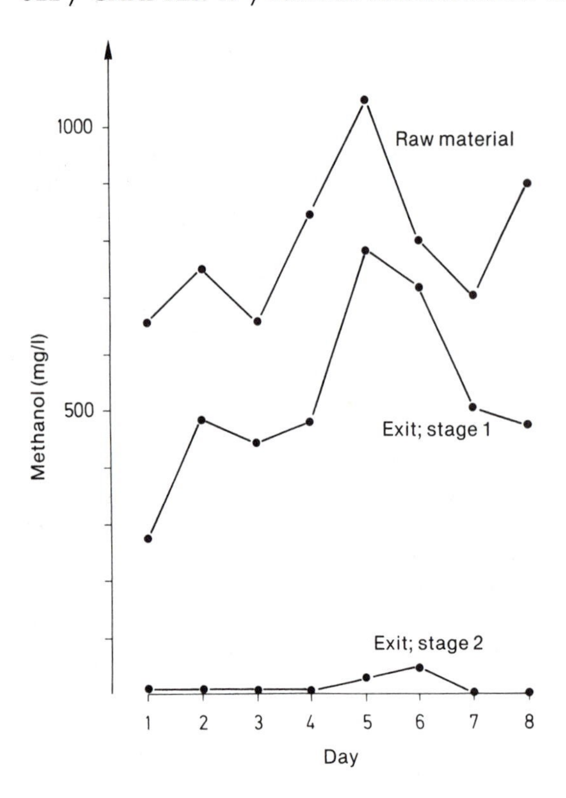

Figure 17.6 Methanol breakdown in a two-stage purification plant using pure oxygen (Mack, 1973)

Table 17.3 Data from a Unox purification plant with pure oxygen

Amount of sewage	20,000 m^3/day
Chemical oxygen demand (COD)-breakdown capacity	61 tons/day
Biological oxygen demand $(BOD)_5$-breakdown capacity	34 tons/day
Oxygen input	
Activation stage 1	51 tons/day
Activation stage 2	7 tons/day
Volume	
Activation stage 1	6200 m^3
Activation stage 2	3200 m^3
Intermediate purification:	4400 m^3
Final purification	4800 m^3
Time required	
Activation stage 1	7.5 h
Activation stage 2	3.8 h
Intermediate purification:	5.3 h
Final purification	5.8 h
Breakdown activity	
Activation stage 1	max. 80% COD max. 93% BOD_5
Activation stage 2	max. 7% COD, total 87% max. 5% BOD_5, total 98%
Sludge load	
Activation stage 1	0.7 kg BOD_5/kg dry weight·d
Activation stage 2	0.2 kg BOD_5/kg dry weight·d
Space requirement	
Activation stage 1	5.6 kg BOD_5/m^3·d
Activation stage 2	0.8 kg BOD_5/m^3·d
Excess sludge	25% of the BOD_5 load

(Lemke and Mack, 1978)

methane yields. Anaerobic installations for sewage treatment are called **digestors**, because the primary process taking place is the digestion of the organic material to soluble and gaseous constituents. In most waste treatment systems, however, even though the methane may be used as a fuel, the main goal has not been the production of methane as an energy source but rather the stabilization of the sludge so that it can be safely disposed of in the environment. In such systems, little attention has been given to the optimization of the methane production process. In one study of 275 sewage treatment systems in western Europe, only a small fraction of the methane produced was distributed to other users, although 75% of the gas produced was used as an energy source within the treatment plants themselves.

In comparison to aerobic processes, anaerobic process have some significant advantages. Energy expenditure for aeration is not required, saving considerably on operating costs. The organic waste is converted almost quantitatively (>90%) to methane and CO_2. Biomass formation is low, so that the cost of sludge disposal is lower. Also, odor problems are decreased since the process is carried out in closed reactors. Finally, the methane produced can be utilized as an energy source.

However, the anaerobic process is not suitable for the more dilute liquid wastes that are primarily treated by the aerobic activated sludge processes.

The methanogenic bacteria constitute a unique group of organisms which are the final and key link in the breakdown of organic matter in the anaerobic food chain. The methanogenic bacteria are able to utilize only a restricted group of substrates for the production of methane, including: acetate, methanol, formate, and $H_2 + CO_2$. In waste disposal systems, about 75% of the methane is derived from acetate and most of the rest from $H_2 + CO_2$. The starting materials of the anaerobic decomposition process are complex polymeric materials such as cellulose, starch, fats, and proteins, none of which the methanogenic bacteria are able to use. The methanogenic bacteria are consequently dependent upon other anaerobic fermentative organisms for the initial breakdown of the substrates and the production of acetate and $H_2 + CO_2$.

Three groups of microorganisms participate in the anaerobic process. The first group breaks down the original organic material (starches, fats, proteins) into organic acids (propionic acid, butyric acid, acetic acid, lactic acid, and valeric acid), alcohol, H_2, and CO_2. Since one of the constituents of the initial load, lignin, is not broken down anaerobically, any lignocelluloses in the starting material are metabolized slowly. The group of bacteria involved in this initial breakdown of organic matter include obligate anaerobes such as the clostridia and facultative anaerobics such as the streptococci and enteric bacteria.

The second group of bacteria convert the longer-chain fatty acids (for instance, propionic and butyric acids) and alcohols to acetic acid, H_2, and CO_2. This conversion is endergonic at pH 7 and can only occur if coupled with exergonic reactions. Thus, the reaction only occurs in mixed cultures. For example, ethanol is converted to acetic acid and H_2 by one bacterium and in a second reaction a methane bacterium utilizes the H_2, thus pulling the reaction:

$$CH_3CH_2OH + H_2O \rightarrow CH_3COO^- + H^+ + 2H_2$$
$$4H_2 + CO_2 \rightarrow CH_4 + 2H_2O$$

The third group of organisms are the extremely anaerobic methanogens. As noted, only a limited range of substrates are used by these bacteria. Most methanogens will use H_2 as energy source and CO_2 as carbon source and electron acceptor. A few species of morphologically diverse groups (cocci, sarcinae, and spirillae) also metabolize formic acid, methanol, or methylamine. The most important organic substrate of methanogens is acetic acid, which is fermented in a reaction that is only weakly favorable energetically:

$$CH_3COO^- + H^+ \rightarrow CH_4 + CO_2$$
$$\Delta G^0 = -37 \text{ kJ}$$

Because of this, these bacteria grow very slowly and are the most critical organisms in the mixed culture of the anaerobic digestor.

Because of the low biomass formed and the low energy yield, the rate of the digestion process is low, the average residence time in the digestor being greater than 20 days. Immobilization of microorganisms in a fixed-bed reactor can be used to increase greatly the biomass concentration. Solutions with very high organic loads (chemical oxygen demand >3 g O_2 consumption/l) can be treated with such fixed-bed reactors within a few hours. Some studies have been carried out using porous sintered glass as carrier material for fixed-bed reactors. In a 1000 l reactor the biochemical oxygen demand (BOD) was reduced from 6.4 kg/m^3 to 1.3 kg/m^3 in 4.8 hours. In another installation for treating high-concentration wastes from a fermentation plant, a two-stage process was used, the first reactor being used for acid production, the second for methanogenesis. Both reactors were 380 m^3 in volume and sand was used as the carrier. Waste volumes of 150–200 m^3/h were treated within 1–1.5 hours with a removal efficiency of 30 kg COD/m^3·day.

In the United States, some studies have been done to produce methane from the biomass of the water hyacinth, a plant which causes considerable problems by excess growth in canals and streams in the warmer parts of the country. Under optimal conditions, about 2 tons dry mass per day and per hectar could be treated, yielding theoretically 800 m³ biogas.

When optimizing methanogenesis, it is important to note that the rate is affected by the nature of the substrate as well as by the temperature (Table 17.4).

With mixed sludges, as much as 600 l biogas per kg dry organic matter have been produced. The gas consists of about 60–70% methane, about 25–30% CO_2, and the rest H_2 and N_2. Developing countries represent the most promising areas for further development of biogas production, since energy is often in short supply and waste disposal can often not be done by the elaborate processes used in developed countries. In such cases, the efficiency of the process is less critical than the energy produced; biogas production in developing countries can often be accomplished in simple installations with little requirement for complicated engineering skills. Biogas can also be used as a motor fuel in those regions where petroleum sources are expensive or in short supply.

Table 17.4 Biogas production in relation to temperature

Substrate	Temperature °C	% Dry weight in input	Time h	l Methane/ g dry weight
Sludge	35–40	2.1	23	0.488⁺
	50	2.1	9	0.488⁺
	60	2.1	12.5	0.488⁺
Sludge	30	3	33–50	0.371
	50	3	20	0.368
Waste and sludge	35	3	30	0.288⁺
	50	3	10	0.369⁺
	60	3	4	0.455⁺
Cow manure	55	2	3	0.16
			9	0.19
	55	6	3	0.16
			9	0.19
	55	10	3	0.10
			9	0.19

⁺=Biogas
(Kandler, 1979)

REFERENCES

Aivasidis, A., H.J. Brandt, A. Joeris, J. Kriese, R. Pick, and C. Wandrey. 1986. Recent developments in process and reactor design for anaerobic waste water treatment. Biotech Forum 3:141–150.

Barnard, G.W. and D.O. Hall. 1983. Energy from renewable resources, pp. 592–625. In: Rehm, H.J. and G. Reed (eds.). Biotechnology 3. Verlag Chemie, Weinheim.

Bishop, P.L. and N. E. Kinner. 1986. Aerobic fixed-film processes, pp. 113–176. In: Rehm, H.J. and G. Reed (eds.). Biotechnology 8. Verlag Chemie, Weinheim.

Chakrabarty, A. 1974. US Patent 3,813,316.

Dohne, E. 1985. Biogasverwertung, Biogasmotoren und Wirtschaftlichkeit. (Economic aspects of biogas utilization and biogas motors.) GIT Fachz. Lab 12/85:1241–1254.

Holt, F. 1986. Environmental management: is there a business? World Biotech Report 1: 89–99. Online, London.

Kandler, O. 1979. Optimierungsmöglichkeiten von Biogasprozessen unter Einsatz thermophiler Mikroorganismen (Optimization of the biogas process using thermophilic microorganisms), pp. 19–31. BMFT (ed.), Biokonversion. 2nd BMFT-Statusseminar "Bioverfahrenstechnik".

Lemke, J.R. and K. Mack. 1978. Biologische Abwasserreinigung mit technisch reinem Sauerstoff in der Kläranlage BAYER AG, Werk Wuppertal-Elberfeld (Biological sewage treatment in activated sludge plants using pure oxygen). Chemie-Technik 7: 193–196.

Mack, K. 1973. Biologische Reinigung von Abwasser mit Sauerstoff (Biological treatment of sewage with oxygen). Wasser, Luft, und Betrieb 17: 108–111.

Magruder, G.C. and J.L. Gaddy. 1981. Production of farm energy from biomass, pp. 269–274. In: Moo-Young, M. (ed.), Advances in biotechnology, vol. II. Pergamon Press, Toronto.

Müller, H.G. 1984. Der Biohochreaktor—eine platz- und energiesparende biologische Abwasserreinigungsanlage. (The vertical bioreactor—a space- and energy-saving sewage-treatment installation.) GIT-Suppl. 4:16–23.

Nyns, E.J. 1986. Biomethanation processes, pp. 207–267. In: Rehm, H.J. and G. Reed (eds.). Biotechnology 8. Verlag Chemie, Weinheim.

Ohta, T. and M. Ikeda. 1981. Rapid microbial deodorization of agricultural and animal wastes, pp. 621–626. In: Moo-Young, M. (ed.) Advances in Biotechnology II. Pergamon Press, Toronto.

Reimann, H., M. Morper, and C.H. Gregor. 1982. Reinigung industrieller Abwässer unter Einsatz von technischem Sauerstoff. (Purification of industrial waste water with oxygen.) Chemie Technik 11:27–32.

Sahm, H. 1984. Anaerobic wastewater treatment. Adv. Biochem. Eng./Biotechnol. 29:83–115.

Shailubhai, K. 1986. Treatment of petroleum industry oil sludge in soil. Reviews. TIBTECH, August 202–206.

Shieh, W.K. and J.D. Keenan. 1986. Fluidized bed biofilm reactor for wastewater treatment. Adv. Biochem. Eng./Biotechnol. 33:131–169.

Slater, J.H. 1986. Microbial inoculants in environmental management. World Biotech Report 1986 1:75–79. Online, London.

Sojka, S.A. 1986. Genetic engineering for dehalogenation and landfill applications. World Biotech Report 1986 1:81–88. Online, London.

Verstreate, W., and E. van Vaerenbergh. 1986. Aerobic activated sludge, pp. 43–112. In: Rehm, H.J. and G. Reed (eds.). Biotechnology 8. Verlag Chemie, Weinheim.

Uhde/Hoechst. Bio-Hochreaktor. No date.

Zlokarnik, M. 1985. Tower-shaped reactors for aerobic biological waste water treatment, pp. 537–569. In: Rehm, H.J. and G. Reed (eds.). Biotechnology 2. Verlag Chemie, Weinheim.

18

Leaching

18.1 INTRODUCTION

Microbial leaching is the process by which metals are dissolved from ore-bearing rocks using microorganisms. At present a number of ores cannot be economically processed with chemical methods due to their low metal content. In addition, large quantities of low-grade ores are produced during the separation of higher-grade ores and are generally discarded in waste heaps. Throughout the world there are vast quantities of such low-grade copper ores that cannot be profitably purified by conventional chemical methods, but that could be processed by microbial leaching. There are also significant quantities of nickel, lead, and zinc ores which could be leached.

Leaching was first discovered as a process occurring in pumps and pipelines installed in mine pits containing acidic water. It was subsequently developed for the recovery of metal from low-grade ores. For many metals, there are now leaching methods which permit extraction from metal sulfides or other ores. The metals are converted to water-soluble metal sulfates with the aid of biochemical oxidation processes.

In commercial applications, copper and uranium have been widely produced through the use of microorganisms. However, there have been difficulties in extrapolating the results from laboratory and pilot-plant studies into practical field conditions. In addition, problems may arise when the large-scale leaching process of a waste dump is improperly managed. Leach fluids containing large quantities of metals and having extremely low pH values (pH 3) can seep from such dumps into nearby natural water supplies and ground waters, causing enormous and lasting damage.

18.2 ORGANISMS FOR LEACHING

The two most commonly used organisms in microbial leaching are *Thiobacillus thiooxidans* and *Thiobacillus ferrooxidans*. A number of others may also be used including: *Thiobacillus concre-*

tivorus, Pseudomonas fluorescens, P. putida, Achromobacter, Bacillus licheniformis, B. cereus, B. luteus, B. polymyxa, B. megaterium, and several thermophilic bacteria including *Thiobacillus thermophilica, Thermothrix thioparus, Thiobacillus* TH1, and *Sulfolobus acidocaldarius.* The heterotrophic organisms listed have not as yet actually been used, but it seems likely that processes will be developed by which metals are extracted from ores with microbially produced organic acids via chelate and salt formation. Because of their more rapid growth rate, the thermophilic bacteria may significantly accelerate the leaching process.

18.3 CHEMISTRY OF MICROBIAL LEACHING

Thiobacillus ferrooxidans is the organism that has been most extensively studied. It is a Gram-negative rod-shaped bacterium which is 0.5–0.8 $\mu m \times 1.0$–2.0 μm in size. An autotrophic aerobe, it can obtain carbon for biosynthesis solely from CO_2 fixation, and obtains its energy from the oxidation of Fe^{2+} to Fe^{3+} or from the oxidation of elemental sulfur and reduced sulfur compounds to sulfate.

$$4\,FeSO_4 + 2\,H_2SO_4 + O_2 \rightarrow 2\,Fe_2(SO_4)_3 + 2\,H_2O \qquad [1]$$

$$2\,S^O + 3\,O_2 + 2\,H_2O \rightarrow 2\,H_2SO_4 \qquad [2]$$

$$2\,FeS_2 + 7\,O_2 + 2\,H_2O \rightarrow 2\,FeSO_4 + 2H_2SO_4 \qquad [3]$$

The oxidation of insoluble sulfur to sulfuric acid, which is also performed by *Thiobacillus thiooxidans,* occurs in the periplasmic space. According to equation 3, iron is dissolved through **"direct bacterial leaching"**.

In addition to this leaching process performed only by microorganisms, there is another process, "indirect, bacterially supported leaching" which takes place slowly in the absence of microbes. The oxidation of pyrite can be used as an example. Pyrite is a common rock mineral that is found in association with many ores. The following equation describes the initial oxidation of pyrite by ferric ions:

$$FeS_2 + Fe_2(SO_4)_3 \rightarrow 3\,FeSO_4 + 2\,S^O \qquad [4]$$

The sulfur which is formed via this process is reoxidized as shown in equation 2.

Examination of leaching dumps always shows the presence of mixtures of *T. thiooxidans* and *T. ferrooxidans.* In pilot-plant reactors (50 liter), leaching can be performed continuously in a cascade series with recycling of the cells and leachate.

Yields such as those in other areas of microbiology can be attained in the laboratory under optimal conditions (temperature control, O_2 and CO_2 adjustment, maintenance of pH around 2–3 and Eh around −300 mV) with very finely ground ores in a tower (percolator), or better yet in fermenters under optimal conditions. However, in field experiments, these conditions and yields cannot be realized due to the high cost.

18.4 COMMERCIAL PROCESSES

Three methods have practical application (Figure 18.1):

- **Slope leaching**. Finely ground ores (up to 100,00 tons) are dumped in large piles down a mountainside and continuously sprinkled with water containing *Thiobacillus.* The water is collected at the bottom and reused after metal extraction and possible regeneration of the bacteria in an oxidation pool.
- **Heap leaching**. The ore is arranged in large heaps and treated as in slope leaching.
- **In-situ leaching**. Water containing *Thiobacillus* is pumped through drilled passages to unextracted ore which remains in its original location in the earth. In most cases, the permeability of the rock must be first increased by subsurface blasting of the rock. The acidic water seeps through the rock and

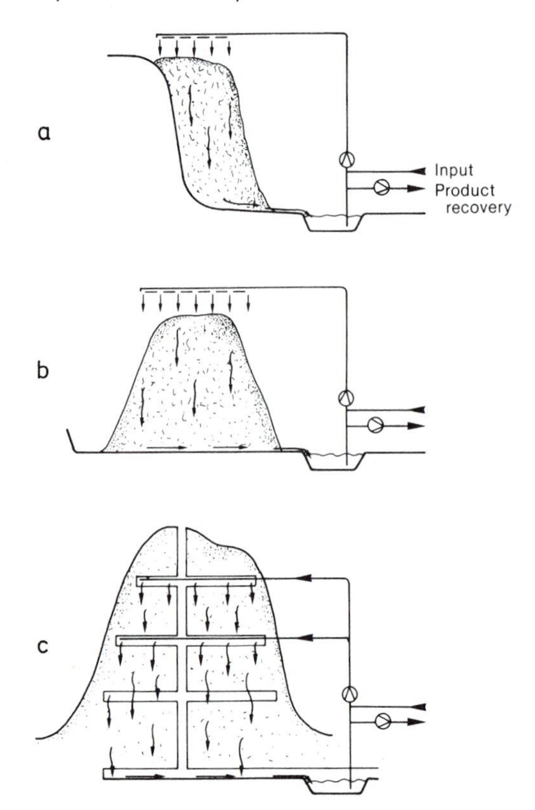

Figure 18.1 Diagram of (a) slope, (b) heap, and (c) in-situ leaching

collects in the bottommost cavity from which it is pumped, the minerals extracted, and the water reused after regeneration of bacteria.

Copper leaching

If chalcocite, chalcopyrite, or covellite are used for the production of copper, several metals are usually found together. For example, chalcopyrite contains 26% copper, 25.9% iron, 2.5% zinc, and 33% sulphur. Chalcopyrite is oxidized as follows:

$$2\,CuFeS_2 + 8\,1/2\,O_2 + H_2SO_4 \rightarrow$$

$$2\,CuSO_4 + Fe_2(SO_4)_3 + H_2O \qquad [5]$$

Covellite is oxidized to copper sulfate:

$$CuS + 2\,O_2 \rightarrow CuSO_4 \qquad [6]$$

Copper leaching plants have been in wide use throughout the world for many years, generally operated as simple heap leaching processes but sometimes as combinations of heap and in-situ leaching. The leaching solution (sulfate/Fe^{3+} solution) carries the microbial nutrients in and the dissolved copper out. The solution is sprinkled over the heap and percolates through the rock pile to the lower level where the copper-rich liquid is collected. The copper-containing solution (up to 0.6g/l) is removed, the copper is precipitated, and the water is reused after readjusting the pH to 2.

Countries in which microbial leaching of copper has been widely used include the United States, Australia, Canada, Mexico, South Africa, Portugal, Spain, and Japan. About 5% of the world copper production is obtained via microbial leaching. A single installation in the United States has produced up to 200 tons of copper per day.

Uranium leaching

Although less uranium than copper is obtained by microbial leaching, the uranium process is more significant economically. Because a thousand tons of uranium ore must be handled to obtain one ton of uranium, in-situ microbial leaching is gaining greater acceptance, since it eliminates the expense of moving such vast amounts of material.

In the uranium leaching process, insoluble tetravalent uranium is oxidized with a hot H_2SO_4/Fe^{3+} solution to soluble hexavalent uranium sulfate.

$$UO_2 + Fe_2(SO_4)_3 \rightarrow UO_2SO_4 + 2\,FeSO_4 \qquad [7]$$

This is an indirect leaching process since the microbial attack is not on the uranium ore directly but on the iron oxidant. Ferric sulfate and sulfuric acid can be produced by *T. ferrooxidans* from the pyrite within the uranium ore.

$$2\ FeS_2 + H_2O + 7.5\ O_2 \rightarrow Fe_2(SO_4)_3 + H_2SO_4 \qquad [8]$$

The pyrite reaction is used for the initial production of the Fe^{3+} leach solution. Pilot plants operate with surface reactors similar to the trickling filters used in sewage.

Optimal uranium leaching conditions are pH 1.5–3.5, 35°C and 0.2% CO_2 in the incoming air. Some thermophilic strains are known which have a temperature optimum of 45–50°C.

In commercial processes, the dissolved uranium is extracted from the leach liquor with organic solvents such as tributylphosphate and the uranium is subsequently precipitated from the organic phase. Adsorption of the uranyl ions with ion exchangers is another possibility. The organic solvents which remain in the water system after extraction may be toxic and hence cause problems when the microbiological system is reused.

In-situ leaching has the disadvantage that the permeability of the rock may be low and the drilled passages may not always allow an adequate supply of nutrients and oxygen to enter deeply into the ore. In such situations the heap system is often still used commercially for leaching of uranium.

Areas where uranium leaching has been carried out include the United States, Canada, and South Africa.

REFERENCES

Atkins, A.S., F.D. Pooley, and C.C. Townsley. 1986. Comparative mineral sulfide leaching in shake flasks, percolation columns, and pachuca reactors using *Thiobacillus ferrooxidans*. Process Biochemistry Febr. 3–10.

Bosecker, K. 1987. Microbial leaching, pp. 551–559. In: Präve, P. (ed.). Handbook of Biotechnology. Oldenbourg Publishers, Munich.

Ebner, H.G. 1977. Metal extraction from industrial waste with *Thiobacilli*, pp. 217–222. In: Schwartz, W. (ed.), Conference on Bacterial Leaching. Verlag Chemie, Weinheim.

Kelly, D.P., P.R. Norris, and C.L. Brierley. 1979. Microbiological methods for the extraction and recovery of metals, pp. 263–308. In: Bull, A.T., D.C. Ellwood, and C. Ratledge (eds.), Microbial technology: Current state, future prospects. Cambridge University Press, Cambridge.

Olsen, G.J. and R.M. Kelly. 1986. Microbiological metal transformations: Biotechnological applications and potential. Biotechn. Progress 2:1–15.

Sanmugasunderam, V., D.W. Duncan, and R.M.R. Branion. 1981. Novel reactor configuration for small scale microbiological leaching, pp. 595–600. In: Moo-Young, M. (ed.), Advances in biotechnology, vol. I. Pergamon Press, Toronto.

Torma, A.E. 1977. The role of *Thiobacillus ferrooxidans* in hydrometallurgical processes. Adv. Biochem. Eng. 6: 1–37.

19

Extracellular polysaccharides

Polysaccharides are used commercially to produce gels, and to thicken and stabilize foods, medicines, and industrial products. They are also used as polymers in the fluids used to force oil to the surface in the tertiary oil recovery process. Although polysaccharides of plant origin (such as starch, alginate, or agar) have been used for many years, microbial polysaccharides have become widely used over the past several decades. Both intracellular polymers such as polybutyric acid and poly-β-hydroxybutyric acid/polyhydroxyvaleric acid copolymer, and extracellular polysaccharides are produced commercially. Around 20 different microbial polysaccharides with market potential have been described, but the largest part of the market is held by xanthan, with production of about 10,000 tons per year. However, microbial polysaccharides are really only a small part of the polysaccharide market as shown by the fact that xanthan holds only about 4% of the market. An important aspect of a polysaccharide, if it is to have market potential, is its rheological properties. Microbial polysaccharides vary greatly in such rheological properties as pseudoplasticity, thixotrophy, and viscoelasticity.

Table 19.1 shows the most important microbial exopolysaccharides produced commercially. In this list, alginate has thus far been produced commercially only from seaweed, but in the future *Azotobacter* may be used. Alginate and xanthan are anionic polysaccharides composed of uronic acid residues.

The **biosynthesis of heteropolysaccharides** is comparable to that of bacterial cell wall components. Proceeding from glucose, the appropriate sugar nucleotide is produced via glucose phosphate, followed by transformation of the glucose into another sugar. At the sugar nucleotide level, further transformations can take place, such as UDP-mannose $\longrightarrow$ UDP-mannuronic acid. The transport of monosaccharides through the membrane and out of the cell takes place after coupling to a C-55 isoprenoid alcohol phosphate. Relatively little is known about the manner in which the polysaccharide polymerase on the out-

Table 19.1 Microbial polysaccharides with commercial uses

Poly-saccha-ride	Structure	Organism
Xanthan	$-\text{Glc } 1 \xrightarrow{\beta} 4 \text{ Glc}$ $\uparrow^{3}_{1}\alpha$ Mann 6-OAc $\uparrow^{2}_{1}\beta$ GlcA $\uparrow^{4}_{1}\beta$ Mann 4,6-OPyruvate	*Xanthomonas campestris*
Alginate	-4 D–MannA $1 \xrightarrow{\beta} 4$D –MannA 1–; -4 L– GulA $1 \xrightarrow{\alpha} 4$ L – GluA	*Pseudomonas aeruginosa, Azotobacter vinelandii*
Curdlan	$-\text{Glc } 1 \xrightarrow{\beta} 3 \text{ Glc} -$	*Alcaligenes*
Sclero-glucan	$-\text{Glc } 1 \xrightarrow{\beta} 3 \text{ Glc} -$ $\uparrow^{6}_{1}\beta$ Glc	*Sclerotium glucanicum, S. delphinii, S. rolfsii*
Pullulan	$\rightarrow 6$ (Glc $1 \overset{\alpha}{-} 4$ Glc $1 \overset{\alpha}{-}$ 4 Glc) $1 \rightarrow$	*Aureobasidium pullulans*
Dextran	$-\text{Glc } 1 \overset{\alpha}{-} 6 \text{ Glc} -$ several $1 \rightarrow 2$, $1 \rightarrow 3$ and $1 \rightarrow 4$ compounds	*Acetobacter* sp., *Leuconostoc mesenteroides, Leuconostoc dextranicum, Streptococcus mutans*

Glc=Glucose, Mann=Mannose, GlcA=Glucuronic acid, MannA=Mannuronic acid, GulA=Guluronic acid, Ac=Acetate

side of the cell membrane acts to bring about chain lengthening and termination and the influence on the development of secondary and tertiary structure. An additional factor influencing the structure of the resulting polysaccharide is degradation by microbial enzymes. It is known for alginate and pullulan, for instance, that at the end of the fermentation enzymes are excreted which degrade these polymers and hence reduce the viscosity of the final product.

Various factors regulate the rate of exopolysaccharide biosynthesis. Molecular oxygen ($>90\%$ saturation) is necessary not only for primary energy metabolism but also for the oxidation of sugar to the corresponding alcohol and for reoxidizing reduced pyridine nucleotides. A carbon/nitrogen ratio of around 10:1 is generally most favorable for optimal yield. For some polysaccharide-forming organisms, *Xanthomonas, Pseudomonas, Azotobacter,* and *Aureobasidium,* growth and product formation can be described quantitatively by the Leudeking-Piret equation:

$$\frac{dP}{dt} = nX + m\frac{dX}{dt}$$

Especially in large-scale fermentations, the marked increase in viscosity which occurs as the polysaccharide concentration increases causes the stirring rate to decrease, resulting in a slowing of the mixing time. This leads to a reduction in mass transfer efficiency and hence to the formation of a polymer of heterogeneous quality.

Isolation of the polymer from the fermentation broth involves precipitation by treatment with an organic solvent, a salt, or an acid. The producing bacteria become entrapped in the precipitating polymer and are carried out of the broth also.

Xanthan

Xanthan was the first heteropolysaccharide to be commercially available. It is produced by *Xanthomonas campestris,* and was developed in the late 1950's and marketed in 1964. The molecular weight of this polymer is around 2–15 $\times$ 10^6 Daltons. As shown in Figure 19.1, the basic repeating unit of xanthan consists of a pentasaccharide containing glucose, mannose, glucuronic acid, acetate, and pyruvate. The number of pyruvate units in the molecule and the molecular weight determines the viscosity of the resulting xanthan gum.

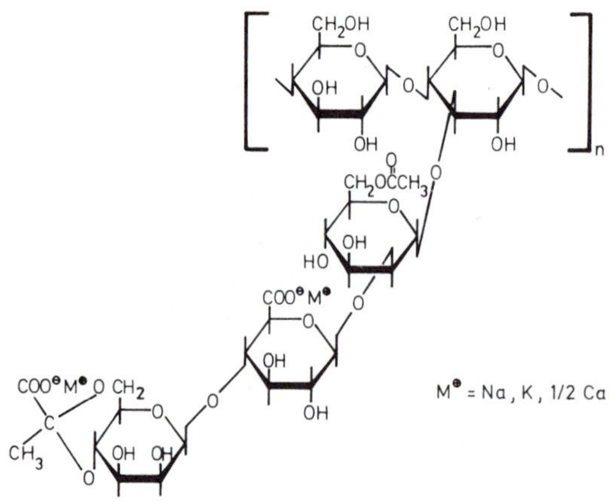

CH2OH CH2OH OH O O OH OH n CH2OCCH3 O OH HO COO M O OH COO M O CH2 C CH3 O OH OH O O
M = Na, K, 1/2 Ca

Figure 19.1 Repeating unit of a xanthan gum

For commercial production, the nutrient solution consists of 4–5% carbohydrate (glucose, sucrose, corn starch hydrolysate), 0.05–0.1% nitrogen (yeast extract, peptone, ammonium nitrate, or urea) and salts, with the pH controlled at 7.0. During the 2-day batch culture, an increase in viscosity begins during the log phase and continues into the resting phase. Large-scale fermentations can have an apparent viscosity up to 20–30,000 centipoise. However, viscosities greater than 10,000 cP may cause technical problems. The efficiency of production of xanthan in weight per weight of carbohydrate used is 70–80% and the yield is 25–30 g/l. Product recovery is carried out by precipitation of the polysaccharide with isopropanol or methanol, a step which also kills the culture. The precipitated xanthan is then dried and ground. Production in continous fermentation has been successfully tested; the effect of the limiting nutrient and the dilution rate on xanthan yield is shown in Table 19.2.

In one actual installation capable of producing 2000 tons of xanthan per year (40 hour fermentation time, yield of 0.7 kg xanthan/kg glucose, 2.5% xanthan final concentration), calculations showed that the production cost per m^3 fermenter volume was markedly lower with the continuous process.

Table 19.2 Influence of limiting nutrient and dilution rate on yield of xanthan

Limitation	Medium	Dilution rate (h^{-1})	Yield (%)
Nitrogen	Complex	0.023	71
Nitrogen	Complex	0.054	82
Sulfur	Synthetic	0.035	60
Sulfur	Synthetic	0.050	32
Phosphate	Synthetic	0.030	40
Magnesium	Synthetilc	0.050	28

Yield based on added glucose
Data of Sutherland (1983)

Alginate

Pseudomonas aeruginosa and *Azotobacter vinelandii* produce alginate from two uronic acids, mannuronic acid and guluronic acid, in proportions of 4:1 to 20:1. Sodium alginate is a commonly used agent for the immobilization of microorganisms (see Chapter 11) and also finds use as an ion-exchange agent.

Curdlan

Alcaligenes faecalis var. *myxogenes* produces two β-1,3 glucans, curdlan and succinoglucan. Succinoglucan contains glucose, about 10% succinic acid, and galactose, whereas curdlan is a simple glucose polymer. Yields of curdlan of 40 g/l are obtained on a defined medium containing 8% glucose in the course of the 80-hour fermentation. Because curdlan is water insoluble, the viscosity does not increase during the fermentation; it only begins to form a gel at temperatures above 54°C. There is an increase in viscosity with succinoglucan, however, so it must be produced from a 4% glucose medium, with an efficiency of production of 35%.

Scleroglucan

Sclerotium glucanicum, *S. delphinii*, *S. rolfsii* and *Helotium* sp. produce scleroglucan, a polysaccharide with glucose units connected primarily in β-1,3- with occasional β-6,1-glycosidic bonds. The organisms grow on hydrolyzed lignocellulose substrates. Scleroglucan is used in food production and is under development for use in tertiary petroleum recovery.

Figure 19.2 shows the kinetics of the production of scleroglucan with *Sclerotium rolfsii* growing on glucose under nitrate limitation. Production (fermentation, purification and possibly enzymatic splitting) of scleroglucans has been shown to be an economically feasible process, provided the conditions of production are optimized.

Pullulan

Pullulan, a glucan with α-1,4- and a few α-1,6-glycosidic bonds, is produced by the fungus *Aureobasidium pullulans* from a 5% glucose solution with 70% efficiency within 5 days. The enzyme pullulanase, produced by bacteria, was discussed in Section 11.2. The physiology of pullulan formation has been intensively studied; Figure 19.3 shows the fermentation kinetics of pullulan production.

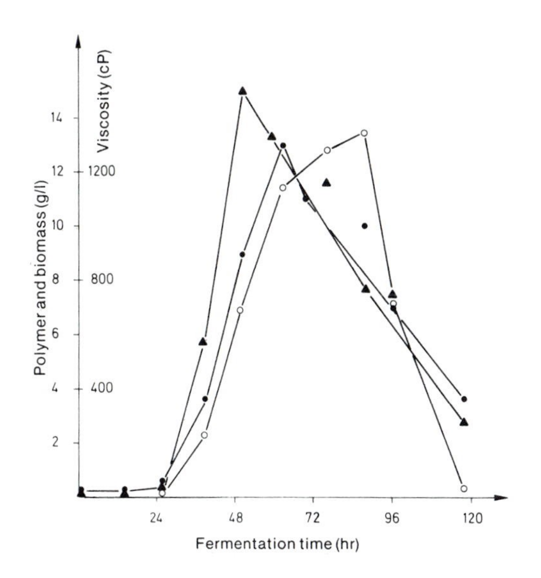

Figure 19.2 Scleroglucan production by *Sclerotium rolfsii* ATCC 15206 during nitrate limitation. • — • biomass; ▲ — ▲ scleroglucan; O — O viscosity (Griffith and Compere, 1978).

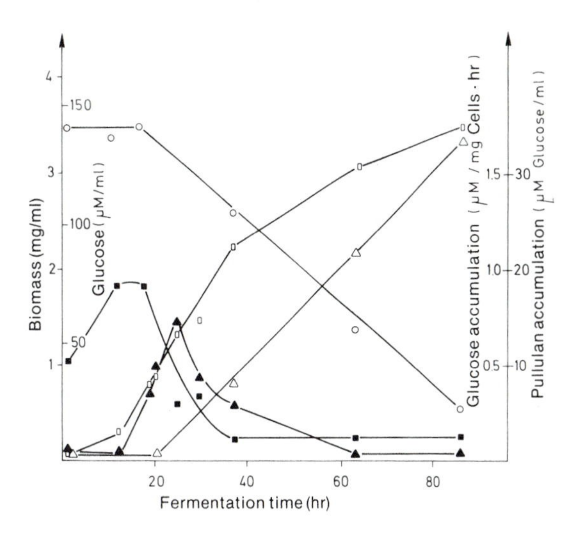

Figure 19.3 Pullulan production by *Aureobasidium pullulans* in a medium with 140mM glucose and 4.5 mM ammonium sulfate. □ — □ cell dry weight (mg/ml); O — O extracellular glucose (μM/ml); △ — △ pullulan (μM anhydroglucose/ml); ■ — ■ ^{14}C-accumulation rate as a measure of glucose assimilation rate (μM/mg cells·h); ▲ — ▲ pullulan formation rate (μM anhydroglucose/mg cells·h) (Catley, 1973).

Dextran

Dextrans are important as blood plasma extenders and are also used in food production. They are a complex group with molecular weights from 15,000–500,000. Most occur as glucans with 1,6-glycosidic bonds, but some dextrans also have 1,2-, 1,3-, or 1,4-glycosidic bonds.

The polysaccharide dextran is produced by the enzyme dextransucrase, a transglucosidase which acts on sucrose (glucose-fructoside), polymerizing the glucose units into dextran and liberating free fructose. This enzyme is produced by many strains of the lactic acid bacterium *Leuconostoc mesenteroides*. In contrast to other exopolysaccharides, dextran is produced extracellularly in the medium. Commercial dextran production can be carried out in a batch process with *L. mesenteroides* with a medium containing inorganic phosphate and an organic nitrogen source. The crude dextran (MW 500,000) obtained through alcohol precipitation is treated with acid for hydrolysis. The desired dextrans (MW 40,000 to 60,000) are subsequently fractionated by ethanol precipitation and dried. The batch process has been further developed into a strictly enzymatic process, in which the cells are separated from their medium, which now contains the enzyme, after fermentation. The extracellular dextransucrase is able to carry out the transformation of sucrose into dextran in the cell-free nutrient solution at pH 5.0–5.2 and a temperature of 25–30°C. The fructose remaining in the solution can be recovered.

REFERENCES

Catley, B.J. 1973. The rate of elaboration of the extracellular polysaccharide, pullulan, during growth of *Pullularia pullulans*. J. Gen. Microbiol. 78: 33–38.

Catley, B.J. 1979. Pullulan synthesis by *Aureobasidium pullulans*, pp. 67–84. In: Berkeley, R.C.W., G.W. Gooday, and D.C. Ellwood (eds.), Microbial polysaccharides and polysaccharases. Academic Press, London.

Compere, A.L. and W.L. Griffith. 1981. Scleroglucan biopolymer production, properties, and economics, pp. 441–446. In: Moo-Young, M. (ed.), Advances in biotechnology, vol. III. Pergamon Press, Toronto.

Griffith, W.L. and A.L. Compere. 1978. Production of a high viscosity glucan by *Sclerotium rolfsii* ATCC 15206. Dev. Ind. Microbiol. 19: 609–617.

Jeanes, A.R. 1978. Dextran bibliography. U.S. Dept. Agric., Res. Serv. Misc. Publ. 1355.

Margaritis, A. and G.W. Pace. 1985. Microbial polysaccharides. pp. 1005–1044. In: Moo-Young, M. (editor), Comprehensive Biotechnology, Volume 3, Pergamon Press, Oxford.

Mian, F.A., T.R. Jarman, and R.C. Righelato. 1978. Biosynthesis of exopolysaccharide by *Pseudomonas aeruginosa*. J. Bacteriol. 134: 418–422.

Sutherland, J.W. 1983. Extracellular polysaccharides. pp. 531–574. In: Rehm, H.J. and G. Reed (editors). Biotechnology, Volume 3, VCH Publishers, Deerfield Beach, FL.

Uttley, N.L. 1986. Polyhydroxybutyrate: a commercial challenge. World Biotech Report 1: 171–177.

20 Other fermentation processes and future prospects

In this chapter we discuss several fermentations that do not fit into the categories established in the previous chapters, including several which are not now of commercial use but may become significant in the future.

Gibberellins

The gibberellins are one class among the five known classes of phytohormones. Gibberellins are used as growth hormones and have found application with barley to improve grain quality in malt production. Sales of plant growth regulators are estimated at $40 million for 1980 in American agriculture, with gross world sales at $118 million.

The gibberellins are derived from gibberellan (Figure 20.1) and are given arbitrary consecutive numbers according to when they were discovered. From the first gibberellin discovery in 1938, many more have been described; by 1975 the number was up to GA_{45}. The most important are GA_3 (also called gibberellic acid) and GA_7. Other gibberellins are also active, e.g. GA_4 from *Sphaceloma manihoticola*. Plants produce only a few gibberellins (GA_1, GA_3, GA_4, and GA_5), and the majority of the rest are produced by fungi. Biosynthesis takes place via isoprenoid units.

Fusarium moniliforme, the imperfect stage of the fungus *Gibberella fujikuroi*, is used for commercial production. This fungus produces the antibiotic bikaverin simultaneously with its production of gibberellins. Six physiologically distinct phases have been characterized for the production process (Figure 20.2).

1. Lag phase
2. Growth phase (24–36 h) without nitrogen limitation, low production of gibberellin
3. Glycine limitation, little cell reproduction, slow gibberellin production
4. Glycine absent, remaining glucose catabolized, strong gibberellin production
5. Glucose content zero, slow accumulation of gibberellins
6. Lysis of cells, increase in pH

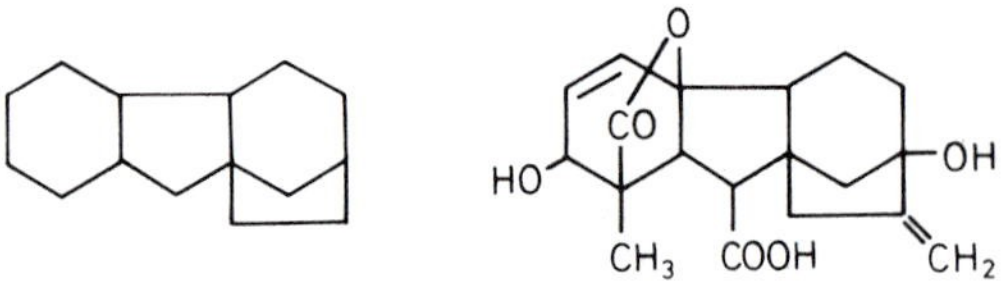

Figure 20.1 Structure of gibberellan (I) and gibberellic acid (II)

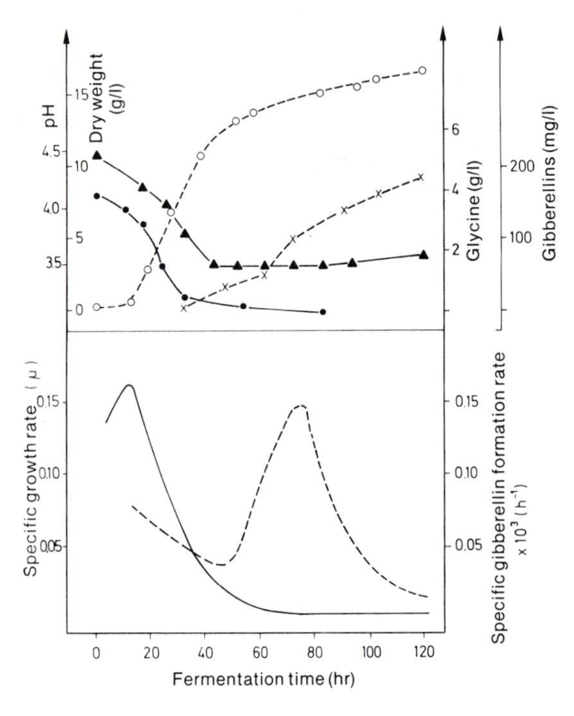

Figure 20.2 Batch fermentation with glycine limitation for the production of gibberellins by *Gibberella fujikuroi*. Upper graph: ▲—▲ pH; •—• glycine; O—O mycelium dry weight; X - - - X gibberellin concentration; Lower graph: - - - specific gibberellin formation rate; —— specific growth rate (Bu'Lock et al., 1974)

A low nitrogen content (<0.2%) and a mixture of different carbon sources are two important conditions for gibberellin production. Yields in production-scale fermenters after 120–150 hours are 1–2 g/l of total gibberellin at 25°C, with the aeration rate at 0.5 vvm. In addition to direct fermentation in submerged culture, some studies have been carried out on the use of surface cultures as well as submerged fed-batch cultures. Continuous processes have been described, but have not become widely used. Table 20.1 shows that in continous culture, gibberellins are formed only at very low dilution rates.

Zearalenone

Zearalenone (Figure 20.3) and its derivatives have estrogenic activity. The various zearalenone derivatives differ by the presence of a keto or hydroxyl group on the 6′ carbon and by the presence or absence of a double bond in the macrolide lactone ring. These substances are formed by *Gibberella zea*, the sexual stage of the fungus *Fusarium roseum gramineαrum*. Production scale today is carried out in a submerged process in 80 m³ fermenters. The yields at 32°C over a 21-day incubation period are 30 g/l. Zearalenones are used as anabolic growth-stimulating compounds in swine and sheep production.

Table 20.1 Gibberellin formation by *Gibberella fujikuroi* under glycine limitation

Dilution rate h^{-1}	0.12	0.09	0.07	0.05	0.02
Mycelium dry weight g/l	7.6	11.6	14.4	15.6	22.5
Glycine content g/l	1.25	0.90	0.49	0.31	0.12
Glucose content g/l	100	92	81	63	53
Gibberellin mg/g dry weight	0	0	0	0.45	2.05

(Bu'Lock et al., 1974)

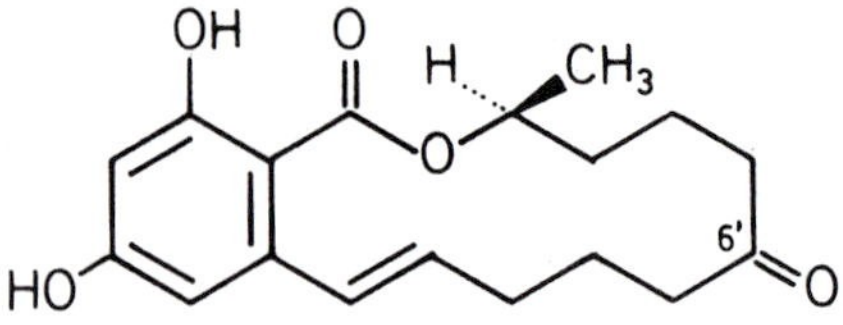

Figure 20.3 Zearalenone

Triglycerides and fatty acids

Fats and oils derived from plant sources find wide use in foods and for industrial applications. Techniques have been developed for the microbial production of fats and oils, but such processes have not been cost-effective for the most significant materials. The fats and oils for food uses mostly contain the longer-chain fatty acids (C_{16}–c_{18}), but uses for those with other chain lengths also exist. Because of the high prices for the fats with shorter-chain fatty acids (C_8 and C_{10}) and for a few of the longer-chain materials, considerable research is underway to produce glycerides with such fatty acids. The proportion of sterols produced in such fermentations must be low (<0.5%) and the fatty acid contents of the cells must be greater than 40%. Only yeasts and filamentous fungi apppear suitable for such processes. In England a process has been developed using species of the fungus genus *Mucor*. Using glucose as the starting material, an oil containing 7% γ-linolenic acid has been produced in 220 m^3 fermenters. γ-Linolenic acid, a C_{18} unsaturated fatty acid with three double bonds (C_6, C_9, and C_{12}) is a starting material in prostaglandin biosynthesis and is also used in dietary foods.

Another approach to the production of desired oils is by the use of specific lipases which will catalyze the exchange of a fatty acid ester of lower value for one of higher value. This approach is currently undergoing research for large-scale production.

Enzyme inhibitors

It has been known for a long time that animals and plants produce enzyme inhibitors. Microorganisms have been studied for enzyme-inhibitor production in Japan since 1965 and worldwide beginning more recently. By 1974, 50 microbial enzyme inhibitors were known. Initially, protease inhibitors were sought, which were usually peptides themselves and were developed for varying applications, including possible use as antitumor agents or as immunosuppressives. Table 20.2 gives examples of several microbially produced protease inhibitors.

Table 20.2 Protease inhibitors

Enzyme inhibitor	Enzyme	Microorganism	Application
Pepstatins	Acid proteases e.g. pepsin	*Streptomyces testaceus, S. argenteolus*	Ulcers
Antipain	Papain	*S. yokosukanensis* *S. michiganensis*	Reduces fever
Chemostatin	Chymotrypsin	*S. hygroscopicus* *S. lavendulae*	Reduces fever
Elastatinal	Elastase	*S. griseoruber*	Pancreatitis
Leupeptin	Trypsin Plasmin		Reduces fever
Bestatin	Amino peptidases	*S. olivoreticuli*	Inhibits cancer
Subtilisin inhibitor	Subtilisin	*S. albogriseolus*	

Figure 20.4 shows the peptide structure of the substance antipain, an inhibitor of papain. In contrast to the relatively simple structure of antipain, the inhibitor of the enzyme subtilisin contains 113 amino acids.

A second group of substances that have been isolated are inhibitors of amylase and other glycoside hydrolases. These substances are important in medicine because they inhibit the breakdown of carbohydrates in the digestive tract. Included here are inhibitors of α-glucosidases, β-glucosidases, cellulases, and dextranases. α-Glu-

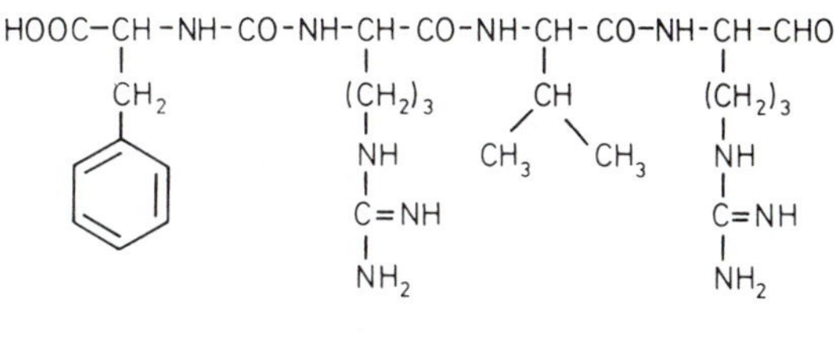

HOOC-CH-NH-CO-NH-CH-CO-NH-CH-CO-NH-CH-CHO
CH$_2$ (phenyl); (CH$_2$)$_3$-NH-C=NH-NH$_2$; CH(CH$_3$)(CH$_3$); (CH$_2$)$_3$-NH-C=NH-NH$_2$

Figure 20.4 Antipain

cosidase inhibitors are widely distributed (Table 20.3). Chemically, such inhibitors are pseudo-oligosaccharides, monosaccharides, cyclopeptides, and polypeptides. Figure 20.5 shows the general formula of a group of α-glucosidase inhibitors which contain a central molecule of an unsaturated cyclitol plus 4,6-dideoxy-4-amino-D-glucopyranose with α-1,4-glucopyranose. This

Table 20.3 α-Glucosidase inhibitors produced by microorganisms

Microorganism	Type of compound	Specificity
Actinomycetes	Proteins Pseudo-oligosaccharides	1,2 plus microbial α-amylases
Streptomyces flavochromogenes	Pseudo-oligosaccharides	1,2,3
S. diastaticus	Pseudo-oligosaccharides	3 plus α-amylases, microbial α-amylases
S. fradiae	Acid polypeptides	Human α-amylases
Bacilli, *S. lavendulae*	Nojirimycin, 1-deoxynojirimycin	2,3 plus β-glucosidases from plants and microorganisms
S. calidus	Glycopeptide	1 plus intestinal maltase and sucrase
S. violaceoruber	Polypeptide	1
Aspergillus niger	Glycopeptide	*Aspergillus* glucoamylase
Cladosporium herbarum	Glycopeptide	1
Cladosporium cladosporioides	Acid glycopeptide	Human α-amylases

1 = pancreatic amylase; 2 = intestinal amylase and disaccharase; 3 = *Rhizopus* glucoamylase.
(Müller, 1986)

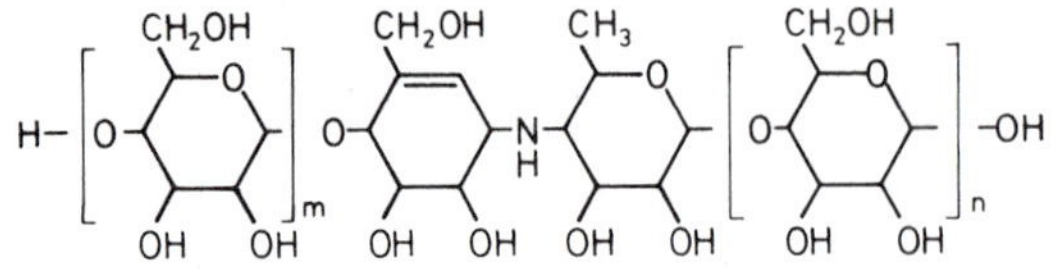

Figure 20.5 α-Glucosidase inhibitor complex. (m+n=1–8)

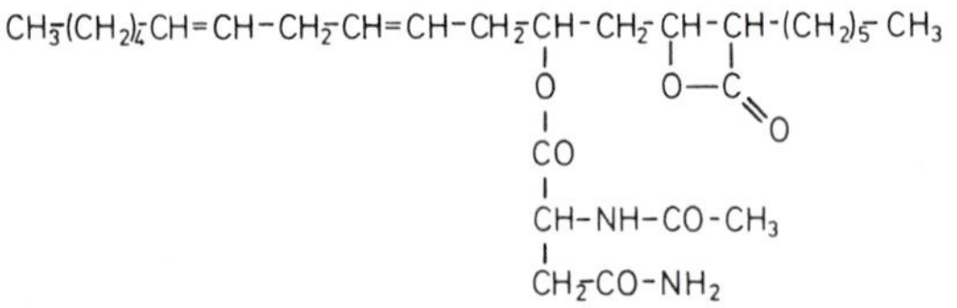

Figure 20.6 Esterastin

group of acarboses are produced by species of the genus *Actinoplanes* using a medium containing maltose and yeast extract. For purification, the products are complexed to an ion-exchange resin and then eluted for further purification.

Nojirimycin (see Figure 13.19) is an example of a substance which has anti-β-glucosidase activity.

Enzyme inhibitors of adrenalin biosynthesis (dopastin, fusaric acid, isoflavone, phenolpicolinic acid, oudenone, and oosponol) have been intensively studied because of their blood-pressure regulating effect. The structure of esterastin, an esterase inhibitor (Figure 20.6) is unusual because it contains a β-lactone ring. β-Lactamase inhibitors such as clavulanic acid are valuable because they overcome bacterial penicillin resistance. Because antibiotics which are active against microbial cell wall synthesis are generally less toxic to humans, penicillinase inhibitors are being sought for possible use in penicillin therapy. Another area of application for enzyme inhibitors concerns the inhibitors of chitin synthesis, such as the polyoxins, which have found limited use as insecticides.

Microbial insecticides

The production of insecticides from microbiological sources is of considerable interest. The emphasis here is not on the production of insecticide-active chemicals, but on the use of microbes which are themselves insect pathogens. In such applications, the living microorganism must be dispersed into the environment. Five specific requirements must be met before living preparations can be used as insecticides:

1. Large-scale production must be feasible
2. Viability must be maintained
3. The organism must not be toxic or pathogenic to other animals and plants
4. The organism must be less expensive than chemical agents.
5. No development of resistance by the target insects.

Bacteria, viruses, fungi, or protozoa can be used as insecticides. They are produced commercially either in large-scale fermentation or in insect hosts.

Five protozoa, *(Nosema locustae, N. algerae, N. pyrausta, Vairimorpha necatrix,* and *Mattesia trogodermae)* have been tested thus far, but have not yet found practical application.

Over 400 fungi are known which attack insects. They are relatively nonspecific in host range but are also less effective than other agents. Those fungi currently under study are summarized in Table 20.4. Many of these fungi can be cultured well in open dishes in semi-solid medium such as rice seeds. Within 15–20 days at 26–29°C, about 2×10^9 conidia per gram are produced.

Table 20.4 Fungi pathogenic against insects

Fungus	Production technique	Status	Countries
Aschersonia sp.	Submerged	Pilot plant	USSR, Holland, UK
Beauveria bassiana	Submerged/ 2 stage	Approved for market	USSR
	Semi-solid medium/ 2 stage	Small scale	USA
Conidiobolus obscurus	Submerged	Pilot plant	France, UK, USA
Culcinomyces clavosporus	Submerged	Small scale	Australia
Entomophthora grylli	In insects	Small scale	USA
Erynia neoaphidis	Submerged In insects	Small scale	UK
Hirsutella thompsonii	Semi-solid	Pilot plant	USA
Lagenidium giganteum	Submerged	Small scale	USA
Metarhizium anisopliae	Semi-solid	Approved for market	Brazil
Nomurea rileyi	Solid	Small scale	USA
Verticillium lecanii	Submerged	Approved for market	UK
Zoophthora radicans	Submerged	Laboratory studies	USA

(Quinlan and Lisanski, 1983)

Over 650 insect viruses have been described. Such viruses offer the best possibilities for practical use as insecticides and several viruses are already being marketed. One disadvantage of viruses for large-scale production is that they must be cultured on living insects, making scale-up difficult.

Three bacteria (*Bacillus popilliae, B. moritae,* and *B. thuringiensis*) are commercially produced. The first of these can only be cultivated in living insects but the latter two can be grown in submerged culture. *B. thuringiensis* is widely cultivated in North America and Europe, with amounts of 2300 tons per year, and is also produced in the USSR and China. About 40 different insects that cause agricultural or forest diseases are treated with *B. thuringiensis*. In addition, control of certain insects which are carriers of human diseases (for example, *Anopheles* sp., malaria; *Simulium damnosum*; river blindness) is also a possibility. The active ingredient of *B. thuringiensis* is a polypeptide toxin (γ-endotoxin) which is produced during the bacterial sporulation process. The cultivation procedure is therefore designed to obtained a high percentage of sporulation and toxin accumulation. The fermentation is carried out in a medium containing starch, corn-steep liquor, casein, and yeast extract. Sporulation occurs after 25–30 hours and the γ-endotoxin accumulates as parasporal crystalline inclusions that amount to around 30% of the cell mass. This toxin is quite selective in its action and is administered by being incorporated into a suitable insect food. Another toxin, the β-exotoxin, is a nucleoside which is produced during the growth phase. It is a nonspecific toxin which rapidly kills caterpillars. Between 10–15 minutes after ingestion of the β-exotoxin, the animals quite feeding and die within 3–5 days.

Cyclosporin A

The cyclosporins are a group of oligopeptides produced by the fungus *Tolypocladium inflatum* which have antifungal, anti-inflammatory, and immunosuppressive properties. The immunosuppressive effect is due to the inhibition of the activation of the T cell response. Cyclosporin A is an eleven-membered cyclic peptide which is used because of its immunosuppressive properties to prevent rejection reactions during organ transplantation. The cyclosporins are produced in a 15-day submerged fermentation from a medium containing 4% glucose, 1% peptone, and mineral salts.

Biochips

A quite different application of biotechnology involves the combination of electronics, biochemistry, and genetics in the development of the concept of the **biocomputer**. Microprocessors of the future may consist of chips in which monolayers of protein are the functional material. Studies with polylysine have given some success. These chips have over 100,000-fold more switches than conventional microprocessors. The appropriate proteins may be produced from microorganisms economically by the use of genetic engineering techniques.

Flavoring substances

We discussed the production of flavor-enhancers in Chapters 9 and 10. Substances with inherent flavor properties are also produced by microorganisms and are of great significance in the manufacture of food and drink. Moreover, microbial enzymes used in food can alter the substrate and thus produce flavoring substances. Intensive studies are being conducted on the use of microbes in food production and on the genetics and biochemistry of aroma production. Among various alcoholic drinks alone, 400 different substances have been identified; of these 118 were esters, 80 aliphatic and aromatic acids, 41 carbonyl compounds, and 38 aliphatic and aromatic alcohols. The aroma components and the percentage distribution vary depending on the strain and substrate solutions.

In addition to the flavor enhancers already discussed (sodium glutamate, Chapter 9, and 5′-nucleotides, Chapter 10) a number of specific flavoring agents can be produced by fermentation. An example is the production of complex flavoring ingredients suitable for cheese production. The most important flavoring substances in cheese are methyl ketones (2-heptanone, 2-nonanone, 2-pentanone) and secondary alcohols with 5–11 carbon atoms. Such flavoring substances can also be produced in separate fermentation processes and then used as supplements to increase the flavor of prepared foods such as salad dressings and cheese pastries. As an example, the fermentation of *Penicillium roqueforti*, the fungus involved in Roquefort cheese production, can be described. Milk, salt, and milk fat are added for the fermentation which is carried out at 24–26°C for three days. During this period, the milk fat is converted by the fungal lipases into methyl ketones and secondary alcohols.

$$\text{Fats} \xrightarrow{\text{Lipases}} \text{Fatty acids} \longrightarrow \beta\text{-Keto acids} \longrightarrow \text{Methylketones} \longrightarrow \text{Secondary alcohols}$$

After growth, the sterilized culture broth contains 4% fat, 5% protein, 8% carbohydrates, 5% salts, and 78% water; this final fermentation product has 6–20 times more cheese aroma than does the cheese itself.

A number of specific microbial products that have distinctive odors are summarized in Table 20.5. In addition, biotransformation of terpenes to terpenols have been used to produce flavor ingredients. For instance, *Penicillium digitatum* is able to convert the terpene R(+)limonene to α-terpineol, a substance with a lilac odor that is used in perfumes.

Table 20.5 Microbial compounds with specific odors

Compound	Structure	Type of odor
Diacetyl	$CH_3-CO-CO-CH_3$	Butter flavor (buttermilk)
Acetaldehyde	CH_3-CHO	Oranges and yogurt
γ-Decalactone	$CH_3-(CH_2)_5-CH$ —— CH_2 (ring: CH–O–C(=O)–CH_2–CH_2)	Peach
Esters		Fruit
Pyrazine		Roast and nut

Biofilters

Biofilters have been used since about 1970 for the removal of odors from air streams arising from sewage treatment plants, animal feed lots, paper mills, tobacco, chocolate and coffee roasting plants. Highly odoriferous off-gasses are purified and moistened by passing through beds containing compost, peat, pine needles, moss, or plastic bodies. Microorganisms growing on the beds are responsible for the removal of much of the odor. During the time which the gas passes through the bed (generally only a few seconds), the odors are absorbed and metabolized to biomass and CO_2. Biofilters themselves only have a compost-like odor. Biofilters may also be used in the future to purify the air from large-scale fermentation installations.

REFERENCES

Anke, T. 1986. Further secondary products of biotechnological interest, pp. 611–628. In: Rehm, H.J. and Reed (eds.). Biotechnology 4. VCH Verlag, Weinheim.

Baldwin, B. 1986. Commercialisation of microbially produced pesticides, pp. 39–49. World Biotech Report 1. Online, London.

Bu'Lock, J.D., R.W. Detroy, Z. Hostalek, and A. Munim-Al-Shakarchi. 1974. Regulation of secondary biosynthesis in *Gibberella fujikuroi*. Trans. Br. Mycol. Soc. 62:377–389.

Gatfield, I.L. 1988. Die enzymatische Bildung von Aromastoffen. (Enzymatic formation of perfume ingredients.) In: Crueger, W. P. Präve, M. Schlingmann, R. Thauer, F. Wagner (eds.). Jahrbuch Biotechnologie 1988/89. Hanser Verlag, München.

Hidy, P.H., R.S. Baldwin, R.L. Greasham, C.L. Keith, and J.R. McMullen. 1977. Zearalenone and some derivatives: Production and biological activities. Adv. Appl. Microbiol. 22:59–82.

Khachatourians, G.G. 1986. Production and use of biological pest control agents. TIBTECH May, 120–124.

Krieg, A. 1986. *Bacillus thuringiensis*, Tagungsbericht. Forum Mikrobiol. 6/86:334–336.

Kumar, P.K.R. and B.K. Lonsane. 1987. Potential of fed-batch culture in solid state fermentation for production of gibberellic acid. Biotech. Letters 9: 179–182.

Lüthy, P. and M.G. Wolfersberger. 1986. The delta-endotoxin of *Bacillus thuringiensis*. Swiss Biotech 4:11–14.

Müller, L. 1986. Microbial glycosidase inhibitors, pp. 531–567. In; Rehm, H.J. and G. Reed (eds.). Biotechnology 4. VCH Verlag, Weinheim.

Quinlan, R.J. and S.G. Lisanski. 1983. Microbial insectisides, pp. 233–254. In: Rehm, H.J. and G. Reed (eds.). Biotechnology 3. VCH Verlag, Weinheim.

Rowe, G.E. and A. Margaritis. 1987. Bioprocess developments in the production of bioinsecticides by *Bacillus thuringiensis*. CRC Critical Rev. Biotechnol. 6:87 ff.

Sinden, K.W. 1987. The production of lipids by fermentation within the EEC. Enzyme Microb. Technol. 9:124–125.

Tholander, P.J. 1987. Biologische Verfahren zur Beseitigung von Geruchsquellen im grosstechnischen Bereich. (Biological technique for the removal of odors in large-scale installations.) BioEng. 1:54–55.

Umezawa, H., T. Takita, and T. Shiba (eds.). 1978. Bioactive peptides produced by microorganisms. John Wiley & Sons, New York.

Vogel, R. 1984. Natürliche Enzym-Inhibitoren. (Natural enzyme inhibitors.) Thieme Verlag, Stuttgart.

Wink, J., A. Lotz, and P. Präve. 1987. Biotechnologisch hergestellte Aromastoffe. (Use of biotechnology for the manufacture of perfume ingredients.) Biotech-Forum 4:235–238.

Index